Radon Series on Computational and Applied Mathematics 2

Managing Editor

Heinz W. Engl (Linz/Vienna)

Editors

Hansjörg Albrecher (Linz)
Karl Kunisch (Graz)
Ronald H. W. Hoppe (Augsburg/Houston)
Ulrich Langer (Linz)
Harald Niederreiter (Singapore)
Christian Schmeiser (Linz/Vienna)

Gröbner Bases
in Symbolic Analysis

Edited by

Markus Rosenkranz
Dongming Wang

Walter de Gruyter · Berlin · New York

Editors

Markus Rosenkranz
Radon Institute for Computational and
Applied Mathematics (RICAM)
Austrian Academy of Sciences
Altenberger Straße 69
A-4040 Linz
Austria
e-mail: Markus.Rosenkranz@oeaw.ac.at

Dongming Wang
School of Science
Beihang University
37 Xueyuan Road
Beijing 100083
China

and

Laboratoire d'Informatique de Paris 6
Université Pierre et Marie Curie – CNRS
104, avenue du Président Kennedy
F-75016 Paris
France
e-mail: Dongming.Wang@lip6.fr

Key words:
Gröbner basis, polynomial ideal, initial value problem, boundary value problem, differential algebra, difference polynomial, elimination theory, differential algebraic equation, invariant theory,
syzygies.

Mathematics Subject Classification 2000:
00B25, 12HXX, 13A50, 13D02, 13HXX, 13N10, 13P10, 16S32, 34A09.

♾ Printed on acid-free paper which falls within the guidelines
of the ANSI to ensure permanence and durability.

Library of Congress Cataloging-in-Publication Data

A CIP catalogue record for this book is available from the Library of Congress.

ISBN 978-3-11-019323-7

Bibliographic information published by the Deutsche Nationalbibliothek

The Deutsche Nationalbibliothek lists this publication in the Deutsche Nationalbibliografie;
detailed bibliographic data are available in the Internet at http://dnb.d-nb.de.

Printed in Germany
Cover design: Martin Zech, Bremen.
Typeset using the authors' TeX files: Matthias Pfizenmaier, Berlin.
Printing and binding: Hubert & Co. GmbH & Co. KG, Göttingen.

Preface

In the spring and summer of 2006, the Radon Institute for Computational and Applied Mathematics (RICAM), in close cooperation with the Research Institute for Symbolic Computation (RISC), organized the *Special Semester on Gröbner Bases and Related Methods*. Directed by Bruno Buchberger (RISC) and Heinz W. Engl (RICAM), the special semester created a unique atmosphere for interdisciplinary cooperation on all aspects of Gröbner bases.

Among the eight workshops of its core program, the D2 Workshop on *Gröbner Bases in Symbolic Analysis* (May 8–17, 2006) provided an inspirational platform for the exchange of ideas on the role of Gröbner bases in differential and difference equations. With great pleasure we present here its main outcome in the form of eleven pertinent survey papers. Reflecting the rich diversity of the field, they cover a broad range of topics in:

> Algebraic Analysis
>
> Involutive and Janet Bases
>
> Differential Algebra
>
> Lie Groups and Invariant Theory
>
> Initial and Boundary Conditions

For the full workshop program, comprising both invited and contributed talks, we refer to the website of the RICAM special semesters at:

> http://www.ricam.oeaw.ac.at/specsem/

We would like to express our gratitude to all the authors for their contributions and cooperation, which are essential for the quality and timely publication of this volume. Finally and most particularly, we extend our warm thanks to Bruno Buchberger: his seminal dissertation from 1965 has created the powerful method of Gröbner bases whose ongoing vitality is witnessed also in this proceedings volume.

Linz and Paris   Markus Rosenkranz
July 2007   Dongming Wang

Contents

Radon Series Comp. Appl. Math **2**, 1–21

Gröbner Bases in Algebraic Analysis:
New Perspectives for Applications

Jean-François Pommaret

Key words. Gröbner basis, homological algebra, differential module, algebraic analysis, elimination theory.

AMS classification. 35A30, 18G10, 18G15.

1   Introduction

Algebraic Analysis, a sophisticated mixture of differential geometry and homological algebra, has been pioneered after 1970 by V. P. Palamodov [14], M. Kashiwara [9] and B. Malgrange [12], successively for the constant and variable coefficients cases. However, the theory of differential modules constituting its core [11] can also be considered as the algebraic dual counterpart of the formal theory of systems of partial differential equations pioneered by D. C. Spencer [25] just before 1970. Surprisingly, it is only in 1990 that such a difficult theory has been applied with quite a success to engineering sciences, namely control theory, thanks to the work of U. Oberst [13]. Among recent works in this direction we may quote the contributions of A. Quadrat [22], J. Wood [27] and E. Zerz [28].

More surprisingly, and even if many people are knowing about their usefulness through explicit calculations, Gröbner bases are often hidden behind elimination techniques not easy to grasp at first sight from tools such as the extension modules that play a central part in modern homological algebra for reasons we shall explain later on. As a byproduct, the very successful workshop organized by RICAM/ RISC can be considered as the first place where people could meet and exchange ideas on the advantages and disadvantages of using Gröbner bases in algebraic analysis.

The purpose of this paper, originating from two lectures at the D2 and D3 workshops, is twofold. First of all, we want to introduce the reader in a way as elementary as possible, through explicit examples, to the main concepts of algebraic analysis and the theory of differential modules. Meanwhile, we shall prove that, in a certain sense,

Gröbner bases are *everywhere* through formal elimination procedures. We shall also present new problems originating from a broad range of engineering applications (continuum mechanics, hydrodynamics, control theory, and many others). However, as a second and *unescapable point*, we shall point out that any recent progress in mathematics has *always* been done along lines becoming more and more intrinsic, that is coordinate-free, even, of course, if computations *must* be done in a particular coordinate system. Therefore, the whole spirit of Gröbner bases must be changed as we shall give examples clearly showing that the underlying principle behind their construction is far from being intrinsic and is superseded by the formal theory of ordinary differential (OD) or partial differential (PD) equations.

At this stage we give hints concerning the types of problems met in engineering applications. In the differential geometric setting, it is clear that all the evolution of the story of OD/PD equations has been done towards more and more intrinsic tools and results. Accordingly, the work of C. Riquier [23] in 1910 has been superseded by that of M. Janet [7] in 1921 which has been directly used, without any quotation, by J.F. Ritt [24] in 1930 while extending differential algebra from OD to PD equations and finally totally superseded by the work of D. C. Spencer [26]. Here come two intermediate tools in their very specific setting.

First, *exterior calculus* has been invented by E. Cartan, at the beginning of the previous century, as a way to generalize *vector calculus* in ordinary three dimensional space geometry (grad, curl and div operators) and has been later on generalized by *tensor calculus*. We should like to react strongly against the usual attractive temptations in all mathematical physics during the last fifty years or so to bring back all types of equations to the exterior derivative d. This is not fair at all as, even if the so-called Poincaré sequence using the fact that $d \circ d = 0$ makes up a *differential sequence*, that is a chain of functional spaces acted on by operators, the composition of two successive ones vanishing, it is well known that, in the Spencer *machinery*, the construction of the Janet or Spencer differential sequences, the only canonical finite-length differential sequences that can be constructed is far from being directly related to the exterior derivative as it only depends on the Spencer operator [18, 19, 26]. At this point, we take the opportunity to notice that the *second* Spencer sequence is nothing else than the Janet sequence for the first order system/operator into which any system/operator can be brought by *reduction* to first order (see later on). Though everyone believes from the OD case to be found in control theory that it is an easy task, we shall show that, in the PD case, it is directly related to delicate computations in the Spencer cohomology of the symbols (roughly the top-order part of the system generalizing the initial of a Gröbner basis) and will lead to a surprising generalization of the so-called *Kalman form* existing in control theory and crucially used in order to check the controllability of a control system.

As a direct byproduct, this is the reason for which all attempts to explain the work of the brothers E. and F. Cosserat [4] in 1909 on the foundation of continuum mechanics/elasticity were done in vain, up to the moment we proved in [18] that the *Cosserat operator is nothing else than the formal adjoint of the Spencer operator* (for an elementary but striking short survey in French, we refer the reader to [17]). Roughly, everybody knows that, if one plies a piece of rubber or a sheet of paper and then carries it in space along a translation or a rotation, nothing is changed in the deformation of the elastic body which has therefore only to do, as everybody believes, with the (differential) invariants of the group of rigid motions. Meanwhile, as only displacemet vectors appear, there is *no way at all*, by duality, to exhibit torques as only forces can be exhibited. Also, if the group should be extended to a bigger group, it should have less invariants. The basic idea of the Cosserat brothers was to create *another machine*(!) only based on a *group acting on a space* in such a way as to exhibit at once forces and torques *together*, both with their corresponding equilibrium equations (see [4], p. 137).

Second, in 1965 B. Buchberger invented Gröbner bases [3], named in honor of his advisor W. Gröbner, whose earlier work [6] provided a source of inspiration. They had apparently been unaware of the work of Janet though it is now known that the two approaches are essentially the same. As we have been among the students of Spencer in 1970, we may say that he did not know at all about the work and even the existence of Janet. However, both Gröbner and Janet approaches suffer from the same lack of intrinsicness and we now explain this point.

As we shall see later on by means of two specific academic examples that will be compared to the modern formal theory of systems of PD equations, let us give to one student, in a room equipped with a computer, a set of polynomials in three indeterminates with rational coefficients, then make a linear change of the indeterminates with a non-degenerate 3×3 matrix with rational coefficients, like the matrix of the permutation $(123) \to (321)$, and give the new set to another student in a different room equipped with the same computer. Then it is clear that the algorithms will proceed along quite different lines and eventually produce different Gröbner bases, even with a different number of elements, that is the second basis will not coincide with the transformation of the first along the above matrix. However, of course, if the bases are used in order to compute intrinsic numbers, such as the differential transcendence degree for instance, these numbers *must* be the same in both cases.

This is the first difficulty that *must* be overcome and we now sketch the second which is quite more subtle. Up to now we have seen changes of indeterminates in polynomial rings. However, in the case of constant coefficients, there is a bijective correspondence between polynomials and differential operators, leading to a bijective correspondence between ideals and systems in one unknown. More precisely, if k (say $\mathbb{Q}$ in the above situation) is a given field of constants for formal commuting derivations $d_1, \ldots, d_n$ and $\chi_1, \ldots, \chi_n$ are indeterminates, we can identify the ring $k[\chi] = k[\chi_1, \ldots, \chi_n]$ of polynomials with the ring $D = k[d] = k[d_1, \ldots, d_n]$ of differential operators in $d_1, \ldots, d_n$ with coefficients in k. If now $\mu = (\mu_1, \ldots, \mu_n)$ is a multi-index of *length* $|\mu| = \mu_1 + \ldots + \mu_n$ with $\mu + 1_i = (\mu_1, \ldots, \mu_{i-1}, \mu_i + 1, \mu_{i+1}, \ldots, \mu_n)$, we may set $\chi_\mu = (\chi_1)^{\mu_1} \ldots (\chi_n)^{\mu_n}$ and similarly $d_\mu = (d_1)^{\mu_1} \ldots (d_n)^{\mu_n}$. Also, for a later use, a multi-index μ is said to be of *class* i if $\mu_1 = 0, \ldots, \mu_{i-1} = 0, \mu_i \neq 0$. Hence using Einstein implicit summation, with any polynomial $a^\mu \chi_\mu$ we may associate the operator $a^\mu d_\mu$ and both can be denoted for simplicity by the same letter P when the interpretation is clear enough, with the same rules of addition and multiplication. Needless to say that the situation drastically changes when k is replaced by a differential field K, that is a field K equipped with n (partial) derivations $\partial_1, \ldots, \partial_n$ such that $\partial_i \partial_j = \partial_j \partial_i, \partial_i(a + b) = \partial_i a + \partial_i b, \partial_i(ab) = (\partial_i a)b + a\partial_i b, \forall i = 1, \ldots, n, \forall a, b \in K$ because in this case D is no longer a commutative ring as we have now $d_i a = \text{ad}_i + \partial_i a$. In this new framework, it becomes clear that any (constant) linear change of the χ_i can be interpreted as a linear change of the d_i corresponding, in the differential geometric framework, to a linear change of the independent variables if we use the basic prolongation formulas for a section, namely $y = f(x) \rightarrow d_i y = \partial_i f(x)$. Finally, everybody is familiar with the effect of linear changes of the unknowns and even changes of the unknowns involving derivatives, provided they are invertible.

Example 1.1 With time derivative indicated by a dot as usual, the *single* second order OD equation $\ddot{y} = 0$ is *equivalent* to the system of *two* first order OD equations $\dot{y}^1 - y^2 = 0, \dot{y}^2 = 0$ by setting $y^1 = y, y^2 = \dot{y}$ and conversely $y = y^1$ in the invertible change $(y) \rightarrow (y^1, y^2)$. In this case, the independent variable is untouched while the numbers of dependent variables and equations are changed. Nevertheless, integrating the only second order equation or the two corresponding first order ones is *equivalent* in a sense coming from the invertible transformation of the unknowns provided that all the equations are satisfied.

However, there is a much more subtle change of the *presentation* of the system and we provide a tricky example showing that even if the independent and dependent variables are untouched, two systems can be again *equivalent*, even if they consist of *two* PD equations of order *four* on one side and *one* PD equation of order *two* on the other side.

Example 1.2 With $n = 2, k = \mathbb{Q}$, let us consider the inhomogeneous formal system with second members

$$Py \equiv d_{22}y = u, \qquad Qy \equiv d_{12}y - y = v$$

where we set as usual $d_{22} = d_{(0,2)}, d_{12} = d_{(1,1)}$ in a coherent way with the multi-index notation introduced above. We obtain at once through crossed derivatives

$$y = d_{11}u - d_{12}v - v$$

and, by substituting, two fourth order compatibility conditions for (u, v), namely

$$\begin{cases} A & \equiv & d_{1122}u - d_{1222}v - d_{22}v - u & = & 0, \\ B & \equiv & d_{1112}u - d_{11}u - d_{1122}v & = & 0, \end{cases}$$

satisfying $d_{12}B + B - d_{11}A = 0$. However, the commutation relation $P \circ Q \equiv Q \circ P$ provides a single compatibility condition for (u, v), namely

$$C \equiv d_{12}u - u - d_{22}v = 0,$$

and we check at once $A = d_{12}C + C, B = d_{11}C$ while $C = d_{22}B - d_{12}A + A$, that is,

$$(A = 0, B = 0) \Leftrightarrow (C = 0).$$

Comparing the two preceding examples, we discover that the concept of *state* in control theory must be revisited if we understand by state the vague idea of *what must be given in order to integrate the system*. Indeed, in the first example, at the origin we need to know $(y(0), \dot{y}(0))$ on one side and $(y^1(0), y^2(0))$ on the other side, that is two numbers in both cases. Nothing similar can be achieved in the second example though we understand intuitively that introducing $C = 0$ is much *better* than introducing $A = 0, B = 0$.

The above situation is far more involved than the one considered by Gröbner bases and we may dream about a tool that should give results independent not only of the change of dependent/independent variables but also of the change of presentation. Henceforth, our main point is to say that *such a tool is already exists* as it is exactly the main purpose of modern homological algebra. As a byproduct, the main difficulty is that it is known from textbooks ([25], Theorem 7.22 (Axioms)) that *the answer needs the computation of the extension modules*, and this led to the birth of algebraic analysis. We understand therefore that, if we want to deal with an intrinsic, that is a coordinate-free and even presentation-free, description of engineering systems like the ones to be found in control theory, it is surely not through Janet or Gröbner techniques. However, this is the reason why algebraic analysis is based on homological algebra on the algebraic side but also on the formal theory of systems of OD/PD equations on the differential geometric side. Again, in our opinion, the second aspect is much more delicate than the first, despite a first outlook, though, nevertheless, we believe that

intrinsicness always competes with complexity,

and a major task for the future will be to overcome this problem.

We are now ready to start the technical part of this paper, which can be considered as a first step towards this direction. For more details and references, we refer the reader to our two-volume book [19] and to its summary [20], coming from an intensive twenty-hours European course.

2   Controllability

In order to motivate the main problem we want to solve and illustrate, we provide three examples:

Example 2.1 Let us consider the following system of OD equations with one independent variable (t=time) and three unknowns (y^1, y^2, y^3), where a is an arbitrary constant parameter and $\dot{y} = dy$ is the usual time derivative:

$$\begin{cases} \dot{y}^1 - ay^2 - \dot{y}^3 & = 0 \\ y^1 - \dot{y}^2 + \dot{y}^3 & = 0 \end{cases}$$

Example 2.2 Let us now consider the following system of PD equations with two independent variables (x^1, x^2) and three unknowns (y^1, y^2, y^3), where again a is an arbitrary constant parameter and we have set for simplicity $y_i^k = d_i y^k$:

$$\begin{cases} y_2^2 + y_2^3 - y_1^3 - y_1^2 - ay^3 & = 0 \\ y_2^1 - y_2^3 - y_1^3 - y_1^2 - ay^3 & = 0 \\ y_1^1 - 2y_1^3 - y_1^2 & = 0 \end{cases}$$

Example 2.3 Let us finally consider the following single PD equation with two independent variables (x^1, x^2) and two unknowns (y^1, y^2), where again a is an arbitrary constant parameter:

$$d_1 y^1 + d_2 y^2 - ax^2 y^1 = 0.$$

Our basic problem will be the following one:

Problem 2.4 How could one *classify* the three families of systems, depending on the value of the parameter involved?

Before providing the general solution at the end of the paper, like in any good crime story, let us get hints from Example 2.1 as, in this case only, we can refer to control theory. A similar problem could be proposed for systems depending on arbitrary functions of the independent variables and could be solved similarly, though with much more effort in general but this is out of the scope of this article (See [18, 19, 20] for more details, in particular the delicate example of the varying double pendulum or treat the three previous examples as an exercise).

From any control textbook, it is well known that any system of OD equations can be put into the so-called *Kalman form*

$$\dot{x} = Ax + Bu, \qquad y = Cx + Du,$$

where $u = (u^1, \ldots, u^p)$ is called the *input* and $x = (x^1, \ldots, x^n)$ is called the *state* (do not confuse with independent variables as we use standard notations in this domain) and $y = (y^1, \ldots, y^m)$ is called the *output*. In this system, A is a constant square $n \times n$ matrix (possibly degenerate), while B, C, D are respectively $n \times p, m \times n, m \times p$ constant rectangular matrices, all the previous matrices having coefficients in a given field of constants k, eventually containing arbitrary parameters like in an electrical circuit with a variable condenser.

As we shall quickly be convinced, that such a form is crucial for studying control systems, we list below its specific properties:

- First order system.
- No zero order equation.
- Differentially independent quations.
- No derivative of input.

A key concept in control theory is the following definition [8, 19].

Definition 2.5 Such a control system is said to be *controllable* if for any given finite time $T \leq \infty$, one can find an input function $u(t)$ such that the corresponding trajectory

$x(t) = x_u(t)$ starting at time $t = 0$ from any given *initial* point $x(0)$ may reach at time $t = T$ any given *final* point $x(T)$.

Surprisingly, such a *purely functional definition* can be tested by a *purely formal test* [8, 19].

KALMAN TEST: The previous system is controllable if and only if

$$\text{rank}(B, AB, \ldots, A^{n-1}B) = n.$$

The matrix involved in the test is called the *controllability matrix* and the rank stabilizes because, according to the well known Cayley-Hamilton theorem, A^n linearly depends on $I = A^0, A = A^1, \ldots, A^{n-1}$. However, it is rarely quoted in control textbooks that *the previous definition is only valid for control systems in Kalman form* (!).

Counterexample 2.6 For a single input/single output (SISO) system $\dot{x} - \dot{u} = 0, y = x$, we have $x = u + cst$ and any $u(t)$ is convenient provided that $u(T) - u(0) = x(T) - x(0)$ given. Accordingly, the apparently technical condition saying that the derivatives of the input must not appear is in fact a crucial condition as otherwise no similar systematic test is known.

Remark 2.7 Of course, if we set $\bar{x} = x - u$, we get the new system $\dot{\bar{x}} = 0$ in Kalman form, which is of course not controllable but this is not helpful because we do not know what kind of intrinsic presentation-free property of the system is hidden behind the definition of controllability. However, even at this stage, *we already know from the introduction that it must (!) be expressed in terms of a certain extension module* .

Of course, according to its definition/test and tradition, controllability seems to highly depend on the choice of the control variables. Example 2.1 will prove that it may not be true at all (!).

Example 2.8 Setting $y^1 = x^1, y^2 = x^2, y^3 = u$ and $\bar{x}^1 = x^1 - u, \bar{x}^2 = x^2 - u$, we obtain the Kalman system $\dot{\bar{x}}^1 = a\bar{x}^2 + au, \dot{\bar{x}}^2 = \bar{x}^1 + u$ which is easily seen to be controllable iff $a \neq 0, a \neq 1$. Similarly, setting $y^1 = x^1, y^2 = u, y^3 = x^2$ and again $\bar{x}^1 = x^1 - u, \bar{x}^2 = x^2 - u$, though with a completely different meaning, we get the new Kalman system $\dot{\bar{x}}^1 = -\bar{x}^1 + (a - 1)u, \dot{\bar{x}}^2 = -\bar{x}^1 - u$ which is again easily seen to be controllable iff $a \neq 0, a \neq 1$.

Hence we arrive to the following striking conjecture [2, 15]:

Conjecture 2.9 *Controllability is a structural property of a control system that does not depend on the meaning of the system variables or even on the presentation of the system.*

The hint towards the formal definition of controllability will come from a careful examination of the non-controllable case. Indeed, when $a = 1$, adding the two OD equations in order to eliminate y^3 and setting $z = y^1 - y^2$, we obtain at once $\dot{z} + z = 0$. Hence, if $z = 0$ at time zero, then $z = 0$ at any time and one cannot escape from the plane $y^1 - y^2 = 0$. Similarly, when $a = 0$, setting $z' = y^1 - y^3$, we get $\dot{z}' = 0$. Accordingly, calling any linear combination of the unknowns and their derivatives an *observable element*, we state the following [15]:

Definition 2.10 An observable element is said to be *free* if it does not satisfy any OD or PD equation for itself; it is called *autonomous* otherwise.

Proposition 2.11 *The Kalman test amounts to the search for autonomous elements. The maximum number of autonomous elements linearly independent over k is equal to the corank of the controllability matrix. In particular, a system is controllable iff any observable element is free.*

Proof. With $x = y$ and thus $\dot{y} = Ay + Bu$, consider the element $z = \lambda y + \mu u + \ldots$. If z were autonomous, it should satisfy at least one OD equation for itself, for example of the type $\dot{z} + \nu z = 0$ and all the derivatives of z should be autonomous too by differentiating sufficiently often. Substituting, we should get $(\lambda A + \nu\lambda)y + (\lambda B + \nu\mu)u + \mu\dot{u} + \ldots = 0$ and z is autonomous iff $\mu = 0$, that is if $z = \lambda y$ as $\dot{u}$ is missing in the Kalman form. Now, if z is autonomous, then $\dot{z} = \lambda\dot{y} = \lambda Ay + \lambda Bu$ is also autonomous as we said and thus $\lambda B = 0$ leading to $\dot{z} = \lambda Ay$. Continuing this way, we get successively $\lambda B = 0, \lambda AB = 0, \ldots, \lambda A^{n-1}B = 0$ and the row vector λ is killed by the controllability matrix. The remainder of the proof is just Cramer's rule in linear algebra.   ☐

If A is a ring containing 1 and having elements $a, b, c, \ldots \in A$, we have $ab, a+b \in A$. If now M is a *module* over A, that is to say roughly a set of elements $x, y, z, \ldots \in M$ such that $ax \in M, \forall a \in A$ and $x + y \in M$, then we may say as usual

Definition 2.12 The torsion submodule of M is the module $t(M) = \{x \in M \mid \exists 0 \neq a \in A, ax = 0\}$. The module M is said to be *torsion-free* if $t(M) = 0$ and M is said to be a *torsion module* if $t(M) = M$.

In the general case of any system of OD or PD equations with unknowns $y = (y^1, \ldots, y^m)$, we may introduce the differential module or D-module M obtained by dividing the D-module $Dy = Dy^1 + \ldots + Dy^m$ by the differential submodule generated by the given equations and all their derivatives [11, 19]. More precisely, studying the formal relations existing between the three possible interpretations

$$\text{SYSTEM} \quad \leftrightarrow \quad \text{OPERATOR} \quad \leftrightarrow \quad \text{MODULE}:$$

we may always introduce a differential operator $\mathcal{D}$ represented by a rectangular matrix with elements in D and write down the system equations symbolically in the form $\mathcal{D}y = 0$. Hence we have $M = Dy/D\mathcal{D}y$, and we obtain the following characterization:

Corollary 2.13 *A control system is controllable iff the corresponding differential module is torsion-free.*

This is still more like a definition than a test; a first step towards a test is based on the following standard definition.

Definition 2.14 If $P = a^\mu d_\mu \in D$, the *formal adjoint* of P is the operator $\mathrm{ad}(P) = \sum(-1)^{|\mu|}d_\mu a^\mu$ that we may transform to normal form by pushing all the derivations to the right using the commutation formula which is indeed exactly equivalent to a formula of integration by part.

It is not so easy to get the relation $\mathrm{ad}(PQ) = \mathrm{ad}(Q)\mathrm{ad}(P)$ where, for simplicity $PQ = P \circ Q$ for the composition of two operators, with of course $\mathrm{ad}(\mathrm{ad}(P)) = P$, and the previous definition can be extended by linearity to $\mathcal{D}$ in order to get $\mathrm{ad}(\mathcal{D})$.

Definition 2.15 We define N from $\mathrm{ad}(\mathcal{D})$ exactly as we defined M from $\mathcal{D}$, that is $N = D\lambda / D\mathrm{ad}(\mathcal{D})\lambda$ when the number of new unknowns λ is equal to the number of OD/PD equations defining $\mathcal{D}$. Of course, N *highly depends on the presentation of M*.

Remark 2.16 In actual engineering practice, we may consider λ as a dual vector density of test functions and we obtain by contraction the equality of volume n-forms (carried out with due care) expressed by the relation:

$$\langle \lambda, \mathcal{D}y \rangle = \langle \mathrm{ad}(\mathcal{D})\lambda, y \rangle + \mathrm{div}(\ldots),$$

where the divergence is in fact the exterior derivative of an $(n-1)$-form according to Stokes formula of integration by part. For more details on this difficult question, see [19].

The link with Gröbner bases will be provided by the elimination steps 3 and 5 in the following purely formal test, first provided in [16, 18] for general systems.

TORSION-FREENESS TEST: The test needs 5 steps:

1) Write the system of OD/PD equations as the kernel of an operator $\mathcal{D}$.

2) Construct the formal adjoint $\mathrm{ad}(\mathcal{D})$.

3) Work out generating compatibility conditions of $\mathrm{ad}(\mathcal{D})$ in the form $\mathrm{ad}(\mathcal{D}_{-1})$.

4) Construct $\mathrm{ad}(\mathrm{ad}(\mathcal{D}_{-1})) = \mathcal{D}_{-1}$.

5) Work out generating compatibility conditions $\mathcal{D}'$ for $\mathcal{D}_{-1}$.

We have $\mathrm{ad}(\mathcal{D} \circ \mathcal{D}_{-1}) = \mathrm{ad}(\mathcal{D}_{-1}) \circ \mathrm{ad}(\mathcal{D}) \equiv 0 \Rightarrow \mathcal{D} \circ \mathcal{D}_{-1} \equiv 0$, and thus $\mathcal{D}$ ranges among the compatibility conditions of $\mathcal{D}_{-1}$, a result symbolically written as $\mathcal{D} \leq \mathcal{D}'$. The differential module M determined by $\mathcal{D}$ is torsion-free iff $\mathcal{D} = \mathcal{D}'$. In that case, with a slight abuse of language, the kernel of $\mathcal{D}$ is parametrized by the image of $\mathcal{D}_{-1}$. Otherwise, any compatibility condition in $\mathcal{D}'$ not in $\mathcal{D}$ provides a torsion element of M and $\mathcal{D}_{-1}$ parametrizes the system determined by $M/t(M)$ or, equivalently, M/tM is the torsion-free differential module determined by $\mathcal{D}'$.

The above test has been crucially used in order to prove that the 10 Einstein second order PD equations to be found in General Relativity could not have a generic parametrized solution like, for example, Maxwell equations in electromagnetism (4-potential) or stress equations in continuum mechanics (Airy function). This negative answer to a conjecture proposed in 1970 by J. Wheeler explains why this result, showing that the Einstein equations cannot be considered as field equations, has not been acknowledged up to now. (For more details, see [16, 20, 28].)

For the reader aware of homological algebra, the above test comparing $\mathcal{D}$ and $\mathcal{D}'$ just amounts to construct $\mathrm{ext}_D^1(N, D)$ in the operator framework, so we get at once a homological consequence (see Remark 3.25 below)

Corollary 2.17 *The module* $\mathrm{ext}_D^1(N, D) = t(M)$ *does not depend on the presentation of* N.

Other extension modules like $\mathrm{ext}_D^2(N, D)$ could be similarly introduced by considering successively $\mathcal{D}_{-2}$ from $\mathcal{D}_{-1}$ exactly as we got $\mathcal{D}_{-1}$ from $\mathcal{D}$ and so on [9, 18, 20].

Example 2.18 Identifying systems with second members with operators as it is done formally in symbolic packages, while taking into account Example 2.1, we successively get:

$$\mathcal{D} \leftrightarrow \dot{y}^1 - a y^2 - \dot{y}^3 = u, y^1 - \dot{y}^2 + \dot{y}^3 = v$$

$$\mathrm{ad}(\mathcal{D}) \leftrightarrow -\dot{\lambda}^1 + \lambda^2 = \mu^1, \dot{\lambda}^2 - a\lambda^1 = \mu^2, \dot{\lambda}^1 - \dot{\lambda}^2 = \mu^3$$

$$\dot{\mu}^1 + \dot{\mu}^2 + \dot{\mu}^3 + (a - 1)\mu^2 + a\mu^3 = -a(a - 1)\lambda^1$$

- $a \neq 0, a \neq 1$:

$$\mathrm{ad}(\mathcal{D}_{-1}) \leftrightarrow \ddot{\mu}^1 + \ddot{\mu}^2 + \ddot{\mu}^3 - a\dot{\mu}^1 - \dot{\mu}^2 - a\mu^3 = \nu$$

$$\mathcal{D}_{-1} \leftrightarrow \ddot{\xi} + a\dot{\xi} = y^1, \ddot{\xi} + \dot{\xi} = y^2, \ddot{\xi} - a\xi = y^3$$

$$\mathcal{D}' = \mathcal{D} \Rightarrow t(M) = 0$$

- $a = 0$:

$$\mathrm{ad}(\mathcal{D}_{-1}) \leftrightarrow \dot{\mu}^1 + \dot{\mu}^2 + \dot{\mu}^3 - \mu^2 = \nu$$

$$\mathcal{D}_{-1} \leftrightarrow -\dot{\xi} = y^1, -\dot{\xi} + \xi = y^2, -\dot{\xi} = y^3$$

$$\mathcal{D}' \leftrightarrow \mathcal{D}, y^1 - y^3 = w$$

The module $t(M)$ is generated by $z' = y^1 - y^3$.

- $a = 1$: The module $t(M)$ is generated by $z'' = y^1 - y^2$.

Example 2.19 With Example 2.2, we get analogously:
- $a \neq 0$: The module $t(M)$ is generated by $z = y^1 - y^2 - 2y^3$.
- $a = 0$: The module $t(M)$ is generated by $z' = y^1 - y^3, z'' = y^2 - y^3$.

Example 2.20 Considering Example 2.3, M is torsion-free for all a, though the generic parametrization may be quite different for $a = 0$ (first order, curl) and for $a \neq 0$ (second order).

As M is *never* torsion-free in Example 2.2 and *always* torsion-free in Exampe 2.3, another concept MUST be introduced in order to classify systems or modules depending on parameters. A first idea is to look at the kernel of $\mathrm{ad}(\mathcal{D})$ or N and we have the following well-known result [10, 19, 21]

Theorem 2.21 *When $\mathcal{D}$ is surjective, if $\mathrm{ad}(\mathcal{D})$ is injective or, equivalently, $N = 0$, then M is a projective module, that is to say a direct summand of a free module. Moreover M is free when D is a principal ideal ring (for example when $D = K[d]$ with $n = 1$) or when $D = k[d_1, \ldots, d_n]$ with k a field of constants (Quillen-Suslin theorem [25]).*

Example 2.22 The kernel of $\mathrm{ad}(\mathcal{D})$ in Example 2.1 is zero if and only if $a \neq 0, a \neq 1$. In particular, for a Kalman type system $-\dot{x} + Ax + Bu = 0$, multiplying on the left by a test row vector, the kernel of the adjoint operator is defined by the OD equations $x \to \dot{\lambda} + \lambda A = 0, u \to \lambda B = 0$. Differentiating the second while tking into acount the first, we get $\lambda AB = 0$ an thus $\lambda A^2 B = 0, \ldots$. Strikingly, the controllability matrix just appears as a way to check the injectivity of the adjoint or its lack of formal integrability (see Definition 3.3 thereafter) by *saturating* the number of zero order equations.

Example 2.23 The kernel of $\mathrm{ad}(\mathcal{D})$ in Example 2.3 is zero if and only if $a \neq 0$. Accordingly, M is torsion-free but not projective and thus not free when $a = 0$ and projective but not free when $a \neq 0$.

3   Purity

The hard section of this paper will be to explain in an intrinsic way the dependence of Example 2.2 on its parameter that we have already exhibited as a pure calculation. For this, we need a few more results on the formal theory of systems of OD/PD equations in order to understand how to generalize the Kalman form from the OD case to the PD case.

If E is a vector bundle over the base manifold X with projection π and local coordinates $(x, y) = (x^i, y^k)$ projecting onto (x^i) for $i = 1, \ldots, n$ and $k = 1, \ldots, m$, identifying a map with its graph, a (local) section $f : U \subset X \to E$ is such that $\pi \circ f = id$ on U and we write $y^k = f^k(x)$ or simply $y = f(x)$. For any change of local coordinates $(x, y) \to (\bar{x} = \varphi(x), \bar{y} = A(x)y)$ on E, the change of section is $y = f(x) \to \bar{y} = \bar{f}(\bar{x})$ such that $\bar{f}^l(\varphi(x)) \equiv A_k^l(x)f^k(x)$. Differentiating with respect to x^i and using new coordinates y_i^k in place of $\partial_i f^k(x)$, we obtain $\bar{y}_r^l \partial_i \varphi^r(x) = A_k^l(x)y_i^k + \partial_i A_k^l(x)y^k$. Prolonging the procedure up to order q, we may construct in this way, by patching coordinates, a vector bundle $J_q(E)$ over X, called the *jet bundle of order q* with local coordinates $(x, y_q) = (x^i, y_\mu^k)$ with $0 \leq |\mu| \leq q$ and $y_0^k = y^k$.

Definition 3.1 (Think of computer algebra systems like Maple.) A *system* of PD equations of order q on E is a vector subbundle $R_q \subset J_q(E)$ locally defined by a constant rank system of linear equations for the jets of order q:

$$R_q \qquad a_k^{\tau\mu}(x)y_\mu^k = 0$$

Its *first prolongation* R_{q+1} will be defined by the equations:

$$R_{q+1} \qquad a_k^{\tau\mu}(x)y_\mu^k = 0, a_k^{\tau\mu}(x)y_{\mu+1_i}^k + \partial_i a_k^{\tau\mu}(x)y_\mu^k = 0$$

Remark 3.2 The first prolongation may not be defined by a system of constant rank as can easily be seen for $xy_x - y = 0 \Rightarrow xy_{xx} = 0$ where the rank drops at $x = 0$.

The next definition will be crucial for our purpose.

Definition 3.3 A system R_q is said to be *formally integrable* if all the R_{q+r} are vector bundles $\forall r \geq 0$ and no new equation of order $q + r$ can be obtained by prolonging the given PD equations more than r times, $\forall r \geq 0$.

Counterexample 3.4 Set $n = 2, m = 1, q = 2$.

$$R_2 \qquad d_{22}y = 0 \qquad d_{12}y - y = 0.$$

We let the reader check that one needs 2 prolongations in order to get $y = 0$ and 2 additional prolongations in order to cancel all jets of order 2.

Counterexample 3.5 Set $n = 4, m = 1, q = 1, K = \mathbb{Q}(x^1, x^2, x^3, x^4)$.

$$R_1 \qquad y_4 - x^3 y_2 - y = 0 \qquad y_3 - x^4 y_1 = 0.$$

Again, the reader will check easily that the subsystem

$$R_1' \subset R_1 \qquad y_4 - x^3 y_2 - y = 0 \qquad y_3 - x^4 y_1 = 0 \qquad y_2 - y_1 = 0,$$

namely the projection of R_2 to R_1, is formally integrable.

Finding an inrinsic test has been achieved by D. C. Spencer in 1970 along coordinate dependent lines sketched by M. Janet [5] in 1920 and W. Gröbner in 1940, as we have already said. The key ingredient, missing explicitly before the modern approach, is provided by the following definition.

Definition 3.6 The family of vector spaces over X given by

$$g_{q+r} \qquad a_k^{\tau\mu}(x)v_{\mu+\nu}^k = 0 \qquad |\mu| = q, |\nu| = r$$

is called the *symbol* of order $q + r$ and only depends on g_q.

The following procedure, *where one may have to change linearly the independent variables if necessary*, is the key step towards the next definition, which is intrinsic even though it must be checked in a particular coordinate system called δ-*regular* [18, 19]:

- *Equations of class n*: Solve the maximum number $\beta = \beta_q^n$ of equations with respect to the jets of order q and class n. Then call $(x^1, \ldots, x^n)$ *multiplicative variables*.

- *Equations of class i*: Solve the maximum number of *remaining* equations with respect to the jets of order q and class i. Then call $(x^1, \ldots, x^i)$ *multiplicative variables* and $(x^{i+1}, \ldots, x^n)$ *non-multiplicative variables*.

- *Remaining equations equations of order $\leq q - 1$*: Call $(x^1, \ldots, x^n)$ *non-multiplicative variables*.

Definition 3.7 A system of PD equations is said to be *involutive* if its first prolongation can be achieved by prolonging its equations only with respect to the corresponding multiplicative variables.

Remark 3.8 For an involutive system, $(y^{\beta+1}, \ldots, y^m)$ can be given arbitrarily and "may" constitute the *input* variables in control theory, though it is not necessary to make such a choice. The intrinsic number $\alpha = m - \beta$ is called the *n-character* and is the module counterpart of the so-called *differential transcendence degree* in differential algebra.

Example 3.9 The system R_1 of counterexample 3.5 is not involutive but the subsystem R_1' is involutive with $\beta = 1$ equation of class 4, 1 equation of class 3 and 1 equation of class 2.

Counterexample 3.10 The system $y_{12} = 0, y_{11} = 0$ is involutive though not in δ-regular coordinates and we need, for example, to effect the change of local coordinates $(x^1, x^2) \rightarrow (x^2, x^1)$.

Though the preceding description was known to Janet (he called it "modules de formes en involution"), surprisingly he never used it explicitly. In any case, such a definition is far from being intrinsic, and we just saw that the hard step will be achieved from the Spencer cohomology that will also play an important part in the so-called *reduction to first order*, a result not so well known today as we shall see.

Let us consider $J_{q+1}(E)$ with jet coordinates $\{y_\lambda^k \mid 0 \leq |\lambda| \leq q + 1\}$ and $J_1(J_q(E))$ with jet coordinates $\{z_\mu^k, z_{\mu,i}^k \mid 0 \leq |\mu| \leq q, i = 1, \ldots, n\}$. At first sight, the canonical inclusion $J_{q+1}(E) \subset J_1(J_q(E))$ seems evident by setting the *two kinds* of equations:

$$z_{\nu,i}^k - z_{\nu+1_i}^k = 0, \qquad\qquad 0 \leq |\nu| \leq q - 1$$

$$z_{\nu+1_j,i}^k - z_{\nu+1_i,j}^k = 0, \qquad\qquad |\nu| = q - 1$$

or using the parametrization $z_{\nu,i}^k = y_{\nu+1_i}^k$ for $|\mu| = q$. However, we shall soon discover that computing the true number of equations of the second kind does involve unexpected algebra.

Let T^* be the cotangent vector bundle of 1-forms on X and $\wedge^s T^*$ be the vector bundle of s-forms on X with usual bases $\{dx^I = dx^{i_1} \wedge \ldots \wedge dx^{i_s}\}$ where we have set $I = (i_1 < \ldots < i_s)$. Also, let $S_q T^*$ be the vector bundle of symmetric q-covariant tensors. We have:

Proposition 3.11 *There exists a map* $\delta : \wedge^s T^* \otimes S_{q+1} T^* \otimes E \to \wedge^{s+1} T^* \otimes S_q T^* \otimes E$ *which restricts to* $\delta : \wedge^s T^* \otimes g_{q+1} \to \wedge^{s+1} T^* \otimes g_q$ *and* $\delta^2 = \delta \circ \delta = 0$.

Proof. Let us introduce the family of s-forms $\omega = \{\omega_\mu^k = v_{\mu,I}^k dx^I\}$ and set $(\delta\omega)_\mu^k = dx^i \wedge \omega_{\mu+1_i}^k$. We obtain at once $(\delta^2\omega)_\mu^k = dx^i \wedge dx^j \wedge \omega_{\mu+1_i+1_j}^k = 0$.   □

The kernel of each δ in the first case is equal to the image of the preceding δ but this may no longer be true in the restricted case and we set:

Definition 3.12 We denote by $H_{q+r}^s(g_q)$ the cohomology at $\wedge^s T^* \otimes g_{q+r}$ of the restricted δ-sequence which only depends on g_q. The symbol g_q is said to be *s-acyclic* if $H_{q+r}^1 = \ldots = H_{q+r}^s = 0, \forall r \geq 0$, involutive if it is n-acyclic and finite type if $g_{q+r} = 0$ becomes trivially involutive for r large enough.

The preceding results will be used in proving the following technical result that will prove to be useful for our purpose [19, 26].

Proposition 3.13 $C_1(E) = J_1(J_q(E))/J_{q+1}(E) \simeq T^* \otimes J_{q-1}(E) \oplus \delta(T^* \otimes S_q T^* \otimes E)$.

Proof. The first commutative and exact diagram:

$$
\begin{array}{ccccccc}
 & 0 & & 0 & & 0 & \\
 & \downarrow & & \downarrow & & \downarrow & \\
0 \to & S_{q+1}T^* \otimes E & \to & T^* \otimes J_q(E) & \to & C_1(E) & \to 0 \\
 & \downarrow & & \downarrow & & \| & \\
0 \to & J_{q+1}(E) & \to & J_1(J_q(E)) & \to & C_1(E) & \to 0 \\
 & \downarrow & & \downarrow & & \downarrow & \\
0 \to & J_q(E) & = & J_q(E) & \to & 0 & \\
 & \downarrow & & \downarrow & & & \\
 & 0 & & 0 & & &
\end{array}
$$

shows that $C_1(E) \simeq T^* \otimes J_q(E)/S_{q+1}T^* \otimes E$. The proof finally depends on the following second commutative and exact diagram by using a (non-canonical) splitting of the right column:

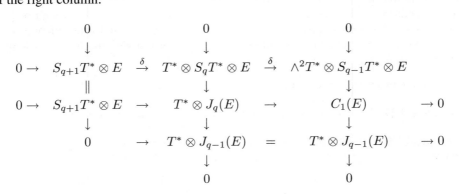

$$\square$$

These non-trivial results can be restricted to the systems and symbols. Accordingly, the inclusion $R_{q+1} \subset J_1(R_q)$ can be considered as a new first order system over R_q, called *first order reduction* or *Spencer form*. One obtains [18, 19]:

Proposition 3.14 *The first order reduction is formally integrable (involutive) whenever R_q is formally integrable (involutive). In that case, the reduction has no longer any zero order equation.*

Counterexample 3.15 In Counterexample 3.5, setting $z^1 = y, z^2 = y_1, z^3 = y_2$, the inclusion $R_2 \subset J_1(R_1)$ is defined by PD equations containing the zero order equation $z^3 - z^2 = 0$. However, the reduction $R'_2 \subset J_1(R'_1)$ no longer contains any zero order equation for $z^1 = y, z^2 = y_1$ and is involutive.

We have therefore set up the problem of "state", even for systems which are not of finite type and *it just remains to modify the Spencer form* in order to generalize the Kalman form to PD equations. Here is the procedure that must be followed in the case of a first order involutive system with no zero order equation, for example like the one we just obtained.

- Look at the equations of class n solved with respect to $y_n^1, \ldots, y_n^\beta$.
- Use integrations by part like:

$$y_n^1 - a(x)y_n^{\beta+1} = d_n(y^1 - a(x)y^{\beta+1}) + \partial_n a(x)y^{\beta+1} = \bar{y}_n^1 + \partial_n a(x)y^{\beta+1}.$$

- Modify $y^1, \ldots, y^\beta$ to $\bar{y}^1, \ldots, \bar{y}^\beta$ in order to *absorb* the various $y_n^{\beta+1}, \ldots, y_n^m$ only *appearing in the equations of class n*.

We have the following unexpected result:

Proposition 3.16 *The new equations of class n only contain $y_i^{\beta+1}, \ldots, y_i^m$ with $0 \le i \le n-1$ while the equations of class $1, \ldots, n-1$ no more contain $y^{\beta+1}, \ldots, y^m$ and their jets.*

Proof. The first assertion comes from the absorption procedure. Now, if y^m or y_i^m should appear in an equation of class $\leq n-1$, prolonging this equation with respect to the non-multiplicative variable x^n should bring y_n^m or y_{in}^m and (here involution is essential) we should get a linear combination of equations of various classes prolonged with respect to $x^1, \ldots, x^{n-1}$ *only*, but this is impossible. $\square$

The same proof as in Proposition 2.11 provides at once:

Corollary 3.17 *Any autonomous element, if it exists, only depends on $\bar{y}^1, \ldots, \bar{y}^\beta$.*

Example 3.18 Coming back to Example 2.2, while setting $\bar{y}^1 = y^1 - y^3, \bar{y}^2 = y^2 + y^3$, we obtain the new first order involutive system:

$$\bar{y}_2^2 - \bar{y}_1^2 - ay^3 = 0, \qquad \bar{y}_2^1 - \bar{y}_1^2 - ay^3 = 0, \qquad \bar{y}_1^1 - \bar{y}_1^2 = 0$$

with two equations of class 2 and one equation of class 1 in which y^3 surprisingly no longer appears. We have already seen in Example 2.19 that autonomous elements depend only on $(\bar{y}^1, \bar{y}^2)$.

We are now in a position to revisit Gröbner bases. First of all, we establish the following approximate "translation":

OLD VOCABULARY	NEW VOCABULARY
leading monomials, initials,…	symbol
passive, coherent,…	formally integrable
complete	involutive

Then it will be sufficient to consider two examples.

Example 3.19 Let $P_1 = (\chi_3)^2, P_2 = \chi_2\chi_3 - (\chi_1)^2, P_3 = (\chi_2)^2$ be three polynomials generating the ideal $\mathfrak{a} = (P_1, P_2, P_3) \subset \mathbb{Q}[\chi_1, \chi_2, \chi_3]$. The corresponding system

$$R_2: \qquad\qquad y_{33} = 0, y_{23} - y_{11} = 0, y_{22} = 0$$

is homogeneous and thus automatically formally integrable but g_2 is not involutive though finite type because $g_4 = 0$ (not evident !!). Elementary computations of ranks of matrices shows that the δ-map

$$0 \rightarrow \wedge^2 T^* \otimes g_3 \xrightarrow{\delta} \wedge^3 T^* \otimes g_2 \rightarrow 0$$

is an isomorphism and thus g_3 is 2-acyclic, a crucial intrinsic property [17, 18, 25] totally absent from any "old" work. Now, denoting the initial by in() while choosing the ordering $\chi_3 > \chi_1 > \chi_2$, we obtain

$$\text{in}(P_1) = (\chi_3)^2, \text{in}(-P_2) = (\chi_1)^2, \text{in}(P_3) = (\chi_2)^2$$

and $\{P_1, P_2, P_3\}$ is a Gröbner basis. However, choosing the ordering $\chi_3 > \chi_2 > \chi_1$, we have now

$$\text{in}(P_1) = (\chi_3)^2, \text{in}(P_2) = \chi_2\chi_3, \text{in}(P_3) = (\chi_2)^2$$

and $\{P_1, P_2, P_3\}$ is not a Gröbner basis because $y_{112} = 0, y_{113} = 0$ AND $y_{1111} = 0$ and a Gröbner basis could be $\{P_1, P_2, P_3, P_4 = (\chi_1)^2\chi_2, P_5 = (\chi_1)^2\chi_3, P_6 = (\chi_1)^4\}$ (!!!).

Example 3.20 Similarly, let $P_1 = (\chi_1)^2, P_2 = \chi_1\chi_3 + \chi_2$ be two polynomials generating the ideal $\mathfrak{a} = (P_1, P_2) \subset \mathbb{Q}[\chi_1, \chi_2, \chi_3]$. The corresponding system

$$R_2: \qquad y_{11} = 0, y_{13} + y_2 = 0$$

is not homogeneous and two prolongations provide the new second order PD equations $y_{12} = 0, y_{22} = 0$, so R_2 is not formally integrable and the problem is to know whether the new system

$$R_2': \qquad y_{11} = 0, y_{13} + y_2 = 0, y_{12} = 0, y_{22} = 0$$

is involutive or not as the given coordinate system is surely not δ-regular. Effecting the permutation $(1, 2, 3) \to (3, 2, 1)$, we get the new system

$$R_2'': \qquad y_{33} = 0, y_{23} = 0, y_{22} = 0, y_{13} + y_2 = 0,$$

which is indeed involutive and a Gröbner basis could be $\{(\chi_3)^2, \chi_2\chi_3, (\chi_2)^2, \chi_1\chi_3 + \chi_2\}$.

We hope to have convinced the reader of how the "modern" approach supersedes the "old" one.

It is only now, at the end of this paper, that we are in position to describe the dependence of the system of Example 2.2 on the parameter, in an intrinsic way through the concept of *purity* [1, 19] that we explain on a simple but illuminating example after recalling that the *characteristic variety* is the algebraic set defined by the radical of the polynomial ideal in $K[\chi] = K[\chi_1, \ldots, \chi_n]$ generated by the $m \times m$ minors of the *characteristic matrix* $(a_k^{\tau\mu}(x)\chi_\mu)$ where $|\mu| = q$ when R_q is involutive.

Example 3.21 The system of two PD equations $y_{22} = 0, y_{12} = 0$ defines a torsion module $t(M) = M$ and the corresponding characteristic variety is defined by the radical of the characteristic ideal, namely $\mathrm{rad}((\chi_2)^2, \chi_1\chi_2) = (\chi_2) \subset \mathbb{Q}[\chi_1, \chi_2]$. The characteristic variety $\chi_2 = 0$ has therefore dimension 1 as χ_1 is arbitrary. If we consider $z' = y_1$, we have $z'_2 = 0$ with the same characteristic variety. However, if we consider $z'' = y_2$, we have now $z''_2 = 0, z''_1 = 0$ and the characteristic variety is the origin $\chi_1 = 0, \chi_2 = 0$ with zero dimension. Hence, we can give a *structure* to $t(M)$ through the nested chain of differential submodules:

$$0 = t_2(M) \subset t_1(M) \subset t_0(M) = t(M) = M$$

in such a way that $t_1(M)/t_2(M)$ or $t_0(M)/t_1(M)$ only contain elements for which the characteristic varieties have dimension 0 or 1 and are therefore called *pure* for this reason.

We cannot enter into the details as it is a long story (Hilbert-Serre theorem involved with a lot of homological algebra and double ext functors [1, 19]) but the procedure is roughly as follows. If we have an involutive reduction to the first order, the set of equations of class $1 + \ldots +$ class $(n - r)$ defines again an involutive system and a differential module M_r with an epimorphism onto M which is defined by the whole set of equations of class $1+\ldots+$ class n. This epimorphism induces a morphism $t(M_r) \to t(M)$ with image $t_r(M)$. Now, $t(M)$ being a torsion module, its n-character is of course zero and it means that any element in $t(M)$ satisfies at least one PD equation in which d_n appears explicitly among $d_1, \ldots, d_n$. As the determination of the classes is done inductively by maximizing the number of equations in each class in decreasing order, starting from class n, any element of $t(M_r)$ satisfies at least one PD equation in which d_{n-r} appears explicitly among $d_1, \ldots, d_{n-r}$. Accordingly, as its image in $t(M)$ satisfies at least these $r+1$ equations (care !) in which $d_n, \ldots, d_{n-r}$ successively appear explicitly, the dimension of the corresponding characteristic variety is $< n - r$ and the codimension $> r$. We have therefore the nested chain of differential submodules:

$$0 = t_n(M) \subseteq t_{n-1}(M) \subseteq \ldots \subseteq t_1(M) \subseteq t_0(M) = t(M) \subseteq M.$$

Again, each element in $t_{r-1}(M)/t_r(M)$ has a characteristic variety of dimension $n-r$, thus of codimension r or is r-*pure*.

Example 3.22 Coming back finally to Example 2.2, the 3×3 characteristic matrix:

$$\begin{pmatrix} 0 & \chi_2 - \chi_1 & \chi_2 - \chi_1 \\ \chi_2 & -\chi_1 & -\chi_1 - \chi_2 \\ \chi_1 & -\chi_1 & -2\chi_1 \end{pmatrix}$$

has zero determinant as first column $-$ second column $+$ third column $= 0$ and the characteristic variety has dimension $n = 2$. It follows that $M/t(M)$ is pure of codimension 0 and in fact free (exercise). We consider two cases:

- $0 \neq a = 1$: $z = y^1 - y^2 - 2y^3 \to z_2 = 0, z_1 = 0$.

$$0 = t_2(M) \subset t_1(M) = t_0(M) = t(M) \subset M$$

- $a = 0$: $z' = y^2 + y^3, z'' = y^1 - y^3 \to z'_2 - z'_1 = 0$.

$$0 = t_2(M) \subset t_1(M) \subset t_0(M) = t(M) \subset M$$

where z' and z'' both have codimension 1 and are identified in the quotient $t_0(M)/t_1(M)$ because $z = z'' - z'$ has codimension 2. In both cases, the only equation of class 1 is $y_1^1 - y_1^2 - 2y_1^3 = 0$ and the corresponding torsion element z generates $t_1(M)$.

Example 3.23 With $n = 4, m = 1, q = 1$, the system $y_4 - x^3 y_2 - y = 0, y_3 - x^4 y_1 = 0$ of Counterexample 3.5 defines a torsion differential module M but is not formally integrable. The subsystem $y_4 - x^3 y_2 - y = 0, y_3 - x^4 y_1 = 0, y_2 - y_1 = 0$ is involutive and defines the same module which is therefore 3-pure.

Example 3.24 (Compare with [19], p. 219) With $n = 3, m = 3, q = 2$, the system $y_{33}^3 - y_1^1 = 0, y_{23}^3 - y_1^2 - y^3 = 0$ is not formally integrable. Adding the second order PD equation $y_{13}^2 - y_{22}^1 + y_3^3 = 0$ provides an involutive subsystem with 2 equations of class 3 and 1 equation of class 2, which defines a differential module of differential rank 1. To see this, one has to make the change of independent variables $\bar{x}^1 = x^1, \bar{x}^2 = x^2, \bar{x}^3 = x^3 + x^1$ in order to get a δ-regular coordinate system. Through the previous test (exercise), the only generating torsion element is $z = y_{33}^2 - y_{23}^1 + y^1$ satisfying $z_1 = 0$. The differential module $M' = M/t(M)$ is projective and thus free according to the Quillen-Suslin theorem. We have $M \simeq t(M) \oplus M'$ and $t(M)$ is 1-pure. (Exercise; *Hint*: Add the PD equation $z = 0$ to the previous system and find a new δ-regular coordinate system in which the system becomes involutive. Find the two compatibility conditions and check that the adjoint of the corresponding operator is injective).

Remark 3.25 One can prove [1, 18] that a finitely generated differential module M of codimension r is r-pure if and only if one has the inclusion of left differential modules:

$$0 \longrightarrow M \longrightarrow \text{ext}_D^r(\text{ext}_D^r(M, D), D).$$

More generally, one may obtain the pure quotient modules through the exact sequences

$$0 \longrightarrow t_r(M) \longrightarrow t_{r-1}(M) \longrightarrow \text{ext}_D^r(\text{ext}_D^r(M, D), D).$$

In particular, taking into account the relation $\text{ext}_D^0(M, D) = \hom_D(M, D)$, a module is 0-pure if and only if it is torsion-free and we have the exact sequence

$$0 \longrightarrow t(M) \longrightarrow M \xrightarrow{\epsilon} \hom_D(\hom_D(M, D), D),$$

where the morphism ϵ is defined by $\epsilon(m)(f) = f(m)$, for all $m \in M$, for all $f \in \hom_D(M, D)$ [19, 20]. This is the main reason for which *any* test for torsion-freeness/controllability *must* be related to double duality, even if this result is breaking many engineering traditions!

4 Conclusion

We have given reasons for revisiting Gröbner bases algorithm within the framework of the formal theory of systems of partial differential equations and its algebraic counterpart, namely the algebraic analysis of such systems. *The key idea is to use systems, operators or modules whenever it is more convenient.* Indeed, certain concepts are natural in one framework but not in the others and torsion is a good example. Finally, we have provided and illustrated the two ways known today for classifying systems depending on parameters.

The first is obtained through the inclusions of various types of (differential) modules:

$$\text{free} \subset \text{projective} \subset \ldots \subset \text{torsion-free}.$$

The second, that must be used for modules that are not torsion-free and thus escape from the preceding classification, is to study the torsion submodule through a nested chain of various submodules, in such a way that the quotient of two successive ones is pure.

Both classifications are based on very delicate results of homological algebra (duality, biduality, extension modules,...). Accordingly, we end this conclusion warning the reader that should find these new techniques far too difficult for engineering applications, that, according to basic hard results of homological algebra, as already said in the introduction:

The use of single or double extension modules is unescapable for applications.

Bibliography

[1] J. E. Bjork, Analytic D-modules and Applications, Kluwer, 1993.

[2] H. Blomberg, Y. Ylinen, Algebraic Theory for Multivariable Linear Systems, Academic Press, 1983.

[3] B. Buchberger, Ein Algorithmus zum Auffinden der Basiselemente des Restklassenringes nach einem Multidimensionalen Polynomideal, PhD thesis (thesis advisor W. Gröbner), University of Innsbruck, Austria, 1965. Journal version: Ein algorithmisches Kriterium für die Lösbarkeit eines algebraischen Gleichungssystems, Aequationes Mathematicae 1970, 4, 374–383. English translation of thesis: An Algorithm for Finding the Basis Elements in the Residue Class Ring Modulo a Zero Dimensional Polynomial Ideal, Journal of Symbolic Computations, Special Issue on logic, Mathematics and Computer Sciences: Interactions, Vol 14, Nb 34, 2006, 475-511.

[4] E. and F. Cosserat, Théorie des Corps Déformables, Hermann, Paris, 1909.

[5] V. P. Gerdt, Y.A. Blinkov, Minimum Involutive Bases, Mathematics and Computers in Simulations, 45, 1998, 543-560

[6] W. Gröbner, Über die Algebraischen Eigenschaften der Integrale von Linearen Differentialgleichungen mit Konstanten Koeffizienten, Monatsh. der Math., 47, 1939, 247-284.

[7] M. Janet, Sur les Systèmes aux dérivées partielles, Journal de Math., 8, 3, 1920, 65-151.

[8] E. R. Kalman, Y. C. Yo, K. S. Narenda, Controllability of Linear Dynamical Systems, Contrib. Diff. Equations, 1, 2, 1963, 189-213.

[9] M. Kashiwara, Algebraic Study of Systems of Partial Differential Equations, Mémoires de la Société Mathématique de France 63, 1995, (Transl. from Japanese of his 1970 Master's Thesis).

[10] E. Kunz, Introduction to Commutative Algebra and Algebraic Geometry, Birkhäuser, 1985.

[11] P. Maisonobe, C. Sabbah, D-Modules Cohérents et Holonomes, Travaux en Cours, 45, Hermann, Paris, 1993.

[12] B. Malgrange, Cohomologie de Spencer (d'après Quillen), Sém. Math. Orsay, 1966.

[13] U. Oberst, Multidimensional Constant Linear Systems, Acta Appl. Math., 20, 1990, 1-175.

[14] V. P. Palamodov, Linear Differential Operators with Constant Coefficients, Grundlehren der Mathematischen Wissenschaften 168, Springer, 1970.

[15] J.-F. Pommaret, Géométrie Différentielle Algébrique et Théorie du Contrôle, C. R. Acad. Sci. Paris, 302, série I, 1986, 547-550.

[16] J.-F. Pommaret, Dualité Différentielle et Applications, C. R. Acad. Sci. Paris, 320, Série I, 1995, 1225-1230.

[17] J.-F. Pommaret, François Cosserat et le Secret de la Théorie Mathématique de l'Elasticité, Annales des Ponts et Chaussées, 82, 1997, 59-66.

[18] J.-F. Pommaret, Partial Differential Equations and Group Theory: New Perspectives for Applications, Kluwer, 1994.

[19] J.-F. Pommaret, Partial Differential Control Theory, Kluwer, 2001. (http://cermics.enpc.fr/~pommaret/home.html).

[20] J.-F. Pommaret, Algebraic Analysis of Control Systems Defined by Partial Differential Equations, in Advanced Topics in Control Systems Theory, Lecture Notes in Control and Information Sciences 311, Chapter 5, Springer, 2005, 155-223.

[21] J.-F. Pommaret, A. Quadrat, Algebraic Analysis of Linear Multidimensional Control Systems, IMA Journal of Mathematical Control and Informations, 16, 1999, 275-297.

[22] A. Quadrat, Analyse Algébrique des Systèmes de Contrôle Linéaires Multidimensionnels, Thèse de Docteur de l'Ecole Nationale des Ponts et Chaussées, 1999 (http://www-sop.inria.fr/cafe/Alban.Quadrat/index.html).

[23] C. Riquier, Les Systèmes d'Equations aux Dérivées Partielles, Gauthier-Villars, Paris, 1910.

[24] J. F. Ritt, Differential Algebra, AMS Coloq. Publ., 33, 1950, and Dover, 1966.

[25] J. J. Rotman, An Introduction to Homological Algebra, Pure and Applied Mathematics, Academic Press, 1979.

[26] D. C. Spencer, Overdetermined Systems of Partial Differential Equations, Bull. Amer. Math. Soc., 75, 1965, 1-114.

[27] J. Wood, Modules and Behaviours in nD Systems Theory, Multidimensional Systems and Signal Processing, 11, 2000, 11-48.

[28] E. Zerz, Topics in Multidimensional Linear Systems Theory, Lecture Notes in Control and Information Sciences 256, Springer, 2000.

Author information

Jean-François Pommaret, CERMICS, Ecole Nationale des Ponts et Chaussées, 6/8 Av. Blaise Pascal, 77455 Marne-la-Vallée Cedex 02, France.
Email: pommaret@cermics.enpc.fr, jean-francois.pommaret@wanadoo.fr

Radon Series Comp. Appl. Math **2**, 23–41

Solving Systems of Linear Partial Difference and Differential Equations with Constant Coefficients Using Gröbner Bases

Ulrich Oberst and Franz Pauer

Key words. Partial difference equation, partial differential equation, Gröbner basis.

AMS classification. 13P10, 39A10, 93C20.

1 Introduction

The main objective of this tutorial paper is to give an elementary presentation of a method developed in [24] to solve systems of linear partial difference equations with constant coefficients using Gröbner bases. This is done in Sections 2, 3 and 5, these sections are self-contained. In Sections 4 and 6 related results of [24] and [28] are described without proofs. The last section contains some historical remarks.

2 The Problem

Let F be a field and let k, ℓ, m, n be positive integers. By $M^{\ell \times m}$ we denote the set of all $\ell \times m$-matrices with entries in a set M.

Definition 2.1 A *system of k linear partial difference equations with constant coefficients for one unknown function in n variables* is given by

(1) a family

$$(R(\alpha))_{\alpha \in \mathbb{N}^n}$$

of columns $R(\alpha) \in F^{k \times 1}$, where only finitely many $R(\alpha)$ are different from 0, and

(2) a family

$$(v(\alpha))_{\alpha \in \mathbb{N}^n}$$

of columns $v(\alpha) \in F^{k \times 1}$.

A *solution* of this system is a family

$$(w(\gamma))_{\gamma \in \mathbb{N}^n}$$

of elements $w(\gamma) \in F$ (i.e. a function $w : \mathbb{N}^n \longrightarrow F$, $\gamma \mapsto w(\gamma)$) such that for all $\beta \in \mathbb{N}^n$

$$\sum_{\alpha \in \mathbb{N}^n} w(\alpha + \beta) R(\alpha) = v(\beta).$$

The system is *homogeneous* if and only if $v(\beta) = 0$, for all $\beta \in \mathbb{N}^n$.

Example 2.2 The problem "Find a function $w : \mathbb{N}^2 \longrightarrow \mathbb{Q}$ such that for all $\beta \in \mathbb{N}^2$

$$
\begin{aligned}
2w((2,1) + \beta) + w(\beta) &= v_1(\beta) \\
3w((1,2) + \beta) + 2w(\beta) &= v_2(\beta)
\end{aligned},
$$

where

$$v(\beta) := \begin{pmatrix} v_1(\beta) \\ v_2(\beta) \end{pmatrix} = \begin{pmatrix} 0 \\ 0 \end{pmatrix} \text{ for } \beta \in \mathbb{N}^2 \setminus \{(1,1), (0,2), (2,3)\}$$

and

$$v(1,1) = \begin{pmatrix} 0 \\ 15 \end{pmatrix}, v(0,2) = \begin{pmatrix} 13 \\ 6 \end{pmatrix}, v(2,3) = \begin{pmatrix} 5 \\ 10 \end{pmatrix}$$ "

is a system of two linear partial difference equations with constant coefficients for one unknown function in two variables. It is defined by the family

$$R(\alpha) := \begin{pmatrix} 0 \\ 0 \end{pmatrix} \text{ for } \alpha \in \mathbb{N}^2 \setminus \{(0,0), (1,2), (2,1)\}$$

and

$$R(0,0) = \begin{pmatrix} 1 \\ 2 \end{pmatrix}, R(2,1) = \begin{pmatrix} 2 \\ 0 \end{pmatrix}, R(1,2) = \begin{pmatrix} 0 \\ 3 \end{pmatrix}$$

and the family $(v(\alpha))_{\alpha \in \mathbb{N}^2}$.

A solution of the system in Definition 2.1 is a function $w : \mathbb{N}^n \longrightarrow F$, $\alpha \mapsto w(\alpha)$. We write $F^{\mathbb{N}^n}$ for the vector space of all functions from $\mathbb{N}^n$ to F. There are other useful interpretations of the function w:

- If we choose n letters $z_1, \ldots, z_n$, the family w can be written as the power series

$$\sum_{\alpha \in \mathbb{N}^n} w(\alpha) z^\alpha \, ,$$

where $z^\alpha := z_1^{\alpha_1} z_2^{\alpha_2} \ldots z_n^{\alpha_n}$. If only finitely many $w(\alpha)$ are different from 0, then $\sum_{\alpha \in \mathbb{N}^n} w(\alpha) z^\alpha$ is a polynomial.

Let $F[[z]] := F[[z_1, \ldots, z_n]]$ be the F-algebra of power series in $z_1, \ldots, z_n$.

- Consider the n-variate polynomial ring $F[s] := F[s_1, \ldots, s_n]$. Then w defines a linear function

$$\varphi : F[s] \longrightarrow F$$

by

$$\varphi(s^\alpha) := w(\alpha) \, , \text{ for all } \alpha \in \mathbb{N}^n \, ,$$

where $s^\alpha := s_1^{\alpha_1} s_2^{\alpha_2} \ldots s_n^{\alpha_n}$.

We denote by $\operatorname{Hom}_F(F[s], F)$ the vector-space of all linear maps from $F[s]$ to F.

The vector spaces $F^{\mathbb{N}^n}$, $F[[z]]$, and $\operatorname{Hom}_F(F[s], F)$ are $F[s]$-modules in a natural way: Let $s^\alpha \in F[s]$, $w \in F^{\mathbb{N}^n}$, $\sum_{\beta \in \mathbb{N}^n} c_\beta z^\beta \in F[[z]]$, and $\varphi \in \operatorname{Hom}_F(F[s], F)$. Then

$$(s^\alpha \circ w)(\beta) := w(\alpha + \beta), \text{ for all } \beta \in \mathbb{N}^n,$$

$$s^\alpha \circ \sum_{\beta \in \mathbb{N}^n} c_\beta z^\beta := \sum_{\beta \in \mathbb{N}^n} c_{\alpha+\beta} z^\beta = \sum_{\beta \in \alpha + \mathbb{N}^n} c_\beta z^{\beta-\alpha} \, ,$$

and

$$(s^\alpha \circ \varphi)(s^\beta) := \varphi(s^{\alpha+\beta}), \text{ for all } \beta \in \mathbb{N}^n.$$

It is easy to verify that the maps

$$F[[z]] \longrightarrow F^{\mathbb{N}^n} \longrightarrow \operatorname{Hom}_F(F[s], F)$$

$$\sum_{\beta \in \mathbb{N}^n} c_\beta z^\beta \mapsto (c_\beta)_{\beta \in \mathbb{N}^n} \mapsto \varphi \text{ (where } \varphi(s^\alpha) = c_\alpha)$$

are isomorphisms of $F[s]$-modules.

Thus, Definition 2.1 could equivalently be formulated as:

Definition 2.3 A *system of k linear partial difference equations with constant coefficients for one unknown function in n variables* is given by

(1') a column

$$\begin{pmatrix} R_1 \\ \vdots \\ R_k \end{pmatrix} \in F[s]^{k \times 1}$$

of polynomials $R_i := \sum_{\alpha \in \mathbb{N}^n} R_i(\alpha) s^\alpha \in F[s]$, and

(2') a column

$$
\begin{pmatrix} v_1 \\ \vdots \\ v_k \end{pmatrix} \in F[[z]]^{k \times 1}
$$

of power series $v_i := \sum_{\alpha \in \mathbb{N}^n} v_i(\alpha) z^\alpha \in F[[z]]$.

A *solution* of this system is a power series

$$
w := \sum_{\alpha \in \mathbb{N}^n} w(\alpha) z^\alpha
$$

such that

$$
R_i \circ w = v_i , \ 1 \leq i \leq k .
$$

Equivalently, we could replace (2') by

(2") a column

$$
\begin{pmatrix} v_1 \\ \vdots \\ v_k \end{pmatrix} \in \mathrm{Hom}_F(F[s], F)^{k \times 1}
$$

of linear functions $v_i : F[s] \longrightarrow F$, $1 \leq i \leq k$.

Then a *solution* of this system is a linear function $\varphi : F[s] \longrightarrow F$ such that

$$
R_i \circ \varphi = v_i , \ 1 \leq i \leq k.
$$

Example 2.4 The data for the system of partial difference equations in Example 2.2 can be written as

$$
R := \begin{pmatrix} R_1 \\ R_2 \end{pmatrix} = \begin{pmatrix} 2s_1^2 s_2 + 1 \\ 3s_1 s_2^2 + 2 \end{pmatrix}
$$

and

$$
v := \begin{pmatrix} v_1 \\ v_2 \end{pmatrix} = \begin{pmatrix} 13z_2^2 + 5z_1^2 z_2^3 \\ 6z_2^2 + 10z_1^2 z_2^3 + 15z_1 z_2 \end{pmatrix} .
$$

A solution is a power series $w := \sum_{\alpha \in \mathbb{N}^2} w(\alpha_1, \alpha_2) z_1^{\alpha_1} z_2^{\alpha_2}$ such that $R_i \circ w = v_i$, $1 \leq i \leq 2$. Or, equivalently, a linear function $\varphi : \mathbb{Q}[s] \longrightarrow \mathbb{Q}$ such that $\varphi(s_1^{\alpha_1} s_2^{\alpha_2} R_i) = v_i(\alpha_1, \alpha_2)$, for all $\alpha \in \mathbb{N}^2$, $1 \leq i \leq 2$.

Remark 2.5 The *inverse system* of the ideal $\langle R_1, \ldots, R_k \rangle$ generated by $R_1, \ldots, R_k$ in $F[s]$ is the set of all linear functions $\varphi : F[s] \longrightarrow F$ such that

$$
\varphi|_{\langle R_1, \ldots, R_k \rangle} = 0 .
$$

This notion was introduced by F. S. Macaulay in [18]. Hence the inverse system is the set of solutions of the homogeneous system of partial difference equations given by $R_1, \ldots, R_k \in F[s]$. Inverse systems have been studied e.g. in [18], [24], [17], [23], and [12].

It is clear that the set of solutions of a homogeneous system is an $F[s]$-submodule of $F^{\mathbb{N}^n}$ resp. $F[[z]]$ resp. $\operatorname{Hom}_F(F[s], F)$. Hence, if w is a solution of a system, then all solutions can be obtained by adding solutions of the homogeneous system to w.

Nevertheless, since a solution is an infinite family, and - even worse - the space of all solutions is in general infinite-dimensional over F, at first view it is not clear how to describe the solutions by finitely many data. We shall describe a family

$$(w(\alpha))_{\alpha \in \mathbb{N}^n} \in F^{\mathbb{N}^n}$$

by an algorithm which permits us to compute $w(\alpha)$ for any $\alpha \in \mathbb{N}^n$.
To *solve* a system of difference equations means

- Decide if there is a solution or not.
- If there are solutions: Determine a subset $\Gamma \subseteq \mathbb{N}^n$ such that for every *initial condition* $x := (x(\gamma))_{\gamma \in \Gamma}$ there is exactly one solution w_x such that $w_x(\gamma) = x(\gamma)$, for all $\gamma \in \Gamma$.
- Give an algorithm to compute $w_x(\alpha)$ for any $\alpha \in \mathbb{N}^n$ and any initial condition x.

Remark 2.6 For $\alpha \in \mathbb{N}^n$, $u \in \mathbb{C}[[z]]$ consider

$$s^\alpha \bullet u := \partial^\alpha u = \frac{\partial^{|\alpha|} u}{\partial z_1^{\alpha_1} \ldots \partial z_n^{\alpha_n}} .$$

If we replace $R_i \circ w$ in Definition 2.3 by $R_i \bullet w$, we get the definition of a *system of k linear partial differential equations with constant coefficients for one unknown function in n variables*. The map

$$\mathbb{C}[[z]] \to \mathbb{C}[[z]], \ \sum_\alpha u(\alpha) z^\alpha \mapsto \sum_\alpha \frac{u(\alpha)}{\alpha!} z^\alpha, \tag{2.1}$$

is an isomorphism of the $\mathbb{C}[s]$−modules $(\mathbb{C}[[z]], \circ)$ and $(\mathbb{C}[[z]], \bullet)$.

Results for systems of partial difference equations ("discrete case") can thus be transferred to systems of partial differential equations ("continuous case").

Example 2.7 The system of partial differential equations corresponding to that of partial difference equations in Examples 2.2 and 2.4 is:

"Find a power series $w \in \mathbb{C}[[z]]$ such that

$$2\frac{\partial^3 w}{\partial z_1^2 \partial z_2} + w = \frac{13}{2} z_2^2 + \frac{5}{12} z_1^2 z_2^3$$

and

$$3\frac{\partial^3 w}{\partial z_1 \partial z_2^2} + 2w = 3 z_2^2 + \frac{5}{6} z_1^2 z_2^3 + 15 z_1 z_2 ."$$

A power series $\sum_{\alpha \in \mathbb{N}^2} w(\alpha) z^\alpha$ is a solution of this system of differential equations if and only if $\sum_{\alpha \in \mathbb{N}^2} \alpha_1! \alpha_2! w(\alpha) z^\alpha$ is so for the system of difference equations in Example 2.2.

3   The Algorithm

Let $R_1, \ldots, R_k \in F[s]$ be polynomials and let I be the ideal generated by them in $F[s]$. Let $\leq$ be a term order on $\mathbb{N}^n$, $\deg(I) \subseteq \mathbb{N}^n$ the set of degrees of all polynomials in I, and

$$\Gamma := \mathbb{N}^n \setminus \deg(I) . \tag{3.1}$$

Then

$$F[s] = I \oplus \bigoplus_{\gamma \in \Gamma} F s^\gamma \tag{3.2}$$

i.e. any polynomial $h \in F[s]$ can uniquely be written as

$$h = h_I + \mathrm{nf}(h) ,$$

where $h_I \in I$ and $\mathrm{nf}(h) \in \bigoplus_{\gamma \in \Gamma} F s^\gamma$. The polynomial $\mathrm{nf}(h)$ is the *normal form* of h with respect to I and $\leq$.

The set $\deg(I)$ and the polynomial $\mathrm{nf}(h)$ can be computed via a Gröbner basis ([4], [5]) of I with respect to $\leq$. Moreover, using Gröbner bases we can compute a system of generators

$$L_i := (L_{i1}, \ldots, L_{ik}) \in F[s]^{1 \times k} , 1 \leq i \leq p ,$$

of the module of syzygies of $(R_1, \ldots, R_k)$, i.e. of the $F[s]$-submodule

$$\{ u \in F[s]^{1 \times k} \mid \sum_{j=1}^{k} u_j R_j = 0 \} \leq F[s]^{1 \times k} .$$

Theorem 3.1   *[28, Th. 3 and Th. 5]*

(1) There exists a function $w \in F^{\mathbb{N}^n}$ such that

$$R_i \circ w = v_i , 1 \leq i \leq k ,$$

  if and only if

$$\text{(S)} \qquad \sum_{j=1}^{k} L_{ij} \circ v_j = 0 , 1 \leq i \leq p .$$

(2) If condition (S) holds then for any initial condition $x : \Gamma \longrightarrow F$ there is a unique function $w \in F^{\mathbb{N}^n}$ such that

$$w|_\Gamma = x \quad \text{and} \quad R_i \circ w = v_i , 1 \leq i \leq k .$$

For any $\alpha \in \mathbb{N}^n$ the element $w(\alpha) \in F$ can be computed as follows:

Compute $c_\gamma \in F$, where $\gamma \in \Gamma$, and $d_{\beta,i} \in F$, where $\beta \in \mathbb{N}^n$, $1 \leq i \leq k$, such that

$$\mathrm{nf}(s^\alpha) = \sum_{\gamma \in \Gamma} c_\gamma s^\gamma$$

and

$$s^\alpha - \mathrm{nf}(s^\alpha) = \sum_{\beta,i} d_{\beta,i} s^\beta R_i \,.$$

Then

$$w(\alpha) = \sum_{\gamma \in \Gamma} c_\gamma x(\gamma) + \sum_{\beta,i} d_{\beta,i} v_i(\beta) \,.$$

Proof. First we show that the existence of a solution w implies condition (S):

$$\sum_{j=1}^k L_{ij} \circ v_j = \sum_{j=1}^k L_{ij} \circ (R_j \circ w) = \Big(\sum_{j=1}^k L_{ij} R_j \Big) \circ w = 0 \,.$$

Now we assume that (S) holds.

If a solution w exists, we consider the associated linear function $\varphi : F[s] \longrightarrow F$, where $\varphi(s^\alpha) = w(\alpha)$. Then

$$w(\alpha) = \varphi(s^\alpha) = \varphi\Big(\sum_{\gamma \in \Gamma} c_\gamma s^\gamma + \sum_{\beta,i} d_{\beta,i} s^\beta R_i \Big) = \sum_{\gamma \in \Gamma} c_\gamma x(\gamma) + \sum_{\beta,i} d_{\beta,i} v_i(\beta) \,.$$

To show the existence of a solution w resp. φ, we define the linear function φ as follows: Since $\{ s^\beta R_i \mid \beta \in \mathbb{N}^n, \, 1 \leq i \leq k \}$ is a system of generators of the F-vector-space I, we may choose a subset B of it which is an F-basis of I. Then define

$$\varphi(s^\gamma) = x(\gamma) \,, \ \gamma \in \Gamma \,,$$

and

$$\varphi(s^\beta R_i) = v_i(\beta) \,, \ \text{for all } s^\beta R_i \in B \,.$$

Now we only have to show that

$$\varphi(s^\alpha R_j) = v_j(\alpha) \,, \ \text{for all } s^\alpha R_j \notin B \,.$$

Let $s^\alpha R_j \notin B$ and let $a_{\beta,i} \in F$ such that

$$s^\alpha R_j = \sum_{s^\beta R_i \in B} a_{\beta,i} s^\beta R_i \,.$$

Then

$$0 = \sum_{s^\beta R_i \in B} a_{\beta,i} s^\beta R_i - s^\alpha R_j$$

$$= \sum_{\substack{i \\ i \neq j}} \Big(\sum_{\substack{\beta \\ s^\beta R_i \in B}} a_{\beta,i} s^\beta \Big) R_i + \Big(\Big(\sum_{\substack{\beta \\ s^\beta R_j \in B}} a_{\beta,j} s^\beta \Big) - s^\alpha \Big) R_j \,.$$

Since (S) holds, it is easy to verify that $\sum_{j=1}^{k} u_j \circ v_j = 0$ for any syzygy u of $(R_1, \ldots, R_k)$. Hence

$$0 = \sum_{\substack{i \\ i \neq j}} \left(\sum_{\substack{\beta \\ s^\beta R_i \in B}} a_{\beta,i} s^\beta \right) \circ v_i + \left(\left(\sum_{\substack{\beta \\ s^\beta R_j \in B}} a_{\beta,j} s^\beta \right) - s^\alpha \right) \circ v_j$$

and

$$0 = \left(\sum_{\substack{i \\ i \neq j}} \left(\sum_{\substack{\beta \\ s^\beta R_i \in B}} a_{\beta,i} s^\beta \right) \circ v_i + \left(\left(\sum_{\substack{\beta \\ s^\beta R_j \in B}} a_{\beta,j} s^\beta \right) - s^\alpha \right) \circ v_j \right)(0)$$

$$= \sum_{\substack{i \\ i \neq j}} \left(\sum_{\substack{\beta \\ s^\beta R_i \in B}} a_{\beta,i} \right) v_i(\beta) + \sum_{\substack{\beta \\ s^\beta R_j \in B}} a_{\beta,j} v_j(\beta) - v_j(\alpha) .$$

Thus

$$v_j(\alpha) = \sum_{\substack{i \\ i \neq j}} \left(\sum_{\substack{\beta \\ s^\beta R_i \in B}} a_{\beta,i} \right) \varphi(s^\beta R_i) + \sum_{\substack{\beta \\ s^\beta R_j \in B}} a_{\beta,j} \varphi(s^\beta R_j) = \varphi(s^\alpha R_j) .$$

$\square$

Example 3.2 We present the solution of the system of linear difference equations in Example 2.2 resp. Example 2.4. Let

$$R := \begin{pmatrix} 2s_1^2 s_2 + 1 \\ 3s_1 s_2^2 + 2 \end{pmatrix}$$

and

$$v := \begin{pmatrix} 13z_2^2 + 5z_1^2 z_2^3 \\ 6z_2^2 + 10z_1^2 z_2^3 + 15z_1 z_2 \end{pmatrix} .$$

The $\mathbb{Q}[s]$-module of syzygies is generated by the element

$$L = (-R_2, R_1) = (-3s_1 s_2^2 - 2, 2s_1^2 s_2 + 1) .$$

We compute $L \circ v = 0$, hence this system is solvable.

A Gröbner basis of the ideal

$$I := \langle R_1, R_2 \rangle = \langle 2s_1^2 s_2 + 1, 3s_1 s_2^2 + 2 \rangle$$

in $\mathbb{Q}[s_1, s_2]$ with respect to the graded lexicographic order ($s_1 > s_2$) is

$$\{4s_1 - 3s_2, 9s_2^3 + 8\}$$

and hence

$$\Gamma = \{(0,0), (0,1), (0,2)\} .$$

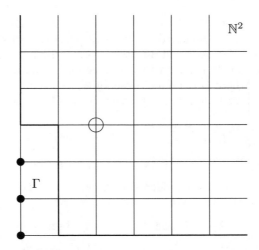

Figure 3.1 The elements $\bullet$ of Γ and $\bigcirc = (2,3) \in \deg(I)$.

Let $\alpha := (2,3)$. Then

$$s_1^2 s_2^3 = -\frac{1}{2} s_2^2 + (-\frac{3}{4} s_1 s_2^4) R_1 + (\frac{1}{2} s_1^2 s_2^3 + \frac{1}{4} s_2^2) R_2 .$$

Hence

$$w(2,3) = -\frac{1}{2} x(0,2) - \frac{3}{4} v_1(1,4) + \frac{1}{2} v_2(2,3) + \frac{1}{4} v_2(0,2)$$

$$= -\frac{1}{2} x(0,2) + \frac{13}{2} .$$

4 Riquier's Decomposition and the Hilbert Function

Let I be an ideal in $F[s]$ and $\deg(I)$ and $\Gamma := \mathbb{N}^n \setminus \deg(I)$ as in the preceding section. We construct disjoint decompositions of Γ, due to Riquier [33], and of $\deg(I)$, due to Janet [14], and apply the first to the computation of the Hilbert function of the ideal I. The details of these considerations are contained in [28] and [26]. In [26] Riquier's decomposition is used for the construction of the canonical state representation of a multidimensional system. In Section 7 we make some further remarks on the history of this decomposition.

Let, more generally, N be an *order ideal* of $\mathbb{N}^n$, i.e. a subset satisfying

$$N = N + \mathbb{N}^n,$$

$\Gamma := \mathbb{N}^n \setminus N$ its complement and

$$\min(N) := \{\alpha \in N \mid \text{there are no } \beta \in N, \gamma \in \mathbb{N}^n \setminus \{0\} \text{ such that } \alpha = \beta + \gamma \}$$

the set of minimal elements of N. The set $\min(N)$ is finite (Dickson's lemma) and satisfies $N = \min(N) + \mathbb{N}^n$. The degree set (with respect to a term order) $N := \deg(I)$ of an ideal I is such an order ideal, and Buchberger's algorithm especially furnishes the finite discrete set $\min(N)$. In the sequel we assume that the order ideal N is given in the form $N = D + \mathbb{N}^n$ where D is a finite subset of $\mathbb{N}^n$. Then $\min(N) = \min(D) \subseteq D$. For a subset S of $\{1, \ldots, n\}$ with its complement $S' := \{1, \ldots, n\} \setminus S$ we *identify* $\mathbb{N}^S$ as subset of $\mathbb{N}^n$ via

$$\mathbb{N}^S = \{x = (x_1, \ldots, x_n) \in \mathbb{N}^n \mid x_i = 0 \text{ for all } i \notin S\} \text{ and}$$

$$\mathbb{N}^n = \mathbb{N}^S \times \mathbb{N}^{S'} \ni x = ((x_i)_{i \in S}, (x_i)_{i \in S'}). \tag{4.1}$$

We are going to construct disjoint decompositions of N and of Γ by induction on n. These decompositions are given by *finite* subsets

$$A_N \subset N \text{ and } A_\Gamma \subset \Gamma \text{ and subsets}$$

$$S(\alpha) \subseteq \{1, \ldots, n\}, \ \alpha \in A_N \uplus A_\Gamma, \text{ such that} \tag{4.2}$$

$$N = \biguplus_{\alpha \in A_N} \left(\alpha + \mathbb{N}^{S(\alpha)}\right) \text{ and } \Gamma = \biguplus_{\alpha \in A_\Gamma} \left(\alpha + \mathbb{N}^{S(\alpha)}\right).$$

The recursive algorithm

Induction beginning: $n = 1$:

(i) If $N = D = \emptyset$, we choose $A_N := \emptyset$, $A_\Gamma = \{0\}$ and $S(0) := \{1\}$ and obtain $\mathbb{N} = \emptyset \uplus (0 + \mathbb{N})$.

(ii) If $N \neq \emptyset$ and $\{d\} := \min(N)$ we define

$$A_N := \{d\}, \ S(d) := \{1\}, \ A_\Gamma := \Gamma = \{0, \ldots, d-1\}, \ S(\alpha) := \emptyset \text{ for } \alpha \in A_\Gamma,$$

$$\text{hence } \mathbb{N} = (d + \mathbb{N}) \uplus \{0\} \uplus \cdots \uplus \{d-1\}.$$

The induction step: For $n > 1$ we assume that the decomposition has already been constructed for order ideals in $\mathbb{N}^{n-1}$. We identify

$$\mathbb{N}^n = \mathbb{N} \times \mathbb{N}^{n-1} \ni \alpha = (\alpha_1, \alpha_{II}), \ \alpha_{II} := (\alpha_2, \ldots, \alpha_n),$$

consider the projection $\text{proj} : \mathbb{N}^n \to \mathbb{N}^{n-1}$, $\alpha = (\alpha_1, \alpha_{II}) \mapsto \alpha_{II}$, and define

$$N_{II} := \text{proj}_{II}(N) := \{\alpha_{II} \mid (\alpha_1, \alpha_{II}) \in N \text{ for some } \alpha_1 \in \mathbb{N}\} \text{ resp.}$$

$$D_{II} := \text{proj}_{II}(D) := \{\alpha_{II} \mid (\alpha_1, \alpha_{II}) \in D \text{ for some } \alpha_1 \in \mathbb{N}\}.$$

Since proj is an epimorphism the representation $N = D + \mathbb{N}^n$ implies $N_{II} = D_{II} + \mathbb{N}^{n-1}$ and in particular that N_{II} is an order ideal in $\mathbb{N}^{n-1}$. Moreover

$$N = \biguplus_{i=0}^{\infty} (\{i\} \times N_{II}(i)) \text{ with}$$

$$N_{II}(i) := \{\alpha_{II} \mid (i, \alpha_{II}) \in N\} = D_{II}(i) + \mathbb{N}^{n-1} \text{ where}$$

$$D_{II}(i) := \{\alpha_{II} \mid (\alpha_1, \alpha_{II}) \in D \text{ for some } \alpha_1 \leq i\}.$$

The $N_{II}(i)$ and $D_{II}(i)$ are increasing, i.e.

$$N_{II}(0) \subseteq N_{II}(1) \subseteq \cdots \subseteq N_{II} = \bigcup_{i=0}^{\infty} N_{II}(i)$$

$$D_{II}(0) \subseteq D_{II}(1) \subseteq \cdots \subseteq D_{II} = \bigcup_{i=0}^{\infty} D_{II}(i).$$

Since D and hence D_{II} are finite the sequences of the $D_{II}(i)$ and hence of the $N_{II}(i) = D_{II}(i) + \mathbb{N}^n$ become stationary and we define

$$k := \min\{i \in \mathbb{N} \mid N_{II}(i) = N_{II}\} = \max\{\alpha_1;\ (\alpha_1, \alpha_{II}) \in \min(D)\}.$$

By induction applied to

$$N_{II}(i) = D_{II}(i) + \mathbb{N}^{n-1} \subset \mathbb{N}^{n-1} = \mathbb{N}^{\{2,\ldots,n\}},\ 0 \le i \le k,$$

we obtain

$$\mathbb{N}^{n-1} = N_{II}(i) \uplus \Gamma_{II}(i) =$$

$$\left[\biguplus_{\alpha_{II} \in B_{II}(i)} (\alpha_{II} + \mathbb{N}^{S_{II}(i,\alpha_{II})}) \right] \uplus \left[\biguplus_{\alpha_{II} \in C_{II}(i)} (\alpha_{II} + \mathbb{N}^{S_{II}(i,\alpha_{II})}) \right]$$

where

$$B_{II}(i) := A_{N_{II}(i)},\ C_{II}(i) := A_{\Gamma_{II}(i)},$$
$$S_{II}(i, \alpha_{II}) \subset \{2, \ldots, n\} \text{ for } \alpha_{II} \in B_{II}(i) \uplus C_{II}(i).$$

With these data in dimension $n - 1$ we define

$$A_N := \{(i, \alpha_{II});\ 0 \le i \le k,\ \alpha_{II} \in B_{II}(i)\}$$
$$A_\Gamma := \{(i, \alpha_{II});\ 0 \le i \le k,\ \alpha_{II} \in C_{II}(i)\}$$
$$A := A_N \uplus A_\Gamma$$
$$S(i, \alpha_{II}) := S_{II}(i, \alpha_{II}) \text{ for } (i, \alpha_{II}) \in A,\ 0 \le i \le k-1$$
$$S(k, \alpha_{II}) := \{1\} \uplus S_{II}(k, \alpha_{II}) \text{ for } (k, \alpha_{II}) \in A.$$

The desired decomposition in dimension n is

$$\mathbb{N}^n = N \uplus \Gamma = \left[\biguplus_{\alpha \in A_N} \left(\alpha + \mathbb{N}^{S(\alpha)} \right) \right] \uplus \left[\biguplus_{\alpha \in A_\Gamma} \left(\alpha + \mathbb{N}^{S(\alpha)} \right) \right]. \tag{4.3}$$

According to Janet [14] the indices i in the set $S(\alpha)$, $\alpha \in A$, resp. the corresponding indeterminates s_i are called *multiplicative* with respect to α and the others are called *non-multiplicative*. Notice that the decomposition (4.3) depends on N only and not on the special choice of the finite set D. The only required computations are comparisons of finitely many vectors in $\mathbb{N}^n$ and $\mathbb{N}^{n-1}$ and reading off first and second components

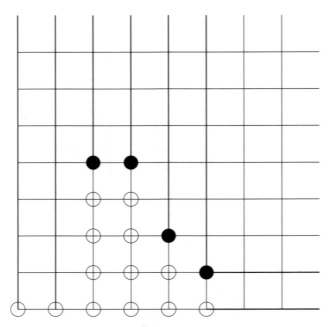

Figure 4.1 The decomposition $\mathbb{N}^2 = N \uplus \Gamma$, the points $\bullet$ of A_N and $\circ$ of A_Γ.

α_1 and α_{II} of vectors α in $\mathbb{N}^n$. These manipulations are simple and easy to implement. Figure 4.1 illustrates the preceding theorem for the order ideal $N = D + \mathbb{N}^2$ with $D = \min(N) := \{(2,4), (4,2), (5,1)\}$.

We denote by $F[s|S(\alpha)]$ resp. $F[[z|S(\alpha)]]$ the polynomial ring resp. power series ring $F[s_i \,;\, i \in S(\alpha)]$ resp. $F[[z_i \,;\, i \in S(\alpha)]]$ in the variables s_i resp. z_i, $i \in S(\alpha) \subseteq \{1, \ldots, n\}$. Application of the preceding decomposition to $N := \deg(I)$ from Section 3 and the direct sum decomposition $F[s] = I \oplus \oplus_{\gamma \in \Gamma} F s^\gamma$ from (3.2) imply

$$\oplus_{\gamma \in \Gamma} F s^\gamma = \oplus_{\alpha \in A_\Gamma} F[s|S(\alpha)] s^\alpha,$$

$$F[s]/I = \oplus_{\alpha \in A_\Gamma} F[s|S(\alpha)] \overline{s^\alpha} \cong \prod_{\alpha \in A_\Gamma} F[s|S(\alpha)] \tag{4.4}$$

$$\text{and } F^\Gamma = \oplus_{\alpha \in A_\Gamma} F[[z|S(\alpha)]] z^\alpha \subset F[[z]]$$

where $\overline{s^\alpha}$ denotes the residue class of s^α in $F[s]/I$. A direct sum decomposition of the factor ring $F[s]/I$ as in (4.4) is called a *Stanley decomposition* in [35] by Sturmfels and White. These authors also construct a Stanley decomposition of the algebra $F[s]/I$ and especially point out its usefulness for the computation of the Hilbert series of the ideal. The decomposition of Γ, due to Riquier, improves Macaulay's result that the residue classes $\overline{s^\gamma}$, $\gamma \in \Gamma$, are an F-basis of the factor ring. The decomposition $F^\Gamma = \oplus_{\alpha \in A_\Gamma} F[[z|S(\alpha)]] z^\alpha$ signifies that the solution w of the Cauchy problem $R \circ w = 0$, $w|\Gamma = x$, depends on $|A_\Gamma|$ power series (functions) in the variables z_i, $i \in S(\alpha)$, $\alpha \in A_\Gamma$.

As observed in [35] Riquier's decomposition also leads to a computation of the

Hilbert function of the ideal I. For this result we need a little preparation. For $\alpha \in \mathbb{N}^n$ define $| \alpha | := \alpha_1 + \ldots + \alpha_n$. This is the total degree of the monomial s^α. For $m \in \mathbb{N}$ the space

$$F[s]_{\leq m} := \oplus \{F s^\alpha; \; \alpha \in \mathbb{N}^n, \; | \alpha | \leq m\}$$

is the finite-dimensional subspace of $F[s]$ of polynomials of total degree at most m. The function

$$HF_I : \mathbb{N} \longrightarrow \mathbb{N}, \; m \mapsto HF_I(m) := \dim_F \left(F[s]_{\leq m} / (F[s]_{\leq m} \cap I)\right),$$

is called the *Hilbert function* of the ideal I. It is a polynomial function for large m, i.e. there is a unique polynomial $HP_I \in \mathbb{Q}[s]$, the *Hilbert polynomial* of I, such that $HF_I(m) = HP_I(m)$ for almost all m. The power series $HS_I := \sum_{m=0}^\infty HF_I(m) t^m$ in one variable t is called the *Hilbert series* of I and is a rational function of t.

Theorem 4.1 *[26, Th. 3.6] Consider a* graded *term order on* $\mathbb{N}^n$ *(i.e.:* $\sum_{i=1}^n \alpha_i < \sum_{i=1}^n \beta_i$ *implies* $\alpha \leq \beta$ *for all* $\alpha, \beta \in \mathbb{N}^n$*). Denote by* $\#(S(\alpha))$ *the number of elements of* $S(\alpha)$*. Then the dimension of* I*, i.e. the Krull dimension of the factor ring* $F[s]/I$*, is*

$$\dim(F[s]/I) = \max_{\alpha \in A_\Gamma} \#(S(\alpha)).$$

The Hilbert function, polynomial and series of the ideal I *are given as*

$$HF_I(m) = \sum \left\{ \binom{m - | \alpha | + \#(S(\alpha))}{\#(S(\alpha))} \; | \; \alpha \in A_\Gamma, \; | \alpha | \leq m \right\}$$

$$HP_I(m) = \sum_{\alpha \in A_\Gamma} \binom{m - | \alpha | + \#(S(\alpha))}{\#(S(\alpha))}$$

$$HS_I = \sum_{\alpha \in A_\Gamma} t^{|\alpha|} (1 - t)^{-\#(S(\alpha))-1}.$$

We find the simplicity and power of Riquier's algorithm remarkable.

5 Difference Equations for ℓ Unknowns

Definition 5.1 A *system of k linear partial difference equations with constant coefficients for ℓ unknown functions in n variables* is given by

(1) a family

$$(R(\alpha))_{\alpha \in \mathbb{N}^n}$$

of matrices $R(\alpha) \in F^{k \times \ell}$, where only finitely many matrices $R(\alpha)$ are different from 0, and

(2) a family

$$(v(\alpha))_{\alpha \in \mathbb{N}^n}$$

of columns $v(\alpha) \in F^{k \times 1}$.

A *solution* of this system is a family

$$(w(\gamma))_{\gamma \in \mathbb{N}^n}$$

of columns $w(\gamma) \in F^{\ell \times 1}$ (i.e. a function $w : \mathbb{N}^n \longrightarrow F^\ell$, $\gamma \mapsto \begin{pmatrix} w_1(\gamma) \\ \vdots \\ w_\ell(\gamma) \end{pmatrix}$) such that

for all $\beta \in \mathbb{N}^n$

$$\sum_{\alpha \in \mathbb{N}^n} R(\alpha)w(\alpha + \beta) = v(\beta),$$

i.e.,

$$\sum_{j=1}^{\ell} \sum_{\alpha \in \mathbb{N}^n} R_{ij}(\alpha)w_j(\alpha + \beta) = v_i(\beta),$$

for $i = 1, \ldots, k$ and all $\beta \in \mathbb{N}^n$.

The system is *homogeneous* if and only if $v(\beta) = 0$, for all $\beta \in \mathbb{N}^n$.

As in the case of one unknown function we can give another interpretation of the data and the solutions:

A function $w : \mathbb{N}^n \longrightarrow F^\ell$, $\alpha \mapsto w(\alpha)$, can be written as a column (with ℓ rows) of power series

$$\begin{pmatrix} \sum_{\alpha \in \mathbb{N}^n} w_1(\alpha)z^\alpha \\ \vdots \\ \sum_{\alpha \in \mathbb{N}^n} w_\ell(\alpha)z^\alpha \end{pmatrix},$$

and the family of matrices $(R(\alpha))_{\alpha \in \mathbb{N}^n}$ can be represented by a matrix $R \in F[s]^{k \times \ell}$, where $R_{ij} := \sum_{\alpha \in \mathbb{N}^n} R_{ij}(\alpha)s^\alpha$, $1 \le i \le k$, $1 \le j \le \ell$. Then a solution is a column

$$\begin{pmatrix} w_1 \\ \vdots \\ w_\ell \end{pmatrix}$$

of power series such that

$$R \circ \begin{pmatrix} w_1 \\ \vdots \\ w_\ell \end{pmatrix} = \begin{pmatrix} v_1 \\ \vdots \\ v_k \end{pmatrix}.$$

Let U be the $F[s]$-submodule generated by the rows $R_{1-}, \ldots, R_{k-}$ of R in $F[s]^{1 \times \ell}$. Let $\le$ be a term order on $(\mathbb{N}^n)^\ell$, $\deg(U) \subseteq (\mathbb{N}^n)^\ell$ the set of degrees of all elements in U, and

$$\Gamma := (\mathbb{N}^n)^\ell \setminus \deg(U).$$

Using the theory of Gröbner bases for $F[s]$-modules we can compute Γ, the normal form of any element in $F[s]^\ell$ and the syzygies for $(R_{1-}, \ldots, R_{k-})$. The results of Theorem 3.1 can be transferred to this case in an obvious way.

6 Convergent Solutions and Solutions of Exponential Type

Let

$$\mathbb{C}\langle z\rangle := \mathbb{C}\langle z_1, \ldots, z_n\rangle$$

be the algebra of (locally) convergent power series (i.e. power series $\sum_\alpha u(\alpha)z^\alpha$ such that there are $C > 0$ and $d_1 > 0, \ldots, d_n > 0$ with $|u(\alpha)| \leq Cd^\alpha$ for all $\alpha \in \mathbb{N}^n$). Consider functions from $\mathbb{N}^n$ to $\mathbb{C}$ as power series. Then the solution of

$$R_i \circ w = v_i, 1 \leq i \leq k, \ w|\Gamma = x$$

is convergent if the data x and $v_i, 1 \leq i \leq k$, are so [28, Theorem 24].

Let

$$O(\mathbb{C}^n; \exp)$$

be the algebra of entire holomorphic functions of *exponential type* (i.e. entire holomorphic functions $b = \sum_{\alpha \in \mathbb{N}^n} b(\alpha)z^\alpha$ on $\mathbb{C}^n$ such that there are $C > 0$ and $d_1 > 0, \ldots, d_n > 0$ with $|b(z)| \leq C\exp(\sum_{i=1}^n d_i|z_i|)$, for all $z \in \mathbb{C}^n$).

The isomorphism (2.1) in Section 1 induces the isomorphism

$$(\mathbb{C}\langle z\rangle, \circ) \cong (O(\mathbb{C}^n; \exp), \bullet) .$$

Hence the Cauchy (or initial value) problem for partial differential equations and entire functions of exponential type can be solved constructively. The solution of the corresponding problem for *locally convergent power series* is due to Riquier [33, Th. d'Existence, p. 254]. This theorem requires that the term order is graded (i.e.: $\sum_{i=1}^n \alpha_i < \sum_{i=1}^n \beta_i$ implies $\alpha \leq \beta$ for all $\alpha, \beta \in \mathbb{N}^n$), as the following example [7, Ch. 1, §2.2] shows:

Consider the two-dimensional heat equation

$$(s_1 - s_2^2) \bullet y = \frac{\partial y}{\partial z_1} - \frac{\partial^2 y}{\partial z_2^2} = 0, \ y \in \mathbb{C}[[z_1, z_2]],$$

where z_1 resp. z_2 are the time resp. the position. As term order on $\mathbb{C}[s_1, s_2]$ take the lexicographic one with $s_1 > s_2$, hence

$$\deg(\langle s_1 - s_2^2\rangle) = (1, 0) + \mathbb{N}^2 \text{ and } \Gamma = \mathbb{N}^2 \setminus \deg(\langle s_1 - s_2^2\rangle) = \{0\} \times \mathbb{N}.$$

For the locally convergent initial condition $y(0, z_2) := (1 - z_2)^{-1} \in \mathbb{C}\langle z_2\rangle$ the unique power series solution $y(z_1, z_2) \in \mathbb{C}[[z_1, z_2]]$ according to Theorem 3.1 satisfies $y(z_1, 0) = \sum_{k=0}^\infty (2k)! z_1^k$ [7, Ch. 1, §2.2] and is not convergent. This example is due to S. Kovalevskaya as one of the referees points out. The exact assumptions on the term order for the validity of Riquier's theorem in the case of systems with ℓ unknown functions as in Section 5 were given by Riquier [33] and reproduced in [28, Assumption 28, p. 287].

According to [13, Th. 8.6.7] the heat equation also has a C^∞-solution y whose support is exactly the half-space $\{z \in \mathbb{R}^2 \mid z_1 \leq 0\}$ and thus satisfies the initial condition

$y(0, z_2) = 0$. In contrast to Theorem 3.1 there is no uniqueness in the C^∞-Cauchy problem. The reason is that $\{0\} \times \mathbb{R}$ is a *characteristic* hyper-plane (line).

7   History

The problem introduced in Section 2 is known as the Cauchy problem for linear systems of partial *difference* equations with constant coefficients over the lattice $\mathbb{N}^n$. Its constructive solution was first developed in [24, §5, Th. 41, Cor. 44] after some inspirations through [11]. The inhomogeneous Cauchy problem $R \circ w = u$, $w|\Gamma = x$, with $u \neq 0$ was solved recursively. The direct algorithmic solution of the inhomogeneous problem according to Theorem 3.1, (2), is less complex and comes from [28, Th. 5]. Extensions of these results to other lattices like $\mathbb{Z}^r$ are contained in [37], [36], [28], [30]. The proof for *locally convergent* power series in $\mathbb{C}\langle z \rangle$ (see Section 5) was given in [28, Th. 24] with some support from [34]. By means of the isomorphism (2.1) (Remark 2.6 in Section 2) this result can be transferred to partial *differential* equations for entire functions of exponential type. The algorithmic solution of Krull zero-dimensional or $\mathbb{C}$-finite dimensional systems was described in [25] and differently in [15]. Systems of partial difference equations over *quasi-Frobenius base rings* instead of fields, for instance over the finite rings $\mathbb{Z}/\mathbb{Z}n$, $n > 0$, and their Galois extensions, are treated in [16].

The history of the solution of systems of partial *differential* equations is highly interesting. An important early result was the *Cauchy-Kovalevskaya* theorem. The by far most important contribution to the field since 1890 was the work of Riquier which built on that of Méray ([21], [22]) and which culminated in the book [33], published in 1910. The introduction of [33] contains a historical account. The book which is still worthwhile, but difficult reading treats the solution of non-linear analytic systems of partial differential equations in spaces of analytic functions. This book was far ahead of its time, did not receive the recognition and distribution it deserved, and many prominent later researchers in the field did not mention it. Essentially Riquier [33, Th. d'Existence, p. 254] proved the analogue of Theorem 3.1 for so-called passive orthonomic, non-linear, analytic systems of partial differential equations, but not fully constructively. The analogues of the conditions $L_i \circ v = 0$ in part (1) of Theorem 3.1 were called the necessary *integrability or compatibility conditions*. However, Riquier's proofs and algorithms for non-linear systems and for linear systems with variable coefficients are valid *generically only*, i.e. they can be applied in many cases, but not in all. For instance, the simple ordinary differential equation $zy'(z) = 1$ has no power series solution although the necessary integrability conditions are trivially satisfied (there are none). The reason that the algorithms of Riquier and later Janet [14] do not apply in all cases is the necessity to use the analytic implicit function at various places to solve a system of differential equations for the highest derivatives, and this is not always possible. Riquier and Janet explicitly point out this difficulty. With a different terminology Riquier was the first to use (the most general) term orders (total orderings with *cotes*), the sets Γ and $\deg(I)$ from Section 2 (*parametric and principal derivatives*, called *standard resp. non-standard* monomials in the algebraic situation [35]), and S-polynomials (*cardinal derivatives*). In 1920 Janet [14] gave a simplified, more algebraic and more algorithmic presentation of Riquier's work and especially developed the theory of involutive division and Janet bases (in today's language) as a very early alternative to Gröbner bases. Riquier's Théorème d'Existence was also exposed by Ritt [34, Ch. IX]. For the linear, constant coefficient case we gave a shorter proof in modern language of Riquier's theorem in [28, Th. 29].

A different and independent version of the theory of analytic systems of partial differential equations is the *Cartan-Kähler theory* and is exposed, for instance, by Malgrange in his recent survey article [20] with many historical comments. Malgrange also stresses that the theory is valid generically only. The first work of E. Cartan in this area was published in 1901, i.e., twenty years after Méray's and ten years after Riquier's fundamental papers, and has been acknowledged, although not always understood, from the start (see the introduction of [20]) in contrast to Riquier's and later Janet's work.

The prominent algebraists Macaulay and later Gröbner did not refer to the important algebraic (in Riquier's case in disguise) work of their contemporaries Riquier and Janet. The revival of the fundamental work of Riquier and Janet, its exposition in modern language and its application in theoretical physics and later in control theory is due to Pommaret [31] (dedicated to Janet) and [32]. Pommaret also pointed out this important work to us in the middle of the 1990s. At the time of writing [24] (1988/89) it was unknown to us. In the 1990s Gerdt et al. [10] took up the work of Janet and Pommaret to develop the modern theory of involutive division and involutive bases.

The analogue of part (1) of Theorem 3.1 for linear systems of partial differential equations with constant coefficients in spaces of C^∞-*functions or distributions* (instead of analytic functions in [33]) was proven by Ehrenpreis [8], Malgrange [19] and Palamodov [29] in the beginning 1960s. That the necessary integrability conditions are also sufficient for the solvability of such a system is called the *Fundamental Principle*. Again, these authors did not mention the work of Riquier or Janet. Compare, however, Malgrange's recent historical remarks [20].

The disjoint decomposition (4.3) in Section 3 of Γ is due to Riquier [33, pp. 143-168]. Janet [14, pp. 74-91] constructed it both for $\deg(I)$ and for Γ and used it for the derivation of his early version of the Gröbner basis theory. We gave a modern and shorter inductive proof in [28, pp. 269-274], but used Buchberger's algorithm for the computation of $\deg(I)$. M. Scheicher has implemented our version of Riquier's algorithm; the program seems to run well, but more experimentation is necessary. A weaker version (in our understanding) of Riquier's decomposition and the ensuing *Stanley decomposition* of $F[s]/I$ was also described by Sturmfels and White in [35, Theorem 2.3, Lemma 2.4], again on the basis of Buchberger's algorithm. Gerdt [10, Decomposition Lemma] derived disjoint decompositions of $\deg(I)$ and Γ by means of involutive division. Various forms of the decomposition are used for the computation of Hilbert functions and polynomials in [31], [32], [10], [26]. The Hilbert function and the Hilbert polynomial are treated in several standard textbooks on Commutative Algebra. Computations by means of Stanley decompositions are described in [35], [1] and [2].

All quoted references treat the more general case of submodules of free polynomial modules instead of polynomial ideals. Further historical comments are given in [27].

Bibliography

[1] J. Apel, *The Theory of Involutive Divisions and an Application to Hilbert Function Computations*, J. Symb. Computation 25 (1998), 683–704

[2] J. Apel, *On a conjecture of R. P. Stanley II: Quotients modulo monomoial ideals*, J. Algebr. Comb. 17 (2003), 57–74

[3] R. Bryant, S. Chern, H. Gardner, P. Griffith, H. Goldshmidt, *Exterior Differential Systems*, MSRI Book Series, Vol. 18, Springer, New York, 1991

[4] B. Buchberger, *Ein Algorithmus zum Auffinden der Basiselemente des Restklassenringes nach einem nulldimensionalen Polynomideal*, Dissertation, Innsbruck, 1965

[5] B. Buchberger, *Ein algorithmisches Kriterium für die Lösbarkeit eines algebraischen Gleichungssystems*, Aequationes Math. 4 (1970), 374–383

[6] D. Cox, J. Little, D. O'Shea, *Ideals, Varieties, and Algorithms*, Springer, New York, 1992

[7] Y. V. Egorov, M. A. Shubin, *Linear Partial Differential Equations. Foundations of the Classical Theory*, in Y. V. Egorov, M. A. Shubin (Editors), *Partial Differential Equations I*, Springer, Berlin, 1992

[8] L. Ehrenpreis, *Fourier Analysis in Several Complex Variables*, Wiley-Interscience Publ., New York, 1970

[9] V. G. Ganzha, E. W. Mayr, E. V. Vorozhtsov (Editors), *Computer Algebra in Scientific Computing*, Springer, Berlin, 1999

[10] V. P. Gerdt, *Completion of Linear Differential Systems to Involution*, in [9], 115–137

[11] J. Gregor, *Convolutional solutions of partial difference equations*, Math. of Control, Signals and Systems 2 (1991), 205–215

[12] W. Heiß, U. Oberst, F. Pauer, *On inverse systems and squarefree decomposition of zero-dimensional polynomial ideals*, J. Symbolic Computation 41 (2006), 261–284

[13] L. Hörmander, *The Analysis of Linear Partial Differential Operators I*, Springer, Berlin, 1983

[14] M. Janet, *Sur les Systèmes d'Équations aux Dérivées Partielles*, J. de Mathématiques Pures et Appliquées 8 (1920), 65–151

[15] V. Lomadze, E. Zerz, *Partial Differential Equations of Krull Dimension Zero*, Proc. Math. Theory of Networks and Systems (MTNS), Perpignan (2000)

[16] P. Lu, M. Liu, U. Oberst, *Linear Recurring Arrays, Linear Systems and Multidimensional Cyclic Codes over Quasi-Frobenius Rings*, Acta Appl. Math. 80 (2004), 175–198

[17] M. Marinari, H. Möller, T. Mora, *On Multiplicities in polynomial system solving*, Trans. Amer. Math. Soc. 348 (1996), 3283–3321

[18] F. S. Macaulay, *The Algebraic Theory of Modular Systems*, Cambridge University Press, Cambridge, 1916

[19] B. Malgrange, *Systèmes différentiels à coefficients constants*, Séminaire Bourbaki 1962/1963, 246.01–246.11

[20] B. Malgrange, *Systèmes Différentiels Involutifs*, Prépublication 636, Institut Fourier, Grenoble, 2004

[21] C. Méray, *Démonstration générale de l'existence des intégrales des équations aux dérivées partielles*, Journal des Mathématiques pures et appliquées, 3e série, t. VI (1880), 235–265

[22] C. Méray, C. Riquier, *Sur la convergence des développements des intégrales ordinaires dun système d' équations différentielles partielles*, Ann. ENS, 3e série, t. VI (1890), 23–88

[23] B. Mourrain, *Isolated points, duality and residues*. Journal of pure and applied algebra 117&118 (1997), 469–493

[24] U. Oberst, *Multidimensional Constant Linear Systems*, Acta Appl. Math. 20 (1990), 1–175

[25] U. Oberst, *Finite Dimensional Systems of Partial Differential or Difference Equations*, Advances in Applied Mathematics 17 (1996), 337–356

[26] U. Oberst, *Canonical State Representations and Hilbert Functions of Multidimensional Systems*, Acta Appl. Math. 94 (2006), 83–135

[27] U. Oberst, *Canonical State Representations and Hilbert Functions of Multidimensional Systems (with historical comments and a discussion of Professor Pommaret's remarks)*, Talk at the conference D3, Linz, May 2006, homepage of the Gröbner-Semester.

[28] U. Oberst, F. Pauer, *The Constructive Solution of Linear Systems of Partial Difference and Differential Equations with Constant Coefficients*, Multidimensional Systems and Signal Processing 12 (2001), 253–308

[29] V. P. Palamodov, *Linear Differential Operators*, Springer, Berlin, 1970

[30] F. Pauer, A. Unterkircher, *Gröbner Bases for Ideals in Laurent Polynomial Rings and their Application to Systems of Difference Equations*, AAECC 9/4 (1999), 271–291

[31] J.-F. Pommaret, *Systems of Partial Differential Equations and Lie Pseudogroups*, Gordon and Breach, New York, 1978

[32] J.-F. Pommaret, *Partial Differential Control Theory, Volume I: Mathematical Tools, Volume II: Control Systems*, Kluwer Academic Publishers, Dordrecht, 2001

[33] C. Riquier, *Les Systèmes d'Équations aux Dérivées Partielles*, Gauthiers - Villars, Paris, 1910

[34] J. F. Ritt, *Differential Equations From The Algebraic Standpoint*, American Mathematical Society, New York, 1932

[35] B. Sturmfels, N. White, *Computing Combinatorial Decompositions of Rings*, Combinatorica 11 (1991), 275–293

[36] S. Zampieri, *A Solution of the Cauchy Problem for Multidimensional Discrete Linear Shift - Invariant Systems*, Linear Algebra Appl. 202(1994), 143–162

[37] E. Zerz, U. Oberst, *The Canonical Cauchy Problem for Linear Systems of Partial Difference Equations with Constant Coefficients over the Complete r-Dimensional Integral Lattice $\mathbb{Z}^r$*, Acta Applicandae Mathematicae 31 (1993), 249–273

Author information

Ulrich Oberst, Institut für Mathematik, Universität Innsbruck, Technikerstraße 13/7, A-6020 Innsbruck, Austria.
Email: Ulrich.Oberst@uibk.ac.at

Franz Pauer, Institut für Mathematik, Universität Innsbruck, Technikerstraße 13/7, A-6020 Innsbruck, Austria.
Email: Franz.Pauer@uibk.ac.at

Radon Series Comp. Appl. Math **2**, 43–73 © de Gruyter 2007

Computation of the Strength
of Systems of Difference Equations
via Generalized Gröbner Bases

Alexander Levin

Key words. Difference field, difference polynomial, difference ideal, difference module, Gröbner basis, dimension polynomial, p-dimensional filtration, reduction, S-polynomial, strength.

AMS classification. 12H10, 39A20.

1 Introduction

In this paper we develop an analog of the classical method of Gröbner bases introduced by B. Buchberger [1] in the case of free modules over a ring of difference operators equipped with several term-orderings. The corresponding technique allows us to prove the existence and outline a method of computation of dimensional polynomials of several variables associated with finitely generated difference modules and difference field extensions. As an important application we obtain a method of computation of the strength of a system of partial difference equations in the sense of A. Einstein [3].

2 Preliminaries

In this section we present some basic concepts and results that are used in the rest of the paper. In what follows, $\mathbf{N}, \mathbf{Z}$, and $\mathbf{Q}$ denote the sets of all non-negative integers, integers, and rational numbers, respectively. As usual, $\mathbf{Q}[t]$ denotes the ring of polynomials in one variable t with rational coefficients. By a ring we always mean an associative ring with a unit. Every ring homomorphism is unitary (maps unit onto unit), every subring of a ring contains the unit of the ring. Unless otherwise indicated, by a module over a ring R we always mean a unitary left R-module.

2.1 Difference Rings and Fields

A *difference ring* is a commutative ring R together with a finite set $\sigma = \{\alpha_1, \dots, \alpha_n\}$ of mutually commuting injective endomorphisms of R into itself. The set σ is called the *basic set* of the difference ring R, and the endomorphisms $\alpha_1, \dots, \alpha_n$ are called *translations*. A difference ring with a basic set σ is also called a σ-*ring*. If $\alpha_1, \dots, \alpha_n$ are automorphisms of R, we say that R is an *inversive difference ring* with the basic set σ. In this case we denote the set $\{\alpha_1, \dots, \alpha_n, \alpha_1^{-1}, \dots, \alpha_n^{-1}\}$ by σ^* and call R a σ^*-ring. If a difference (σ-) ring R is a field, it is called a *difference* (or σ-) *field*. If R is inversive, it is called an *inversive difference field* or a σ^*-*field*.

Let R be a difference (inversive difference) ring with a basic set σ and R_0 a subring of R such that $\alpha(R_0) \subseteq R_0$ for any $\alpha \in \sigma$ (respectively, for any $\alpha \in \sigma^*$). Then R_0 is called a *difference* or σ- (respectively, *inversive difference* or σ^*-) *subring* of R, while the ring R is said to be a *difference* or σ- (respectively, *inversive difference* or σ^*-) *overring* of R_0. In this case the restriction of an endomorphism α_i on R_0 is denoted by the same symbol α_i. If R is a difference (σ-) or an inversive difference (σ^*-) field and R_0 a subfield of R which is also a σ- (respectively, σ^*-) subring of R, then R_0 is said to be a *difference* or σ- (respectively, *inversive difference* or σ^*-) *subfield* of R; R, in turn, is called a *difference* or σ- (respectively, *inversive difference* or σ^*-) *field extension* or a σ-(respectively, σ^*-) *overfield* of R_0. In this case we also say that we have a σ-(or σ^*-) field extension R/R_0.

If R is a difference ring with a basic set σ and J is an ideal of the ring R such that $\alpha(J) \subseteq J$ for any $\alpha \in \sigma$, then J is called a *difference* (or σ-) *ideal* of R. If a prime (maximal) ideal P of R is closed with respect to σ (that is, $\alpha(P) \subseteq P$ for any $\alpha \in \sigma$), it is called a *prime* (respectively, *maximal*) *difference* (or σ-) *ideal* of R.

A σ-ideal J of a σ-ring R is called *reflexive* if for any translation α, the inclusion $\alpha(a) \in J$ $(a \in R)$ implies $a \in J$. A reflexive σ-ideal is also called a σ^*-ideal (if R is inversive and J is a σ^*-ideal of R, then $\alpha(J) = J$ for any $\alpha \in \sigma^*$).

If R is a difference ring with a basic set $\sigma = \{\alpha_1, \dots, \alpha_n\}$, then T_σ (or T if the set σ is fixed) will denote the free commutative semigroup with identity generated by $\alpha_1, \dots, \alpha_n$. Elements of T_σ will be written in the multiplicative form $\alpha_1^{k_1} \dots \alpha_n^{k_n}$ $(k_1, \dots, k_n \in \mathbf{N})$ and considered as injective endomorphisms of R (which are the corresponding compositions of the endomorphisms of σ). If the σ-ring R is inversive, then Γ_σ (or Γ if the set σ is fixed) will denote the free commutative group generated by the set σ. It is clear that elements of the group Γ_σ (written in the multiplicative form $\alpha_1^{i_1} \dots \alpha_n^{i_n}$ where $i_1, \dots, i_n \in \mathbf{Z}$) act on R as automorphisms and T_σ is a subsemigroup of Γ_σ.

For any $a \in R$ and for any $\tau \in T_\sigma$, the element $\tau(a)$ is called a *transform* of a. If the σ-ring R is inversive, then an element $\gamma(a)$ $(a \in R, \gamma \in \Gamma_\sigma)$ is also called a transform of a.

If J is a σ-ideal of a σ-ring R, then $J^* = \{a \in R \,|\, \tau(a) \in J \text{ for some } \tau \in T_\sigma\}$ is a reflexive σ-ideal of R contained in any reflexive σ-ideal of R containing J. The ideal J^* is called the *reflexive closure* of the σ-ideal J.

Let R be a difference ring with a basic set σ and $S \subseteq R$. Then the intersection of all σ-ideals of R containing S is denoted by $[S]$. Clearly, $[S]$ is the smallest σ-ideal of R containing S; as an ideal, it is generated by the set $T_\sigma S = \{\tau(a) | \tau \in T_\sigma, a \in S\}$. If

$J = [S]$, we say that the σ-ideal J is generated by the set S called a *set of σ-generators* of J. If S is finite, $S = \{a_1, \ldots, a_k\}$, we write $J = [a_1, \ldots, a_k]$ and say that J is a *finitely generated σ-ideal* of the σ-ring R. (In this case elements $a_1, \ldots, a_k$ are said to be σ-*generators* of J.)

If R is an inversive difference (σ-) ring and $S \subseteq R$, then the inverse closure of the σ-ideal $[S]$ is denoted by $[S]^*$. It is easy to see that $[S]^*$ is the smallest σ^*-ideal of R containing S; as an ideal, it is generated by the set $\Gamma_\sigma S = \{\gamma(a) | \gamma \in \Gamma_\sigma, a \in S\}$. If S is finite, $S = \{a_1, \ldots, a_k\}$, we write $[a_1, \ldots, a_k]^*$ for $I = [S]^*$ and say that I is a *finitely generated σ^*-ideal* of R. (In this case, elements $a_1, \ldots, a_k$ are said to be σ^*-generators of I.)

Let R be a difference ring with a basic set σ, R_0 a σ-subring of R and $B \subseteq R$. The intersection of all σ-subrings of R containing R_0 and B is called the σ-*subring of R generated by the set B over R_0*, it is denoted by $R_0\{B\}$. (As a ring, $R_0\{B\}$ coincides with the ring $R_0[\{\tau(b) | b \in B, \tau \in T_\sigma\}]$ obtained by adjoining the set $\{\tau(b) | b \in B, \tau \in T_\sigma\}$ to the ring R_0). The set B is said to be the set of σ-*generators* of the σ-ring $R_0\{B\}$ over R_0. If this set is finite, $B = \{b_1, \ldots, b_k\}$, we say that $R' = R_0\{B\}$ is a finitely generated difference (or σ-) ring extension (or overring) of R_0 and write $R' = R_0\{b_1, \ldots, b_k\}$. If R is a σ-field, R_0 a σ-subfield of R and $B \subseteq R$, then the intersection of all σ-subfields of R containing R_0 and B is denoted by $R_0\langle B \rangle$ (or $R_0\langle b_1, \ldots, b_k \rangle$ if $B = \{b_1, \ldots, b_k\}$ is a finite set). This is the smallest σ-subfield of R containing R_0 and B; it coincides with the field $R_0(\{\tau(b) | b \in B, \tau \in T_\sigma\})$. The set B is called a set of σ-*generators* of the σ-field $R_0\langle B \rangle$ over R_0.

Let R be an inversive difference ring with a basic set σ, R_0 a σ^*-subring of R and $B \subseteq R$. Then the intersection of all σ^*-subrings of R containing R_0 and B is the smallest σ^*-subring of R containing R_0 and B. This ring coincides with the ring $R_0[\{\gamma(b) | b \in B, \gamma \in \Gamma_\sigma\}]$; it is denoted by $R_0\{B\}^*$. The set B is said to be a *set of σ^*-generators* of $R_0\{B\}^*$ over R_0. If $B = \{b_1, \ldots, b_k\}$ is a finite set, we say that $S = R_0\{B\}^*$ is a finitely generated inversive difference (or σ^*-) ring extension (or overring) of R_0 and write $S = R_0\{b_1, \ldots, b_k\}^*$.

If R is a σ^*-field, R_0 a σ^*-subfield of R and $B \subseteq R$, then the intersection of all σ^*-subfields of R containing R_0 and B is denoted by $R_0\langle B \rangle^*$. This is the smallest σ^*-subfield of R containing R_0 and B; it coincides with the field $R_0(\{\gamma(b) | b \in B, \gamma \in \Gamma_\sigma\})$. The set B is called a *set of σ^*-generators of the σ^*-field extension $R_0\langle B \rangle^*$ of R_0*. If B is finite, $B = \{b_1, \ldots, b_k\}$, we write $R_0\langle b_1, \ldots, b_k \rangle^*$ for $R_0\langle B \rangle^*$.

In what follows we often consider two or more difference rings $R_1, \ldots, R_p$ with the same basic set $\sigma = \{\alpha_1, \ldots, \alpha_n\}$. Formally speaking, it means that for every $i = 1, \ldots, p$, there is some fixed mapping ν_i from the set σ into the set of all injective endomorphisms of the ring R_i such that any two endomorphisms $\nu_i(\alpha_j)$ and $\nu_i(\alpha_k)$ of R_i commute ($1 \leq j, k \leq n$). We shall identify elements α_j with their images $\nu_i(\alpha_j)$ and say that elements of the set σ act as mutually commuting injective endomorphisms of the ring R_i ($i = 1, \ldots, p$).

Let R_1 and R_2 be difference rings with the same basic set $\sigma = \{\alpha_1, \ldots, \alpha_n\}$. A ring homomorphism $\phi : R_1 \to R_2$ is called a *difference* (or σ-) *homomorphism* if $\phi(\alpha(a)) = \alpha(\phi(a))$ for any $\alpha \in \sigma, a \in R_1$. Clearly, if $\phi : R_1 \to R_2$ is a σ-homomorphism of inversive difference rings, then $\phi(\alpha^{-1}(a)) = \alpha^{-1}(\phi(a))$ for any $\alpha \in \sigma$, $a \in R_1$. If a σ-homomorphism is an isomorphism (endomorphism, automorphism, etc.), it is called a difference (or σ-) isomorphism (respectively, difference (or σ-) endomorphism, difference (or σ-) automorphism, etc.). If R_1 and R_2 are two σ-overrings of the same σ-ring R_0 and $\phi : R_1 \to R_2$ is a σ-homomorphism such that $\phi(a) = a$ for any $a \in R_0$, we say that ϕ is a difference (or σ-) homomorphism over R_0 or that ϕ leaves the ring R_0 fixed. It is easy to see that the kernel of any σ-homomorphism of σ-rings $\phi : R \to R'$ is an inversive σ-ideal of R. Conversely, let g be a surjective homomorphism of a σ-ring R onto a ring S such that Ker g is a σ^*-ideal of R. Then there is a unique structure of a σ-ring on S such that g is a σ-homomorphism. In particular, if I is a σ^*-ideal of a σ-ring R, then the factor ring R/I has a unique structure of a σ-ring such that the canonical surjection $R \to R/I$ is a σ-homomorphism. In this case R/I is said to be the *difference* (or σ-) *factor ring* of R by the σ^*-ideal I.

If a difference (inversive difference) ring R with a basic set σ is an integral domain, then its quotient field $Q(R)$ can be naturally considered as a σ-(respectively, σ^*-) overring of R. (We identify an element $a \in R$ with its canonical image $\frac{a}{1}$ in $Q(R)$.) In this case $Q(R)$ is said to be the *quotient difference* or σ-*field* of R. Clearly, if the σ-ring R is inversive, then its quotient σ-field $Q(R)$ is also inversive. Furthermore, if a σ-field K contains an integral domain R as a σ-subring, then K contains the quotient σ-field $Q(R)$.

2.2 Difference and Inversive Polynomials. Algebraic Difference Equations

Let R be a difference ring with a basic set $\sigma = \{\alpha_1, \ldots, \alpha_n\}$, T_σ the free commutative semigroup generated by σ, and $U = \{u_\lambda | \lambda \in \Lambda\}$ a family of elements from some σ-overring of R. We say that the family U is *transformally* (or σ-*algebraically*) *dependent* over R, if the family $T_\sigma(U) = \{\tau(u_\lambda) | \tau \in T_\sigma, \lambda \in \Lambda\}$ is algebraically dependent over R (that is, there exist elements $v_1, \ldots, v_k \in T_\sigma(U)$ and a non-zero polynomial $f(X_1, \ldots, X_k)$ with coefficients from R such that $f(v_1, \ldots, v_k) = 0$). Otherwise, the family U is said to be *transformally* (or σ-*algebraically*) *independent* over R or a family of *difference* (or σ-) *indeterminates* over R. In the last case, the σ-ring $R\{(u_\lambda)_{\lambda \in \Lambda}\}$ is called the *algebra of difference* (or σ-) *polynomials* in the difference (or σ-) indeterminates $\{(u_\lambda)_{\lambda \in \Lambda}\}$ over R. If a family consisting of one element u is σ-algebraically dependent over R, the element u is said to be *transformally algebraic* (or σ-*algebraic*) over the σ-ring R. If the set $\{\tau(u) | \tau \in T_\sigma\}$ is algebraically independent over R, we say that u is *transformally* (or σ-) *transcendental* over the ring R.

Let R be a σ-field, L a σ-overfield of R, and $S \subseteq L$. We say that *the set S is σ-algebraic over R* if every element $a \in S$ is σ-algebraic over R. If every element of L is σ-algebraic over R, we say that L is a σ-*algebraic field extension* of the σ-field R.

Proposition 2.1 ([2, Chapter 2, Theorem I], [6, Proposition 3.3.7]). *Let R be a difference ring with a basic set σ and I an arbitrary set. Then there exists an algebra of σ-polynomials over R in a family of σ-indeterminates with indices from the set I. If S and S' are two such algebras, then there exists a σ-isomorphism $S \to S'$ that leaves the ring R fixed. If R is an integral domain, then any algebra of σ-polynomials over R is an integral domain.*

The algebra of σ-polynomials over the σ-ring R can be constructed as follows. Let $T = T_\sigma$ and let S be the polynomial R-algebra in the set of indeterminates $\{y_{i,\tau}\}_{i \in I, \tau \in T}$ with indices from the set $I \times T$. For any $f \in S$ and $\alpha \in \sigma$, let $\alpha(f)$ denote the polynomial from S obtained by replacing every indeterminate $y_{i,\tau}$ that appears in f by $y_{i,\alpha\tau}$ and every coefficient $a \in R$ by $\alpha(a)$. We obtain an injective endomorphism $S \to S$ that extends the original endomorphism α of R to the ring S (this extension is denoted by the same letter α). Setting $y_i = y_{i,1}$ (where 1 denotes the identity of the semigroup T) we obtain a set $\{y_i | i \in I\}$ whose elements are σ-algebraically independent over R and generate S as a σ-ring extension of R. Thus, $S = R\{(y_i)_{i \in I}\}$ is an algebra of σ-polynomials over R in a family of σ-indeterminates $\{y_i | i \in I\}$.

Let R be an inversive difference ring with a basic set σ, $\Gamma = \Gamma_\sigma$, I a set, and S^* a polynomial ring in the set of indeterminates $\{y_{i,\gamma}\}_{i \in I, \gamma \in \Gamma}$ with indices from the set $I \times \Gamma$. If we extend the automorphisms $\beta \in \sigma^*$ to S^* setting $\beta(y_{i,\gamma}) = y_{i,\beta\gamma}$ for any $y_{i,\gamma}$ and denote $y_{i,1}$ by y_i, then S^* becomes an inversive difference overring of R generated (as a σ^*-overring) by the family $\{(y_i)_{i \in I}\}$. Obviously, this family is σ^*-*algebraically independent* over R, that is, the set $\{\gamma(y_i) | \gamma \in \Gamma, i \in I\}$ is algebraically independent over R. (Note that a set is σ^*-algebraically dependent (independent) over an inversive σ-ring if and only if this set is σ-algebraically dependent (respectively, independent) over this ring.) The ring $S^* = R\{(y_i)_{i \in I}\}^*$ is called the *algebra of inversive difference* (or σ^*-) *polynomials* over R in the set of σ^*-indeterminates $\{(y_i)_{i \in I}\}$. It is easy to see that S^* is the inversive closure of the ring of σ-polynomials $R\{(y_i)_{i \in I}\}$ over R. Furthermore, if a family $\{(u_i)_{i \in I}\}$ from some σ^*-overring of R is σ^*-algebraically independent over R, then the inversive difference ring $R\{(u_i)_{i \in I}\}^*$ is naturally σ-isomorphic to S^*. Any such overring $R\{(u_i)_{i \in I}\}^*$ is said to be an algebra of inversive difference (or σ^*-) polynomials over R in the set of σ^*-indeterminates $\{(u_i)_{i \in I}\}$. We obtain the following analog of Proposition 2.1.

Proposition 2.2 ([6, Proposition 3.4.4]). *Let R be an inversive difference ring with a basic set σ and I an arbitrary set. Then there exists an algebra of σ^*-polynomials over R in a family of σ^*-indeterminates with indices from the set I. If S and S' are two such algebras, then there exists a σ^*-isomorphism $S \to S'$ that leaves the ring R fixed. If R is an integral domain, then any algebra of σ^*-polynomials over R is an integral domain.*

Let R be a σ-ring, $R\{(y_i)_{i \in I}\}$ an algebra of difference polynomials in a family of σ-indeterminates $\{(y_i)_{i \in I}\}$, and $\{(\eta_i)_{i \in I}\}$ a set of elements from some σ-overring of R. Since the set $\{\tau(y_i) | i \in I, \tau \in T_\sigma\}$ is algebraically independent over R, there exists a unique ring homomorphism $\phi_\eta : R[\tau(y_i)_{i \in I, \tau \in T_\sigma}] \to R[\tau(\eta_i)_{i \in I, \tau \in T_\sigma}]$ that maps every $\tau(y_i)$ onto $\tau(\eta_i)$ and leaves R fixed. Clearly, ϕ_η is a surjective σ-homomorphism of $R\{(y_i)_{i \in I}\}$ onto $R\{(\eta_i)_{i \in I}\}$; it is called the *substitution* of $(\eta_i)_{i \in I}$ for $(y_i)_{i \in I}$. Similarly, if R is an inversive σ-ring, $R\{(y_i)_{i \in I}\}^*$ an algebra of σ^*-polynomials over R and $(\eta_i)_{i \in I}$ a family of elements from a σ^*-overring of R, one can define a surjective σ-homomorphism $R\{(y_i)_{i \in I}\}^* \to R\{(\eta_i)_{i \in I}\}^*$ that maps every y_i onto η_i and leaves the ring R fixed. This homomorphism is also called the substitution of $(\eta_i)_{i \in I}$ for $(y_i)_{i \in I}$. (It will be always clear whether we talk about substitutions for difference or inversive difference polynomials.) If g is a σ- or σ^*- polynomial, then its image under a substitution of $(\eta_i)_{i \in I}$ for $(y_i)_{i \in I}$ is denoted by $g((\eta_i)_{i \in I})$. The kernel of a substitution ϕ_η is an inversive difference ideal of the σ-ring $R\{(y_i)_{i \in I}\}$ (or the σ^*-ring $R\{(y_i)_{i \in I}\}^*$); it is called the *defining difference* (or σ-) *ideal* of the family $(\eta_i)_{i \in I}$ over R. If R is a σ- (or σ^*-) field and $(\eta_i)_{i \in I}$ is a family of elements from some σ- (respectively, σ^*-) overfield S, then $R\{(\eta_i)_{i \in I}\}$(respectively, $R\{(\eta_i)_{i \in I}\}^*$) is an integral domain (it is contained in the field S). It follows that the defining σ-ideal P of the family $(\eta_i)_{i \in I}$ over R is a reflexive prime difference ideal of the ring $R\{(y_i)_{i \in I}\}$ (respectively, of the ring of σ^*-polynomials $R\{(y_i)_{i \in I}\}^*$). Therefore, the difference field $R\langle(\eta_i)_{i \in I}\rangle$ can be treated as the quotient σ-field of the σ-ring $R\{(y_i)_{i \in I}\}/P$. (In the case of inversive difference rings, the σ^*-field $R\langle(\eta_i)_{i \in I}\rangle^*$ can be considered as a quotient σ-field of the σ^*-ring $R\{(y_i)_{i \in I}\}^*/P$.)

Let K be a difference field with a basic set σ and s a positive integer. By an *s-tuple over K* we mean an s-dimensional vector $a = (a_1, \ldots, a_s)$ whose coordinates belong to some σ-overfield of K. If the σ-field K is inversive, the coordinates of an s-tuple over K are supposed to lie in some σ^*-overfield of K. If each a_i $(1 \le i \le s)$ is σ-algebraic over the σ-field K, we say that the *s-tuple a is σ-algebraic over K*.

Definition 2.3 Let K be a difference (inversive difference) field with a basic set σ and let R be the algebra of σ- (respectively, σ^*-) polynomials in finitely many σ- (respectively, σ^*-) indeterminates $y_1, \ldots, y_s$ over K. Furthermore, let $\Phi = \{f_j | j \in J\}$ be a set of σ- (respectively, σ^*-) polynomials from R. An s-tuple $\eta = (\eta_1, \ldots, \eta_s)$ over K is said to be a solution of the set Φ or a solution of the system of *algebraic difference equations* $f_j(y_1, \ldots, y_s) = 0$ $(j \in J)$ if Φ is contained in the kernel of the substitution of $(\eta_1, \ldots, \eta_s)$ for $(y_1, \ldots, y_s)$. In this case we also say that η annuls Φ. (If Φ is a subset of a ring of inversive difference polynomials, the system is said to be a system of algebraic σ^*-equations.) A system of algebraic difference equations Φ is called *prime* if the reflexive difference ideal generated by Φ in the ring of σ (or σ^*-) polynomials is prime.

As we have seen, if one fixes an s-tuple $\eta = (\eta_1, \ldots, \eta_s)$ over a σ-field F, then all σ-polynomials of the ring $K\{y_1, \ldots, y_s\}$, for which η is a solution, form a reflexive prime difference ideal. It is called the *defining σ-ideal* of η. If η is an s-tuple over a σ^*-field K, then all σ^*-polynomials g of the ring $K\{y_1, \ldots, y_s\}^*$ such that $g(\eta_1, \ldots, \eta_s) = 0$ form a prime σ^*-ideal of $K\{y_1, \ldots, y_s\}^*$. This ideal is called the *defining σ^*-ideal* of η over K.

Let Φ be a subset of the algebra of σ-polynomials $K\{y_1, \ldots, y_s\}$ over a σ-field K. An s-tuple $\eta = (\eta_1, \ldots, \eta_s)$ over K is called a *generic zero* of Φ if for any σ-polynomial $f \in K\{y_1, \ldots, y_s\}$, the inclusion $f \in \Phi$ holds if and only if $f(\eta_1, \ldots, \eta_s) = 0$. If the σ-field K is inversive, then the notion of a generic zero of a subset of $K\{y_1, \ldots, y_s\}^*$ is defined similarly.

Two s-tuples $\eta = (\eta_1, \ldots, \eta_s)$ and $\zeta = (\zeta_1, \ldots, \zeta_s)$ over a σ- (or σ^*-) field K are called *equivalent* over K if there is a σ-homomorphism $K\langle\eta_1, \ldots, \eta_s\rangle \to K\langle\zeta_1, \ldots, \zeta_s\rangle$ (respectively, $K\langle\eta_1, \ldots, \eta_s\rangle^* \to K\langle\zeta_1, \ldots, \zeta_s\rangle^*$) that maps each η_i onto ζ_i and leaves the field K fixed.

Proposition 2.4 ([2, Chapter 2, Theorem VII], [6, Proposition 3.3.7]). *Let R denote* *the algebra of σ-polynomials $K\{y_1, \ldots, y_s\}$ over a difference field K with a basic set σ.*

(i) *A set $\Phi \subsetneqq R$ has a generic zero if and only if Φ is a prime reflexive σ-ideal of R. If $(\eta_1, \ldots, \eta_s)$ is a generic zero of Φ, then $K\langle \eta_1, \ldots, \eta_s \rangle$ is σ-isomorphic to the quotient σ-field of R/Φ.*

(ii) *Any s-tuple over K is a generic zero of some prime reflexive σ-ideal of R.*

(iii) *If two s-tuples over K are generic zeros of the same prime reflexive σ-ideal of R, then these s-tuples are equivalent.*

2.3 Ring of Difference Operators. Difference Modules

Let R be a difference ring with a basic set $\sigma = \{\alpha_1, \ldots, \alpha_n\}$ and T the free commutative semigroup generated by the elements $\alpha_1, \ldots, \alpha_n$. If $\tau = \alpha_1^{k_1} \ldots \alpha_n^{k_n} \in T$ $(k_1, \ldots, k_n \in \mathbf{N})$, then the number $\operatorname{ord} \tau = \sum_{\nu=1}^{n} k_\nu$ is called the *order* of τ. Furthermore, for any $r \in \mathbf{N}$, the set $\{\tau \in T | \operatorname{ord} \tau \leq r\}$ is denoted by $T(r)$.

Definition 2.5 An expression of the form $\sum_{\tau \in T} a_\tau \tau$, where $a_\tau \in R$ for any $\tau \in T$ and only finitely many elements a_τ are different from 0, is called a difference (or σ-) operator over the difference ring R. Two σ-operators $\sum_{\tau \in T} a_\tau \tau$ and $\sum_{\tau \in T} b_\tau \tau$ are considered to be equal if and only if $a_\tau = b_\tau$ for all $\tau \in T$.

The set of all σ-operators over the σ-ring R can be equipped with a ring structure if we set $\sum_{\tau \in T} a_\tau \tau + \sum_{\tau \in T} b_\tau \tau = \sum_{\tau \in T} (a_\tau + b_\tau)\tau$, $a \sum_{\tau \in T} a_\tau \tau = \sum_{\tau \in T} (aa_\tau)\tau$, $(\sum_{\tau \in T} a_\tau \tau)\tau_1 = \sum_{\tau \in T} a_\tau (\tau \tau_1)$, $\tau_1 a = \tau_1(a)\tau_1$ for any $\sum_{\tau \in T} a_\tau \tau$, $\sum_{\tau \in T} b_\tau \tau \in \mathcal{D}$, $a \in R$, $\tau_1 \in T$, and extend the multiplication by distributivity. The ring obtained in this way is called *the ring of difference* (or σ-) *operators over R*; it will be denoted by $\mathcal{D}$.

The order of a non-zero σ-operator $A = \sum_{\tau \in T} a_\tau \tau \in \mathcal{D}$ is defined as the number $\operatorname{ord} A = \max\{\operatorname{ord} \tau | a_\tau \neq 0\}$. We also set $\operatorname{ord} 0 = -1$.

Let $\mathcal{D}_r = \{A \in \mathcal{D} | \operatorname{ord} A \leq r\}$ for any $r \in \mathbf{N}$ and let $\mathcal{D}_r = 0$ for any $r \in \mathbf{Z}, r < 0$. Then the ring $\mathcal{D}$ can be treated as a filtered ring with the ascending filtration $(\mathcal{D}_r)_{r \in \mathbf{Z}}$. Below, while considering $\mathcal{D}$ as a filtered ring, we always mean this filtration.

Definition 2.6 Let R be a difference ring with a basic set σ and $\mathcal{D}$ the ring of σ-operators over R. Then a left $\mathcal{D}$-module is called a difference R-module or a σ-R-module. In other words, an R-module M is a difference (or σ-) R-module, if the elements of σ act on M in such a way that $\alpha(x + y) = \alpha(x) + \alpha(y)$, $\alpha(\beta x) = \beta(\alpha x)$, and $\alpha(ax) = \alpha(a)\alpha(x)$ for any $x, y \in M; \alpha, \beta \in \sigma; a \in R$.

If R is a difference (σ-) field, then a σ-R-module M is also called a difference vector space over R or a vector σ-R-space.

We say that a difference R-module M is finitely generated, if it is finitely generated as a left $\mathcal{D}$-module. By a filtered σ-R-module we always mean a left $\mathcal{D}$-module equipped with an exhaustive and separated filtration. Thus, a filtration of a σ-R-module M is an ascending chain $(M_r)_{r \in \mathbf{Z}}$ of R-submodules of M such that $\mathcal{D}_r M_s \subseteq M_{r+s}$ for all $r, s \in \mathbf{Z}$, $M_r = 0$ for all sufficiently small $r \in \mathbf{Z}$, and $\bigcup_{r \in \mathbf{Z}} M_r = M$.

A filtration $(M_r)_{r \in \mathbf{Z}}$ of a σ-R-module M is called *excellent* if all R-modules M_r ($r \in \mathbf{Z}$) are finitely generated and there exists $r_0 \in \mathbf{Z}$ such that $M_r = \mathcal{D}_{r-r_0} M_{r_0}$ for any $r \in \mathbf{Z}, r \geq r_0$.

The following result was obtained in [7] (see also [6, Theorem 6.2.5]).

Theorem 2.7 *Let R be an Artinian difference ring with a basic set of translations $\sigma = \{\alpha_1, \ldots, \alpha_n\}$, let $\mathcal{D}$ be the ring of σ-operators over R, and let $(M_r)_{r \in \mathbf{Z}}$ be an excellent filtration of a σ-R-module M. Then there exists a polynomial $\psi(t) \in \mathbf{Q}[t]$ with the following properties.*

(i) *$\psi(r) = l_R(M_r)$ for all sufficiently large $r \in \mathbf{Z}$, that is, there exists $r_0 \in \mathbf{Z}$ such that the last equality holds for all integers $r \geq r_0$. ($l_R(M_r)$ denotes the length of the R-module M_r, that is, the length of a composition series of M_r.)*

(ii) *$\deg \psi(t) \leq n$ and the polynomial $\psi(t)$ can be written as*

$$\psi(t) = \sum_{i=0}^{n} c_i \binom{t+i}{i}$$

where $c_0, c_1, \ldots, c_n \in \mathbf{Z}$. (As usual, $\binom{t+i}{i}$ denotes the polynomial $(t+i)(t+i-1)\ldots(t+1)/i! \in \mathbf{Q}[t]$ that takes integer values for all sufficiently large integer values of t.)

(iii) *If R is a difference field, then the integers $d = \deg \psi(t)$, c_n and c_d (if $d < n$) do not depend on the choice of the excellent filtration of M. Furthermore, c_n is equal to the maximal number of elements of M linearly independent over the ring $\mathcal{D}$.*

The polynomial $\psi(t)$ whose existence is established by Theorem 2.7 is called the *difference (σ-) dimension polynomial* or the *characteristic polynomial* of the σ-R-module M associated with the excellent filtration $(M_r)_{r \in \mathbf{Z}}$. If R is a difference field, then the integers d, c_n, and c_d are called the *difference type*, *difference dimension*, and *typical difference dimension* of M, respectively. A number of results on difference dimension polynomials and methods of their computation can be found in [6, Chapters 6–9].

2.4 Partitions of the Basic Set and Multidimensional Filtrations of Difference Modules

Let K be a difference field with a basic set $\sigma = \{\alpha_1, \ldots, \alpha_n\}$, T the commutative semigroup of all power products $\alpha_1^{k_1} \ldots \alpha_n^{k_n}$ ($k_1, \ldots, k_n \in \mathbf{N}$) and $\mathcal{D}$ the ring of σ-operators over K. Let us fix a partition of the set σ into a disjoint union of its subsets:

$$\sigma = \sigma_1 \cup \cdots \cup \sigma_p \tag{2.1}$$

where $p \in \mathbf{N}$, and $\sigma_1 = \{\alpha_1, \ldots, \alpha_{n_1}\}$, $\sigma_2 = \{\alpha_{n_1+1}, \ldots, \alpha_{n_1+n_2}\}, \ldots,$ $\sigma_p = \{\alpha_{n_1+\cdots+n_{p-1}+1}, \ldots, \alpha_n\}$ ($n_i \geq 1$ for $i = 1, \ldots, p; n_1 + \cdots + n_p = n$).

If $\tau = \alpha_1^{k_1} \ldots \alpha_n^{k_n} \in T$ ($k_1, \ldots, k_n \in \mathbf{N}$), then the numbers $\mathrm{ord}_i \tau = \sum_{\nu=n_1+\cdots+n_{i-1}+1}^{n_1+\cdots+n_i} k_\nu$ ($1 \leq i \leq p$) are called the *orders* of τ with respect to σ_i (we assume that $n_0 = 0$, so the indices in the sum for $\mathrm{ord}_1\tau$ change from 1 to n_1). As before, the order of the element τ is defined as $\mathrm{ord}\,\tau = \sum_{\nu=1}^{n} k_i = \sum_{i=1}^{p} \mathrm{ord}_i\tau$.

Below we shall consider p orders $<_1, \ldots, <_p$ on the set T that are defined as follows: $\tau = \alpha_1^{k_1} \ldots \alpha_n^{k_n} <_i \tau' = \alpha_1^{l_1} \ldots \alpha_n^{l_n}$ if and only if the vector

$$(\mathrm{ord}_i\tau, \mathrm{ord}\,\tau, \mathrm{ord}_1\tau, \ldots, \mathrm{ord}_{i-1}\tau, \mathrm{ord}_{i+1}\tau, \ldots, \mathrm{ord}_p\tau, k_{n_1+\cdots+n_{i-1}+1}, \ldots, k_{n_1+\cdots+n_i},$$

$$k_1, \ldots, k_{n_1+\cdots+n_{i-1}}, k_{n_1+\cdots+n_i+1}, \ldots, k_n)$$

is less than the vector

$$(\mathrm{ord}_i\tau', \mathrm{ord}\,\tau', \mathrm{ord}_1\tau', \ldots, \mathrm{ord}_{i-1}\tau', \mathrm{ord}_{i-1}\tau', \ldots, \mathrm{ord}_p\tau', l_{n_1+\cdots+n_{i-1}+1}, \ldots,$$

$$l_{n_1+\cdots+n_i}, l_1, \ldots, l_{n_1+\cdots+n_{i-1}}, l_{n_1+\cdots+n_i+1}, \ldots, l_n)$$

with respect to the lexicographic order on $\mathbf{N}^{n+p+1}$. It is easy to see that the set T is well-ordered with respect to each of the orders $<_1, \ldots, <_p$.

If $r_1, \ldots, r_p$ are non-negative integers, then $T(r_1, \ldots, r_p)$ will denote the set of all elements $\tau \in T$ such that $\mathrm{ord}_i\tau \leq r_i$ $(i = 1, \ldots, p)$. The vector K-subspace of $\mathcal{D}$ generated by the set $T(r_1, \ldots, r_p)$ will be denoted by $\mathcal{D}_{r_1,\ldots,r_p}$.

Setting $\mathcal{D}_{r_1,\ldots,r_p} = 0$ for any $(r_1, \ldots, r_p) \in \mathbf{Z}^p \setminus \mathbf{N}^p$, we obtain a family $\{\mathcal{D}_{r_1,\ldots,r_p} | (r_1, \ldots, r_p) \in \mathbf{Z}^p\}$ of vector K-subspaces of $\mathcal{D}$ which is called the *standard p-dimensional filtration* of the ring $\mathcal{D}$. It is easy to see that $\mathcal{D}_{r_1,\ldots,r_p} \subseteq \mathcal{D}_{s_1,\ldots,s_p}$ if $(r_1, \ldots, r_p) \leq_P (s_1, \ldots, s_p)$, where $\leq_P$ denotes the product order on $\mathbf{Z}^p$ (recall that this is a partial order on $\mathbf{Z}^p$ such that $(a_1, \ldots, a_p) \leq_P (b_1, \ldots, b_p)$ if and only if $a_i \leq b_i$ for $i = 1, \ldots, p$). Furthermore, $\mathcal{D}_{i_1,\ldots,i_p}\mathcal{D}_{r_1,\ldots,r_p} = \mathcal{D}_{r_1+i_1,\ldots,r_p+i_p}$ for any $(r_1, \ldots, r_p)$, $(i_1, \ldots, i_p) \in \mathbf{N}^p$.

Definition 2.8 Let M be a vector σ-K-space (that is, a left $\mathcal{D}$-module). A family $\{M_{r_1,\ldots,r_p} | (r_1, \ldots, r_p) \in \mathbf{Z}^p\}$ is said to be a p-dimensional filtration of M if the following four conditions hold:

(i) $M_{r_1,\ldots,r_p} \subseteq M_{s_1,\ldots,s_p}$ for any p-tuples $(r_1, \ldots, r_p), (s_1, \ldots, s_p) \in \mathbf{Z}^p$ such that $(r_1, \ldots, r_p) \leq_P (s_1, \ldots, s_p)$.

(ii) $\bigcup_{(r_1,\ldots,r_p)\in\mathbf{Z}^p} M_{r_1,\ldots,r_p} = M$.

(iii) There exists a p-tuple $(r_1^{(0)}, \ldots, r_p^{(0)}) \in \mathbf{Z}^p$ such that $M_{r_1,\ldots,r_p} = 0$ if $r_i < r_i^{(0)}$ for at least one index i $(1 \leq i \leq p)$.

(iv) $\mathcal{D}_{r_1,\ldots,r_p} M_{s_1,\ldots,s_p} \subseteq M_{r_1+s_1,\ldots,r_p+s_p}$ for any $(r_1, \ldots, r_p), (s_1, \ldots, s_p) \in \mathbf{Z}^p$.

If every vector K-space $M_{r_1,\ldots,r_p}$ is finite-dimensional and there exists an element $(h_1, \ldots, h_p) \in \mathbf{Z}^p$ such that $\mathcal{D}_{r_1,\ldots,r_p} M_{h_1,\ldots,h_p} = M_{r_1+h_1,\ldots,r_p+h_p}$ for any $(r_1, \ldots, r_p) \in \mathbf{N}^p$, the p-dimensional filtration $\{M_{r_1,\ldots,r_p} | (r_1, \ldots, r_p) \in \mathbf{Z}^p\}$ is called **excellent**.

It is easy to see that if $z_1, \ldots, z_k$ is a finite system of generators of a vector σ-K-space M, then $\{\sum_{i=1}^{k} \mathcal{D}_{r_1,\ldots,r_p} z_i | (r_1, \ldots, r_p) \in \mathbf{Z}^p\}$ is an excellent p-dimensional filtration of M.

2.5   Numerical Polynomials

A polynomial $f(t_1, \ldots, t_p)$ in p variables $t_1, \ldots, t_p$ $(p \geq 1)$ with rational coefficients is called *numerical* if $f(t_1, \ldots, t_p) \in \mathbf{Z}$ for all sufficiently large $t_1, \ldots, t_p \in \mathbf{Z}$, i.e., there exists an element $(s_1, \ldots, s_p) \in \mathbf{Z}^p$ such that $f(r_1, \ldots, r_p) \in \mathbf{Z}$ as soon as $(r_1, \ldots, r_p) \in \mathbf{Z}^p$ and $r_i \geq s_i$ for all $i = 1, \ldots, p$.

It is clear that every polynomial with integer coefficients is numerical. As an example of a numerical polynomial in p variables with non-integer coefficients one can consider the polynomial $\prod_{i=1}^{p} \binom{t_i}{m_i}$ where $m_1, \ldots, m_p \in \mathbf{N}$.

If f is a numerical polynomial in p variables and $p > 1$, then $\deg f$ and $\deg_{t_i} f$ $(1 \leq i \leq p)$ will denote the total degree of f and the degree of f relative to the variable t_i, respectively. The following theorem proved in [5] gives the "canonical" representation of a numerical polynomial in several variables.

Theorem 2.9 *Let* $f(t_1, \ldots, t_p)$ *be a numerical polynomial in* $t_1, \ldots, t_p$, *and let* $\deg_{t_i} f = m_i$ $(m_1, \ldots, m_p \in \mathbf{N})$. *Then the polynomial* $f(t_1, \ldots, t_p)$ *can be represented in the form*

$$f(t_1, \ldots t_p) = \sum_{i_1=0}^{m_1} \cdots \sum_{i_p=0}^{m_p} a_{i_1 \ldots i_p} \binom{t_1 + i_1}{i_1} \cdots \binom{t_p + i_p}{i_p} \tag{2.2}$$

with integer coefficients $a_{i_1 \ldots i_p}$ *uniquely defined by the numerical polynomial.*

In what follows (until the end of the section), we deal with subsets of the set $\mathbf{N}^m$ where the positive integer m is represented as a sum of p nonnegative integers $m_1, \ldots, m_p$ $(p \in \mathbf{N}, p \geq 1)$. In other words, we assume that a partition $(m_1, \ldots, m_p)$ of the number m is fixed.

If $\mathcal{A} \subseteq \mathbf{N}^m$, then for any $r_1, \ldots, r_p \in \mathbf{N}$, $\mathcal{A}(r_1, \ldots, r_p)$ will denote the subset of $\mathcal{A}$ that consists of all m-tuples $(a_1, \ldots, a_m)$ such that $a_1 + \cdots + a_{m_1} \leq r_1$, $a_{m_1+1} + \cdots + a_{m_1+m_2} \leq r_2, \ldots, a_{m_1+\cdots+m_{p-1}+1} + \cdots + a_m \leq r_p$. Furthermore, we shall associate with the set $\mathcal{A}$ a set $V_{\mathcal{A}} \subseteq \mathbf{N}^m$ that consists of all m-tuples $v = (v_1, \ldots, v_m) \in \mathbf{N}$ that are not greater than or equal to any m-tuple from $\mathcal{A}$ with respect to the product order on $\mathbf{N}^m$. Clearly, an element $v = (v_1, \ldots, v_m) \in \mathbf{N}^m$ belongs to $V_{\mathcal{A}}$ if and only if for any element $(a_1, \ldots, a_m) \in \mathcal{A}$, there exists $i \in \mathbf{N}, 1 \leq i \leq m$, such that $a_i > v_i$.

The following two theorems proved in [5] generalize the well-known Kolchin's result on univariate numerical polynomials of subsets of $\mathbf{N}^m$ (see [4, Chapter 0, Lemma 17]) and give the explicit formula for the numerical polynomials in p variables associated with a finite subset of $\mathbf{N}^m$.

Theorem 2.10 ([5, Theorem 4]). *Let* $\mathcal{A}$ *be a subset of* $\mathbf{N}^m$ *where* $m = m_1 + \cdots + m_p$ *for some nonnegative integers* $m_1, \ldots, m_p$ $(p \geq 1)$. *Then there exists a numerical polynomial* $\omega_{\mathcal{A}}(t_1, \ldots, t_p)$ *with the following properties:*

  (i) $\omega_{\mathcal{A}}(r_1, \ldots, r_p) = \mathrm{Card}\, V_{\mathcal{A}}(r_1, \ldots, r_p)$ *for all sufficiently large* $(r_1, \ldots, r_p) \in \mathbf{N}^p$ *(as usual,* $\mathrm{Card}\, M$ *denotes the number of elements of a finite set* M).

  (ii) *The total degree of the polynomial* $\omega_{\mathcal{A}}$ *does not exceed* m *and* $\deg_{t_i} \omega_{\mathcal{A}} \leq m_i$ *for all* $i = 1, \ldots, p$.

(iii) deg $\omega_\mathcal{A} = m$ *if and only if the set $\mathcal{A}$ is empty. In this case*

$$\omega_\mathcal{A}(t_1,\ldots,t_p) = \prod_{i=1}^{p} \binom{t_i + m_i}{m_i}.$$

(iv) $\omega_\mathcal{A}$ *is a zero polynomial if and only if* $(0,\ldots,0) \in \mathcal{A}$.

Definition 2.11 The polynomial $\omega_\mathcal{A}(t_1,\ldots,t_p)$ whose existence is stated by Theorem 2.7 is called the *dimension polynomial* of the set $\mathcal{A} \subseteq \mathbf{N}^m$ associated with the partition $(m_1,\ldots,m_p)$ of m. If $p = 1$, the polynomial $\omega_\mathcal{A}$ is called the Kolchin polynomial of the set $\mathcal{A}$.

Theorem 2.12 *Let $\mathcal{A} = \{a_1,\ldots,a_n\}$ be a finite subset of $\mathbf{N}^m$ where m is a positive integer and $m = m_1 + \cdots + m_p$ for some nonnegative integers $m_1,\ldots,m_p$ $(p \geq 1)$. Let $a_i = (a_{i1},\ldots,a_{im})$ $(1 \leq i \leq n)$ and for any $l \in \mathbf{N}$, $0 \leq l \leq n$, let $\Gamma(l,n)$ denote the set of all l-element subsets of the set $\mathbf{N}_n = \{1,\ldots,n\}$. Furthermore, for any $\xi \in \Gamma(l,p)$, let $\bar{a}_{\xi h} = \max\{a_{ih} | i \in \xi\}$ $(1 \leq h \leq m)$ and $b_{\xi j} = \sum_{h \in \sigma_j} \bar{a}_{\xi h}$. (If $\xi = \emptyset$, we assume that $\bar{a}_{\xi h} = 0$ for $h = 1,\ldots,m$, so $b_{\xi j} = 0$ for $j = 1,\ldots,p$.) Then*

$$\omega_\mathcal{A}(t_1,\ldots,t_p) = \sum_{l=0}^{n} (-1)^l \sum_{\xi \in \Gamma(l,n)} \prod_{j=1}^{p} \binom{t_j + m_j - b_{\xi j}}{m_j}. \qquad (2.3)$$

It is clear that if $\mathcal{A}$ is any subset of $\mathbf{N}^m$ and $\mathcal{A}'$ is the set of all minimal elements of the set $\mathcal{A}$ with respect to the product order on $\mathbf{N}^m$, then the set $\mathcal{A}'$ is finite and $\omega_\mathcal{A}(t_1,\ldots,t_p) = \omega_{\mathcal{A}'}(t_1,\ldots,t_p)$. Thus, Theorem 2.12 gives an algorithm that allows one to find a numerical polynomial associated with any subset of $\mathbf{N}^m$ (and with a given representation of m as a sum of p non-negative integers): one should first find the set of all minimal points of the subset and then apply Theorem 2.12.

3 Difference Dimension Polynomials and Strength of Systems of Difference Equations

The following two theorems proved, respectively, in [7] and [6, Theorem 6.4.8] introduce the concepts and provide some description of dimension polynomials associated with finitely generated difference and inversive difference field extensions.

Theorem 3.1 *Let K be a difference field of zero characteristic with a basic set $\sigma = \{\alpha_1,\ldots,\alpha_n\}$, let T be the free commutative semigroup generated by σ, and $T(r) = \{\tau = \alpha_1^{k_1}\ldots\alpha_n^{k_n} \in T | \operatorname{ord}\tau = \sum_{i=1}^{n} k_i \leq r\}$ for any $r \in \mathbf{N}$. Furthermore, let $L = K\langle\eta_1,\ldots,\eta_s\rangle$ be a difference field extension of K generated by a finite set $\eta = \{\eta_1,\ldots,\eta_s\}$. Then there exists a numerical polynomial $\phi_\eta(t)$ such that*

(i) $\phi_\eta(r) = \operatorname{trdeg}_K K(\{\tau\eta_j | \tau \in T(r), 1 \leq j \leq s\})$ *for all sufficiently large $r \in \mathbf{Z}$.*

(ii) $\deg \phi_\eta(t) \leq n$ and $\phi_\eta(t)$ can be written as $\phi_\eta(t) = \sum_{i=0}^{n} a_i \binom{t+i}{i}$ where $a_0, \ldots, a_n \in \mathbf{Z}$;

(iii) The degree d of the polynomial $\phi_\eta(t)$ and the coefficients a_n and a_d do not depend on the choice of the system of generators η (clearly, $a_d \neq a_n$ if and only if $d < n$, that is, $a_n = 0$). Moreover, a_n is equal to the difference transcendence degree of L over K, i. e., to the maximal number of elements $\xi_1, \ldots, \xi_k \in L$ such that the set $\{\tau(\xi_i)|\tau \in T, 1 \leq i \leq k\}$ is algebraically independent over K (this characteristic of the σ-field extension is denoted by $\sigma\text{-trdeg}_K L$).

Theorem 3.2 Let K be an inversive difference field of zero characteristic with a basic set $\sigma = \{\alpha_1, \ldots, \alpha_n\}$. Let Γ be the free commutative group generated by the set σ, and for any $r \in \mathbf{N}$, let $\Gamma(r)$ denote the set of all elements $\gamma = \alpha_1^{k_1} \ldots \alpha_n^{k_n} \in \Gamma$ ($k_1, \ldots, k_n \in \mathbf{Z}$) such that $\sum_{i=1}^{n} |k_i| \leq r$. Furthermore, let L be an inversive difference field extension of K generated by a finite set $\eta = \{\eta_1, \ldots, \eta_p\}$.

Then there exists a polynomial $\psi_{\eta|K}(t)$ in one variable t with rational coefficients (called a **difference dimension polynomial** of the σ^*-field extension L/K) such that

(i) $\psi_{\eta|K}(r) = \text{trdeg}_K K(\{\gamma(\eta_j)|\gamma \in \Gamma(r), 1 \leq j \leq p\})$ for all sufficiently large integers r.

(ii) $\deg \psi_{\eta|K} \leq n$ and the polynomial $\psi_{\eta|K}(t)$ can be written as $\psi_{\eta|K}(t) = \sum_{i=0}^{n} a_i 2^i \binom{t+i}{i}$ where $a_0, \ldots, a_n \in \mathbf{Z}$.

(iii) The degree d of the polynomial $\psi_{\eta|K}$ and the coefficients a_n and a_d do not depend on the choice of the system of σ^*-generators η of the extension L/K. Furthermore, the coefficient a_n is equal to the difference transcendence degree of L over K.

Let K be a difference (inversive difference) field with a basic set σ and R an algebra of σ- (respectively, σ^*-) polynomials in difference indeterminates $y_1, \ldots, y_s$ over K. Let P be a prime reflexive difference ideal of R and $\eta = (\eta_1, \ldots, \eta_s)$ a generic zero of P. Then the dimension polynomial $\phi_{\eta|K}(t)$ (respectively, $\psi_{\eta|K}(t)$) associated with the σ- (σ^*-) field extension $K\langle \eta_1, \ldots, \eta_s \rangle/K$ (respectively, $K\langle \eta_1, \ldots, \eta_s \rangle^*/K$) is called the σ- (respectively, σ^*-) dimension polynomial of the ideal P. It is denoted by $\phi_P(t)$ (respectively, $\psi_P(t)$).

The difference dimension polynomial of a prime system of algebraic difference equations has an interesting interpretation as a measure of strength of a system of such equations in the sense of A. Einstein. The concept of the strength of a system of functional equations in finite differences is an analog of the corresponding characteristic of a system of partial differential equations introduced and studied by A. Einstein. In his work [3] A. Einstein defined the strength of a system of partial differential equations governing a physical field as follows: "... the system of equations is to be chosen so that the field quantities are determined as strongly as possible. In order to apply this principle, we propose a method which gives a measure of strength of an equation system. We expand the field variables, in the neighborhood of a point $\mathcal{P}$, into a Taylor series (which presupposes the analytic character of the field); the coefficients of these series, which are the derivatives of the field variables at $\mathcal{P}$, fall into sets according to the degree of differentiation. In every such degree there appear, for the first time, a set

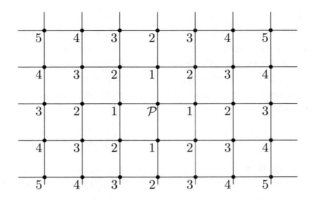

Figure 3.1

of coefficients which would be free for arbitrary choice if it were not that the field must satisfy a system of differential equations. Through this system of differential equations (and its derivatives with respect to the coordinates) the number of coefficients is restricted, so that in each degree a smaller number of coefficients is left free for arbitrary choice. The set of numbers of "free" coefficients for all degrees of differentiation is then a measure of the "weakness" of the system of equations, and through this, also of its "strength"."

Considering a system of equations in finite differences over a field of of functions in several real variables, one can use A. Einstein's approach to define the concept of strength of such a system as follows. Let

$$A_i(f_1, \ldots, f_s) = 0 \quad (i = 1, \ldots, p) \tag{3.1}$$

be a system of equations in finite differences with respect to s unknown grid functions $f_1, \ldots, f_s$ in n real variables $x_1, \ldots, x_n$ with coefficients in some functional field K over the field of real numbers $\mathbf{R}$. We also assume that the difference grid, whose nodes form the domain of considered functions, has equal cells of dimension $h_1 \times \cdots \times h_n$ ($h_1, \ldots, h_n \in \mathbf{R}$) and fills the whole space $\mathbf{R}^n$. As an example, one can consider a field K consisting of a zero function and fractions of the form u/v where u and v are grid functions defined almost everywhere and vanishing at a finite number of nodes. (As usual, we say that a grid function is defined almost everywhere if there are only finitely many nodes where it is not defined.)

Let us fix some node $\mathcal{P}$ and say that *a node $\mathcal{Q}$ has order i* (with respect to $\mathcal{P}$) if the shortest path from $\mathcal{P}$ to $\mathcal{Q}$ along the edges of the grid consists of i steps (by a step we mean a path from a node of the grid to a neighbor node along the edge between these two nodes). For example, the orders of the nodes in the two-dimensional case are as follows (a number near a node shows the order of this node, see Figure 3.1).

Let us consider the values of the unknown grid functions $f_1, \ldots, f_s$ at the nodes whose order does not exceed r ($r \in \mathbf{N}$). If $f_1, \ldots, f_s$ should not satisfy any system of equations (or any other condition), their values at nodes of any order can be chosen arbitrarily. Because of the system in finite differences (and equations obtained from the equations of the system by transformations of the form $f_j(x_1, \ldots, x_n) \mapsto$ $f_j(x_1 + k_1 h_1, \ldots, x_n + k_n h_n)$ with $k_1, \ldots, k_n \in \mathbf{Z}$, $1 \leq j \leq s)$, the number of independent values of the functions $f_1, \ldots, f_s$ at the nodes of order $\leq r$ decreases. This number, which is a function of r, is considered as a "measure of strength" of the system in finite differences (in the sense of A. Einstein). We denote it by S_r.

With the above conventions, suppose that the transformations α_j of the field of coefficients K defined by

$$\alpha_j f(x_1, \ldots, x_n) = f(x_1, \ldots, x_{j-1}, x_j + h_j, \ldots, x_n) \quad (1 \leq j \leq n)$$

are automorphisms of this field. Then K can be considered as an inversive difference field with the basic set $\sigma = \{\alpha_1, \ldots, \alpha_n\}$. Furthermore, assume that the replacement of the unknown functions f_i by σ^*-indeterminates y_i ($i = 1, \ldots, s$) in the ring $K\{y_1, \ldots, y_s\}^*$ leads to a prime system of algebraic σ^*-equations (then the original system of equations in finite differences is also called *prime*). The difference dimension polynomial $\psi(t)$ of the latter system is said to be the *difference dimension polynomial of the given system in finite differences*.

Clearly, $\psi(r) = S_r$ for any $r \in \mathbf{N}$, so the difference dimension polynomial of a prime system of equations in finite differences is the measure of strength of such a system in the sense of A. Einstein.

The natural generalization of the concept of strength arises in the case when one evaluates the maximal number of values of unknown grid functions that can be chosen arbitrarily in a region which is not symmetric with respect to a fixed node $\mathcal{P}$. More precisely, using the previous settings, let us denote the automorphisms α_i^{-1} by α_{n+i} ($1 \leq i \leq n$), add to the system (3.1) n new equations $\alpha_i \alpha_{n+i} f - f = 0$ ($1 \leq i \leq n$) and divide the set $\{\alpha_1, \ldots, \alpha_{2n}\}$ into p disjoint subsets $\sigma_1, \ldots, \sigma_p$. Then for any $r_1, \ldots, r_p \in \mathbf{N}$, the transcendence degree of the field $K(\{\alpha_1^{k_1} \ldots \alpha_{2n}^{k_{2n}} f_j \mid k_1, \ldots, k_{2n} \in \mathbf{N}, 1 \leq j \leq s, \sum_{\nu \in \sigma_i} k_i \leq r_i \text{ for } i = 1, \ldots, p\})$ over K is a function of $r_1, \ldots, r_p$ which can be naturally treated as a generalized strength of the given system of equations in finite differences. Theorem 3.3 below shows that this characteristic is a polynomial function of $r_1, \ldots, r_p$.

In what follows, for any permutation $(j_1, \ldots, j_p)$ of the set $\{1, \ldots, p\}$, we define a lexicographic order $\leq_{j_1, \ldots, j_p}$ on $\mathbf{N}^p$ such that $(r_1, \ldots, r_p) \leq_{j_1, \ldots, j_p} (s_1, \ldots, s_p)$ if and only if either $r_{j_1} < s_{j_1}$ or there exists $k \in \mathbf{N}$, $1 \leq k \leq p-1$, such that $r_{j_\nu} = s_{j_\nu}$ for $\nu = 1, \ldots, k$ and $r_{j_{k+1}} < s_{j_{k+1}}$. If $\Sigma \subseteq \mathbf{N}^p$, then Σ' will denote the set $\{e \in \Sigma \mid e \text{ is a maximal element of } \Sigma \text{ with respect to one of the } p! \text{ lexicographic orders } \leq_{j_1, \ldots, j_p}\}$. For example, if $\Sigma = \{(4, 1, 0), (2, 2, 1), (3, 1, 2), (4, 0, 2), (1, 3, 1), (3, 3, 0), (3, 2, 1), (1, 1, 6), (3, 1, 0), (1, 2, 0)\} \subseteq \mathbf{N}^3$, then $\Sigma' = \{(4, 1, 0), (4, 0, 2), (3, 3, 0), (1, 3, 1), 1(3, 1, 2), (1, 1, 6)\}$.

Theorem 3.3 *Let K be an inversive difference field with a basic set $\sigma = \{\alpha_1, \ldots, \alpha_n\}$, let $T = T_\sigma$, and let a partition $\sigma = \sigma_1 \cup \cdots \cup \sigma_p$ of the set σ into a disjoint union of its subsets be fixed (we use the notation of section 2.4). Let $L = K\langle \eta_1, \ldots, \eta_s \rangle$ be a difference (σ-) field extension of K generated by a finite set $\eta = \{\eta_1, \ldots, \eta_s\}$. Then there exists a polynomial $\phi_{\eta|K}(t_1, \ldots, t_p)$ in p variables with rational coefficients such that*

(i) $\phi_\eta(r_1, \ldots, r_p) = \mathrm{trdeg}_K K(\{\tau\eta_i \mid \tau \in T(r_1, \ldots, r_p), 1 \leq j \leq s\})$ *for all suffi-ciently large* $(r_1, \ldots, r_p) \in \mathbf{N}^p$.

(ii) $\deg_{t_i} \phi_\eta \leq n_i$ $(1 \leq i \leq p)$, *so that* $\deg \phi \leq n$ *and the polynomial* $\phi_\eta(t_1, \ldots, t_p)$ *can be represented as*

$$\phi_\eta(t_1, \ldots, t_p) = \sum_{i_1=0}^{n_1} \cdots \sum_{i_p=0}^{n_p} a_{i_1 \ldots i_p} \binom{t_1 + i_1}{i_1} \cdots \binom{t_p + i_p}{i_p}$$

where $a_{i_1 \ldots i_p} \in \mathbf{Z}$ *for all* $i_1, \ldots, i_p$.

(iii) *The degree* d *of the polynomial* $\phi_\eta(t)$, *the coefficient* $a_{n_1 \ldots n_p}$, *p-tuples* $(j_1, \ldots, j_p)$ $\in \Sigma'$, *the corresponding coefficients* $a_{j_1 \ldots j_p}$, *and the coefficients of the terms of total degree* d *do not depend on the choice of the system of σ-generators η of L over K. Furthermore,* $a_{n_1 \ldots n_p} = \sigma\text{-trdeg}_K L$.

The numerical polynomial $\phi_\eta(t_1, \ldots, t_p)$ whose existence is established by Theorem 3.3 is called a *dimension polynomial of the difference field extension L/K associated with the given system of generators η and with the given partition of the basic set σ into p disjoint subsets $\sigma_1, \ldots, \sigma_p$.*

Applying the proof of [6, Theorem 6.4.11] one obtains (with the notation of the last theorem) that the module of differentials $\Omega_K(L)$ associated with the difference field extension L/K is a σ-L-module generated by the differentials $d\eta_1, \ldots, d\eta_s$ $(d\eta_i(D) = D(\eta_i)$ for any K-derivation $D : L \to L$, and $d(\alpha_j(\eta_i)) = \alpha_j(d\eta_i)$ for all $i = 1, \ldots, s$; $j = 1, \ldots, n$. Furthermore, $\Omega_K(L)$ has a natural excellent p-dimension filtration $\{(\Omega_K(L))_{r_1 \ldots r_p} \mid r_1, \ldots, r_p \in \mathbf{N}\}$, where $(\Omega_K(L))_{r_1 \ldots r_p}$ is a vector L-space generated by the set $\{d\eta \mid \eta \in K(\{\tau\eta_i \mid \tau \in T(r_1, \ldots, r_p), 1 \leq j \leq s\})\}$, and $\dim_L \Omega_K(L)_{r_1 \ldots r_p} = \phi_\eta(r_1, \ldots, r_p)$ for all sufficiently large $r_1, \ldots, r_p \in \mathbf{Z}$. It follows that the proof of Theorem 3.3 and the computation of the dimension polynomial $\phi_\eta(t_1, \ldots, t_p)$ can be reduced to the proof of the corresponding existence theorem on a multivariable dimension polynomial of a finitely generated difference module and the computation of such a polynomial. In the next section we obtain this result and a method of computation of multivariable dimension polynomials via a generalized Gröbner basis technique in free difference modules with several term orderings.

4   Gröbner Bases for Several Orderings

Let K be a difference field with a basic set $\sigma = \{\alpha_1, \ldots, \alpha_n\}$, T the commutative semi-group of all power products $\alpha_1^{k_1} \ldots \alpha_n^{k_n}$ $(k_1, \ldots, k_n \in \mathbf{N})$, $\mathcal{D}$ the ring of σ-operators over K, and

$$\sigma = \sigma_1 \cup \cdots \cup \sigma_p \tag{4.1}$$

a partition of the set σ into a disjoint union of its subsets $(p \in \mathbf{N}, \sigma_1 = \{\alpha_1, \ldots, \alpha_{n_1}\}$, $\sigma_2 = \{\alpha_{n_1+1}, \ldots, \alpha_{n_1+n_2}\}, \ldots, \sigma_p = \{\alpha_{n_1+\cdots+n_{p-1}+1}, \ldots, \alpha_n\}$, $n_i \geq 1$ for $i = 1, \ldots, p$, and $n_1 + \cdots + n_p = n$.)

In what follows we keep the notation of section 2.4. In particular, we consider p well-orderings $<_1, \ldots, <_p$ of the set T defined in that section, the standard p-dimensional filtration of $\mathcal{D}$ and p-dimensional filtrations of vector σ-K-spaces.

By a *free vector σ-K-space* we mean a vector σ-K-space which is a free module over the ring of difference operators $\mathcal{D}$. If a free vector σ-K-space E is generated (as a free $\mathcal{D}$-module) by a finite system of elements $\{e_1, \ldots, e_m\}$, we say that E is a finitely generated free vector σ-K-space and call $e_1, \ldots, e_m$ *free generators* of E. In this case the elements of the form τe_ν ($\tau \in T, 1 \leq \nu \leq m$) are called *terms* while the elements of the semigroup T are called *monomials*. The set of all terms is denoted by Te. It is easy to see that this set generates E as a vector space over the field K.

By the order of a term τe_ν with respect to σ_i ($1 \leq i \leq p$) we mean the order of the monomial τ with respect to σ_i. A term $u' = \tau' e_\mu$ is said to be a *multiple* of a term $u = \tau e_\nu$ if $\mu = \nu$ and τ' is a multiple of τ in the semigroup T. In this case we write $u | u'$. (Clearly, $u | u'$ if and only if there exists $\tau'' \in T$ such that $u' = \tau'' u$.)

The least common multiple of two terms $u = \tau_1 e_i$ and $v = \tau_2 e_j$ is defined as follows:

$$\mathrm{lcm}(u, v) = \begin{cases} 0, & \text{if } i \neq j, \\ \mathrm{lcm}(\tau_1, \tau_2) e_i, & \text{if } i = j. \end{cases}$$

We shall consider p orderings of the set Te that correspond to the orderings of the semigroup T introduced at the beginning of this section. These ordering are denoted by the same symbols $<_1, \ldots, <_p$ and defined as follows: if $\tau e_\mu, \tau' e_\mu \in Te$, then $\tau e_\mu <_i \tau' e_\nu$ if and only if $\tau <_i \tau'$ in T or $\tau = \tau'$ and $\mu < \nu$.

Since the set Te is a basis of the vector K-space E, every non-zero element $f \in E$ has a unique (up to the order of the terms in the sum) representation in the form

$$f = a_1 \tau_1 e_{i_1} + \cdots + a_l \tau_l e_{i_l} \tag{4.2}$$

where $\tau_1, \ldots, \tau_l$ are distinct elements of T, $a_1, \ldots, a_l$ are non-zero elements of K, and $1 \leq i_1, \ldots, i_l \leq m$.

Definition 4.1 Let f be a non-zero element of the $\mathcal{D}$-module E written in the form *(4.2.)* and let $\tau_\nu e_{i_\nu}$ ($1 \leq \nu \leq p$) be the greatest term of the set $\{\tau_1 e_{i_1}, \ldots, \tau_l e_{i_l}\}$ with respect to the order $<_j$ ($1 \leq j \leq p$). Then the term $\tau_\nu e_{i_\nu}$ is called the j-leader of the element f; it is denoted by $u_f^{(j)}$. (Of course, it is possible that $u_f^{(j)} = u_f^{(j')}$ for some distinct numbers j and j'.) The non-negative integer $\mathrm{ord}_i u_f^{(i)}$ is called the ith order of f and denoted by $\mathrm{ord}_i f$ ($i = 1, \ldots, p$). The coefficient of $u_f^{(j)}$ in f is said to be the j-leading coefficient of f; it is denoted by $lc_j(f)$.

Definition 4.2 Let $f, g \in E$, $g \neq 0$, and let $k, i_1, \ldots, i_r$ be distinct elements of the set $\{1, \ldots, p\}$. Then f is said to be $(<_k, <_{i_1}, \ldots, <_{i_r})$-**reduced** with respect to g if $f = 0$ or f does not contain any multiple $\tau u_g^{(k)}$ ($\tau \in T$) such that $\mathrm{ord}_{i_\nu} (\tau u_g^{(i_\nu)}) \leq \mathrm{ord}_{i_\nu} u_f^{(i_\nu)}$ ($\nu = 1, \ldots, r$).

An element $f \in E$ is said to be $(<_k, <_{i_1}, \ldots, <_{i_r})$-reduced with respect to a set $G \subseteq E$, if f is $(<_k, <_{i_1}, \ldots, <_{i_r})$-reduced with respect to every element of G.

Let us consider $p - 1$ new symbols $z_1, \ldots, z_{p-1}$ and the free commutative semigroup Γ of all power products $\gamma = \alpha_1^{k_1} \ldots \alpha_n^{k_n} z_1^{l_1} \ldots z_{p-1}^{l_{p-1}}$ with non-negative integer exponents. Let $\Gamma e = \{\gamma e_j | \gamma \in \Gamma, 1 \leq j \leq m\} = \Gamma \times \{e_1, \ldots, e_n\}$. For any non-zero element $f \in E$, let $d_i(f) = \operatorname{ord}_i u_f^{(i)} - \operatorname{ord}_i u_f^{(1)}$ $(2 \leq i \leq p)$ and let $\rho : E \to \Gamma e$ be defined by $\rho(f) = z_1^{d_2(f)} \ldots z_{p-1}^{d_p(f)} u_f^{(1)}$.

Let N be a $\mathcal{D}$-submodule of E. A finite set of non-zero elements $G = \{g_1, \ldots, g_r\} \subseteq N$ is called a **Gröbner basis of N with respect to the orders** $<_1, \ldots, <_p$ if for any $0 \neq f \in N$, there exists $g_i \in G$ such that $\rho(g_i)|\rho(f)$ in Γe.

It is clear that every Gröbner basis of N with respect to the orders $<_1, \ldots, <_p$ is a Gröbner basis of N in the usual sense. Therefore, every Gröbner basis of N with respect to the orders $<_1, \ldots, <_p$ generates N as a left D-module. A set $G = \{g_1, \ldots, g_r\} \subseteq E$ is said to be a *Gröbner basis with respect to the orders* $<_1, \ldots, <_p$ if G is a Gröbner basis of $N = \sum_{i=1}^r \mathcal{D}g_i$ with respect to $<_1, \ldots, <_p$.

Definition 4.3 Given $f, g, h \in E$, with $g \neq 0$, we say that the element f $(<_k, <_{i_1}, \ldots, <_{i_l})$-**reduces** to h **modulo** g in one step and write $f \xrightarrow[<_k, <_{i_1}, \ldots, <_{i_l}]{g} h$ if and only if f contains some term w with a coefficient a such that $u_g^{(k)}|w$,

$$h = f - a \left(\frac{w}{u_g^{(k)}} (lc_k(g)) \right)^{-1} \frac{w}{u_g^{(k)}} g$$

and $\operatorname{ord}_{i_\nu} \frac{w}{u_g^{(k)}} u_g^{(i_\nu)} \leq \operatorname{ord}_{i_\nu} u_f^{(i_\nu)}$ $(1 \leq \nu \leq l)$.

Definition 4.4 Let $f, h \in E$ and let $G = \{g_1, \ldots, g_r\}$ be a finite set of non-zero elements of E. We say that f $(<_k, <_{i_1}, \ldots, <_{i_l})$-**reduces** to h **modulo** G and write $f \xrightarrow[<_k, <_{i_1}, \ldots, <_{i_l}]{G} h$ if and only if there exists a sequence of elements $g^{(1)}, g^{(2)}, \ldots, g^{(q)} \in G$ and a sequence of elements $h_1, \ldots, h_{q-1} \in E$ such that

$$f \xrightarrow[<_k, <_{i_1}, \ldots, <_{i_l}]{g^{(1)}} h_1 \xrightarrow[<_k, <_{i_1}, \ldots, <_{i_l}]{g^{(2)}} \cdots \xrightarrow[<_k, <_{i_1}, \ldots, <_{i_l}]{g^{(q-1)}} h_{q-1} \xrightarrow[<_k, <_{i_1}, \ldots, <_{i_l}]{g^{(q)}} h.$$

Theorem 4.5 *With the above notation, let $G = \{g_1, \ldots, g_r\} \subseteq E$ be a Gröbner basis with respect to the orders $<_1, \ldots, <_p$ on Γe. Then there exist elements $g \in E$ and $Q_1, \ldots, Q_r \in \mathcal{D}$ such that $f - g = \sum_{i=1}^r Q_i g_i$ and g is $(<_1, \ldots, <_p)$-reduced with respect to the set G.*

Proof. If f is $(<_1, \ldots, <_p)$-reduced with respect to G, the statement is obvious (one can set $g = f$). Suppose that f is not $(<_1, \ldots, <_p)$-reduced with respect to G. Let $u_i^{(j)} = u_{g_i}^{(j)}$ $(1 \leq i \leq r, 1 \leq j \leq p)$ and let a_i be the coefficient of the term $u_i^{(1)}$ in g_i $(i = 1, \ldots, r)$. In what follows, a term w_h, that appears in an element $h \in E$, will be called the G-leader of h if w_h is the greatest (with respect to the order $<_1$) term among all terms $\tau u_i^{(1)}$ $(\tau \in T, 1 \leq i \leq r)$ that appear in h and satisfy the condition $\operatorname{ord}_j(\tau u_i^{(j)}) \leq \operatorname{ord}_j u_h^{(j)}$ for $j = 2, \ldots, p$.

Let w_f be the G-leader of the element f and let c_f be the coefficient of w_f in f. Then $w_f = \tau u_i^{(1)}$ for some $\tau \in T$, $1 \leq i \leq r$, such that $\mathrm{ord}_j(\tau u_i^{(j)}) \leq \mathrm{ord}_j u_f^{(j)}$ for $j = 2, \ldots, p$. Without loss of generality we may assume that i corresponds to the maximum (with respect to the order $<_1$) 1-leader $u_i^{(1)}$ in the set of all 1-leaders of elements of G. Let us consider the element $f' = f - c_f(\tau(a_i))^{-1}\tau g_i$. Obviously, f' does not contain w_f and $\mathrm{ord}_j(u_{f'}^{(j)}) \leq \mathrm{ord}_j u_f^{(j)}$ for $j = 2, \ldots, p$. Furthermore, f' cannot contain any term of the form $\tau' u_i^{(1)}$ ($\tau' \in T$, $1 \leq i \leq r$), that is greater than w_f (with respect to $<_1$) and satisfies the condition $\mathrm{ord}_j(\tau' u_i^{(j)}) \leq \mathrm{ord}_j u_{f'}^{(j)}$ for $j = 2, \ldots, p$. Indeed, if the last inequality holds, then $\mathrm{ord}_j(\tau' u_i^{(j)}) \leq \mathrm{ord}_j u_f^{(j)}$, so that the term $\tau' u_i^{(1)}$ cannot appear in f. This term cannot appear in τg_i either, since $u_{\tau g_i}^{(1)} = \tau u_{g_i}^{(1)} = w_f <_j \tau' u_i^{(1)}$. Thus, $\tau' u_i^{(1)}$ cannot appear in f', whence the G-leader of f' is strictly less (with respect to the order $<_1$) than the G-leader of f. Applying the same procedure to f' and continuing in the same way, we obtain an element $g \in E$ such that $f - g$ is a linear combination of elements $g_1, \ldots, g_r$ with coefficients in $\mathcal{D}$ and g is $(<_1, \ldots, <_p)$-reduced with respect to G. This completes the proof. $\square$

The process of reduction described in the proof of the last theorem can be realized by the following algorithm (that can be used for the reduction with respect to any finite set of elements of the free $\mathcal{D}$-module E).

Algorithm 4.6 $(f, r, g_1, \ldots, g_r; g; Q_1, \ldots, Q_r)$
Input: $f \in E$, a positive integer r, $G = \{g_1, \ldots, g_r\} \subseteq E$ where $g_i \neq 0$ for $i = 1, \ldots, r$
Output: Element $g \in E$ and elements $Q_1, \ldots, Q_r \in \mathcal{D}$ such that
$$g = f - (Q_1 g_1 + \cdots + Q_r g_r) \text{ and } g \text{ is reduced with respect to } G$$
Begin $Q_1 := 0, \ldots, Q_r := 0, g := f$
While there exist i, $1 \leq i \leq r$, and a term w, that appears in g with a non-zero coefficient $c(w)$, such that $u_{g_i}^{(1)} | w$ and $\mathrm{ord}_j(\frac{w}{u_{g_i}^{(1)}} u_{g_i}^{(j)}) \leq \mathrm{ord}_j u_g^{(j)}$ for $j = 2, \ldots, p$

do $z :=$ the greatest (with respect to $<_1$) of the terms w that satisfy the above conditions.
     $k :=$ the smallest number i for which $u_{g_i}^{(1)}$ is the greatest (with respect to $<_1$) 1-leader of an element $g_i \in G$ such that $u_{g_i}^{(1)} | z$ and $\mathrm{ord}_j(\frac{z}{u_{g_i}^{(1)}} u_{g_i}) \leq \mathrm{ord}_j u_g^{(j)}$ for $j = 2, \ldots, p$.
$$Q_k := Q_k + c(z) \left(\frac{z}{u_{g_k}^{(1)}} (lc_1(g_k)) \right)^{-1} \frac{z}{u_{g_k}^{(1)}} g_k$$
$$g := g - c(z) \left(\frac{z}{u_{g_k}^{(1)}} (lc_1(g_k)) \right)^{-1} \frac{z}{u_{g_k}^{(1)}} g_k$$

End

The following example illustrates Algorithm 4.6.

Example 4.7 Let $K = \mathbf{Q}(x, y)$ be the field of rational fractions in two variables x and y over $\mathbf{Q}$ treated as a difference field with basic set $\sigma = \{\alpha_1, \alpha_2\}$ where $\alpha_1 f(x, y) =$

$f(x + 1, y)$ and $\alpha_2 f(x, y) = f(x, y + 1)$ for every $f(x, y) = \mathbf{Q}(x, y)$. We consider a partition $\sigma = \sigma_1 \bigcup \sigma_2$ of the set σ where $\sigma_1 = \{\alpha_1\}$ and $\sigma_2 = \{\alpha_2\}$.

Let $\mathcal{D}$ be the ring of σ-operators over K, let E be the free left $\mathcal{D}$-module with free generators e_1 and e_2, and let $G = \{g_1, g_2\}$ where

$$g_1 = x^2 \alpha_1 \alpha_2 e_1 + x \alpha_2^2 e_2 - xy \alpha_1 e_1 + y^2 e_2,$$

$$g_2 = 2x \alpha_1^2 \alpha_2 e_2 + \alpha_1 \alpha_2^2 e_1 + y \alpha_2 e_1.$$

We are going to to apply Algorithm 4.6 to reduce the element

$$f = x \alpha_1^2 \alpha_2^2 e_1 + y \alpha_1 \alpha_3^2 e_2 - xy \alpha_1 \alpha_2 e_1$$

with respect to G.

Notice that $u_{g_1}^{(1)} = \alpha_1 \alpha_2 e_1$, $u_{g_1}^{(2)} = \alpha_2^2 e_2$, $u_{g_2}^{(1)} = \alpha_1^2 \alpha_2 e_2$, and $u_{g_2}^{(2)} = \alpha_1 \alpha_2^2 e_1$. Furthermore, g_1 and g_2 are $(<_1, <_2)$-reduced with respect to each other. Indeed, $u_{g_2}^{(1)}$ divides no term of g_1, so g_1 is $(<_1, <_2)$-reduced with respect to g_2. The only term of g_2 which is a multiple of $u_{g_1}^{(1)}$ is $w = \alpha_1 \alpha_2^2 e_1$. Since $w = \alpha_2 u_{g_1}^{(1)}$ and $\mathrm{ord}_2(\alpha_2 u_{g_1}^{(2)}) = \mathrm{ord}_2(\alpha_2^3) = 3 > \mathrm{ord}_2 u_{g_2}^{(2)} = 2$, we obtain that g_2 is $(<_1, <_2)$-reduced with respect to g_1. (If any two elements of a set $\Sigma \subseteq E$ are $(<_1, <_2)$-reduced with respect to each other, we say that the set Σ is $(<_1, <_2)$-autoreduced. The theory of such sets is developed in [10]. Notice that one can consider a similar theory for subsets of a ring of differential polynomials, see [9]).

According to Algorithm 4.7 the first step of the $(<_1, <_2)$-reduction of the element f with respect to G is as follows.

$$
\begin{aligned}
f \to f_1 &= f - x \frac{1}{(x+1)^2} \alpha_1 \alpha_2 g_1 \\
&= f - \frac{x}{(x+1)^2} \left[(x+1)^2 \alpha_1^2 \alpha_2^2 e_1 + (x+1) \alpha_1 \alpha_2^3 e_2 \right. \\
&\qquad\qquad \left. - (x+1)(y+1) \alpha_1^2 \alpha_2 e_1 + (y+1)^2 \alpha_1 \alpha_2 e_2 \right] \\
&= \left(y - \frac{x}{x+1} \right) \alpha_1 \alpha_2^3 e_2 + \frac{x(y+1)}{x+1} \alpha_1^2 \alpha_2 e_1 - \frac{x(y+1)^2}{(x+1)^2} \alpha_1 \alpha_2 e_2.
\end{aligned}
$$

(Notice that $\mathrm{ord}_2(\alpha_1 \alpha_2 u_{g_1}^{(2)}) \leq \mathrm{ord}_2 u_f^{(2)}$; both orders are equal to 3.)

Since $u_{f_1}^{(1)} = \alpha_1^2 \alpha_2 e_1 = \alpha_1 u_{g_1}^{(1)}$ and the order of $\alpha_1 u_{g_1}^{(2)} = \alpha_1 \alpha_2^2 e_2$ with respect to σ_2 does not exceed $\mathrm{ord}_2(u_{f_1}^{(2)}) = 3$, the second step of the $(<_1, <_2)$-reduction is

$$
f_1 \to f_2 = f_1 - \frac{x(y+1)}{x+1} \frac{1}{(x+1)^2} \alpha_1 g_1 = \frac{xy(y+1)}{(x+1)^2} \alpha_1^2 e_1 +
$$

$$
\left(y - \frac{x(y+1)}{x+1} \right) \alpha_1 \alpha_2^3 e_2 - \frac{x(y+1)}{(x+1)^2} \alpha_1 \alpha_2^2 e_2 - \frac{x(y+1)^2}{(x+1)^2} \alpha_1 \alpha_2 e_2 - \frac{xy^2(y+1)}{(x+1)^3} \alpha_1 e_2.
$$

Since no term of f_2 is divisible by $u_{g_1}^{(1)} = \alpha_1 \alpha_2 e_1$ or by $u_{g_2}^{(1)} = \alpha_1^2 \alpha_2 e_2$, the element f_2 is $(<_1, <_2)$-reduced with respect to G.

The proof of Theorem 4.5 shows that if G is a Gröbner basis of a $\mathcal{D}$-submodule N of E, then a reduction step described in Definition 4.3 (with some $g \in G$) can be applied to every non-zero element $f \in N$. As a result of such a step, we obtain an element of N whose G-leader is strictly less (with respect to $<_1$) than the G-leader of f. This observation leads to the following statement.

Proposition 4.8 *Let* $G = \{g_1, \ldots, g_r\}$ *be a Gröbner basis of a $\mathcal{D}$-submodule N of E with respect to the orders* $<_1, \ldots, <_p$. *Then*

(i) $f \in N$ *if and only if* $f \xrightarrow{\ G\ }_{<_1,<_2,\ldots,<_p} 0$.

(ii) *If* $f \in N$ *and f is* $(<_1, <_2, \ldots, <_p)$-*reduced with respect to G, then* $f = 0$.

Definition 4.9 Let f and g be two elements in the free $\mathcal{D}$-module E and let $k \in \{1, \ldots, p\}$. Then the element

$$
S_k(f, g) = \left(\frac{\mathrm{lcm}\left(u_f^{(k)}, u_g^{(k)} \right)}{u_f^{(k)}} (lc_k(f)) \right)^{-1} \frac{\mathrm{lcm}\left(u_f^{(k)}, u_g^{(k)} \right)}{u_f^{(k)}} f
$$
$$
- \left(\frac{\mathrm{lcm}\left(u_f^{(k)}, u_g^{(k)} \right)}{u_g^{(k)}} (lc_k(g)) \right)^{-1} \frac{\mathrm{lcm}\left(u_f^{(k)}, u_g^{(k)} \right)}{u_g^{(k)}} g
$$

is called the k**th S-polynomial** of f and g.

With the above notation, one can obtain the following two statements that generalize the corresponding properties of the classical Gröbner basis.

Proposition 4.10 *Let* $f, g_1, \ldots, g_r \in E$ $(r \geq 1)$ *and let* $f = \sum_{i=1}^{r} c_i \omega_i g_i$ *where* $\omega_i \in T$, $c_i \in K$ $(1 \leq i \leq r)$. *Let* $k \in \{1, \ldots, p\}$ *and for any* $\nu, j \in \{1, \ldots, r\}$, *let* $u_{\nu j}^{(k)} = \mathrm{lcm}(u_{g_\nu}^{(k)}, u_{g_j}^{(k)})$. *Furthermore, suppose that* $\omega_1 u_{g_1}^{(k)} = \cdots = \omega_r u_{g_r}^{(k)} = u$ *for some* $k \in \{1, \ldots, p\}$, $u_f^{(k)} <_k u$ *and there is a nonempty set* $I \subseteq \{1, \ldots, p\} \setminus \{k\}$ *such that* $\omega_i u_{g_i}^{(l)} \leq_l u_f^{(l)}$ *for all* $i \in \{1, \ldots, r\}, l \in I$. *Then there exist elements* $c_{\nu j} \in K$ $(1 \leq \nu \leq s, 1 \leq j \leq t)$ *such that*

$$
f = \sum_{\nu=1}^{s} \sum_{j=1}^{t} c_{\nu j} \theta_{\nu j} S_k(g_\nu, g_j)
$$

where $\theta_{\nu j} = \frac{u}{u_{\nu j}^{(k)}}$ *and*

$$
\theta_{\nu j} u_{S_k(g_\nu, g_j)}^{(k)} <_k u, \qquad \theta_{\nu j} u_{S_k(g_\nu, g_j)}^{(l)} \leq_l u_f^{(l)} \qquad (1 \leq \nu \leq s,\ 1 \leq j \leq t,\ l \in I).
$$

Proof. For every $i = 1, \ldots, r$, let $d_i = lc_k(\omega_i g_i) = \omega_i(lc_k(g_i))$. Since $\omega_1 u_{g_1}^{(k)} = \cdots = \omega_r u_{g_r}^{(k)} = u$ and $u_f^{(k)} <_k u$, $\sum_{i=1}^{r} c_i d_i = 0$. Let $h_i = d_i^{-1} \omega_i g_i = (\omega_i(lc_k(g_i)))^{-1} \omega_i g_i$ $(i =$

$1, \ldots, r$). Then $lc_k(h_i) = 1$ and

$$f = \sum_{i=1}^{r} c_i \omega_i g_i = \sum_{i=1}^{r} c_i d_i h_i = c_1 d_1 (h_1 - h_2) + (c_1 d_1 + c_2 d_2)(h_2 - h_3)$$

$$+ \cdots + (c_1 d_1 + \cdots + c_{r-1} d_{r-1})(h_{r-1} - h_r).$$

(The last sum should end with the term $(c_1 d_1 + \cdots + c_r d_r) h_r$ in order to represent an identity, but this term is equal to zero.)

For every $\nu, j \in \{1, \ldots, r\}$, $\nu \neq j$, let $\tau_{\nu j} = \frac{u_{\nu j}^{(k)}}{u_{g_\nu}^{(k)}}$, $\gamma_{\nu j} = \frac{u_{\nu j}^{(k)}}{u_{g_j}^{(k)}}$, and $\theta_{\nu j} = \frac{u}{u_{\nu j}^{(k)}}$ (since $u_{g_\nu}^{(k)} \mid u$ and $u_{g_j}^{(k)} \mid u$, the term $u_{\nu j}^{(k)}$ divides u). Then

$$\theta_{\nu j} S_k(g_\nu, g_j) = \theta_{\nu j}[(\tau_{\nu j}(lc_k(g_\nu)))^{-1} \tau_{\nu j} g_\nu - (\gamma_{\nu j}(lc_k(g_j)))^{-1} \gamma_{\nu j} g_j]$$

$$= [\theta_{\nu j}(\tau_{\nu j}(lc_k(g_\nu)))]^{-1} \frac{u}{u_\nu^{(k)}} g_\nu - [\theta_{\nu j}(\gamma_{\nu j}(lc_k(g_j)))]^{-1} \frac{u}{u_j^{(k)}} g_j$$

$$= [\omega_\nu(lc_k(g_\nu))]^{-1} \omega_\nu g_\nu [\omega_j(lc_k(g_j))]^{-1} \omega_j g_j = h_\nu - h_j.$$

It follows that $f = c_1 d_1 \theta_{12} S_k(g_1, g_2) + (c_1 d_1 + c_2 d_2) \theta_{23} S_k(g_2, g_3) + \cdots + (\sum_{i=1}^{r-1} c_i d_i) \theta_{r-1,r} S_k(g_{r-1}, g_r)$, $\theta_{i,i+1} u_{S_k(g_i, g_{i+1})}^{(k)} = u_{\theta_{i,i+1}}^{(k)} S_k(g_i, g_{i+1}) = u_{h_i - h_{i+1}}^{(k)} <_k u$ (since $u_{h_i}^{(k)} = u_{h_{i+1}}^{(k)}$ and $lc_k(h_i) = lc_k(h_{i+1}) = 1$), and $\theta_{i,i+1} u_{S_k(g_i, g_{i+1})}^{(l)} \leq_l u_f^{(l)}$ for any $i = 1, \ldots, r-1$, $l \in I$. This completes the proof. $\square$

Theorem 4.11 *With the above notation, let $G = \{g_1, \ldots, g_r\}$ be a Gröbner basis of a $\mathcal{D}$-submodule N of E with respect to each of the following sequences of orders: $<_p$, $<_{p-1}, <_p, \ldots, <_{k+1}, \ldots, <_p$ ($1 \leq k \leq p-1$). Furthermore, suppose that*

$$S_k(g_i, g_j) \xrightarrow[<_k, <_{k+1}, \ldots, <_p]{G} 0 \text{ for any } g_i, g_j \in G.$$

Then G is a Gröbner basis of N with respect to $<_k, <_{k+1}, \ldots, <_p$.

Proof. First, let us prove that under the conditions of the theorem every element $f \in N$ can be represented as

$$f = \sum_{i=1}^{r} h_i g_i \qquad (4.3)$$

where $h_1, \ldots, h_r \in \mathcal{D}$,

$$\max_{<_k} \left\{ u_{h_i}^{(k)} u_{g_i}^{(k)} \mid 1 \leq i \leq r \right\} = u_f^{(k)} \qquad (4.4)$$

and

$$\mathrm{ord}_j \left(u_{h_i}^{(j)} u_{g_i}^{(j)} \right) \leq \mathrm{ord}_j u_f^{(j)} \qquad (1 \leq i \leq r, \; 1 \leq j \leq p). \qquad (4.5)$$

(The symbol $\max_{<_k}$ in (4.4) means the maximum with respect to the term order $<_k$.)

We proceed by induction on $p - k$. If $p - k = 0$, that is $k = p$, our statement is a classical result of the theory of Gröbner bases (the fact that we consider modules over $\mathcal{D}$ rather than modules over a polynomial ring does not play an essential role). Let $k < p$ and let $f \in N$. By the induction hypothesis, f can be written as

$$f = \sum_{i=1}^{r} H_i g_i \tag{4.6}$$

where $H_1, \ldots, H_r \in \mathcal{D}$,

$$\max_{<_{k+1}} \left\{ u_{H_i}^{(k+1)} u_{g_i}^{(k+1)} \mid 1 \le i \le r \right\} = u_f^{(k+1)} \tag{4.7}$$

and

$$\mathrm{ord}_j \left(u_{H_i}^{(j)} u_{g_i}^{(j)} \right) \le \mathrm{ord}_j u_f^{(j)} \quad (j = k+2, \ldots, p;\ i = 1, \ldots, r). \tag{4.8}$$

Let us choose among all representations of f in the form (4.6) with conditions (4.7) and (4.8) a representation with the smallest (with respect to $<_k$) possible term

$$u = \max_{<_k} \left\{ u_{H_i}^{(k)} u_{g_i}^{(k)} \mid 1 \le i \le r \right\}.$$

Setting $d_i = lc_k(H_i)\,(1 \le i \le r)$ and breaking the sum in (4.6) in two parts, we can write

$$f = \sum_{i=1}^{r} H_i g_i = \sum_{u_{H_i}^{(k)} u_{g_i}^{(k)} = u} H_i g_i + \sum_{u_{H_i}^{(k)} u_{g_i}^{(k)} <_k u} H_i g_i$$

or

$$f = \sum_{u_{H_i}^{(k)} u_{g_i}^{(k)} = u} d_i u_{H_i}^{(k)} g_i + \sum_{u_{H_i}^{(k)} u_{g_i}^{(k)} = u} \left(H_i - d_i u_{H_i}^{(k)} \right) g_i + \sum_{u_{H_i}^{(k)} u_{g_i}^{(k)} <_k u} H_i g_i. \tag{4.9}$$

Notice that if $u = u_f^{(k)}$, then the expansion (4.9) satisfies conditions (4.3) - (4.5). Indeed, in this case $u_f^{(k+1)} = \max_{<_{k+1}} \{ u_{H_i}^{(k+1)} u_{g_i}^{(k+1)} \mid 1 \le i \le r \}$ (see (4.7)), hence $\max\{\mathrm{ord}_{k+1}(u_{H_i}^{(k+1)} u_{g_i}^{(k+1)}) \mid 1 \le i \le r\} \le \mathrm{ord}_{k+1} u_f^{(k+1)}$ and $\mathrm{ord}_j(u_{H_i}^{(j)} u_{g_i}^{(j)}) \le \mathrm{ord}_j u_f^{(j)}$ for $j = k+2, \ldots, p;\ i = 1, \ldots, r$ (see (4.8)).

Suppose that $u_f^{(k)} <_k u$. Since $u = \max_{<_k} \{ u_{H_i}^{(k)} u_{g_i}^{(k)} \mid 1 \le i \le r \}$, we have $u_{H_i - d_i u_{H_i}^{(k)}}^{(k)} <_k u\,(1 \le i \le r)$, so the expansion (4.9) implies that the k-leader of the first sum in (4.9) does not exceed u with respect to $<_k$. Furthermore, it is clear that $u_{H_i}^{(k)} u_{g_i}^{(k)} = u$ for any term in the sum

$$\tilde{f} = \sum_{u_{H_i}^{(k)} u_{g_i}^{(k)} = u} d_i u_{H_i}^{(k)} g_i \tag{4.10}$$

and for every $j = k+1, \ldots, p$,

$$\mathrm{ord}_j u_{\tilde{f}}^{(j)} \le \max_{i \in I} \left\{ \mathrm{ord}_j \left(u_{H_i}^{(k)} u_{g_i}^{(j)} \right) \right\} \le \max_{i \in I} \left\{ \mathrm{ord}_j \left(u_{H_i}^{(j)} u_{g_i}^{(j)} \right) \right\} \le \mathrm{ord}_j u_f^{(j)}$$

where I denotes the set of all indices $i \in \{1, \ldots, r\}$ that appear in (4.10).

Let $u_{\nu j}^{(k)} = \mathrm{lcm}(u_{g_\nu}^{(k)}, u_{g_j}^{(k)})$ for any $\nu, j \in I, \nu \neq j$ and let $\tau_{\nu j} = \frac{u}{u_{\nu j}^{(k)}} \in T$. (Since $u = u_{H_i}^{(k)} u_{g_i}^{(k)}$ for every $i \in I$, $u_{\nu j}^{(k)} | u$.)

By Proposition 4.10, there exist elements $c_{\nu j} \in K$ such that

$$\tilde{f} = \sum_{\nu, j} c_{\nu j} \tau_{\nu j} S_k(g_\nu, g_j) \tag{4.11}$$

where

$$u_{\tau_{\nu j} S_k(g_\nu, g_j)}^{(k)} <_k u_{\tilde{f}}^{(k)} = u$$

and

$$\mathrm{ord}_j u_{\tau_{\nu j} S_k(g_\nu, g_j)}^{(j)} \leq \mathrm{ord}_j u_{\tilde{f}}^{(j)} \quad (j = k+1, \ldots, p).$$

Since $S_k(g_\nu, g_j) \xrightarrow[<_k, <_{k+1}, \ldots, <_p]{G} 0$, there exist $q_{i\nu j} \in \mathcal{D}$ such that

$$S_k(g_\nu, g_j) = \sum_{i=1}^{r} q_{i\nu j} g_i$$

and

$$u_{q_{i\nu j}}^{(k)} u_{g_i}^{(k)} \leq_k u_{S_k(g_\nu, g_j)}^{(k)},$$

$$\mathrm{ord}_l \left(u_{q_{i\nu j}}^{(l)} u_{g_i}^{(l)} \right) \leq_l \mathrm{ord}_l u_{S_k(g_\nu, g_j)}^{(l)}$$

for $l = k+1, \ldots, p$. Thus, for any indices ν, j in the sum (4.11) we have

$$\tau_{\nu j} S_k(g_\nu, g_j) = \sum_{i=1}^{r} (\tau_{\nu j} q_{i\nu j}) g_i$$

where

$$u_{\tau_{\nu j} q_{i\nu j}}^{(k)} u_{g_i}^{(k)} = \tau_{\nu j} u_{q_{i\nu j}}^{(k)} u_{g_i}^{(k)} \leq_k \tau_{\nu j} u_{S_k(g_\nu, g_j)}^{(k)} <_k u.$$

It follows that

$$\tilde{f} = \sum_{\nu, j} c_{\nu j} \sum_{i=1}^{r} (\tau_{\nu j} q_{i\nu j}) g_i = \sum_{i=1}^{r} \left(\sum_{\nu, j} c_{\nu j} \tau_{\nu j} q_{i\nu j} \right) g_i = \sum_{i=1}^{r} \tilde{H}_i g_i \tag{4.12}$$

where

$$\tilde{H}_i = \sum_{\nu, j} c_{\nu j} \tau_{\nu j} q_{i\nu j} \quad (1 \leq i \leq r)$$

and

$$u_{\tilde{H}_i}^{(k)} u_{g_i}^{(k)} <_k u \quad (1 \leq i \leq r).$$

Furthermore, for any $l = k + 1, \ldots, p$, we have

$$\operatorname{ord}_l \left(u_{\tilde{H}_i}^{(l)} u_{g_i}^{(l)} \right) \leq \max_{\nu, j} \left\{ \operatorname{ord}_l \left(\tau_{\nu j} u_{\tilde{H}_i}^{(l)} u_{g_i}^{(l)} \right) \right\}$$

$$\leq \max_{\nu, j} \left\{ \max \left\{ \operatorname{ord}_l \left(\tau_{\nu j} \frac{u_{\nu j}^{(k)}}{u_{g_\nu}^{(k)}} u_{g_\nu}^{(l)} \right), \operatorname{ord}_l \left(\tau_{\nu j} \frac{u_{\nu j}^{(k)}}{u_{g_j}^{(k)}} u_{g_j}^{(l)} \right) \right\} \right\}$$

$$= \max_{\nu, j} \left\{ \max \left\{ \operatorname{ord}_l \left(\frac{u}{u_{g_\nu}^{(k)}} u_{g_\nu}^{(l)} \right), \operatorname{ord}_l \left(\frac{u}{u_{g_j}^{(k)}} u_{g_j}^{(l)} \right) \right\} \right\}$$

$$\leq \operatorname{ord}_l u = \operatorname{ord}_l u_{\tilde{f}}^{(k)} \leq \operatorname{ord}_l u_{\tilde{f}}^{(l)},$$

so that representation (4.12) satisfies the condition

$$\operatorname{ord}_l \left(u_{\tilde{H}_i}^{(l)} u_{g_i}^{(l)} \right) \leq \operatorname{ord}_l u_{\tilde{f}}^{(l)} \tag{4.13}$$

for $i = 1, \ldots, r$. Substituting (4.12) into (4.9) we obtain

$$f = \sum_{i=1}^{r} \tilde{H}_i g_i + \sum_{u_{H_i}^{(k)} u_{g_i}^{(k)} = u} \left(H_i - d_i u_{H_i}^{(k)} \right) g_i + \sum_{u_{H_i}^{(k)} u_{g_i}^{(k)} <_k u} H_i g_i \tag{4.14}$$

where, denoting each $H_i - d_i u_{H_i}^{(k)}$ in the second sum by H_i', we have the following conditions:

(i) $u_{\tilde{H}_i}^{(k)} u_{g_i}^{(k)} <_k u$

(ii) $u_{H_i'}^{(k)} u_{g_i}^{(k)} <_k u$ for any term with index i in the second sum in (4.14).

(iii) $u_{H_i}^{(k)} u_{g_i}^{(k)} <_k u$ for any term with index i in the third sum in (4.14).

Also, for every $l = k+1, \ldots, p$, the inequality (4.13) implies $\operatorname{ord}_l \left(u_{\tilde{H}_i}^{(l)} u_{g_i}^{(l)} \right) \leq \operatorname{ord}_l u_{\tilde{f}}^{(l)}$. Therefore,

$$\operatorname{ord}_l \left(u_{\tilde{H}_i}^{(l)} u_{g_i}^{(l)} \right) \leq \max \left\{ \operatorname{ord}_l \left(u_{H_i}^{(k)} u_{g_i}^{(l)} \right) \right\} \leq \max \left\{ \operatorname{ord}_l \left(u_{H_i}^{(l)} u_{g_i}^{(l)} \right) \right\} \leq \operatorname{ord}_l u_{\tilde{f}}^{(l)}$$

where the maxima are taken over the set of all indices i that appear in the first sum in (4.9). Furthermore, inequality (4.8) implies that for every index i in the second sum in (4.14), one has

$$\operatorname{ord}_l \left(u_{H_i'}^{(l)} u_{g_i}^{(l)} \right) \leq \operatorname{ord}_l \left(u_{H_i}^{(k)} u_{g_i}^{(l)} \right) \leq \operatorname{ord}_l u_{\tilde{f}}^{(l)} \quad (l = k+1, \ldots, p)$$

and for every index i in the third sum in (4.14) we have

$$\operatorname{ord}_l \left(u_{H_i}^{(l)} u_{g_i}^{(l)} \right) \leq \operatorname{ord}_l u_{\tilde{f}}^{(l)} \quad (l = k+1, \ldots, p).$$

Thus, (4.14) is a representation of f in the form (4.6) with conditions (4.7) and (4.8) such that if one writes (4.14) as $f = \sum_{i=1}^{r} \tilde{H}_i f_i$ (combining the sums in (4.14)), then $\max\{u_{\tilde{H}_1'}^{(k)}u_{g_1}^{(k)},\ldots,u_{\tilde{H}_r'}^{(k)}u_{g_r}^{(k)}\} <_k u$ and one has conditions of the types (4.7) and (4.8). We have arrived at a contradiction with our choice of representation (4.6) of f with conditions (4.7), (4.8) and the smallest (with respect to $<_k$) possible value of $\max\{u_{\tilde{H}_i'}^{(k)}u_{g_i}^{(k)} \mid 1 \leq i \leq r\} = u$. Thus, every element $f \in N$ can be represented in the form (4.3) with conditions (4.4) and (4.5). □

The last theorem allows one to construct a Gröbner basis of a $\mathcal{D}$-module $N \subseteq E$ with respect to $<_1,\ldots,<_p$ starting with a Gröbner basis of N with respect to $<_p$.

One should note that the developed technique is an essential improvement of the characteristic set approach used in [8] for the proof of a theorem on bivariate difference-differential dimension polynomials.

Theorem 4.12 *Let $\mathcal{D}$ be the ring of difference (σ-) operators over a difference (σ-) field K, M a vector σ-K-space generated (as a left $\mathcal{D}$-module) by a finite set $\{f_1,\ldots,f_m\}$, and E a free left $\mathcal{D}$-module with free generators $e_1,\ldots,e_m$. Let $\pi : E \longrightarrow M$ be the natural $\mathcal{D}$-epimorphism ($\pi(e_i) = f_i$ for $i = 1,\ldots,m$), $N = Ker\,\pi$, and $G = \{g_1,\ldots,g_d\}$ a Gröbner basis of N with respect to $<_1,\ldots,<_p$. Furthermore, for any $(r_1,\ldots,r_p) \in \mathbf{Z}^p$, let $M_{r_1\ldots r_p} = \sum_{i=1}^m \mathcal{D}_{r_1\ldots r_p}f_i$ and let*

$$V_{r_1\ldots r_p} = \left\{u \in Te \mid \mathrm{ord}_i u \leq r_i \text{ for } i = 1,\ldots,p, \text{ and } u \neq \tau u_g^{(1)} \text{ for any } \tau \in T, g \in G\right\},$$

$$W_{r_1\ldots r_p} = \left\{u \in Te \setminus V_{r_1\ldots r_p} \mid \mathrm{ord}_i u \leq r_i \text{ for } i = 1,\ldots,p \text{ and for every } \tau \in T, g \in G\right.$$
$$\left. \text{such that } u = \tau u_g^{(1)}, \text{ there exists } i \in \{2,\ldots,p\} \text{ such that } \mathrm{ord}_i \tau u_g^{(i)} > r_i\right\},$$

and $U_{r_1\ldots r_p} = V_{r_1\ldots r_p} \bigcup W_{r_1\ldots r_p}$.

Then for any $(r_1,\ldots,r_p) \in \mathbf{N}^p$, the set $\pi(U_{r_1\ldots r_p})$ is a basis of the vector K-space $M_{r_1\ldots r_p}$.

Proof. Let us prove, first, that every element τf_i ($1 \leq i \leq m$, $\tau \in T(r_1,\ldots,r_p)$), which does not belong to $\pi(U_{r_1\ldots r_p})$, can be written as a finite linear combination of elements of $\pi(U_{r_1\ldots r_p})$ with coefficients in K (so that the set $\pi(U_{r_1\ldots r_p})$ generates the vector K-space $M_{r_1\ldots r_p}$). Indeed, since $\tau f_i \notin \pi(U_{r_1\ldots r_p})$, $\tau e_i \notin U_{r_1\ldots r_p}$ whence $\tau e_i = \tau' u_{g_j}^{(1)}$ for some $\tau' \in T$, $1 \leq j \leq d$, such that $\mathrm{ord}_\nu(\tau' u_{g_j}^{(\nu)}) \leq r_\nu$ ($\nu = 2,\ldots,p$). Let us consider the element $g_j = a_j u_{g_j}^{(1)} + \ldots$ ($a_j \in K, a_j \neq 0$), where dots are placed instead of the sum of the other terms of g_j with non-zero coefficients (obviously, those terms are less than $u_{g_j}^{(1)}$ with respect to the order $<_1$). Since $g_j \in N = Ker\,\pi$, $\pi(g_j) = a_j\pi(u_{g_j}^{(1)}) + \cdots = 0$, whence $\pi(\tau' g_j) = a_j\pi(\tau' u_{g_j}^{(1)}) + \cdots = a_j\pi(\tau e_i) + \cdots = a_j\tau f_i + \cdots = 0$, so that τf_i is a finite linear combination with coefficients in K of some elements $\tilde{\tau}_l f_l$ ($1 \leq l \leq m$) such that $\tilde{\tau}_l \in T(r_1,\ldots,r_p)$ and $\tilde{\tau}_l e_l <_1 \tau' u_{g_j}^{(1)}$. ($\mathrm{ord}_1\tilde{\tau}_l \leq r_1$, since $\tilde{\tau}_l e_l <_1 \tau e_i$ and $\tau \in T(r_1,\ldots,r_p)$; $\mathrm{ord}_\nu\tilde{\tau}_l \leq r_\nu$ ($\nu = 2,\ldots,p$), because $\tilde{\tau}_l e_l \leq_\nu u_{\tau' g_j}^{(\nu)} = \tau' u_{g_j}^{(\nu)}$ and $\mathrm{ord}_\nu(\tau' u_{g_j}^{(\nu)}) \leq r_\nu$.) Thus, we can apply the induction on τe_j ($\tau \in T, 1 \leq j \leq m$) with respect to the order $<_1$ and obtain that every element τf_i ($\tau \in T(r_1,\ldots,r_p), 1 \leq j \leq m$) can be written as a finite linear combination of elements of $\pi(U_{r_1\ldots r_p})$ with coefficients in the field K.

Now, let us prove that the set $\pi(U_{r_1\ldots r_p})$ is linearly independent over K. Suppose that $\sum_{i=1}^k a_i\pi(u_i) = 0$ for some $u_1,\ldots,u_k \in U_{r_1\ldots r_p}$, $a_1,\ldots,a_k \in K$. Then $h = \sum_{i=1}^k a_iu_i$ is an element of N which is reduced with respect to G. Indeed, if a term $u = \tau e_j$ appears in h (so that $u = u_i$ for some $i = 1,\ldots,k$), then either u is not a multiple of any $u_{g_\nu}^{(1)}$ ($1 \leq \nu \leq d$) or $u = \tau u_{g_\nu}^{(1)}$ for some $\tau \in T$, $1 \leq \nu \leq d$, such that $\operatorname{ord}_\mu(\tau u_{g_\nu}^{(\mu)}) > r_\mu \geq \operatorname{ord}_\mu u_h^{(\mu)}$ for some μ, $2 \leq \mu \leq p$. By Proposition 4.8, $h = 0$, whence $a_1 = \cdots = a_k = 0$. This completes the proof of the theorem. $\square$

Now we are ready to prove the main result of this section, the theorem on a multivariable dimension polynomial associated with a difference vector space with an excellent p-dimension filtration.

Theorem 4.13 *Let K be a difference field with a basic set $\sigma = \{\alpha_1,\ldots,\alpha_n\}$ and let $\mathcal{D}$ be the ring of difference operators over K equipped with the standard p-dimensional filtration corresponding to partition (4.1) of the set σ. Furthermore, let $n_i = \operatorname{Card}\sigma_i$ ($i = 1,\ldots,p$) and let $\{M_{r_1\ldots r_p}|(r_1,\ldots,r_p) \in \mathbf{Z}^p\}$ be an excellent p-dimensional filtration of a vector σ-K-space M. Then there exists a polynomial $\phi(t_1,\ldots,t_p) \in \mathbf{Q}[t_1,\ldots,t_p]$ such that*

(i) $\phi(r_1,\ldots,r_p) = \dim_K M_{r_1\ldots r_p}$ for all sufficiently large $(r_1,\ldots,r_p) \in \mathbf{Z}^p$;

(ii) $\deg_{t_i} \phi \leq n_i$ ($1 \leq i \leq p$), so that $\deg \phi \leq n$ and the polynomial $\phi(t_1,\ldots,t_p)$ can be represented as

$$\phi(t_1,\ldots,t_p) = \sum_{i_1=0}^{n_1} \cdots \sum_{i_p=0}^{n_p} a_{i_1\ldots i_p}\binom{t_1+i_1}{i_1}\cdots\binom{t_p+i_p}{i_p}$$

where $a_{i_1\ldots i_p} \in \mathbf{Z}$ for all $i_1,\ldots,i_p$.

Proof. Since the p-dimensional filtration $\{M_{r_1\ldots r_p}|(r_1,\ldots,r_p) \in \mathbf{Z}^p\}$ is excellent, there exists an element $(h_1,\ldots,h_p) \in \mathbf{Z}^p$ such that $\mathcal{D}_{r_1,\ldots,r_p}M_{h_1,\ldots,h_p} = M_{r_1+h_1,\ldots,r_p+h_p}$ for any $(r_1,\ldots,r_p) \in \mathbf{N}^p$. Furthermore, $M_{h_1,\ldots,h_p}$ is a finite-dimensional vector space over K and any basis of this vector space generates M as a left $\mathcal{D}$-module, so $M = \sum_{i=1}^m \mathcal{D}y_i$ for some elements $y_1,\ldots,y_m \in M_{h_1,\ldots,h_p}$.

Let E be a free $\mathcal{D}$-module with a basis $e_1,\ldots,e_m$, let N be the kernel of the natural σ-epimorphism $\pi : E \to M$ ($\pi(e_i) = y_i$ for $i = 1,\ldots,m$), and let the set $U_{r_1\ldots r_p}$ ($r_1,\ldots,r_p \in \mathbf{N}$) be the same as in the assumptions of Theorem 4.12. Furthermore, let $G = \{g_1,\ldots,g_d\}$ be a Gröbner basis of N with respect to $<_1,\ldots,<_p$. By Theorem 4.12, for any $r_1,\ldots,r_p \in \mathbf{N}$, $\pi(U_{r_1,\ldots,r_p})$ is a basis of the vector K-space $M_{r_1,\ldots,r_p}$. Therefore, $\dim_K M_{r_1,\ldots,r_p} = \operatorname{Card}\pi(U_{r_1,\ldots,r_p}) = \operatorname{Card}U_{r_1,\ldots,r_p}$. (It was shown in the second part of the proof of Theorem 4.12 that the restriction of the mapping π on $U_{r_1,\ldots,r_p}$ is bijective).

Let $U'_{r_1,\ldots,r_p} = \{w \in U_{r_1,\ldots,r_p}|w$ is not a multiple of any element $u_{g_i}^{(1)}$ ($1 \leq i \leq d$)$\}$ and let $U''_{r_1,\ldots,r_p} = \{w \in U_{r_1,\ldots,r_p}|$ there exists $g_j \in G$ and $\tau \in T$ such that $w = \tau u_{g_j}^{(1)}$

and $\mathrm{ord}_\nu(\tau u_{g_j}^{(\nu)}) > r_\nu$ for some ν, $2 \leq \nu \leq p\}$. Then $U_{r_1,\ldots,r_p} = U'_{r_1,\ldots,r_p} \bigcup U''_{r_1,\ldots,r_p}$ and $U'_{r_1,\ldots,r_p} \bigcap U''_{r_1,\ldots,r_p} = \emptyset$, hence

$$\mathrm{Card}\, U_{r_1,\ldots,r_p} = \mathrm{Card}\, U'_{r_1,\ldots,r_p} + \mathrm{Card}\, U''_{r_1,\ldots,r_p}.$$

By Theorem 2.10, there exists a numerical polynomial $\omega(t_1,\ldots,t_p)$ in p variables $t_1,\ldots,t_p$ such that $\omega(r_1,\ldots,r_p) = \mathrm{Card}\, U'_{r_1,\ldots,r_p}$ for all sufficiently large $(r_1,\ldots,r_p) \in \mathbf{N}^p$. In order to express $\mathrm{Card}\, U''_{r_1,\ldots,r_p}$ in terms of $r_1,\ldots,r_p$, let us set $a_{ij} = \mathrm{ord}_i u_{g_j}^{(1)}$ and $b_{ij} = \mathrm{ord}_i u_{g_j}^{(i)}$ for $i = 1,\ldots,p$; $j = 1,\ldots,d$. Clearly, $a_{1j} = b_{1j}$ and $a_{ij} \leq b_{ij}$ for $i = 1,\ldots,p$; $j = 1,\ldots,d$. Furthermore, for any $\mu = 1,\ldots,p$ and for any integers $k_1,\ldots,k_\mu$ such that $2 \leq k_1 < \cdots < k_\mu \leq p$, let $V_{j;k_1,\ldots,k_\mu}(r_1,\ldots,r_p) = \{\tau u_{g_j}^{(1)}|\mathrm{ord}_i\tau \leq r_i - a_{ij}$ for $i = 1,\ldots,p$ and $\mathrm{ord}_\nu\tau > r_\nu - b_{\nu j}$ if and only if ν is equal to one of the numbers $k_1,\ldots,k_\mu\}$.

Then $\mathrm{Card}\, V_{j;k_1,\ldots,k_\mu}(r_1,\ldots,r_p) = \phi_{j;k_1,\ldots,k_\mu}(r_1,\ldots,r_p)$, where $\phi_{j;k_1,\ldots,k_\mu}(t_1,\ldots,t_p)$ is a numerical polynomial in p variables $t_1,\ldots,t_p$ defined by the formula

$$\phi_{j;k_1,\ldots,k_\mu}(t_1,\ldots,t_p) = \binom{t_1 + n_1 - b_{1j}}{n_1}\cdots\binom{t_{k_1-1} + n_{k_1-1} - b_{k_1-1,j}}{n_{k_1-1}}$$

$$\left[\binom{t_{k_1} + n_{k_1} - a_{k_1,j}}{n_{k_1}} - \binom{t_{k_1} + n_{k_1} - b_{k_1,j}}{n_{k_1}}\right]\binom{t_{k_1+1} + n_{k_1+1} - b_{k_1+1,j}}{n_{k_1+1}}\cdots$$

$$\binom{t_{k_\mu-1} + n_{k_\mu-1} - b_{k_\mu-1,j}}{n_{k_\mu-1}}\left[\binom{t_{k_\mu} + n_{k_\mu} - a_{k_\mu,j}}{n_{k_\mu}} - \binom{t_{k_\mu} + n_{k_\mu} - b_{k_\mu,j}}{n_{k_\mu}}\right]$$

$$\cdots\binom{t_p + n_p - b_{pj}}{n_p}. \tag{4.15}$$

(Statement (iii) of Theorem 2.10 shows that $\mathrm{Card}\,\{\tau \in T|\mathrm{ord}_1\tau \leq r_1,\ldots, \mathrm{ord}_p\tau \leq r_p\} = \prod_{i=1}^p \binom{r_i+n_i}{n_i}$ for any $r_1,\ldots,r_p \in \mathbf{N}$.) It is easy to see that $\deg_{t_i} \phi_{j;k_1,\ldots,k_\mu} \leq n_i$ for $i = 1,\ldots,p$.

Now, for any $j = 1,\ldots,d$, let $V_j(r_1,\ldots,r_p) = \{\tau u_{g_j}^{(1)}|\mathrm{ord}_i\tau \leq r_i - a_{ij}$ for $i = 1,\ldots,p$ and there exists $\nu \in \mathbf{N}$, $2 \leq \nu \leq p$, such that $\mathrm{ord}_\nu\tau > r_\nu - b_{\nu j}\}$. Then the combinatorial principle of inclusion and exclusion implies that $\mathrm{Card}\, V_j(r_1,\ldots,r_p) = \phi_j(r_1,\ldots,r_p)$, where $\phi_j(t_1,\ldots,t_p)$ is a numerical polynomial in p variables $t_1,\ldots,t_p$ defined by the formula

$$\phi_j(t_1,\ldots,t_p) = \sum_{k_1=1}^p \phi_{j;k_1}(t_1,\ldots,t_p) - \sum_{1\leq k_1<k_2\leq p}\phi_{j;k_1,k_2}(t_1,\ldots,t_p) + \cdots$$

$$+(-1)^{\mu-1}\sum_{1\leq k_1<\cdots<k_\mu\leq p}\phi_{j;k_1,\ldots,k_\mu}(t_1,\ldots,t_p) + \cdots + (-1)^{p-1}\phi_{j;2,\ldots,p}(t_1,\ldots,t_p).$$

It is easy to see that $\deg_{t_i} \phi_j(t_1,\ldots,t_p) \leq n_i$ for $i = 1,\ldots,p$.

Applying the principle of inclusion and exclusion once again we obtain that

$$\text{Card}\, U''_{r_1\ldots r_p} = \text{Card}\, \bigcup_{j=1}^{d} V_j(r_1,\ldots,r_p) = \sum_{j=1}^{d} \text{Card}\, V_j(r_1,\ldots,r_p)$$

$$- \sum_{1\le j_1 < j_2 \le d} \text{Card}\, \left(V_{j_1}(r_1,\ldots,r_p)\bigcap V_{j_2}(r_1,\ldots,r_p) \right) + \ldots$$

$$+ (-1)^{d-1}\text{Card}\, \bigcap_{\nu=1}^{d} V_{j_\nu}(r_1,\ldots,r_p),$$

so it is sufficient to prove that for any $s = 1,\ldots,d$ and for any indices $j_1,\ldots,j_s$, $1 \le j_1 < \cdots < j_s \le d$,

$$\text{Card}\, \left(V_{j_1}(r_1,\ldots,r_p)\bigcap\cdots\bigcap V_{j_s}(r_1,\ldots,r_p) \right) = \phi_{j_1,\ldots,j_s}(r_1,\ldots,r_p)$$

where $\phi_{j_1,\ldots,j_s}(t_1,\ldots,t_p)$ is a numerical polynomial in p variables $t_1,\ldots,t_p$ such that $\deg_{t_i} \phi_{j_1,\ldots,j_s} \le n_i$ for $i = 1,\ldots,p$. It is clear that the intersection $V_{j_1}(r_1,\ldots,r_p)\bigcap\cdots \bigcap V_{j_s}(r_1,\ldots,r_p)$ is not empty (therefore, $\phi_{j_1,\ldots,j_s} \ne 0$) if and only if the leaders $u_{g_{j_1}}^{(1)},\ldots, u_{g_{j_s}}^{(1)}$ contain the same indeterminate e_i ($1 \le i \le m$). Let us consider such an intersection

$$V_{j_1}(r_1,\ldots,r_p)\bigcap\cdots\bigcap V_{j_s}(r_1,\ldots,r_p),$$

let $v(j_1,\ldots,j_s) = \text{lcm}(u_{j_1}^{(1)},\ldots,u_{j_s}^{(1)})$, and let $v(j_1,\ldots,j_s) = \gamma_\nu u_{g_{j_\nu}}$ ($1 \le \nu \le s$; $\gamma_\nu \in T$). Then $V_{j_1}(r_1,\ldots,r_p)\bigcap\cdots\bigcap V_{j_s}(r_1,\ldots,r_p)$ is the set of all terms $u = \tau v(j_1,\ldots,j_s)$ such that $\text{ord}_i u \le r_i$ (that is, $\text{ord}_i\tau \le r_i - \text{ord}_i v(j_1,\ldots,j_s)$) for $i = 1,\ldots,p$, and for any $l = 1,\ldots,s$, there exists at least one index $\nu \in \{2,\ldots,p\}$ such that $\text{ord}_\nu(\tau\gamma_l u_{g_{j_l}}^{(\nu)}) > r_\nu$ (i. e., $\text{ord}_\nu\tau > r_\nu - \text{ord}_\nu v(j_1,\ldots,j_s) - \text{ord}_\nu u_{g_{j_l}}^{(\nu)} + \text{ord}_\nu u_{g_{j_l}}^{(1)}$). Denoting $\text{ord}_i v(j_1,\ldots,j_s)$ by $c_{j_1,\ldots,j_s}^{(i)}$ ($1 \le i \le p$) and applying the principle of inclusion and exclusion one more time, we obtain that Card $\bigcap_{\mu=1}^{s} V_{j_\mu}(r_1,\ldots,r_p)$ is an alternating sum of terms of the form

$$\text{Card}\, W(j_1,\ldots,j_s; k_{11},k_{12},\ldots,k_{1q_1},k_{21},\ldots,k_{sq_s}; r_1,\ldots,r_p)$$

where $W(j_1,\ldots,j_s; k_{11},k_{12},\ldots,k_{1q_1},k_{21},\ldots,k_{sq_s}; r_1,\ldots,r_p) = \{\tau \in T | \text{ord}_i\tau \le r_i - c_{j_1,\ldots,j_s}^{(i)}$ for $i = 1,\ldots,p$, and for any $l = 1,\ldots,s$, $\text{ord}_k\tau > r_k - c_{j_1,\ldots,j_s}^{(k)} + a_{kj_l} - b_{kj_l}$ if and only if $k = k_{li}$ for some $i = 1,\ldots,q_l\}$ ($q_1,\ldots,q_s$ are some positive integers from the set $\{1,\ldots,p\}$ and $\{k_{i\mu}|1 \le i \le s, 1 \le \mu \le q_s\}$ is a family of integers such that $2 \le k_{i1} < k_{i2} < \cdots < k_{iq_i} \le p$ for $i = 1,\ldots,s$).

Thus, it is sufficient to show that Card $W(j_1,\ldots,j_s; k_{11},\ldots,k_{sq_s}; r_1,\ldots,r_p)$ $= \psi_{k_{11},\ldots,k_{sq_s}}^{j_1,\ldots,j_s}(r_1,\ldots,r_p)$ where $\psi_{k_{11},\ldots,k_{sq_s}}^{j_1,\ldots,j_s}(t_1,\ldots,t_p)$ is a numerical polynomial in p variables $t_1,\ldots,t_p$ such that $\deg_{t_i} \psi_{k_{11},\ldots,k_{sq_s}}^{j_1,\ldots,j_s} \le n_i$ ($i = 1,\ldots,p$). But this is almost evident: as in the process of evaluation of the value of Card $V_{j;k_1,\ldots,k_q}(r_1,\ldots,r_p)$ (when we used Theorem 2.10 (iii) to obtain formula (4.15)), we see that Card $W(j_1,\ldots,j_s; k_{11},\ldots,k_{sq_s}; r_1,\ldots,r_p)$ is a product of terms of the form $\binom{r_\nu + n_\nu - c_{j_1,\ldots,j_s}^{(\nu)} - S_\nu}{n_\nu}$ (such a term corresponds to a number $\nu \in \{1,\ldots,p\}$ that is different from all $k_{i\mu}$ ($1 \leq i \leq s$, $1 \leq \mu \leq q_s$); S_ν is defined as $\max\{b_{\nu j_l} - a_{\nu j_l}|1 \leq l \leq s\}$) and terms of the form $\left[\binom{r_\nu + n_\nu - c_{j_1,\ldots,j_s}^{(\nu)}}{n_\nu} - \binom{r_\nu + n_\nu - c_{j_1,\ldots,j_s}^{(\nu)} - S_\nu'}{n_\nu}\right]$ (such a term appears in the product if $\nu = k_{i\mu}$ for some i, μ; if $k_{i_1\mu_1}, \ldots, k_{i_e\mu_e}$ are all elements of the set $\{k_{i\mu}|1 \leq i \leq s, 1 \leq \mu \leq q_s\}$ that are equal to ν ($1 \leq e \leq s$, $1 \leq i_1 < \cdots < i_e \leq s$), then S_ν' is defined as $\min\{b_{\nu j_{i_\lambda}} - a_{\nu j_{i_\lambda}}|1 \leq \lambda \leq l\}$). The corresponding numerical polynomial $\psi_{k_{11},\ldots,k_{sq_s}}^{j_1,\ldots,j_s}(t_1,\ldots,t_p)$ is a product of p "elementary" numerical polynomials, each of which is equal to either $\binom{z_\nu + n_\nu - c_{j_1,\ldots,j_s}^{(\nu)} - S_\nu}{n_\nu}$ or $\left[\binom{t_\nu + n_\nu - c_{j_1,\ldots,j_s}^{(\nu)}}{n_\nu} - \binom{t_\nu + n_\nu - c_{j_1,\ldots,j_s}^{(\nu)} - S_\nu'}{n_\nu}\right]$ ($1 \leq \nu \leq p$). Since the degree of such a product with respect to any variable t_i ($1 \leq i \leq p$) does not exceed n_i, this completes the proof of the theorem. □

Definition 4.14 The polynomial $\phi(t_1, \ldots, t_p)$, whose existence is established by Theorem 4.13, is called a dimension (or $(\sigma_1, \ldots, \sigma_p)$-dimension) polynomial of the vector σ-K-space M associated with the p-dimensional filtration $\{M_{r_1 \ldots r_p}|(r_1, \ldots, r_p) \in \mathbf{Z}^p\}$.

Example 4.15 With the notation of Theorem 4.13, let $n = 2$, $\sigma_1 = \{\alpha_1\}, \sigma_2 = \{\alpha_2\}$, and let a σ-K-module M be generated by one element x that satisfies the defining equation

$$\sum_{i=0}^{p} a_i \alpha_1^i x + \sum_{j=1}^{q} b_j \alpha_2^j x = 0,$$

where $p, q \geq 1$, $a_i, b_j \in K$ ($1 \leq i \leq p$, $1 \leq j \leq q$), $a_p \neq 0$, $b_q \neq 0$. In other words, M is a factor module of a free $\mathcal{D}$-module $E = \mathcal{D}e$ with a free generator e by its $\mathcal{D}$-submodule $N = \mathcal{D}(\sum_{i=0}^{p} a_i \alpha_1^i + \sum_{i=1}^{q} b_j \alpha_2^j)e$.

It is easy to see that the set consisting of a single element $g = (\sum_{i=0}^{p} a_i \alpha_1^i + \sum_{i=1}^{q} b_j \alpha_2^j)e$ is a Gröbner basis of N with respects to the orders $<_1, <_2$. In this case, the proof of Theorem 4.13 shows that the (σ_1, σ_2)-dimension polynomial of M associated with the natural bifiltration $(M_{rs} = \sum_{i=0}^{r} \sum_{j=0}^{s} K\alpha_1^i \alpha_2^j x)_{r,s \in \mathbf{N}}$ is as follows:

$$\phi(t_1, t_2) = \left[\binom{t_1 + 1}{1}\binom{t_2 + 1}{1} - \binom{t_1 + 1 - p}{1}\binom{t_2 + 1}{1}\right]$$
$$+ \binom{t_1 + 1 - p}{1}\left[\binom{t_2 + 1}{1} - \binom{t_2 + 1 - q}{1}\right]$$
$$= qt_1 + pt_2 + p + q - pq.$$

(With the notation of the proof, the polynomial in the first brackets gives Card U_{rs}' while the polynomial $\binom{t_1}{1}\left[\binom{t_2+1}{1} - \binom{t_2-1}{1}\right]$ gives Card U_{rs}'' for all sufficiently large $(r, s) \in \mathbf{N}^2$.) Note that the obtained dimension expresses the strength of the difference equation $\sum_{i=0}^{p} a_i \alpha_1^i y + \sum_{j=1}^{q} b_j \alpha_2^j y = 0$ with respect to the difference indeterminate y (the right-hand side of this equation is treated as a σ-polynomial in the ring $K\{y\}$.)

We conclude with a theorem that gives invariants of a multivariable dimension polynomial of a difference module.

Let K be a difference field and let partition (4.1) of its basic set σ be fixed. As we have seen, if M is a finitely generated vector σ-K-space, then every finite set of generators of M over the ring of σ-operators $\mathcal{D}$ produces an excellent p-dimensional filtration of M and therefore a dimension polynomial of M associated with this filtration. Generally speaking, different finite systems of generators of M over $\mathcal{D}$ produce different $(\sigma_1, \ldots, \sigma_p)$-dimension polynomials (we leave the corresponding example to the reader as an exercise), however every dimension polynomial carries certain integers that do not depend on the system of generators. These integers, that characterize the vector σ-K-space M, are called *invariants* of a dimension polynomial. In what follows, we describe some of the invariants.

Theorem 4.16 *Let K be a difference (σ-) field, M a finitely generated σ-K-vector space, $\{M_{r_1 \ldots r_p} | (r_1, \ldots, r_p) \in \mathbf{Z}^p\}$ an excellent p-dimensional filtration of M, and $\phi(t_1, \ldots, t_p) = \sum_{i_1=0}^{n_1} \cdots \sum_{i_p=0}^{n_p} a_{i_1 \ldots i_p} \binom{t_1+i_1}{i_1} \ldots \binom{t_p+i_p}{i_p}$ the dimension polynomial associated with this filtration. Let $\Sigma_\phi = \{(i_1, \ldots, i_p) \in \mathbf{N}^p \,|\, 0 \le i_k \le n_k \ (k = 1, \ldots, p) \text{ and } a_{i_1 \ldots i_p} \ne 0\}$.*

Then $d = \deg \phi$, $a_{n_1 \ldots n_p}$, the elements $(k_1, \ldots, k_p) \in \Sigma'_\phi$, the corresponding coefficients $a_{k_1 \ldots k_p}$, and the coefficients of the terms of total degree d do not depend on the choice of the excellent filtration.

Proof. Let $\{M_{r_1 \ldots r_p} | (r_1, \ldots, r_p) \in \mathbf{Z}^p\}$ and $\{M'_{r_1 \ldots r_p} | (r_1, \ldots, r_p) \in \mathbf{Z}^p\}$ be two excellent p-dimensional filtrations of the same finitely generated σ-K-vector space M and let $\phi(t_1, \ldots, t_p)$ and $\phi'(t_1, \ldots, t_p)$ be dimension polynomials associated with these excellent filtrations, respectively. Then there exists an element $(s_1, \ldots, s_p) \in \mathbf{N}^p$ such that $M_{r_1 \ldots r_p} \subseteq M'_{r_1+s_1, \ldots, r_p+s_p}$ and $M'_{r_1 \ldots r_p} \subseteq M_{r_1+s_1, \ldots, r_p+s_p}$ for all sufficiently large $(r_1, \ldots, r_p) \in \mathbf{Z}^p$. It follows that there exist $u_1, \ldots, u_p \in \mathbf{Z}$ such that

$$\phi(t_1, \ldots, t_p) \le \phi'(t_1 + s_1, \ldots, t_p + s_p) \tag{4.16}$$

and

$$\phi'(t_1, \ldots, t_p) \le \phi(t_1 + s_1, \ldots, t_p + s_p) \tag{4.17}$$

for all integer (and real) values of $t_1, \ldots, t_p$ such that $t_1 \ge u_1, \ldots, t_p \ge u_p$. If we set $t_i = t$ $(1 \le i \le p)$ in (4.16) and (4.17) and let $t \to \infty$ we obtain that $\phi(t_1, \ldots, t_p)$ and $\phi'(t_1, \ldots, t_p)$ have the same degree d and the same coefficient of the monomial $t_1^{n_1} \ldots t_p^{n_p}$. If $(k_1, \ldots, k_p) \in \Sigma'_\phi$ is the maximal element of Σ_ϕ with respect to the lexicographic order $\le_{j_1, \ldots, j_p}$, then we set $t_{j_p} = t$, $t_{j_{p-1}} = 2^{t_{j_p}} = 2^t, \ldots, t_{j_1} = 2^{t_{j_2}}$ and let $t \to \infty$ in (4.16) and (4.17). We obtain that $(k_1, \ldots, k_p)$ is the maximal element of $\Sigma_{\phi'}$ with respect to $\le_{j_1, \ldots, j_p}$ and the coefficients of $t_1^{k_1} \ldots t_p^{k_p}$ in the polynomials $\phi(t_1, \ldots, t_p)$ and $\phi'(t_1, \ldots, t_p)$ are equal. Finally, let us order the terms of the total degree d in ϕ and ϕ' using the lexicographic order $\le_{p, p-1, \ldots, 1}$, set $w_1 = t$, $w_2 = 2^{w_1} = 2^t, \ldots, w_p = 2^{w_{p-1}}$, $T = 2^{w_p}$, $t_i = w_i T$ $(1 \le i \le p)$ and let $t \to \infty$. Then the inequalities (4.16) and (4.17) immediately imply that the polynomials $\phi(t_1, \ldots, t_p)$ and $\phi'(t_1, \ldots, t_p)$ have the same coefficients of the terms of total degree d. $\square$

A substantial part of this work was written during the Special Semester on Gröbner Bases and Related Methods organized by the Radon Institute for Computational and

Applied Mathematics and the Research Institute for Symbolic Computation (Linz, Austria, 2006). I would like to thank the organizers of the Semester for their support and the atmosphere of productive scientific collaboration that stimulated my work on this paper.

Bibliography

[1] Buchberger, B. Ein Algorithmus zum Auffinden der Basiselemente des Restklassenringes nach einem nulldimensionalen Polynomideal. *PhD. Thesis. Univ. of Innsbruck*, Austria, 1965.

[2] Cohn, R. M. Difference Algebra. *Interscience*, New York, 1965.

[3] Einstein, A. The Meaning of Relativity. Appendix II (Generalization of gravitation theory), 4th edn. Princeton, 1953, 133–165.

[4] Kolchin, E. R. Differential Algebra and Algebraic Groups. *Academic Press*, New York, 1973.

[5] Kondrateva, M. V., Levin, A. B., Mikhalev, A. V., Pankratev, E. V. Computation of Dimension Polynomials. *Internat. J. Algebra and Comput.*, 2 (1992), 117–137.

[6] Kondrateva, M.V.; Levin, A. B.; Mikhalev, A. V.; Pankratev, E. V. Differential and Difference Dimension Polynomials. *Kluwer Academic Publishers*, Dordrecht, 1998.

[7] Levin, A. B. Characteristic Polynomials of Filtered Difference Modules and Difference Field Extensions. *Russian Math Surv.*, 33 (1978), 165–166.

[8] Levin, A. B. Reduced Grobner Bases, Free Difference-Differential Modules and Difference-Differential Dimension Polynomials. *J. Symbolic Comput.*, 30, no. 4 (2000), 357–382.

[9] Levin, A. B. Multivariable Dimension Polynomials and New Invariants of Differential Field Extensions. *Internat. Journal of Math. and Math. Sciences*, 27, no. 4 (2001), 201–214.

[10] Levin, A. B. Multivariable Difference Dimension Polynomials. *Journal of Mathematical Sciences*, 131, no. 6 (2005), 6060–6082.

Author information

Alexander Levin, Department of Mathematics, The Catholic University of America, Washington, D. C. 20064, United States of America.
Email: levin@cua.edu

Editors' Note
in Memory of Giuseppa Carrà Ferro

Giuseppa Carrà Ferro left us on March 22, 2007 — forever. The editors were very much depressed by the sad and unbelievable news. Carrà Ferro not only attended the Workshop on Gröbner Bases in Symbolic Analysis and submitted her paper for publication, but also helped referee two papers for the volume. She must have worked very hard with the paper and the reviews while struggling with cancer. In fact, she delivered her last lecture and completed her course just a few days before she passed away.

On the occasion of Carrà Ferro's death, the editors wish to include her paper for publication in this volume even though it is no longer possible to have the paper thoroughly revised and improved by the author herself. With the support of the two referees and the consent of Alfredo Ferro, husband of Carrà Ferro, the editors took the version she originally submitted and made a number of editorial changes, corrections, and improvements, taking most of the referees' constructive comments and suggestions into account. The resulting version is what the reader can see on the following pages. This paper of Carrà Ferro is published exceptionally in order that her unfinished work does not get lost, in memory of her untimely death, and in honor of her numerous contributions to differential algebra and computational algebra. The editors also wish to condole with Alfredo Ferro and his family on Giuseppa's death and to thank him sincerely for providing the information on which the following biography of Carrà Ferro is based.

Giuseppa Carrà Ferro was born on August 1, 1952 in Catania, Italy. She studied Mathematics at the University of Catania, obtained her Master (Laurea) degree there in July 1974, and held positions at the same university from October 1974 to August 1979. She was a visiting researcher, working with Ellis Kolchin, at Columbia University in New York from September 1979 to October 1980. After returning to the University of Catania, she was an assistant professor there until November 1993 and then an associate professor before she became a full professor of algebra in November 2003.

Carrà Ferro is a pioneer in computational differential algebra, an area to which her scientific activities were mostly devoted. She gave the first definition of differential Gröbner bases and designed an algorithm for computing the differential dimension of a differential polynomial ideal even if its basis is infinite. Carrà Ferro's work was much influenced by E. Kolchin's group, into which she brought her knowledge of computer algebra acquired from the strong Italian scientific community.

Author of more than 40 journal and conference papers, Carrà Ferro made original contributions also to several other subjects of research including commutative algebra, algebraic geometry, differential equations, algorithms and computational complexity, graph theory, automated theorem proving in differential geometry, and applications of computer algebra techniques to systems of differential equations for mathematical physics, control theory, robotics, and circuits design.

Radon Series Comp. Appl. Math **2**, 77–108

A Survey on Differential Gröbner Bases

Giuseppa Carrà Ferro

Key words. Gröbner basis, ranking, differential term ordering, differential Gröbner basis.

AMS classification. 12HXX, 13NXX.

1 Introduction

Gröbner bases of polynomial ideals in rings of polynomials with coefficients in a field, and more generally in a unique factorization domain, have a fundamental place in many branches of mathematical research. In fact, Buchberger's algorithm [11] with its optimizations is one of the most studied subjects in computer algebra. This algorithm and its extension to a large class of noncommutative polynomial rings as in Apel and Lassner [1], Kandri Rody and Weispfenning [35], Mora [43], Ufnarovski [77] and Takayama [72] allow one to make effective computations. They have been successfully applied in many other branches of mathematics such as algebraic geometry, automated theorem proving in geometry and control theory. Since a big part of the study of algebraic differential equations is nothing else than the study of polynomials in a finite number of differential variables and their derivatives, the notion of Gröbner basis was extended first to the case of Weyl algebras in Galligo [24] and Takayama [72], and to Lie algebras in Apel and Lassner [1]. Finally, the notion of differential Gröbner basis in Carrà Ferro [14] and the notion of differential standard basis in Ollivier [49] are the extension of such notions to the case of differential polynomial rings with coefficients in a universal differential field. Mansfield and Fackerell [41] also introduced a notion of differential Gröbner basis but their notion differs from the one treated here as it uses only Ritt and Kolchin's notion of reduction and Gröbner bases in the polynomial case.

Every differential ideal generated by linear differential polynomials always has a differential Gröbner basis and this coincides with the characteristic set of the ideal.

Supported by Italian MIUR Prin Project: Algebra Commutativa, Combinatoria e Computazionale.

Furthermore, if differential ideals generated by homogeneous linear differential polynomials are considered, then the differential Gröbner basis is precisely the standard basis of a left $A_m(K)$-modules over the Weyl algebra $A_m(K)$ or standard basis of a set of differential operators, when the corresponding differential operators are considered.

It is well known in algebra that polynomial rings are a particular case of differential polynomial rings and many results from polynomial ring theory and in particular from algebraic geometry do not extend to the differential case. So many results about Gröbner bases in polynomial rings cannot be extended to the differential case. For example, while a polynomial ideal always has a finite Gröbner basis with respect to a fixed term ordering, there exist polynomial differential ideals that do not have any finite differential Gröbner basis with respect to a fixed differential term ordering as shown in Carrá Ferro [14, 49]. Indeed, Zobnin [87, 88] conjectured that there exist polynomial differential ideals that do not admit a finite differential Gröbner basis with respect to any term ordering. So, while the membership problem is always solvable in the algebraic case, in the differential case it is not always solvable using differential Gröbner bases for finitely generated ideals and it is undecidable in the general case (see [23]).

The aim of this survey is to present the known results about differential Gröbner bases. Since the definition of such bases is strongly related to the definition of differential term ordering, recent results about this subject will be presented. In particular, results about classification of rankings in Carrà Ferro and Sit [21], Reid and Rust [59], Rust [59], and recent results on general differential term ordering in Zobnin [86] and about particular classes of differential term orderings in Weispfenning [80] will be presented.

A characterization of differential polynomials f so that $\{f\}$ is a differential Gröbner basis of the differential ideal $[f]$ was shown by the author in [17] and extended in the ordinary case by Zobnin in [87, 88]. Other related results about differential Gröbner bases will also be presented.

2   Gröbner Bases

Here some well-known facts about Gröbner bases of ideals in polynomial rings are introduced.

Let K be a field of characteristic zero, i.e., a field such that the field $\mathbb{Q}$ of rational numbers is contained in K, and let $R = K[Y_1, \ldots, Y_m]$. Let $\mathbb{N}_0 = \{0, 1, \ldots, n, \ldots\}$ and let $T_R = \{Y_1^{a_1} \cdots Y_m^{a_m} : (a_1, \ldots, a_m) \in \mathbb{N}_0^m\}$ be the set of terms of R. If $t = Y_1^{a_1} \cdots Y_m^{a_m} \in T_R$, then the degree of t is $\deg(t) = \sum_{i=1}^m a_i$.

Definition 2.1 A *term ordering* σ on T_R is a total order such that

$$\forall t' \in T_R \setminus \{1\}, \quad t_1 <_\sigma t_2 \Longrightarrow t_1 t' <_\sigma t_2 t' \quad \text{(translation property);} \tag{2.1}$$

$$\forall t \in T_R \setminus \{1\}, \quad 1 <_\sigma t \quad \text{(positivity property).} \tag{2.2}$$

It is well known that every term ordering is a well ordering. Term orderings are studied in Bayer [4] and Robbiano [63]; furthermore they are completely classified in Mora and Robbiano [45].

If the map $\log$ from T_R to $\mathbb{N}_0^m$ is defined by $\log(Y_1^{a_1} \cdots Y_m^{a_m}) = (a_1, \ldots, a_m)$, then it is an order preserving monoid isomorphism. So every term ordering on the first monoid will be identified with a term ordering on the second one.

Let R^* (respectively K^*) be equal to $R \setminus \{0\}$ (respectively $K \setminus \{0\}$) and let $f \in R^*$, $f = \sum_{i=1}^r c_i t_i$, $c_i \in K^*$ and $t_i \in T_R$ for all $i = 1, \ldots, r$.

If σ is a term ordering on T_R and $f \in R$, then $M_\sigma(f)$ is the monomial $c_j t_j$ iff $t_i <_\sigma t_j$ for all $i \neq j$, $i = 1, \ldots, r$, while $T_\sigma(f) = t_j = M_\sigma(f)/c_j$. Moreover, $\mathrm{lc}_\sigma(f) = M_\sigma(f)/T_\sigma(f) = c_j$ and $\mathrm{Supp}(f) = \{t \in T_R : f = \sum ct \text{ and } c \in K^*\}$.

$M_\sigma(f)$ is the *leading monomial* of f, $T_\sigma(f)$ is the *leading term* of f and $\mathrm{lc}_\sigma(f)$ is the *leading coefficient* of f with respect to σ.

Definition 2.2 Let σ be a term ordering on T_R and let I be an ideal in R. Define

$$M_\sigma(I) = (M_\sigma(f) : f \in I), \tag{2.3}$$

$$T_\sigma(I) = (T_\sigma(f) : f \in I). \tag{2.4}$$

Definition 2.3 ([12]) Let $I = (g_1, \ldots, g_s)$ be an ideal in R and let σ be a term ordering on T_R. Then $G = \{g_1, \ldots, g_s\}$ is a Gröbner (or standard) basis of I with respect to σ on T_R iff either

$$M_\sigma(I) = (M_\sigma(g_1), \ldots, M_\sigma(g_s)) \tag{2.5}$$

or

$$T_\sigma(I) = (T_\sigma(g_1), \ldots, T_\sigma(g_s)). \tag{2.6}$$

Definition 2.4 Let $f, g \in R^*$ and let σ be a term ordering on T_R. f is said to be *reduced with respect to* g iff no monomial in f is a multiple of $M_\sigma(g)$. More generally, if $G = \{g_1, \ldots, g_s\} \subseteq R^*$, then f is said to be *reduced with respect to* G iff it is reduced with respect to every g_i, $i = 1, \ldots, s$.

Remark 2.5 Let $f, g \in R^*$. If f is not reduced with respect to g, then there exist polynomials q and r such that $f = qg + r$ and r is reduced with respect to g. In this case, f *reduces to* r *with respect to* g and r is called the *normal form of* f *with respect to* g. More generally, if f is not reduced with respect to G, then there exist polynomials q_i with $i = 1, \ldots, s$ and r such that $f = \sum_{i=1}^s q_i g_i + r$ and r is reduced with respect to G. In this case, f *reduces to* r *with respect to* G and r is called a *normal form of* f *with respect to* G.

Definition 2.6 Let $f, g \in R^*$ and let σ be a term ordering.

$$\mathrm{lc}_\sigma(g) \frac{\mathrm{lcm}(T_\sigma(f), T_\sigma(g))}{T_\sigma(f)} f - \mathrm{lc}_\sigma(f) \frac{\mathrm{lcm}(T_\sigma(f), T_\sigma(g))}{T_\sigma(g)} g \tag{2.7}$$

is the *S-polynomial* of f and g and it will be denoted by $S(f, g)$.

Proposition 2.7 ([4, 11]) $G = \{g_1, \ldots, g_s\}$ *is a Gröbner basis of the ideal I in R with respect to the term ordering σ on T_R iff*

$$\forall i, j = 1, \ldots, s, \quad S(g_i, g_j) \text{ reduces to zero with respect to } G. \tag{2.8}$$

Proposition 2.8 ([4, 11]) *Let I be a nonzero ideal in R and let σ be a term ordering on T_R. There exists a unique set $G = \{g_1, \ldots, g_s\} \subset I \setminus \{0\}$ such that*

$$\forall i = 1, \ldots, r, \quad g_i = t_i + R_i, \quad with \ t_i \in T_R; \tag{2.9}$$

$$\{t_1, \ldots, t_r\} \ minimally \ generates \ M_\sigma(I); \tag{2.10}$$

$$\forall i = 1, \ldots, r, \quad t_i = M_\sigma(g_i); \tag{2.11}$$

$$\forall i = 1, \ldots, r, \quad \mathrm{Supp}(R_i) \cap M_\sigma(I) = \emptyset. \tag{2.12}$$

G is a basis of I and it is called the reduced Gröbner basis *of I with respect to σ.*

Proposition 2.9 *Let I be a nonzero ideal in R and let σ be a term ordering. The following statements are equivalent:*

$$G = \{g_1, \ldots, g_s\} \ is \ a \ Gröbner \ basis \ of \ I \ with \ respect \ to \ a \ term \ ordering \ \sigma; \tag{2.13}$$

$$A \ polynomial \ f \in I \ iff \ f \ reduces \ to \ 0 \ with \ respect \ to \ G; \tag{2.14}$$

$$A \ polynomial \ f \in R^* \ has \ a \ unique \ normal \ form \ with \ respect \ to \ G. \tag{2.15}$$

3 Gröbner Bases and Linear Homogeneous Differential Equations

We assume that everything, which is undefined, is as in Kolchin [36] and Ritt [62].

Let $\Delta = \{\delta_1, \ldots, \delta_m\}$ be a set of derivation operators on a ring R, such that $\delta_i \delta_j = \delta_j \delta_i$ for all $i, j = 1, \ldots, m$.

Let R be a *differential* ring (or Δ-*ring*) of characteristic zero, i.e., a commutative ring with unit and with a set Δ of commuting derivation operators, such that the field $\mathbb{Q}$ of rational numbers is contained in R.

Let $C_R = \{a \in R: \delta_i(a) = 0 \text{ for all } i = 1, \ldots, m\}$ be the *ring of constants* of R and let $R_\Delta = R[\delta_1, \ldots, \delta_n]$ be the Δ-ring of differential operators generated by the set Δ with coefficients in R.

Example 3.1 $R = C^\infty(\mathbb{R}^m)$ and the field of meromorphic functions on a domain of $\mathbb{C}^m$ with the usual derivations $\delta_i = \partial/\partial X_i$, $i = 1, \ldots, m$, are differential rings.

Remark 3.2 $\mathbb{Q} \subseteq C_R$ because $\mathbb{Q} \subseteq R$. If $R \neq C_R$, then R_Δ is not commutative, but rather for all $a \in R$ and all $i = 1, \ldots, m$ the following property is satisfied: $\delta_i a = a\delta_i + \delta_i(a)$.

Let $\Theta = \{\theta^a = \delta_1^{a_1} \cdots \delta_m^{a_m} : a = (a_1, \ldots, a_m) \in \mathbb{N}_0^m\}$ be the monoid of derivatives generated by Δ. Θ can be identified with the set of terms of R_Δ and the map log from Θ to $\mathbb{N}^m$ is an isomorphism of monoids. The order of θ^a is $\mathrm{ord}(\theta^a) = a_1 + \cdots + a_m = |a|$.

3.1   Standard Bases of Left Ideals in the Weyl Algebras

A first extension of the concept of Gröbner basis to the differential case is given by the extension to the case of Weyl algebras, which are very important in microlocal analysis as shown in Maisonobe [40].

Definition 3.3 Let K be a field of characteristic zero. A *Weyl algebra* $A_m(K)$ is the noncommutative polynomial ring with coefficients in K and in the $2m$ variables $Y_1, \ldots, Y_m, D_1, \ldots, D_m$ such that

$$\forall\, i,j = 1, \ldots, m, \quad [D_i, D_j] = [Y_i, Y_j] = 0; \tag{3.1}$$

$$[D_i, Y_j] = \delta_{ij} = \text{ the Kronecker symbol} \tag{3.2}$$

and it will be denoted by $K\langle Y_1, \ldots, Y_m, D_1, \ldots, D_m \rangle$.

Example 3.4 If $D_i = \partial/\partial Y_i$ for all $i = 1, \ldots, m$, then the corresponding Weyl algebra $K\langle Y_1, \ldots, Y_m, \partial/\partial Y_1, \ldots, \partial/\partial Y_m \rangle$ is nothing else than the ring of differential operators with coefficients in $K[Y_1, \ldots, Y_m]$.

In what follows, Y^a denotes $Y_1^{a_1} \cdots Y_m^{a_m}$, while D^b denotes $D_1^{b_1} \cdots D_m^{b_m}$.

It is well known as in Björk [5] that every element of $A_m(K)$ can be written in a unique way as $\sum k_{ab} Y^a D^b$ with $k_{ab} \in K$. Moreover, $A_m(K)$ is a simple ring, i.e., (0) and $A_m(K)$ are the unique two-sided ideals.

Let σ be a term ordering on $\mathbb{N}^{2m}$. It induces a term ordering on the set $LT_{A_m(K)} = \{Y^a D^b : a, b \in \mathbb{N}_0^m\}$ of *left terms* of $A_m(K)$.

If $D = \sum k_{ab} Y^a D^b \in A_m(K)$, then $\mathrm{Supp}(D) = \{Y^a D^b \in LT_{A_m(K)} : k_{ab} \neq 0\}$.

Definition 3.5 Let $D = \sum k_{ab} Y^a D^b$ with $k_{ab} \neq 0$ and let σ be a term ordering on $\mathbb{N}_0^{2m}$.

$$M_\sigma(D) = k_{ab} Y^a D^b \text{ iff } Y^a D^b = \max_\sigma \{Y^a D^b : Y^a D^b \in \mathrm{Supp}(D)\}$$

and $T_\sigma(D) = M_\sigma(D)/k_{ab}$, while $\mathrm{lc}_\sigma(D) = M_\sigma(D)/T_\sigma(D) = k_{ab}$.

$$\exp_\sigma(D) = \max_\sigma \{(a,b) \in \mathbb{N}^{2m} : Y^a D^b \in \mathrm{Supp}(D)\}$$

is the exponential of D with respect to σ.

Of course, $\exp_\sigma(M_\sigma(D)) = \exp_\sigma(T_\sigma(D))$.

Definition 3.6 Let $I = (D_{(1)}, \ldots, D_{(r)})$ be a left ideal in $A_m(K)$. Define

$$E_\sigma(I) = (\exp_\sigma(D) : D \in I) \tag{3.3}$$

and

$$ES_\sigma(I) = \{(a_i, b_i) \in \mathbb{N}_0^{2m} : i = 1, \ldots, r\}, \tag{3.4}$$

where $(a_i, b_i) = \exp_\sigma(D_{(i)})$.

It is not too hard to show that

$$E_\sigma(I) = \bigcup_{i=1}^{r} \{(a_i, b_i) + \mathbb{N}_0^{2m}\}. \tag{3.5}$$

Definition 3.7 Let $I = (D_{(1)}, \ldots, D_{(r)})$ be a left ideal in $A_m(K)$. $\{D_{(1)}, \ldots, D_{(r)}\}$ is a Gröbner basis of I with respect to σ on $LT_{A_m(K)}$ iff

$$E_\sigma(I) = (\exp_\sigma(D_{(1)}), \ldots, \exp_\sigma(D_{(r)})). \tag{3.6}$$

Definition 3.8 Let $D = \sum k_{ab} Y^a D^b$ and $D' = \sum k_{a'b'} Y^{a'} D^{b'}$ be nonzero elements of $A_m(K)$ and let σ be a term ordering. D is said to be *reduced with respect to* D' iff $\exp_\sigma(Y^a D^b) \neq \exp_\sigma(M_\sigma(D')) + c$ for all $c \in \mathbb{N}^{2m}$ and $Y^a D^b \in \mathrm{Supp}(D)$. More generally, D is said to be *reduced with respect to the elements* $D_{(1)}, \ldots, D_{(r)}$ in $A_m(K)$ iff it is reduced with respect to every $D_{(i)}$, $i = 1, \ldots, r$.

Remark 3.9 Let $D, D' \in A_m(K)$ and let σ be a term ordering. If D is not reduced with respect to D', then there exist unique operators Q and R such that $D = QD' + R$ and R is reduced with respect to D'. In this case, D reduces to R with respect to D'.

Proposition 3.10 *The following two statements are equivalent:*

(a) $\{D_{(1)}, \ldots, D_{(r)}\}$ *is a Gröbner basis of the left ideal I in $A_m(K)$ with respect to the term ordering σ;*

(b) *a differential operator D reduces to zero with respect to $D_{(1)}, \ldots, D_{(r)}$ iff $D \in I$.*

It is easy to show that the notion of S-operator and the algorithm to construct the standard basis of a left ideal I are essentially the same as the analogous ones for commutative polynomial rings as shown in Castro [22], Galligo [24], Noumi [46], Oaku [47] and Takayama [72].

Some results about the complexity, when the number m of variables is small, can be extended from the algebraic case to the case of Weyl algebras, while some results about the complexity of the resolution of differential linear systems with polynomial coefficients were obtained in Grigoriev [28, 29, 30, 31].

3.2   Some Preliminaries on Rankings

Now we introduce the notion of rankings. Let $\mathbb{N}_p = \{1, \ldots, p\}$ and let $\mathbb{N}(m, p) = \mathbb{N}_0^m \times \mathbb{N}_p$.

Definition 3.11 A *ranking* on $\mathbb{N}(m, p)$ is a total order O such that

$$\forall a, a', b \in \mathbb{N}_0^m, \ \forall i, j = 1, \ldots, p, \quad (a, i) <_O (a', j) \implies (a + b, i) <_O (a' + b, j)$$

$$\text{(translation property);} \quad (3.7)$$

$$\forall a \in \mathbb{N}_0^m, \quad (0, i) <_O (a, i) \quad \text{(positivity property).} \tag{3.8}$$

It is easy to show that every ranking is a well ordering. Let

$$M = R \oplus R \oplus \cdots \oplus R = R^p = K[Y_1, \ldots, Y_m]^p$$

be the direct sum of p copies of $R = K[Y_1, \ldots, Y_m]$. Let

$$T_M = T_R^p = \{Y_1^{a_1} \cdots Y_n^{a_m} e_i : (a_1, \ldots, a_m) \in \mathbb{N}_0^m \text{ and } i = 1, \ldots, p\}$$

be the set of all terms in M.

The map $\log$ from T_M to $\mathbb{N}(m, p)$ is an isomorphism of direct sum of monoids. By using such isomorphism a ranking is nothing else than a term ordering on the set of the terms of M.

By using this definition of ranking, the definitions of leading monomial, leading term, leading coefficient, reduction, S-polynomial and standard basis extend to M. So the notion of Gröbner basis extends to the case of finitely generated submodules of the free module M as in Bayer [4] and Möller and Mora [42].

Example 3.12 Let σ be a term ordering on $\mathbb{N}_0^m$. Let $(a, i) <_O (a', j)$ when either $a <_\sigma a'$ or $a = a'$ and $i < j$ for all $a \in \mathbb{N}_0^m$ and all $i, j = 1, \ldots, p$. O is a ranking on $\mathbb{N}(m, p)$.

Example 3.13 Let $(a, i) <_O (a', j)$ when either $i < j$ or $i = j$ and $(a, i) <_{\sigma_i} (a', i)$ with respect to some term ordering σ_i on $\mathbb{N}_0^m$ for all $a \in \mathbb{N}_0^m$ and all $i, j = 1, \ldots, p$. O is a ranking on $\mathbb{N}(m, p)$. O is called a *lexicographic* ranking.

Of course, every ranking on $\mathbb{N}(m, p)$ induces a term ordering on each subset $\mathbb{N}_0^m \times i$ with $i = 1, \ldots, p$, that can be identified with $\mathbb{N}_0^m$.

If we consider M as a free $K[Y_1, \ldots, Y_m]$-module, then it can be identified with the set of all linear homogeneous polynomials in the ring $K[Y_1, \ldots, Y_m, Z_1, \ldots, Z_p]$ of polynomials in $m + p$ variables of degree one in the Z_j's.

By using the map $\log$, every term of M can be identified either with the $(m + 1)$-tuple $(a, i) \in \mathbb{N}(m, p)$ or with the $(m + p)$-tuple $(a, 0, \ldots, 0, 1, 0, \ldots, 0)$ with 1 in the $(m + i)$-th place for all $a \in \mathbb{N}_0^m$, furthermore the term ordering $\sigma_i = \sigma_j$ for all $i, j = 1, \ldots, p$ by definition of R-module.

3.3   Standard Bases of Left $A_m(K)$-modules

Let $\mathbb{N}_p = \{1, \ldots, p\}$ and let $\mathbb{N}(m, p) = \mathbb{N}_0^m \times \mathbb{N}_p$.

Let O be a ranking on $\mathbb{N}(2m, p)$. By using this definition of ranking, the definitions of reduction, S-polynomial and standard basis extend to the free left $A_m(K)$-module $A_m(K)^p$ for every p as in Oaku [48], Takayama [73] and Saito, Sturmfels and Takayama [67].

So such notions can be extended to the systems of linear partial homogeneous differential equations in the m independent variables $Y_1, \ldots, Y_m$ and in the dependent variables $Z_1, \ldots, Z_p$, i.e., in the variables $Z_1, \ldots, Z_p$ and with coefficients in $K[Y_1, \ldots, Y_m]$.

Remark 3.14 The existence of Gröbner bases of left ideals of the Weyl algebras and the extension of the Buchberger algorithm are a particular case of such extension. The existence and the algorithm to construct Gröbner bases can be established in the case of finitely generated K-algebras $R = K\langle Y_1, \ldots, Y_m \rangle$ such that $Y_i Y_j - c_{ij} Y_j Y_i = f_{ij}(Y_1, \ldots, Y_m)$ for every $i < j$, some $c_{ij} \in K$ and commuting polynomials f_{ij} as in Mora [43] .

In fact, Weyl algebras are given in the case $c_{ij} = 1$ for each $i < j$ and they are a particular case of the enveloping algebras of Lie algebras, that are given in the case $c_{ij} = 1$ and f_{ij} linear for all $i < j$ as in Apel and Lassner [1] .

These last ones are a particular case of the algebras of resolvable type, that are given in the case $c_{ij} \neq 0$ as in Kandri-Rodi and Weispfenning [35].

A first generalization to the differential case is shown in the following section.

3.4   Differential Operators

As in the case of Weyl algebras, every differential operator $D \in R_\Delta$ can be written in a unique way as $\sum_{|a| \leq d} c_a \theta^a$.

Definition 3.15 ([32]) Let $D = \sum_{|a| \leq d} c_a \theta^a$ be a differential operator in R_Δ. The *order* of D is the least upper bound of the integers $|a|$ such that $c_a \neq 0$.

Given a term ordering σ on $\mathbb{N}_0^m$, $M_\sigma(D)$, $T_\sigma(D)$ and $\mathrm{lc}_\sigma(D)$ can be defined in a unique way as in the case of a commutative polynomial ring in m variables. In a similar way, the concepts of reduction, Gröbner basis and the corresponding Buchberger algorithm can be extended. In some particular cases, for example, when R is the ring of holomorphic functions in one variable, some particular term orderings are considered in Maisonobe [40], while Gröbner bases of differential-difference modules are studied in Zhou and Winkler [85].

As an alternative way, the notions of *pseudo-reduction* in the differential case and *chain* in Ritt, the notion of *autoreduced set* in Kolchin and the notion of *coherent set* in Rosenfeld can be used. In fact, the concepts of autoreduced and coherent subset of a differential ideal and of Gröbner basis coincides in this case, since every differential operator D can be identified with the corresponding linear homogeneous differential polynomial $D(Z)$, that will be introduced in the next section.

Another alternative way is given by the theory of Janet based on the concepts of involutive and formally integrable systems. This theory is based on the notion of *symbol* associated to a system of differential operators

$$\left\{ D_h = \sum_{|a_h| \leq d} c_{a_h} \delta_1^{a_1} \cdots \delta_m^{a_m}, \ h = 1, \ldots, r \right\} \tag{3.9}$$

which is nothing else than the dual of the vector space, defined by the system of linear equations

$$\left\{ P_h = \sum_{|a_h| \leq d} c_{a_h} Y_1^{a_1} \cdots Y_m^{a_m} = 0, \ h = 1, \ldots, r \right\} \tag{3.10}$$

in the vector space of all terms in $\{Y_1, \ldots, Y_m\}$ of degree less than or equal to $d = \max\{\text{order of } D_h\colon h = 1, \ldots, r\}$.

Prolongations of the system until order $d + k$ are considered, in order that the corresponding set of symbols becomes *involutive*, i.e., the dual of the vector space defined by the system

$$\{tP_h = 0 : t \text{ term in } \{Y_1, \ldots, Y_m\}, \deg(t) \le k, h = 1, \ldots, r\} \tag{3.11}$$

in the set of all terms of degree less than or equal to $d + k$ has a required dimension.

Further prolongations of an involutive system make it *formally integrable*, i.e., allow to put the system in the required form for the existence of analytical solutions as in Janet [33], Pommaret [55] and Topunov [76].

The relation between involutive and formally integrable systems and Gröbner bases are studied by many people, for example in Apel [3], Carrà Ferro and Duzhin [18], Gerdt and Blinkov [25, 26], Gerdt [27], Pommaret [56, 57], Reid [58], Schwarz [68], Seiler [71], Zharkov and Blinkov [83], Zharkov [84].

Such notions extend to the free left R_Δ-module R_Δ^p, by using the rankings for every m and p. In other words, they extend to the case of systems of linear homogeneous partial differential equations.

There are still many open problems about the relations between Gröbner bases and either involutive or formally integrable systems and the relations between either the characteristic set or the coherent set for a system and either involutive or formally integrable systems of differential equations.

4   Differential Term Orderings

Let K be a differential field of characteristic zero and let Δ and Θ be as before.

Definition 4.1 $S = K\{Z_1, \ldots, Z_p\} = K[\theta Z_i \colon \theta \in \Theta \text{ and } i = 1, \ldots, p]$ is called the *ring of differential polynomials* in the differential indeterminates $Z_1, \ldots, Z_p$.

$K\{Z_1, \ldots, Z_p\}$ is a commutative polynomial ring in an infinite number of variables. Every algebraic ordinary $(m = 1)$ and partial $(m > 1)$ differential equation can be written as $f = 0$, where f is a differential polynomial.

Remark 4.2 $K[\delta_1, \ldots, \delta_m]^p = K[\delta_1, \ldots, \delta_m]Z_1 \oplus \cdots \oplus K[\delta_1, \ldots, \delta_m]Z_p$ as $K[\delta_1, \ldots, \delta_m]$-module can be identified with the subset of all linear homogeneous differential polynomials in $K\{Z_1, \ldots, Z_p\}$.

Definition 4.3 A *differential monomial* is a monomial

$$m = a(\theta_{11}Z_1)^{a_{11}} \cdots (\theta_{1r(1)}Z_1)^{a_{1r(1)}} \cdots (\theta_{p1}Z_p)^{a_{p1}} \cdots (\theta_{pr(p)}Z_p)^{a_{pr(p)}} \tag{4.1}$$

in $S = K\{Z_1, \ldots, Z_p\}$. A *differential term* is a differential monomial with $a = 1$.

T_S will denote the set of all differential terms in S.

We shall consider only some particular ideals in $K\{Z_1, \ldots, Z_p\}$, that are called differential ideals and correspond to systems of algebraic differential equations in the variables $Z_1, \ldots, Z_p$.

Definition 4.4 An ideal I in $S = K\{Z_1, \ldots, Z_p\}$ is *differential* (or a Δ-*ideal*) iff

$$\forall a \in I, \forall \theta \in \Theta, \quad \theta a \in I. \tag{4.2}$$

If $A \subset S$, then $[A] = (\theta a: a \in A$ and $\theta \in \Theta)$. A differential ideal I is *finitely differentially generated* iff $I = [A]$ for some finite set A.

4.1   Classification of Rankings

In this section, some results about rankings are introduced. Everything which is undefined is as in Carrá Ferro and Sit [21].

Definition 4.5 Let A be an $(m + p) \times (m + p)$-matrix of rank $m + p$ with entries in $\mathbb{R}$. Let $\mu_1, \ldots, \mu_m, \nu_1, \ldots, \nu_p \in \mathbb{Z}$ and suppose that A has $(\mu_1, \ldots, \mu_n, \nu_1, \ldots, \nu_p)^t$ as first column. Let $(a, 0, \ldots, 0, 1, 0, \ldots, 0)$ with 1 in the $(m + i)$-th place and let $(a', 0, \ldots, 0, 1, 0, \ldots, 0)$ with 1 in the $(m + j)$-th place. Define an order $<_O$ on $\mathbb{R}^{m+p}$ by

$$(a, 0, \ldots, 0, 1, 0, \ldots, 0) <_O (a', 0, \ldots, 0, 1, 0, \ldots, 0)$$

iff

$$((a, 0, \ldots, 0, 1, 0, \ldots, 0) \times A)^t <_{\text{lex}} ((a', 0, \ldots, 0, 1, 0, \ldots, 0) \times A)^t.$$

It is easy to show that O is a total order on the set of terms of M and it satisfies property (3.7) of rankings. O is called a *permissible ordering* as in Kolchin [36, p. 72]. O satisfies property (3.8) of rankings, i.e., it is a ranking, when the first column of the matrix A has nonnegative entries. The elements of the first column of A are called *weights*.

The ranking of Example 3.12 is permissible, while the one in Example 3.13 is not permissible when $\sigma_i \neq \sigma_j$ for some i and j.

Permissible rankings were studied in Riquier [60], Janet [33], Kolchin [36] and Thomas [74, 75] in order to have good total orderings compatible with the derivations on the set of derivatives of some variables as will be shown in the next sections.

Definition 4.6 A ranking O on $\mathbb{N}(m, p)$ is said to be *compatibile with the structure of* R-*module on* M (or a *Riquier ranking*) iff there exists a term ordering σ on the set of terms of $R = K[Y_1, \ldots, Y_m]$, such that

$$\forall a, b \in T_R \text{ and } \forall t \in T_M, \quad a <_\sigma b \Longrightarrow at <_O bt; \tag{4.3}$$

$$\forall a \in T_R \text{ and } \forall t, t' \in T_M, \quad t <_O t' \Longrightarrow at <_O at'. \tag{4.4}$$

The ranking of Example 3.12 is compatible with the structure of R-module of M.

On the other hand, there are rankings that are compatible with the structure of R-module of M and they are not permissible, as shown in Carrà Ferro and Sit [13]. Such rankings are determined by $(m+p) \times (m+p)$-matrices B of rank $m+p$ with entries in $\mathbb{R}$, such that the first nonzero element in each row of B is positive as in Reid and Rust [59]. In fact, all such rankings are characterized by inducing the same term ordering σ on each $\mathbb{N}_0^m \times i$. Other results can be found in Caboara and Silvestri [13].

Of course, there are permissible rankings and rankings like the one in Example 3.13 that are not compatible with the structure of R-module of M.

A characterization of rankings and the the notion of cut of an ordered group can be found in Carrà Ferro and Sit [21]. Another characterization of rankings can be found in Reid and Rust [59], where cuts and matrices are used.

There are still many open problems about rankings. For example, it is not known if every ranking on $\mathbb{N}(m, p)$ is a restriction of a ranking on $\mathbb{R}(m, p) = \mathbb{R}^m \times \mathbb{N}_p$. Furthermore, it is still open to determine a characterization of rankings through a minimal number of linear systems of equations and inequations.

4.1.1   Rankings of Variables

Definition 4.7 Let $\{Z_1, \ldots, Z_p\}$ be a set of differential variables. The set V of all variables in $S = K\{Z_1, \ldots, Z_p\}$ is

$$V = \{\theta Z_i \colon \theta \in \Theta \text{ and } i = 1, \ldots, p\}.$$

Definition 4.8 A ranking O is *integrated* iff

$$\forall (\theta Z_{i1}, \theta' Z_{i2}), \, \exists \, \theta'' \in \Theta, \quad \text{such that } \theta'' \theta Z_{i1} >_O \theta' Z_{i2}. \qquad (4.5)$$

A ranking O is *sequential* iff

$$\forall \theta Z_i, \quad \text{the set } \{\theta' Z_j \in V : \theta' Z_j <_O \theta Z_i\} \text{ is finite.} \qquad (4.6)$$

A ranking O is *orderly* if

$$\theta Z_i <_O \theta' Z_j \text{ whenever } \operatorname{ord}(\theta) < \operatorname{ord}(\theta'). \qquad (4.7)$$

It is not too hard to show that each sequential order is integrated. Of course, every orderly ranking is sequential.

Rankings are also studied in Riquier [60], Thomas [74, 75], Reid and Rust [59], Rust [64] and Ovchinnikov and Zobnin [52].

4.2   Differential Term Orderings

Let K be a differential field of characteristic zero and let Δ and Θ be as before. A reasonable definition of differential Gröbner basis of a differential ideal requires the choice of a term ordering σ on the set T_S of differential terms, that must be compatible with the derivations. Of course, the derivative of a differential monomial is not a differential monomial, unless it has degree one.

$V = \{\theta Z_i \colon \theta \in \Theta, \, i = 1, \ldots, p\}$ is the set of all variables in $S = K\{Z_1, \ldots, Z_p\}$. If the map $\log$ from V to $\mathbb{N}(m, p)$ is defined by $\log(\theta Z_i) = \log(\delta_1^{a_1} \cdots \delta_m^{a_m} Z_i) = (a_1, \ldots, a_m, i)$, then it is an isomorphism of a direct sum of monoids. So every ranking O on $\mathbb{N}(m, p)$ determines in a unique way a total order on V compatible with the derivations.

4.2.1   Admissible Orderings

The notion of differential term ordering on the set T_S was first introduced by Carrà Ferro [14] and Ollivier [49] and then by Weispfenning [80] in a more restricted way. A more general one was introduced recently in Zobnin [86].

First we introduce the more general definition of admissible ordering given by Zobnin [86].

Definition 4.9 ([86]) An *admissible ordering* σ on T_S is a total order such that

$$\forall t \in T_S \setminus \{1\}, \quad 1 <_\sigma t \quad \text{(positivity property)}; \tag{4.8}$$

$$\forall t' \in T_S \setminus \{1\}, \quad t_1 <_\sigma t_2 \implies t_1 t' <_\sigma t_2 t' \quad \text{(translation property)}; \tag{4.9}$$

$$\text{the restriction of } \sigma \text{ to } V \text{ is a ranking (restriction property)}. \tag{4.10}$$

In Zobnin [86] it is proved that every admissible ordering is a well order.

If σ is an admissible ordering on T_S and $f \in S$, then $M_\sigma(f)$, $T_\sigma(f)$, $\mathrm{lc}_\sigma(f)$ and $\mathrm{Supp}(f)$ are defined as in the polynomial case.

Example 4.10 Given a ranking O on the set of variables, the corresponding lexicographic, deglexicographic, degrevlexicographic and weighted-degrevlexicographic orderings are admissible orderings.

Definition 4.11 ([38]) Let $m = 1$ and let $t = \prod_{i=0}^{k} (\delta^i Z_1)^{\alpha_i}$. t is called an α-term with respect to an integer s if $a_{i-1} + \alpha_i < s$ for all $i = 1, \ldots, k$. If a monomial is not an α-term, then it is said to be a β-term.

An admissible ordering σ such that the leading term $T_\sigma(\delta^m (\delta^k Z_1)^q)$ is a β-term for all q and $k \geq 0$ is a pure β-ordering.

Example 4.12 The degrevlexicographic and weighted-degrevlexicographic orderings are pure β-orderings.

Definition 4.13 ([87, 88]) An admissible ordering σ is said to be Δ-*stable* iff

$$\forall t, t' \in T_S, \, \forall j = 1, \ldots, m, \quad t <_\sigma t' \implies T_\sigma(\delta_j t) \leq_\sigma T_\sigma(\delta_j t'). \tag{4.11}$$

An admissible ordering σ is said to be *strictly* Δ-*stable* iff

$$\forall t, t' \in T_S, \, \forall j = 1, \ldots, m, \quad t <_\sigma t' \implies T_\sigma(\delta_j t) <_\sigma T_\sigma(\delta_j t'). \tag{4.12}$$

Example 4.14 The lexicographic and deglexicographic orderings are strictly Δ-stable. The degrevlexicographic and weighted-revlexicographic orderings are Δ-stable but they are not strictly Δ-stable. In fact, if $p = m = 1$, $\delta = \delta_1$ and $<$ is the degrevlexicographic ordering, then $(\delta^i(Z_1))^2 > \delta^{i-1}(Z_1)\delta^{i+1}(Z_1)$ but

$$\delta(\delta^i(Z_1))^2 = 2\delta^i(Z_1)\delta^{i+1}(Z_1) > \delta(\delta^{i-1}(Z_1)\delta^{i+1}(Z_1)).$$

Definition 4.15 ([88]) Let $m = 1$ and let $\delta = \delta_1$. An admissible ordering σ is said to be δ-*lexicographic* if it satisfies one of the following equivalent conditions:

$$\forall \text{ term } t \neq 1, \quad T_\sigma(\delta t) = T_{\text{lex}}(t); \tag{4.13}$$

$$\forall 0 < i \leq j, \quad \delta^i Z_1 \delta^j Z_1 <_\sigma \delta^{i-1} Z_1 \delta^{j-1} Z_1; \tag{4.14}$$

$$\forall t \neq 1, \quad \text{all terms in } \delta t \text{ are compared lexicographically.} \tag{4.15}$$

Remark 4.16 The lexicographic and deglexicographic orderings are δ-lexicographic. No β-ordering (e.g., an ordering σ where $T_\sigma(\delta t) = T_{\text{revlex}}(\delta t)$ for all terms t) is δ-lexicographic. Any strictly δ-stable ordering is δ-lexicographic.

4.2.2 Differential Term Orderings

The first definition of differential term ordering on T_S, i.e., a term ordering compatible with derivatives, is the following one.

Definition 4.17 ([14, 49]) A *differential term ordering* σ on T_S is a total order such that

$$\forall t, t', t'' \in T_S, \quad t <_\sigma t' \implies t''t <_\sigma t''t' \quad \text{(translation property)}; \tag{4.16}$$

$$\forall t \in T_S \setminus \{1\}, \quad 1 <_\sigma t \quad \text{(positivity property)}; \tag{4.17}$$

$$\forall t \in T_S, \forall \theta \in \Theta, \quad t <_\sigma T_\sigma(\theta t) \quad \text{(domination of differentiation)}; \tag{4.18}$$

$$\forall t, t' \in T_S, \forall \theta \in \Theta, \quad t <_\sigma t' \implies T_\sigma(\theta t) <_\sigma T_\sigma(\theta t'). \tag{4.19}$$

Remark 4.18 (4.18) and (4.19) in Definition 4.17 are obvious conditions. (4.18) and (4.19) imply condition (4.10), but there are admissible term orderings in which (4.18) and (4.19) do not hold. For example, (4.19) is equivalent to the strong Δ-stability (4.12). (4.10) implies (4.18) as in Zobnin [87, 88].

Remark 4.18 clarifies the definition of differential term ordering given in Weispfenning [80], where conditions (4.18) and (4.19) are substituted by the following conditions.

Definition 4.19 ([80])

$$\forall i, j = 1, \ldots, p, \forall \theta, \theta', \theta'' \in \Theta, \quad \theta Z_i <_\sigma \theta' Z_j \implies \theta'' \theta Z_i <_\sigma \theta'' \theta' Z_j; \tag{4.20}$$

$$\forall i, j = 1, \ldots, p, \forall \theta \in \Theta, \quad Z_i <_\sigma \theta Z_j; \tag{4.21}$$

$$\forall\, i,j = 1,\ldots,p,\ \forall\,\theta \in \Theta, \quad \theta Z_i <_\sigma \theta' Z_j \iff \theta Z_1 <_\sigma \theta' Z_1; \qquad (4.22)$$

$$\forall\, i,j = 1,\ldots,p,\ \forall\,\theta,\theta' \in \Theta,\ \forall\, t \in T_S \setminus \{1\},$$
$$\theta Z_i <_\sigma \theta' Z_j \quad \text{and} \quad t <_\sigma \theta' Z_j \implies \theta Z_i t <_\sigma \theta' Z_j. \qquad (4.23)$$

Remark 4.20 Property (4.21) in Definition 4.19 of Weispfenning implies that a differential term ordering on T_S induces a ranking O on the set V, that is a Riquier ranking. Properties (4.21), (4.22) and (4.23) imply that the derivatives determine the ordering on monomials in a much stronger way than multiplication. Furthermore, they allow to show that σ is a term ordering and they provide a characterization of all differential term orderings by systems of good linear forms with coefficients in $\mathbb{R}[\alpha]$. The property (4.18) with the properties of an admissible ordering implies the property (4.21).

The definition of differential term ordering as in Carrà Ferro [14] and Ollivier [49] implies that the induced ranking on V is not a Riquier ranking in general, as will be shown in the next subsection.

Remark 4.21 Any admissible ordering σ that is an extension of a ranking does not satisfy property (4.22). In fact, if $Z_1 <_\sigma Z_2$ and $\theta_1 Z_1 <_\sigma \theta_2 Z_2$, then $\theta_2 Z_1 <_\sigma \theta_1 Z_2$. By (4.22) we should have $\theta_2 Z_1 <_\sigma \theta_1 Z_1$.

The deglexicographic term ordering does not satisfy (4.23). In fact, if X and Y are variables with $X < Y$, then if $U = X$, we have $UX = X^2 > Y$, that contradicts (4.23).

We have also the following definition in Weispfenning [80].

Definition 4.22 ([80]) A differential term ordering σ is *lexicographic with respect to multiplication* iff

$$\forall\, i = 1,\ldots,p,\ \forall\,\theta \in \Theta,\ \forall\, t \in T_S, \quad t <_\sigma \theta Z_i \implies t^2 <_\sigma \theta Z_i. \qquad (4.24)$$

Example 4.23 Let $m = 1$ and $p = 2$. If

$$t = Z_1^{a_{10}}(\delta Z_1)^{a_{11}} \cdots (\delta^{r(1)} Z_1)^{a_{1r(1)}} Z_2^{a_{20}}(\delta Z_2)^{a_{21}} \cdots (\delta^{r(2)} Z_2)^{a_{2r(2)}}$$

and

$$t' = Z_1^{a_{b0}}(\delta Z_1)^{a_{b1}} \cdots (\delta^{r(1)} Z_1)^{b_{1r(1)}} Z_2^{b_{20}}(\delta Z_2)^{b_{21}} \cdots (\delta^{r(2)} Z_2)^{b_{2r(2)}},$$

then

$$t <_\sigma t' \iff (a_{2r(2)},\ldots,a_{20},a_{1r(1)},\ldots,a_{10}) < (b_{2r(2)},\ldots,b_{20},b_{1r(1)},\ldots,b_{10})$$

with respect to the lexicographic order. σ is a differential term ordering as in Definition 4.19 and lexicographic with respect to the product.

Now we introduce a differential term ordering, whenever a ranking O is known.

Differential Term Orderings Compatible with Rankings

Let O be a ranking on the set $V = \{\theta Z_i : \theta \in \Theta, i = 1, \dots, p\}$.

Definition 4.24 A differential term ordering σ on T_S is compatible with the ranking O on V iff $\theta Z_i <_\sigma \theta' Z_j$ whenever $\theta Z_i <_O \theta' Z_j$ for all $\theta, \theta' \in \Theta$ and $i, j = 1, \dots, p$.

If $k \geq 0$, then let $\Theta_k = \{\theta \in \Theta : \mathrm{ord}(\theta) \leq k\}$, $S_k = K[\theta Z_j : \theta \in \Theta_k, j = 1, \dots, p]$ and let T_{S_k} be the corresponding set of terms. Of course, $T_{S_k} = T_S \cap S_k$.

The following definition is given in Carrà Ferro [14] and Ollivier [49].

Definition 4.25 A differential term ordering σ on T_S is called the *canonical differential term ordering associated with O* iff for every $k \in \mathbb{N}_0$:

$$\forall \theta, \theta' \in \Theta_k, \ \forall i, j = 1, \dots, p, \quad \theta Z_i <_\sigma \theta' Z_j \ \text{whenever} \ \theta Z_i <_O \theta' Z_j; \qquad (4.25)$$

$$\text{the restriction } \sigma_k \text{ of } \sigma \text{ to } T_{S_k} \text{ is a term ordering on } T_{S_k}; \qquad (4.26)$$

$$\forall r \in \mathbb{N}_0, \quad \text{the restriction of } \sigma_{k+r} \text{ to } T_{S_k} \text{ is equal to } \sigma_k. \qquad (4.27)$$

Every σ_k is nothing else than the inverse lexicographic order on T_{S_k} defined by

$$\prod_{l=1}^{s} \theta_l Z_{j(l)}^{a_l} <_{\sigma_k} \prod_{l=1}^{s} \theta_l Z_{j(l)}^{b_l}$$

with $\theta_s Z_{j(s)} >_O \theta_{s-1} Z_{j(s-1)} >_O \cdots >_O \theta_1 Z_{j(1)}$ iff $(a_s, \dots, a_1) < (b_s, \dots, b_1)$ with respect to the lexicographic order.

Remark 4.26 Every differential term ordering as in Carrà Ferro [14], Ollivier [49] and [80] is a well ordering.

Remark 4.27 The canonical differential term ordering σ associated with O is not the unique differential term ordering on T_S compatible with O and having the properties (4.18) and (4.19) of the differential term orderings. For example, if we define $t <_\tau t'$ iff either the degree of t is less than the degree of t' or t and t' have the same degree and $t <_\sigma t'$, then τ is a term ordering compatible with O as in Ollivier [49, lemma 5]. On the other hand, if $m = p = 1$, then there is only one ranking O and σ is the unique differential term ordering on T_S following the definition given in Weispfenning [80]. Finally, σ is lexicographic with respect to multiplication by its own definition.

Example 4.28 Let $m = p = 2$ and let O be the lexicographic ranking on $\mathbb{N}(2, 2)$ as in Example 3.13, where σ_1 is the lexicographic order with $(1, 0) >_{\sigma_1} (0, 1)$ and σ_2 is the lexicographic order with $(0, 1) >_{\sigma_2} (1, 0)$, i.e., $Z_1 <_O \delta_2 Z_1 <_O \delta_1 Z_1 <_O Z_2 <_O \delta_1 Z_2 <_O \delta_2 Z_2$. Let σ be the canonical differential term ordering on T_S as in Carrà Ferro [14] and Ollivier [49]. Then σ is not a differential term ordering as in Weispfenning's definition, since property (4.21) is not satisfied.

Example 4.29 Let $m = 1$ and $p = 2$. If

$$t = Z_1^{a_{10}}(\delta Z_1)^{a_{11}} \cdots (\delta^{1r(1)} Z_1)^{a_{1r(1)}} Z_2^{a_{20}}(\delta Z_2)^{a_{21}} \cdots (\delta^{2r(2)} Z_2)^{a_{2r(2)}}$$

and

$$t' = Z_1^{a_{b0}}(\delta Z_1)^{a_{b1}} \cdots (\delta^{1r(1)} Z_1)^{b_{1r(1)}} Z_2^{b_{20}}(\delta Z_2)^{b_{21}} \cdots (\delta^{2r(2)} Z_2)^{b_{2r(2)}},$$

then $t <_\sigma t'$ iff

$$(a_{2r(2)}, \ldots, a_{20}, a_{1r(1)}, \ldots, a_{10}) < (b_{2r(2)}, \ldots, b_{20}, b_{1r(1)}, \ldots, b_{10})$$

with respect to the lexicographic order. σ is a differential term ordering in the sense of Weispfenning and lexicographic with respect to the product.

5 Differential Gröbner Bases

Here the notion and some important properties of differential Gröbner and differential standard bases are introduced. More details can be found in Carrà Ferro [14], Ollivier [49], Pankratiev [53, 54] and Zobnin [87, 88].

5.1 Definition of Differential Gröbner Basis

Now the definition of differential Gröbner basis is given by trying to extend the corresponding property in the polynomial case. If σ is a differential term ordering on T_S and $f \in S$, then $M_\sigma(f)$, $T_\sigma(f)$, $\mathrm{lc}_\sigma(f)$ and $\mathrm{Supp}(f)$ are defined as before.

Definition 5.1 Let I be a Δ-ideal in S and let σ be a differential term ordering on T_S. Define

$$M_\sigma(I) = (M_\sigma(\theta f) : f \in I, \theta \in \Theta) \tag{5.1}$$

and

$$T_\sigma(I) = (T_\sigma(\theta f) : f \in I, \theta \in \Theta). \tag{5.2}$$

Differential Gröbner bases were first introduced in [14] when σ is the canonical differential term ordering compatible with a ranking O, while differential standard bases were first introduced in [49] when σ is a term ordering associated to a ranking O.

Definition 5.2 ([14, 49]) Let $I = [g_1, \ldots, g_s]$ be a Δ-ideal in S. $\{f_j : j \in J\}$ is a *differential standard basis* of I with respect to a differential term ordering σ on T_S iff one of the following equivalent statements holds:

$$M_\sigma(I) = (M_\sigma(\theta f_j) : \theta \in \Theta, j \in J); \tag{5.3}$$

$$T_\sigma(I) = (T_\sigma(\theta f_j) : \theta \in \Theta, j \in J). \tag{5.4}$$

If J is finite, i.e., $J = \{1, \ldots, r\}$, then $\{f_1, \ldots, f_r\}$ is a *differential Gröbner basis* of I.

Remark 5.3 The definition of differential standard basis can be extended to the case when σ is an admissible ordering on T_S as in Zobnin [87, 88].

Definition 5.4 ([61]) Let $f \in K\{Z_1, \ldots, Z_p\} \setminus K$ and let O be a ranking. The highest ranking derivative θZ_j present in f is called the *leader* of f and it is denoted by u_f. If $d = \deg_{u_f} f$, then the coefficient of u_f^d in f is called the *initial* of f and it is denoted by I_f. $S_f = \partial f / \partial u_f$ is called the *separant* of f. The *order* of f is $\mathrm{ord}(f) = \mathrm{ord}(u_f)$.

Definition 5.5 ([36, 61]) Let $g \in K\{Z_1, \ldots, Z_p\} \setminus K$. A differential polynomial $f \in K\{Z_1, \ldots, Z_p\}$ is said to be *R-reduced* with respect to g if f is free of every proper derivative of u_g and $\deg_{u_g} f < \deg_{u_g} g$. More generally, if $G = \{g_1, \ldots, g_s\}$, f is said to be *R-reduced* with respect to G if f is R-reduced with respect to g_i for all $i = 1, \ldots, s$.

Definition 5.6 Let $f, g \in S \setminus \{0\}$ and let σ be an admissible ordering on T_S. f is said to be Δ-*reduced* with respect to g iff no monomial in f is a multiple of some $M_\sigma(\theta g)$, for all $\theta \in \Theta$. More generally, if $G = \{g_1, \ldots, g_s\}$ are in $S \setminus \{0\}$, f is said to be Δ-*reduced* with respect to G iff it is Δ-reduced with respect to every g_i, $i = 1, \ldots, r$.

Remark 5.7 The R-reduction implies the Δ-reduction, when σ is the canonical differential term ordering associated with the ranking O. The converse is not true. In fact, θZ_j is Δ-reduced with respect to the monomial $Z \theta Z_j$ for all θ and j, but it is not R-reduced.

Remark 5.8 Let $f, g \in S \setminus \{0\}$ and let σ be an admissible ordering on T_S. If f is not Δ-reduced with respect to g, then there exist differential polynomials $q_1, \ldots, q_t$ and r such that $f = \sum_{i=1}^{t} q_i \theta_i g + r$ and r is Δ-reduced with respect to g. In this case, we say that f Δ-reduces to r with respect to g and r is called the *normal form of f with respect to g*. More generally, if $G = \{g_1, \ldots, g_s\}$ and f is not Δ-reduced with respect to G, then there exist differential polynomials $q_{1(j)}, \ldots, q_{t(j)}$ with $j = 1, \ldots, s$ and r such that $f = \sum_{j=1}^{s} \sum_{l=1}^{t(j)} q_l \theta_l g_j + r$ and r is Δ-reduced with respect to G. In this case, we say that f Δ-reduces to r with respect to G and r is called a *normal form of f with respect to G*. The existence of the differential polynomial r follows from the finiteness of the reduction algorithm as in the polynomial case, because an admissible ordering is a well ordering.

Proposition 5.9 *Let I be a Δ-ideal in S and let σ be a differential term ordering on T_S. $\{f_1, \ldots, f_r\}$ is a differential Gröbner basis of I with respect to σ iff*

$$\forall \theta, \theta' \in \Theta, \, \forall i, j = 1, \ldots, r,$$

$$S(\theta f_i, \theta' f_j) \ \Delta\text{-reduces to zero with respect to } \ f_1, \ldots, f_r. \tag{5.5}$$

Proposition 5.10 *Let I be a Δ-ideal in S and let σ be a differential term ordering on T_S. If I has a differential Gröbner basis with respect to σ, then there exists a unique set $\{f_1, \ldots, f_r\} \subset I \setminus \{0\}$ such that*

$$\forall i = 1, \ldots, r, \quad f_i = t_i + R_i, \text{ with } t_i \in T_S; \tag{5.6}$$

$$\{M_\sigma(\theta t_1), \ldots, M_\sigma(\theta t_r) : \theta \in \Theta\} \quad \text{minimally generates } M_\sigma(I); \qquad (5.7)$$

$$\forall i = 1, \ldots, r, \quad t_i = M_\sigma(f_i); \qquad (5.8)$$

$$\forall i = 1, \ldots, r, \quad \text{Supp}(R_i) \cap M_\sigma(I) = \emptyset; \qquad (5.9)$$

$$\forall \theta \in \Theta, \forall j \neq i, \quad f_i \neq \theta f_j. \qquad (5.10)$$

$\{f_1, \ldots, f_r\}$ *is a differential Gröbner basis of I and it is called the* reduced differential Gröbner basis *of I with respect to* σ.

We have similar propositions in the case of differential standard bases, i.e., infinite differential Gröbner bases.

Remark 5.11 A differential ideal always has a differential standard basis with respect to a differential term ordering σ associated with ranking O and more generally with respect to an admissible ordering. On the other hand, there are many examples of differential ideals that do not have a differential Gröbner basis with respect to any differential term ordering σ associated with a ranking O, as will be shown in the next section.

Proposition 5.12 *Let I be a nonzero ideal in R and let σ be a differential term ordering. The following statements are equivalent:*

$F = \{f_1, \ldots, f_r\}$ *is a differential Gröbner basis of I with respect to* σ; (5.11)

A differential polynomial $f \in I$ iff f Δ-reduces to 0 with respect to F; (5.12)

A differential polynomial $f \in R^*$ has a unique normal form with respect to F. (5.13)

Proof. The proof is the same as in the algebraic case, since F is finite and σ is a well ordering. $\square$

Proposition 5.12 implies that the membership problem of a differential polynomial to a differential ideal has a solution whenever there exists a differential Gröbner basis with respect to some admissible ordering. Unfortunately, the nonexistence of such bases leaves the problem still open in the case of nonradical differential ideals.

Another approach to the study of ideals of differential polynomials is the study of the *characteristic set* of a differential ideal with respect to a fixed ranking O as in the Wu-Ritt theory as in Ritt [61, 62], Wu [81, 82] and Wang [78, 79].

Such tools were used recently for solving the membership problem in the case of radical differential ideals, by using the Gröbner-Rosenfeld techniques in Boulier [7, 8], Boulier, Lazard, Ollivier and Petitot [6], Bouziane, Kandri Rody and Maarouf [9, 10], Maarouf, Kandri Rody and Ssafini [39], Hubert [34], Ovchinnikov [51] and Sadik [65, 66] and isobaric polynomials in Gallo, Mishra and Ollivier [23].

Remark 5.13 Let $I = [f_1, \ldots, f_r]$ be a Δ-ideal in S with f_i linear for all $i = 1, \ldots, r$ and let σ be the canonical differential term ordering on T_S associated to a ranking O. I always has a differential Gröbner basis G with respect to σ and G is also a characteristic set of I with respect to O. In fact, the linearity of the differential polynomials f_i's implies that the notion of R-reduction coincides with the notion of Δ-reduction.

By using a Gröbner-Ritt-Kolchin variant of the Ritt-Kolchin algorithm, that substitutes in this algorithm the notion of R-reduction with the notion of reduction as in the definition of Buchberger, the notion of differential Gröbner basis was introduced in Mansfield and Fackerell [41] in the following way. Of course, this notion is not equal to the notion of differential Gröbner basis as in Definition 5.2 before.

Definition 5.14 ([41]) Let K be a differential field and let O be a ranking on the set of variables $V = \{\theta Z_i \colon \theta \in \Theta$ and $i = 1, \ldots, p\}$. Let F be a finite set of differential polynomials in $K\{Z_1, \ldots, Z_p\}$ and let I be a differential ideal such that $F \subseteq I$. Let G be the output of the Gröbner-Ritt-Kolchin algorithm with input F. G is a differential Gröbner basis if every $f \in I$ R-reduces to zero with respect to G.

Recently, in Zobnin [87, 88] it is shown that in the case $m = p = 1$ the differential ideal $[Z_1^2]$ has $[Z_1^2]$ as differential Gröbner basis with respect to the degrevlexicographic β-ordering induced by the unique ranking O on the set of variables $\{Z_1, \delta Z_1, \ldots\}$, while it has no finite standard differential Gröbner basis with respect to any differential term ordering induced by O as in Carrà Ferro [14] and Ollivier [49].

In Zobnin [87, 88], it is conjectured that in the case $m = p = 1$ (e.g., the ordinary case in one differential variable) the differential ideal $[Z_1 \delta Z_1]$ has no differential Gröbner basis.

5.2 Existence of Differential Gröbner Bases

5.2.1 Differential Gröbner Bases of $[f]$

Here some results about differential Gröbner bases of differential ideals generated by only one differential polynomial are shown. Moreover, some definitions in the ordinary case (e.g., $m = 1$) as in Zobnin [87, 88] are introduced, in order to extend some result.

Definition 5.15 ([88]) Let $m = 1$. An admissible ordering σ on T_S is δ_1-*fixed* if for every $f \in K\{Z_1\} \setminus K$ there exist $t \in T_S$, $k, r \in \mathbb{N}_0$, such that $T_\sigma(f) = t\delta_1^{r+k} Z_1$ for all $k \geq k_0$.

Remark 5.16 Any δ_1-lexicographic admissible ordering is δ_1-fixed as shown in Zobnin [87, 88].

Definition 5.17 Let $m = 1$ and let σ be an admissible ordering on T_S. A differential polynomial f is σ-*quasi-linear* if $\deg(T_\sigma(f)) = 1$.

Example 5.18 $f = Z_1^2 - \delta_1 Z_1$ is quasi-linear with respect to the lexicographic order, e.g., the canonical differential term ordering associated to the unique orderly ranking on the set $V = \{\delta_1^h Z_1 : h \in \mathbb{N}_0\}$.

Definition 5.19 Let $m = 1$ and let σ be an admissible ordering on T_S. σ is *concordant with quasi-linearity* if the derivative of a quasi-linear differential polynomial is quasi-linear.

Remark 5.20 Lexicographic, deglexicographic and degrevlexicographic orderings are concordant with quasi-linearity. More generally, any δ_1-lexicographic admissible ordering is concordant with quasi-linearity as in Zobnin [88].

Let $K^* = K \setminus \{0\}$. Let O be a ranking on $\mathbb{N}(m,p)$ and let σ be the associated canonical differential term ordering on T_S.

All lemmas and theorems in this subsection were proved in Carrà Ferro [17]. Their proofs hold in the more general case of differential polynomials in p differential variables $Z_1, \ldots, Z_p$, whenever a canonical differential term ordering associated with a ranking O is considered, because the proofs depend only on the leaders of the differential polynomials.

Lemma 5.21 *Let* $f = \sum_{h=0}^{d} I_h u_f^h \in K\{Z_1, \ldots, Z_p\}$. *Then* $S(f, \delta_i f)$ *and* $S(\delta_i f, \delta_j f)$ *reduce to zero with respect to* $f, \delta_1 f, \ldots, \delta_m f$ *for all* $i, j = 1, \ldots, m$ *iff*

$$f = I_d(A + u_f)^d,$$

where $I_d \in K^*$ *and* A *is* Δ-*reduced with respect to* u_f.

Lemma 5.22 *Let* $f = a(A + u_f)^d \in K\{Z_1, \ldots, Z_p\}$, *where* A *is* Δ-*reduced with respect to* u_f *and* $a \in K^*$. *Then* $S(\delta_i f, \delta_i^2 f)$ Δ-*reduces to zero with respect to* $\{\theta f: \operatorname{ord}(\theta) \leq 2\}$ *iff* $d = 1$.

Theorem 5.23 *Let* $f \in K\{Z_1, \ldots, Z_p\}$. *The differential polynomial* f *is a differential Gröbner basis of the differential ideal* $[f]$ *with respect to the canonical differential term ordering* σ *on* T_S *associated to the ranking* O *on* $\mathbb{N}(m,p)$ *iff*

$$f = I_0 + I_1 u_f,$$

with $I_1 \in K^*$.

Remark 5.24 Theorem 5.23 implies that f is a differential Gröbner basis of the ideal $[f]$ with respect to the canonical differential term ordering σ associated to a ranking iff f is quasi-linear with respect to σ, e.g., f has the normal form in the sense of mathematical analysis.

Furthermore, Theorem 5.23 implies that the differential ideal $[f]$ is prime, because f is irreducible and its initial is an element of the differential field K as in Kolchin [36].

Remark 5.25 There are examples of differential polynomials f, such that they satisfy conditions of Lemma 5.21, but f is not the differential Gröbner basis of $[f]$, see for example $f = Z_1^2$.

In order to characterize the differential Gröbner bases, we need some preliminary facts about the differential dimension of a differential ideal.

5.2.2 The Differential Dimension of a Differential Ideal

Here we introduce some facts about the differential dimension. Let $\Delta = \{\delta_1, \ldots, \delta_m\}$ and let K be a universal Δ-field.

Roughly speaking, the differential dimension of a differential ideal I is either the maximal number d of variables, such that no differential polynomial in these variables lies in I, or in terms of the corresponding system of partial differential equations the number of arbitrary functions in m variables, that appear in the general solution of the system.

Definition 5.26 If P is a prime differential ideal in $K\{Z_1, \ldots, Z_p\}$, then the Δ-$\dim_K P$ is equal to the Δ-tr.degree of q.f.(S/P) over K. If I is a differential ideal in $K\{Z_1, \ldots, Z_p\}$, then Δ-$\dim_K I = \max\{\Delta$-$\dim_K P$: P minimal prime differential ideal over $I\}$.

Since a universal Δ-field is differentially algebraically closed, the definition above is equivalent to the following definition.

Definition 5.27 Let K be a differentially algebraically closed field and let I be a differential ideal in $K\{Z_1, \ldots, Z_p\}$. Then Δ-$\dim_K I = \max\{h : K\{Z_{i1}, \ldots, Z_{ih}\} \cap I = (0)\}$.

If I is a differential ideal, then

$$\forall k \in \mathbb{N}_0, \quad I(k) = I \cap K[Z_1, \ldots, Z_p, \ldots, \theta Z_1, \ldots, \theta Z_p : \mathrm{ord}(\theta) \leq k]. \quad (5.14)$$

It is well known that if P is a prime differential ideal in $S = K\{Z_1, \ldots, Z_p\}$ with Δ-$\dim_K P = d$ and O is an orderly ranking on $\mathbb{N}(m, p)$, then $\dim_K P(t) = h_\Delta(P)(t)$ is a numerical function, the so-called *differential dimension Hilbert function of P*. Furthermore, there exists $t_0 \in \mathbb{N}_0$ such that

$$h_\Delta(P)(t) = H_\Delta(P)(t) = d\binom{t+m}{m} + \sum_{j=0}^{\tau} a_j \binom{t+j}{j}$$

for all $t \geq t_0$, where $d = -1, 0, \ldots, m-1$ for every $P \neq 0$, $\tau \leq m-1$, $a_j \in \mathbb{Z}$ for all j and a_τ is a positive integer.

$H_\Delta(P)(t)$ is called the *differential dimension polynomial of P*, d is called the *differential dimension of P*, τ is called the *Δ-type of P* and a_τ is called the *typical differential dimension of P*.

Remark 5.28 a_τ is the number of arbitrary functions in τ variables appearing in the general solution of the system of partial differential equations corresponding to the differential ideal P. If $p = 1$, i.e., we have systems of partial differential equations in only one differential variable Z, then Δ-$\dim_K P = 0$.

It is possible to define in the same way the differential dimension Hilbert function and the differential Hilbert polynomial of a differential ideal I according to [36, Proposition 5, p. 170].

Other properties and results about differential dimension polynomials can be found in Kondrateva, Levin, Mikhalev and Pankratev [37].

While the differential dimension d of P is invariant up to differential isomorphisms of S/P, the polynomial $H_\Delta(P)(t)$ is invariant up to linear transformations of the derivation operators $\delta_1, \ldots, \delta_m$, $i = 1, \ldots, m$, over the field of constants.

By [36, Theorem 6, p. 115], given a prime differential ideal P in S and an orderly ranking on $\mathbb{N}(m, p)$, if $\{f_1, \ldots, f_r\}$ is a characteristic set of P with respect to O, then for each $t \in \mathbb{N}_0$

$$h_\Delta(P)(t) = \sharp\{(v_1, \ldots, v_m) \in \mathbb{N}_0^m : \delta_1^{v_1} \cdots \delta_m^{v_m} Z_j < u_{f_h} \text{ with respect to the product}$$
$$\text{order for all } j = 1, \ldots, p \text{ and some } h = 1, \ldots, r \text{ with } \textstyle\sum_{i=1}^m v_i \le t\}.$$

So, if

$$E_j(t) = \sharp\{\log(\theta) \colon \theta Z_j <_{f_h} \text{ with respect to the product}$$
$$\text{order for some } h = 1, \ldots, r \text{ and } \operatorname{ord}(\theta) \le t\}$$

for all $j = 1, \ldots, p$, then $h_\Delta(P)(t) = \sum_{j=1}^p E_j(t)$.

The proof of the theorem as above is based on [36, Proposition 7, p. 100 and Proposition 8, p. 101], which are still true when $\mathbb{Q} \subseteq K$, i.e., char. $K = 0$ and O is a ranking on $(Z_1, \ldots, Z_p)$. In fact, Proposition 7 is true when O is a ranking on $\mathbb{N}(m, p)$, as pointed out on p. 100 there. The proof of Proposition 8 is still true when O is not an orderly ranking, because it depends on the degree of the Hilbert polynomial of a finitely generated $K[\delta_1, \ldots, \delta_m]$-module. Since orderly rankings are nothing else than degree preserving term orderings on such $K[\delta_1, \ldots, \delta_m]$-modules and the degree of the Hilbert polynomial with respect to a fixed term ordering of an ideal in $K[\delta_1, \ldots, \delta_m]$ is independent of the term ordering by [16], the proof of Proposition 8 in [36] is true for all rankings.

The following lemma in Carrá Ferro [17] shows a relation among the classical notion of a characteristic set of a differential ideal with respect to a ranking O and a differential Gröbner basis of the same ideal.

Lemma 5.29 *Let O be a ranking and let $\{g_1, \ldots, g_s\}$ be the reduced differential Gröbner basis of a differential ideal I with respect to the canonical differential term ordering σ associated with O. Let $\{f_1, \ldots, f_r\}$ with $u_{f_1} <_O \cdots <_O u_{f_r}$ be a characteristic set of I with respect to O. Then for each $i = 1, \ldots, r$ there exists a $g_{j(i)}$, such that $u_{g_{j(i)}} = u_{f_i}$.*

5.2.3 Characterization of Differential Gröbner Bases

Here a characterization of differential Gröbner bases is shown. This follows from results in Carrà Ferro [15, 16, 17] and some extensions in Zobnin [87, 88] from the ordinary case.

The following theorem is proved in [17].

Theorem 5.30 *Let $p = 1$. Let O be a ranking and let $\{g_1, \ldots, g_s\}$ be a set of differential polynomials, such that $\operatorname{ord}(u_{g_1}) \le \cdots \le \operatorname{ord}(u_{g_s}) = k_0$. If $\{g_1, \ldots, g_s\}$ is the* *reduced differential Gröbner basis of a differential ideal I with respect to the canonical differential term ordering σ associated with the ranking O, then*

$$\forall i \neq j, \forall \theta \in \Theta, \quad g_j \neq \theta g_i; \tag{5.15}$$

$\{\theta_i g_i : \theta_i \in \Theta, \mathrm{ord}(\theta_i g_i) \leq k, i = 1, \ldots, s\}$ *is the reduced Gröbner basis of*

$$(\theta_i g_i : \theta_i \in \Theta, \mathrm{ord}(\theta_i g_i) \leq k, i = 1, \ldots, s) \ \text{with respect to} \ \sigma_k; \tag{5.16}$$

there exists at least one $g_j = I_{j0} + I_{j1}\theta_j(Z_1)$, with

$$I_{j0} \ \text{R-reduced with respect to} \ u_{g_j} = \theta_j Z \ \text{and} \ I_{j1} \in K^*. \tag{5.17}$$

The following examples are very useful.

Example 5.31 Let $f = (\delta Z_1)^3 - 3Z_1$. Then

$$\{(\delta Z_1)^3 - 3Z_1, (\delta Z_1)^2\delta^2 Z_1 - \delta Z_1, 3Z_1\delta^2 Z_1 - (\delta Z_1)^2, \delta Z_1(\delta^2 Z_1)^2 - \delta^2 Z_1, \delta^3 Z_1 + (\delta^2 Z_1)^3\}$$

is a differential Gröbner basis of $[f]$ with respect to σ, and

$$\{(\delta Z_1)^3 - 3Z, 3Z_1\delta^2 Z_1 - (\delta Z_1)^2, \delta Z_1(\delta^2 Z_1)^2 - \delta^2 Z_1, \delta^3 Z_1 + (\delta^2 Z_1)^3\}$$

is the reduced differential Gröbner basis with respect to σ.

Example 5.32 Let $f = \delta_1 Z_1 \delta_2 Z_1 - 1$ and let $\delta_2 Z_1 >_\sigma \delta_1 Z_1$. Then

$$\{\delta_1 Z_1 \delta_2 Z_1 - 1, \delta_1 Z_1 \delta_1 \delta_2 Z_1 + \delta_1^2 Z_1 \delta_2 Z_1, \delta_1 Z_1 \delta_2^2 Z_1 + \delta_1 \delta_2 Z_1 \delta_2 Z_1,$$
$$\delta_1 \delta_2 Z_1 + \delta_1^2 Z_1 (\delta_2 Z_1)^2, \delta_2^2 Z_1 + \delta_1 \delta_2 Z_1 (\delta_2 Z_1)^2\}$$

is a differential Gröbner basis of $[f]$ with respect to σ, and

$$\{\delta_1 Z_1 \delta_2 Z_1 - 1, \delta_1 \delta_2 Z_1 + \delta_1^2 Z_1 (\delta_2 Z_1)^2, \delta_2^2 Z_1 - \delta_1^2 Z_1 (\delta_2 Z_1)^4\}$$

is the reduced differential Gröbner basis with respect to σ.

Theorem 5.33 *Let $m = p = 1$ and let $\{g_1, \ldots, g_s\}$ be the reduced differential Gröbner basis of a differential ideal I with respect to the canonical differential term ordering σ associated with O. If $\mathrm{ord}(g_1) \leq \cdots \leq \mathrm{ord}(g_s) = k_0$ and a_0 is the typical differential dimension of I, then $a_0 = \dim_K I(k)$ for all $k \geq k_0$.*

Proof. By (5.15), (5.16) and (5.17) of Theorem 5.30, we have

$$I(k)\backslash I(k-1) = \{\delta^{k-k_0} g_s\} \quad \text{and} \quad M_\sigma(I(k))\backslash M_\sigma(I(k-1)) = \{\delta^k Z_1\}$$

for all $k \geq k_0$. It follows that the Hilbert polynomial $H(I(k))(t)$ of the ideal $I(k)$ coincides with the Hilbert polynomial $H(I(k+1))(t)$ for all $t \geq t_0$. In particular, they have the same degree and then by [16] $I(k)$ and $I(k-1)$ have the same dimension, since the degree of the Hilbert polynomial of the monomial ideal corresponding to a Gröbner basis of a polynomial ideal coincides with its dimension. On the other hand, since $m = p = 1$, $\dim_K I(s) = h_\Delta(I)(s) = a_0$ for all $s \geq s_0$, being $\Delta\text{-}\dim_K I = 0$. It follows that $a_0 = \dim_K I(k_0)$ and $s_0 = k_0$. $\qquad\square$

The following theorems in Zobnin [87, 88] are extensions of the above results in the ordinary case $m = p = 1$.

Theorem 5.34 *Let $m = p = 1$ and let σ be a δ_1-fixed admissible ordering. If a differential ideal I has a differential Gröbner basis with respect to σ, then it contains a quasi-linear differential polynomial.*

Theorem 5.35 *Let $m = p = 1$ and let σ be an admissible ordering that is concordant with quasi-linearity. If a differential ideal I contains a quasi-linear differential polynomial, then it has a differential Gröbner basis with respect to σ.*

The above theorems give a characterization of differential Gröbner bases in the case $m = p = 1$ when the admissible term ordering σ is δ_1-fixed and concordant with quasi-linearity, for example when σ is δ_1-lexicographic.

Corollary 5.36 *Let $m = p = 1$ and let σ be a δ_1-lexicographic admissible ordering. A differential ideal I has a differential Gröbner basis with respect to σ iff it contains a quasi-linear differential polynomial.*

Remark 5.37 There are examples of differential polynomials f, such that they do not have any differential Gröbner basis with respect to any differential term ordering as in Definition 4.17 but they have a differential Gröbner basis with respect to the degrevlexicographic order associated to a ranking O as in Zobnin [87, 88].

Example 5.38 Let $m = p = 1$. The differential ideal $[Z_1^2]$ has no differential Gröbner basis with respect to any differential term ordering on the set of differential terms of $S = K\{Z_1\}$. Z_1^2 is the differential Gröbner basis with respect to the revlexicographic admissible ordering associated to the unique ranking O on the set $V = \{\delta^h Z_1 : h \in \mathbb{N}_0\}$.

Corollary 5.39 *Let f be a differential polynomial such that the degree of each monomial in f is greater than one. If O is a ranking, then $[f]$ has no differential Gröbner basis with respect to the canonical differential term ordering σ associated with O.*

The counterexample in Gallo, Mishra and Ollivier [23] can be considered as consequence of the above corollary.

Remark 5.40 The statement of the above corollary does not give sufficient conditions on the existence of differential Gröbner bases as in the following example.

Example 5.41 Let $f = Z_1^2 - 2Z_1 + 1 = (Z_1 - 1)^2$. Then

$$\delta_i f = 2Z_1 \delta_i Z_1 - 2\delta_i Z_1,$$
$$\delta_i^2 f = 2Z_1 \delta_i^2 Z_1 + 2(\delta_i Z_1)^2 - 2\delta_i^2 Z_1,$$
$$S(\delta_i f, \delta_i^2 f) = (2\delta_i Z_1)^3$$

for all $i = 1, \ldots, m$. Since $[2(\delta_i Z_1)^3]$ has no differential Gröbner basis and $[2(\delta_i Z_1)^3] \subset [f]$ for all $i = 1, \ldots, m$, $[f]$ has no differential Gröbner basis.

The results in Corollary 5.39 are extended in the following corollaries to a more general case in Zobnin [87, 88].

Corollary 5.42 *Let* $m = 1$ *and let* σ *be a* δ_1-*fixed admissible ordering. Let* $I = [f_1, \ldots, f_r]$ *be a differential ideal. If the degree of each term in* $f_1, \ldots, f_r$ *is greater than* 1, *then* I *has no differential Gröbner basis with respect to* σ.

Corollary 5.43 *Let* $m = 1$, *let* σ *be a strongly* δ_1-*stable admissible ordering and let* $f \in K\{Z_1\}$. *The reduced differential Gröbner basis of* $[f]$ *with respect to* σ *consists of* f *itself iff* f *is a quasi-linear differential polynomial.*

5.2.4 Sufficient Conditions for the Existence of Differential Gröbner Bases

Here we show some sufficient conditions for the existence of a differential Gröbner basis related to some properties of the differential polynomials.

First of all we introduce the separant ideal, as in Zobnin [87, 88].

Definition 5.44 *Let* σ *be an admissible ordering and let* I *be a differential ideal in* $K\{Z_1, \ldots, Z_p\}$. *The* ideal of separants of I *is the ideal* $S_I = (S_f : f \in I)$, *where the separant* S_f *is as defined in Definition 5.4.*

Theorem 5.45 *Let* $m = 1$ *and let* σ *be an admissible ordering. Let* I *be a differential ideal. Then* $S_I = 1$ *iff* I *contains a quasi-linear polynomial.*

The above theorem gives a characterization of a differential Gröbner bases in the case $m = p = 1$ whenever σ is δ_1-lexicographic, for example if σ is the canonical differential term ordering associated to a ranking O.

Corollary 5.46 *Let* $m = 1$, *let* σ *be a* δ_1-*lexicographic admissible ordering and let* I *be a differential ideal.* I *has a differential Gröbner basis with respect to* σ *iff* $S_I = 1$.

The following theorem and corollary show some sufficient conditions for the existence of differential Gröbner bases.

Theorem 5.47 *Let* $m = 1$. *Let* O *be a ranking and let* $f = I_0 + I_1 u_f$. *Then* $[f]$ *has a reduced differential Gröbner basis with respect to the canonical differential term ordering* σ *associated with* O *if* $(I_0, I_1) = 1$.

Proof. Note that $f = I_0 + I_1 u_f$ implies $\delta f = \delta(I_0) + \delta(I_1)u_f + I_1\delta u_f$. Let $a = \mathrm{lc}_\sigma(f)$. Then $a = \mathrm{lc}_\sigma(I_1)$ by its own definition and thus $a = \mathrm{lc}_\sigma(\delta f)$. Let

$$g = S(f, \delta f) = \delta u_f f - u_f \delta f = I_0 \delta u_f - \delta(I_0)u_f - \delta(I_1)u_f^2.$$

If $(I_0, I_1) = 1$, then there exist $A, B \in K\{Z_1\}$, such that $AI_0 + BI_1 = 1$. The differential polynomial

$$h = Ag + B\delta f = \delta u_f - A\delta(I_1)u_f^2 + (B\delta(I_1) - A\delta(I_0))u_f + B\delta(I_0) \in [f].$$

Since $S_h = 1$, $S_{[f]} = 1$ and $[f]$ has a reduced differential Gröbner basis with respect to the canonical differential term ordering σ associated with O by Corollary 5.46. It is easy to show that $\delta f - I_1 h$ is a multiple of f and thus it δ-reduces to zero with respect to f. So $[f, h]$ is a reduced differential Gröbner basis of $[f]$ with respect to the canonical differential term ordering σ associated with O. □

Corollary 5.48 *Let $m = 1$. Let O be a ranking, let $f = \sum_{j=0}^d I_j u_f$ and let $\delta f = \sum_{j=0}^d \delta(I_j) u_f + S_f \delta u_f$. $[f]$ has a reduced differential Gröbner basis with respect to the canonical differential term ordering σ associated with O if $(\sum_{j=0}^d \delta(I_j) u_f, S_f) = 1$.*

Proof. This follows by Theorem 5.47. □

Once again the conditions of the previous theorem and corollary are only sufficient for the existence of a differential Gröbner basis of $[f]$.

Example 5.49 Let $m = p = 1$ and let $f = Z_1^2 - Z_1$. Then

$$\delta f = 2Z_1 \delta Z_1 - \delta Z_1, \quad S(f, \delta f) = -Z_1 \delta Z_1 \quad \text{and} \quad h = \delta f + 2S(f, \delta f) = -\delta Z_1.$$

So $[f] = [Z_1^2 - Z_1, \delta Z_1]$ and $\{Z_1^2 - Z_1, \delta Z_1\}$ is the reduced differential Gröbner basis of $[f]$.

5.3   Algorithms for Differential Gröbner Bases

Let $I = [f_1, \ldots, f_r]$ be a finitely differentially generated differential ideal in the differential ring $S = K\{Z_1, \ldots, Z_p\}$. In [14] the following recursive algorithm was introduced for the calculation of a differential Gröbner basis, when the canonical differential term ordering σ associated to a ranking O is considered.

Let $I_k = (\theta f_j: \operatorname{ord}(\theta f_j) \le k$ and $j = 1, \ldots, r)$ in

$$S_k = K[Z_1, \ldots, Z_p, \ldots, \theta Z_1, \ldots, \theta Z_p: \operatorname{ord}(\theta) \le k]$$

for all $k \in \mathbb{N}_0$. Let $\{g_1, \ldots, g_{s(0)}, g_{s(0)+1}, \ldots, g_{s(1)}, \ldots, g_{s(k-1)+1}, \ldots, g_{s(k)}\}$ be the reduced Gröbner basis of I_k with respect to the term ordering σ_k for all k. Then $I_k \setminus I_{k-1} = \{g_{s(k-1)+1}, \ldots, g_{s(k)}\}$.

Given I_k as above, the following algorithm finds a new ideal $D(I_k)$ in the same polynomial ring, which is complete with respect to the derivatives.

The Interior Operator Algorithm D

Input: $m, p, k, O, \sigma, I = [f_1, \ldots, f_r]$.

Output: $D(I_k)$.

Step 1. Let I_k be $(\theta f_j: \operatorname{ord}(\theta f_j) \le k$ and $j = 1, \ldots, r)$.

Step 2. Let $\{g_1, \ldots, g_{s(k)}\}$ be the reduced Gröbner basis of I_k with respect to σ_k.

Step 3. Compute θg_j such that $\operatorname{ord}(\theta g_j) \le k$ for all $j = 1, \ldots, s(k)$.

Step 4. Put $(I_k)_1 := (I_k, \theta g_j : \text{ord}(\theta g_j) \le k)$ Find the reduced Gröbner basis of $(I_k)_1$ with respect to σ_k and check whether $I_k = (I_k)_1$. If the latter holds, then return $D(I_k) = I_k$ and stop; else go to step 5;

Step 5. Put $I_k := (I_k)_1$ and go to step 3.

Remark 5.50 The correctness and finiteness of the above algorithm is shown in [14]. The number of loops depends on the degrees of the input differential polynomials and the order k and bounds for this number are still unknown.

The ideal $D(I_k)$ is complete with respect to derivatives until order k, since it contains all derivatives of differential polynomials in it until order k and it can be found by using any known algorithm for Gröbner bases of polynomial ideals.

Note that $I_k \subseteq D(I_k) \subseteq I(k)$ for all k by its own definition.

The following theorem shows a relation between the operator D and the differential Gröbner bases.

Theorem 5.51 *Let* $I = [f_1, \ldots, f_r]$ *be a differential ideal in* $K\{Z_1, \ldots, Z_p\}$. *Let* σ *be the canonical differential term ordering associated with a ranking* O. *Let* $k_0 = \max\{\text{ord}(f_j): j = 1, \ldots, r\}$ *and let* $k \ge k_0 + 1$. *Let*

$$D(I_k) = (h_1, \ldots, h_{t(0)}, \ldots, h_{t(k-1)}, h_{t(k-1)+1}, \ldots, h_{t(k)}),$$

with $D(I_k) \cap S_{k-1} = (h_1, \ldots, h_{t(0)}, \ldots, h_{t(k-1)})$ *and* $h_j = I_{oj} + I_{1j}u_{h_j}$ *with* $I_{1j} \in K^*$ *for all* $j = t(k-1)+1, \ldots, t(k)$. *Let* $u_{h_j} = \theta_j Z_{l(j)}$ *for all* j *and suppose that* $S(\theta_{kj'}h_j, \theta_{kj}h_{j'})$ Δ-reduces to zero with respect to

$$\{h_1, \ldots, h_{t(0)}, \ldots, h_{t(k-1)}, h_{t(k-1)+1}, \ldots, h_{t(k)}\},$$

whenever $l(j) = l(j')$ *and* $\theta_{kj'}, \theta_{kj'}$ *are the derivatives of minimal order for which* $\theta_{kj'}\theta_j = \theta_{kj}\theta_{j'}$. *Then* $\{h_1, \ldots, h_{t(0)}, \ldots, h_{t(k-1)}, h_{t(k-1)+1}, \ldots, h_{t(k)}\}$ *is a differential Gröbner basis of the differential ideal* I *with respect to* σ.

Proof. Let $k_0 = \max\{\text{ord}(f_j): j = 1, \ldots, r\}$ and let $k \ge k_0 + 1$. Then

$$I = [D(I_k)] = J = [h_1, \ldots, h_{t(0)}, \ldots, h_{t(k-1)}, h_{t(k-1)+1}, \ldots, h_{t(k)}],$$

by definition of I and $D(I_k)$. Moreover, $J_k = D(I_k)$, while

$$J_{k+1} = (D(I_k), \delta_i h_j : i = 1, \ldots, p, \ j = t(k-1)+1, \ldots, t(k)).$$

If u_{h_j} and $u_{h'_j}$ are derivatives of distinct differential variables Z_j's, then they are coprime by hypothesis and by definition of reduced Gröbner basis. It follows that $u_{\delta_i h_j}$ and $u_{\delta_{i'}h'_j}$ are coprime and $S(\delta_i h_j, \delta_{i'}h'_j)$ reduces to zero. Furthermore, $S(h_j, \delta_{i'}h'_j)$ reduces to zero for all $j = 1, \ldots, t(k)$ and $j' = t(k-1)+1, \ldots, t(k)$, because their leading terms are coprime. If $u_{h_j} = \theta Z_{l(j)}$ and $u_{h'_j} = \theta_{j'}Z_{l(j)}$, then either $\delta_i u_{h_j}$ and $\delta_{i'}u_{h'_j}$ are coprime or not. If they are coprime, we can repeat the above proof. If they are not coprime, then $\delta_i u_{h_j} = \delta_i \theta_j Z_{l(j)} = \delta_{i'} u_{h'_j} = \delta_{i'}\theta_{j'}Z_{l(j)}$. By hypothesis, $S(\delta_i u_{h_j}, \delta_{i'}u_{h'_j})$ Δ-reduces to zero with respect to $\{h_1, \ldots, h_{t(0)}, \ldots, h_{t(k-1)}, h_{t(k-1)+1}, \ldots, h_{t(k)}\}$ and

$S(h_j, \delta_{i'} h'_j)$ reduces to zero as above for all j, j'. So $J_{k+1} = D(J_{k+1})$. Since the same proof can be repeated in the case of

$$J_{k+h}(D(I_k), \theta h_j : \text{ord}(\theta) \leq h, j = t(k-1)+1, \ldots, t(k))$$

for all $h \in \mathbb{N}$, $J_{k+h} = D(J_{k+h})$ for all such h and the theorem is proved.   □

In order to find a differential Gröbner basis, we can use the above algorithm recursively.

A Differential Gröbner Basis Algorithm

Input: $m, p, O, \sigma, I = [f_1, \ldots, f_r]$.

Output: a differential Gröbner basis of I.

Step 1. Let $k_0 := \max\{\text{ord}(f_j) : j = 1, \ldots, r\}$ and let $k := k_0 + 1$.

Step 2. Compute $D(I_k) = (h_1, \ldots, h_{t(0)}, \ldots, h_{t(k-1)}, h_{t(k-1)+1}, \ldots, h_{t(k)})$ and check whether $h_{t(k-1)+1}, \ldots, h_{t(k)}$ satisfy Theorem 5.51. If the latter holds, then return $G := \{h_1, \ldots, h_{t(0)}, \ldots, h_{t(k-1)}, h_{t(k-1)+1}, \ldots, h_{t(k)}\}$ and stop; else go to step 3.

Step 3. Put $k := k + 1$ and go to step 2.

If the above algorithm stops after a finite number of loops, then we have a differential Gröbner basis of the ideal I.

If the ideal I has a differential Gröbner basis $G = \{g_1, \ldots, g_s\}$, that we can always suppose reduced, then $\{g_1, \ldots, g_s\} \subseteq D(I_{k_0+h})$ for all $h \geq h_0$ and

$$I(k_0 + h) = (\theta g_j : \text{ord}(\theta g_j) \leq k_0 + h, j = 1, \ldots, s) \subseteq D(I_{k_0+h}) \subseteq I(k_0 + h),$$

i.e.,
$$D(I_{k_0+h}) = (\theta g_j : \text{ord}(\theta g_j) \leq k_0 + h, j = 1, \ldots, s)$$

and $\{\theta g_j : \text{ord}(\theta g_j) \leq k_0 + h, j = 1, \ldots, s\}$ is the reduced Gröbner basis of $D(I_{k_0+h})$ with respect to σ_{k_0+h}. So the algorithm stops again after a finite number of loops when $p = 1$ by Theorem 5.30.

If the ideal I has no finite differential Gröbner basis, then the algorithm does not stop.

It is still an open problem to extend Theorem 5.30 to the general case $m \geq 1$ and $p \geq 1$, which is equivalent to characterizing a differential Gröbner basis by the sufficient condition of Theorem 5.51.

Another completion algorithm can be found in Ollivier [49], where an optimization strategy is shown. Almost nothing is known about the complexity of such algorithms.

Bibliography

[1] J. Apel and W. Lassner, *An extension of Buchberger's algorithm and calculations in envelopping fields of Lie algebras*, J. Symb. Comput. 6 (1988), pp. 361–370.

[2] J. Apel and W. Lassner, *Computation of reduced Gröbner bases and syzigies in envelopping algebras*, Proc. SYMSAC '86, 1986.

[3] J. Apel, *A Gröbner approach to involutive bases*, J. Symb. Comput. 6 (1995), pp. 441–457.

[4] D. Bayer, *The Division Algorithm and the Hilbert Scheme*, Ph.D. Thesis, Harvard University, 1982.

[5] J. Björk, *Rings of Differential Operators*, Amsterdam, North Holland Math. Library, vol. 21, Amsterdam, 1979.

[6] F. Boulier, D. Lazard, F. Ollivier and M. Petitot, *Representation for the radical of a finitely generated differential ideal*, Proc. ISSAC '95, ACM Press, New York, 1995, pp. 158–166.

[7] F. Boulier, *Étude et implantation de quelques algorithmes en algèbre différentielle*, Ph.D. Thesis, Universite de Lille, 1994.

[8] F. Boulier, *Some improvements of a lemma of Rosenfeld*, Preprint, Proc. IMACS '96, 1996.

[9] D. Bouziane, A. Kandri-Rody and H. Maarouf, *Computing representations for radicals of finitely generated differential ideals*, submitted to J. Symb. Comput., 1998.

[10] D. Bouziane, A. Kandri-Rody and H. Maarouf, *Unmixed-dimensional decomposition of a finitely generated perfect differential ideals*, J. Symb. Comput. 31 (2001), pp. 631–649.

[11] B. Buchberger, *A theoretical basis for the reduction of polynomials to canonical form*, ACM SIGSAM Bull. 10(3) (1976), pp. 19–29.

[12] B. Buchberger, *Some properties of Gröbner bases for polynomial ideals*, ACM SIGSAM Bull. 10 (1976), pp. 19–24.

[13] M. Caboara and M. Silvestri, *Compatible module orderings*, ISSAC '96 Poster Session Abstracts, 1996, pp. 17–22.

[14] G. Carrá Ferro, *Gröbner bases and differential algebra*, LNCS 356, Springer, 1987, pp. 129–140.

[15] G. Carrá Ferro, *Some remark on the differential dimension*, LNCS 357, Springer, 1988, pp. 152–163.

[16] G. Carrá Ferro, *Gröbner bases and Hilbert schemes*, Computational Aspects of Commutative Algebra, Academic Press, London, 1989, pp. 85–96.

[17] G. Carrá Ferro, *Differential Gröbner bases in one variable and in the partial case*, Mathematical Computing Modelling, Pergamon Press, 25 (1997), pp. 1–10.

[18] G. Carrá Ferro and S. Duzhin, *Differential algebraic and differential geometric approach to the study of involutive symbols*, Proc. Modern Group Analysis: Advanced Analytical and Computational Methods (Catania, 1992), Kluwer, 1993, pp. 93–99.

[19] G. Carrá Ferro and V. Gerdt, *Extended characteristic sets of finitely generated differential ideals*, Computer Algebra in Scientific Computing (CASC, V. Ganzha, E. Mayr and E. Vorozhtsov, eds.), 2002, pp. 29–36.

[20] G. Carrá Ferro and V. P. Gerdt, *An improved Kolchin-Ritt algorithm*, Programmirovanie 2 (2003), pp. 35–40.

[21] G. Carrá Ferro and W. Sit, *On term orderings and rankings*, Computational Algebra, Marcel Dekker, New York, 1994, pp. 31–77.

[22] F. Castro, *Calculus effectifs pour les ideaux d'operateurs differentiells*, Geometrie Algebrique et Application III Travaux en Cours, 24, 1987, pp. 1–19.

[23] G. Gallo, B. Mishra and F. Ollivier, *Some constructions in rings of differential polynomials*, LNCS 539, Springer, 1991, pp. 171–182.

[24] A. Galligo, *Some algorithmic questions on ideals of differential operators*, LNCS 204, Springer, 1985, pp. 413–421.

[25] V. P. Gerdt and Y. A. Blinkov, *Involutive bases of polynomial ideals*, Math. Comp. Simul. 45 (1998), pp. 519–542.

[26] V. P. Gerdt and Y. A. Blinkov, *Minimal involutive bases*, Math. Comp. Simul. 45 (1998), pp. 543–560.

[27] V. P. Gerdt, *Gröbner bases and involutive methods for algebraic and differential equations*, Computer Algebra in Science and Engineering, World Scientific, Singapore, 1995, pp. 117–137.

[28] D. Grigoriev, *Complexity of quantifier elimination in the theory of ordinary differential equations*, LNCS 378, Springer, 1989, pp. 11–25.

[29] D. Grigoriev, *Complexity of irreducibility testing for a system of linear ordinary differential equations*, Proc. ISSAC '90, ACM Press, New York, 1990, pp. 225–230.

[30] D. Grigoriev, *Complexity of solving systems of linear equations over the ring of differential operators*, Proc. MEGA '90, Birkhäuser, Boston, 1991, pp. 195–202.

[31] D. Grigoriev, NC solving a system of linear ordinary differential equations in several unknowns, Preprint, 1994.

[32] L. Hormander, *Linear Partial Differential Operators*, Springer, Berlin, 1963.

[33] M. Janet, *Sur les systèmes d'équations aux dérivées partielles*, J. Math. 3 (1920), pp. 65–151.

[34] E. Hubert, *Essential components of an algebraic differential equation*, J. Symb. Comput. 28 (1999), pp. 657–680.

[35] A. Kandri Rody and V. Weispfenning, *Noncommutative Gröbner bases in algebras of solvable type*, J. Symb. Comput. 9 (1990), pp. 1–26.

[36] E. Kolchin, *Differential Algebra and Algebraic Groups*, Academic Press, New York, 1973.

[37] M. V. Kondrateva, A. B. Levin, A. V. Mikhalev and E. V. Pankratev, *Differential and Difference Dimension Polynomial*, Kluwer, Dordrecht, 1999.

[38] H. Levi, *On the structure of differential polynomials and on their theory of ideals*, Trans. AMS 51 (1942), pp. 532–568.

[39] H. Maarouf, A. Kandri Rody and M. Ssafini, *Triviality and dimension of a system of algebraic differential equations*, J. Automated Reasoning 20 (1998), pp. 365–385.

[40] P. Maisonobe, *D-modules: an overview towards effectivity*, Computer Algebra and Differential Equations, Cambridge Univ. Press, 1994, pp. 21–56.

[41] E. L. Mansfield and E. D. Fackerell, *Differential Gröbner bases*, Preprint, Macquarie University, 1992.

[42] H. Möller and T. Mora, *New constructive methods in classical ideal theory*, J. Algebra 100 (1986), pp. 138–178.

[43] T. Mora, *Seven variations on Gröbner bases*, Preprint, 1988.

[44] T. Mora, *Gröbner bases in non-commutative algebras*, Proc. ISSAC '88, LNCS 358, Springer, 1988, pp. 150–161.

[45] T. Mora and L. Robbiano, *The Gröbner fan of an ideal*, J. Symb. Comput. 6 (1988) pp. 183–208.

[46] N. Noumi, *Wronskian determinants and the Gröbner representation of a linear differential equation*, Algebraic Analysis, Academic Press, Boston, 1988, pp. 549–569.

[47] T. Oaku, *Computation of the characteristic variety and the singular locus of a system of differential equations with polynomial coefficients*, Japan J. Indust. Appl. Math. 11 (1994), pp. 485–497.

[48] T. Oaku, *Gröbner bases for D-modules on a non-singular affine algebraic variety*, Tohoku Math. J. 48 (1996), pp. 575–600.

[49] F. Ollivier, *Standard bases of differential ideals*, LNCS 508, Springer, 1990, pp. 304–321.

[50] O. J. Ore, *Formale Theorie der linearen Differentialgleichungen I*, J. reine angew. Math. 167 (1932), pp. 221–234.

[51] A. Ovchinnikov, *Characterizable radical differential ideals and some properties of characteristic sets*, Programming and Computer Software 30 (2004), pp. 141–149.

[52] A. Ovchinnikov and A. Zobnin, *Classification and applications of monomial orderings and the properties of differential orderings*, Proc. CASC '02 (V. G. Ganzha, E. W. Mayr and E. V. Vorozhtsov, eds.), 2002, pp. 237–252.

[53] E. V. Pankratiev, *Some approaches to construction of standard bases in commutative and differential algebra*, Proc. CASC '02 (V. G. Ganzha, E. W. Mayr and E. V. Vorozhtsov, eds.), 2002, pp. 265–268.

[54] E. V. Pankratiev, *Some approaches to construction of differential Groebner bases*, Proc. Calculemus '02, 2002, pp. 50–55.

[55] F. Pommaret, *Differential Galois Theory*, Gordon and Breach, New York, 1985.

[56] F. Pommaret, *Effective methods for systems of algebraic partial differential equations*, Proc. MEGA '90, Birkhäuser, Boston, 1990, pp. 411–426.

[57] F. Pommaret, *Partial Differential Equations and Group Theory: New Perspective for Applications*, Kluwer, Dordrecht, 1994.

[58] G. J. Reid, *Algorithms for reducing a system of PDE's to standard form, determining the dimension of its solution space and calculating its Taylor series solution*, Eur. J. Appl. Math. 2 (1991), pp. 293–318.

[59] G. J. Reid and C. Rust, *Rankings of partial derivatives*, Proc. ISSAC '97, ACM Press, New York, 1997, pp. 9–16.

[60] C. H. Riquier, *Les systèmes d'équations aux dérivées partielles*, Gauthier-Villars, Paris, 1910.

[61] J. F. Ritt, *Differential Equations from the Algebraic Standpoint*, AMS Coll. Publ. 14, New York, 1932.

[62] J. F. Ritt, *Differential Algebra*, AMS Coll. Publ. 33, New York, 1950.

[63] L. Robbiano, *Term orderings on the polynomial rings*, Proc. EUROCAL '85, LNCS 204, Springer, 1985, pp. 513–517.

[64] C. Rust, *Rankings of Derivatives for Elimination Algorithms and Formal Solvability of Analytic Partial Differential Equations*, Ph.D. Thesis, University of Chicago, 1998.

[65] B. Sadik, *The complexity of formal resolution of linear partial differential equations*, Proc. AAECC-11 (Paris, 1995), LNCS 948, Springer, 1995, pp. 408–414.

[66] B. Sadik, *A bound for the order of characteristic set elements of an ordinary prime differential ideal and some applications*, Appl. Alg. Eng. Comm. Comp. 10 (2000), pp. 251–268.

[67] M. Saito, B. Sturmfels and N. Takayama, *Gröbner Deformations of Hypergeometric Differential Equations*, Algorithms and Computation in Mathematics, vol. 6, Springer, Berlin, 2000.

[68] F. Schwarz, *Monomial orderings and Gröbner bases*, ACM SIGSAM Bull. 25 (1991), pp. 10–23.

[69] F. Schwarz, *An algorithm for determining the size of symmetry groups*, Computing 49 (1992), pp. 95–115.

[70] F. Schwarz, *Reduction and completion algorithms for partial differential equations*, Proc. IS-SAC '92, ACM Press, New York, 1992, pp. 49–56.

[71] W. M. Seiler, *Indices and solvability for general systems of differential equations*, Computer Algebra in Scientific Computing (CASC '99, V. Ganzha, E. Mayr and E. Vorozhtsov, eds.), Springer, Berlin, 1999, pp. 365–385.

[72] N. Takayama, *Gröbner bases and the problem of contigous relations*, Japan J. Appl. Math. 6 (1989), pp. 147–160.

[73] N. Takayama, *An algorithm for constructing the integral of a module*, Proc. ISSAC '90, ACM Press, New York, 1990, pp. 206–211.

[74] J. M. Thomas, *Riquier's existence theorems*, Ann. Math. 30 (1929), pp. 285–321.

[75] J. M. Thomas, *Riquier's existence theorems*, Ann. Math. 35 (1934), pp. 306–311.

[76] V. L. Topunov, *Reducing systems of linear differential equations to a passive form*, Acta Appl. Math. 16 (1989), pp. 191–206.

[77] V. Ufnarovski, *Introduction to noncommutative Gröbner bases theory*, Gröbner Bases and Applications (B. Buchberger and F. Winkler, eds.), Cambridge Univ. Press, 1998, pp. 259–280.

[78] D. Wang, *An elimination method for differential polynomial systems I*, Syst. Sci. Math. Sci. 9 (1996), pp. 216–228.

[79] D. Wang, *Elimination Methods*, Springer, Wien New York, 2001.

[80] V. Weispfenning, *Differential term-orders*, Proc. ISAAC '93, ACM Press, New York, 1993, pp. 245–253.

[81] W.-T. Wu, *Some recent advances in mechanical theorem proving of geometries*, Contemporary Mathematics 29 (1984), pp. 235–241.

[82] W.-T. Wu, *Basic principles of mechanical theorem proving in elementary geometries*, J. Automated Reasoning 2 (1986), pp. 219–252.

[83] A. Y. Zharkov and Y. A. Blinkov, *Involution approach to solving systems of algebraic equations*, Proc. IMACS '93 (Lille, France), 1993, pp. 11–16.

[84] A. Y. Zharkov, *Solving zero-dimensional involutive systems*, Algorithms in Algebraic Geometry and Applications, Birkhäuser, Basel, 1996, pp. 389–399.

[85] M. Zhou and F. Winkler, *Gröbner bases in differential-difference modules*, Proc. ISSAC '06, ACM Press, New Yrok, 2006, pp. 353–360.

[86] A. Zobnin, *Essential properties of admissible orderings and rankings*, Contributions to General Algebra 14 (2004), pp. 205–221.

[87] A. Zobnin, *On standard bases in differential polynomal rings*, Preprint, 2004, to appear in Journal of Mathematical Science.

[88] A. Zobnin, *On admissible orderings and finiteness criteria for differential standard bases*, Proc. ISSAC '05, ACM Press, New Yrok, 2005, pp. 365–372.

Radon Series Comp. Appl. Math **2**, 109–137

Differential Elimination and
Biological Modelling

François Boulier

Key words. Differential algebra, elimination theory, applications, biology.

AMS classification. 12H05, 37N25, 62P10.

1   Introduction

This paper describes applications of a *computer algebra* method, *differential elimination*, to applied mathematics problems mostly borrowed from biology. The two considered applications are related to the *parameters estimation* (chapter 3) and the *model reduction* (chapter 4) problems. In both cases, differential elimination can be viewed as a preparation to numerical treatments. Those numerical treatments are, at least partly, sketched in this paper in order to put some light on the real limitations of the applications. Together with the applications, the paper introduces two *implementations* of the differential elimination algorithms: the *diffalg* package, which is embedded in the MAPLE computer algebra software and the *BLAD* libraries [4] which are standalone open source C libraries. The *diffalg* package is designed to be manipulated interactively and can be used very quickly and easily by casual readers. The *BLAD* libraries are designed to provide differential elimination for scientific software independent of any computer algebra system. They are probably better suited than *diffalg* to the development of software dedicated to the described applications. Using the *BLAD* libraries implies however to write a C program. For this reason, in this paper, examples are illustrated with *diffalg* rather than with *BLAD*.

The author would like to thank Marc Lefranc for his comments and his advices.

2   Differential Elimination

The three next sections can be read in any order and provide three different introductions to differential elimination: Section 2.1 provides historical notes, Section 2.2 presents it more algebraically, through the differential ideal membership problem while Section 2.3 introduces it through software. For a wider survey on differential equations and computer algebra, see [68].

2.1   Historical Introduction

Differential elimination is an algorithmic subtheory of *differential algebra* (see Section 2.2 for mathematical definitions). It solves the membership problem for radical differential ideals[1].

The membership problem for polynomial ideals was one of the main problems of commutative algebra. It was solved by Bruno Buchberger in [16], thanks to the theory of Gröbner bases. Similarly, the membership problem for differential ideals is one of the main problems of differential algebra. It is proven undecidable in general [33]. It is still open for finitely generated differential ideals. It is only solved in the special case of radical differential ideals.

The development of differential elimination was undertaken by Ritt who developed the concept of *characteristic sets*. In his book, Ritt gave an algorithm to decompose the radical of any finitely generated differential ideal as an intersection of finitely many differential prime ideals presented by characteristic sets[2]. Ritt's algorithm relies on factorizations over towers of algebraic extensions of the base field of the polynomials and does not cover the case of partial differential polynomials. Abraham Seidenberg designed in [66] an elimination algorithm for systems of differential polynomials which only relies on addition, multiplication and the equality test with zero in the base field of the polynomials. However, Seidenberg's method is not convenient: it takes as input a differential polynomial, a differential system and decides if the polynomial belongs to the radical of the differential ideal generated by the system. It does not provide a description of this radical differential ideal. It also involves some useless operations (e.g. computation of *preparation polynomials*). To cover the case of partial differential systems, Seidenberg developed an analogue of the S-polynomials theory of the Gröbner bases theory. However, the proof of his [66, Theorem VI] seems to be incomplete. A few years later, Azriel Rosenfeld fixed and generalized Seidenberg's Theorem VI in [62, Lemma] but did not provide any algorithm. In his book, Kolchin generalized "Rosenfeld's lemma" and described a generalized method [43, Section IV.9]. However, Kolchin's method involves some non effective steps: his approach cannot be treated as an algorithm. Later, Wu Wen-Tsün described in [70] an algorithm to decompose a given system of differential polynomials as finitely many characteristic sets but the characteristic sets in the sense of Wu are weaker than those of Ritt and are not sufficient

---

[1] In this paper, differential ideals always refer to differential polynomial ideals.

[2] The intersection may be redundant. Surprisingly, the inclusion problem of two differential prime ideals presented by characteristic sets is still open while the equality test is straightforward [43, Chapter IV, Problem 3].

(without any extra process) to decide membership in the radical of the differential ideal generated by the system. Dongming Wang developed Wu's method in [75].

Giuseppa Carra-Ferro and François Ollivier developed the concept of *differential Gröbner bases* in [19, 57] but the bases they define do not need to be finite. Elizabeth Mansfield developed another concept of *differential Gröbner bases* in [50] but Mansfield's bases do not solve the membership problem in differential ideals. Greg Reid developed the concept of *reduced involutive forms* together with an algorithm in [59]. This concept applies more generally to systems of analytic differential equations. In this setting, no satisfactory analogue of the Rosenfeld's lemma is however available.

The author developed the so-called *Rosenfeld-Gröbner* algorithm in [7] from the papers of Seidenberg and Rosenfeld. He used Gröbner bases to convert Rosenfeld's lemma into an algorithm[3]. *Rosenfeld-Gröbner* gathers as input a differential system and a *ranking*. It represents the radical of the differential ideal generated by the input system as a finite intersection of radical differential ideals presented by characteristic sets (in the sense of Ritt). It solves the membership problem to radical differential ideals (ordinary or with partial derivatives). It only relies on addition, multiplication and the equality test with zero in the base field of the polynomials. The algorithm described in [7] was much improved, theoretically and practically, by a lemma[4] due to Daniel Lazard[5] [9, Lemma 2]. See [13] for a survey on Lazard's lemma. Some variants of *Rosenfeld-Gröbner* were published afterwards [48, 40, 14, 41].

2.2   Algebraic Introduction

Differential algebra is an algebraic theory for differential equations (ordinary or with partial derivatives) which was founded by Joseph Fels Ritt in the first half of the twentieth century. Ritt was much impressed by the development of commutative algebra and wanted to achieve a similar theory for differential equations. He summarized the work of his team in [61]. One of his students, Ellis Robert Kolchin, developed still further Ritt's theory and summarized his results and that of his team in [43]. See [18] for a survey. A *differential ring* (resp. field) is a ring (resp. field) R endowed with a derivation (this paper is restricted to the case of a single derivation but the theory is more general) i.e. a unitary mapping $R \to R$ such that (denoting $\dot{a}$ the derivative of a):

$$\overset{\cdot}{\widehat{(a+b)}} = \dot{a} + \dot{b}, \quad \overset{\cdot}{\widehat{(a\,b)}} = \dot{a}\,b + a\,\dot{b}.$$

Observe that, theoretically, the derivation is an abstract operation. For legibility, one views it as the derivation w.r.t. the time t. Algorithmically, one is led to manipulate finite subsets of some *differential polynomial ring* $R = K\{U\}$ where K is the differential field of coefficients (in practice, $K = \mathbb{Q}$ or $K = \mathbb{Q}(t)$) and U is a finite set of

---

[3] Gröbner bases are no more involved in current implementations of *Rosenfeld-Gröbner*. Instead, a variant [10, 12, *RegCharacteristic*] of *LexTriangular* [46, 52] is used.

[4] Lazard's lemma is a non differential lemma which implies, when combined to Rosenfeld's lemma, that the differential ideals presented by characteristic sets are necessarily radical.

[5] There was a gap in the proof of "Lazard's lemma" in [9] which was fixed for the first time by Sally Morrison in [54, 55].

dependent variables[6]. The elements of R, the *differential polynomials* are just polynomials in the usual sense, built over the infinite set, denoted ΘU, of all the derivatives of the dependent variables.

A famous example of Ritt [61, Section II.4]. The left-hand side of the ordinary differential equation $\dot{u}^2 - 4\,u = 0$ is a differential polynomial of the differential polynomial ring $R = \mathbb{Q}\{u\}$. Its analytic solutions are the zero function $u(t) = 0$ and the family of parabolas $u(t) = (t + c)^2$ where c is an arbitrary constant.

Definition 2.1 A *differential ideal* of a differential ring R is an ideal of R, stable under the action of the derivation.

The study of the radical of the differential ideal generated[7] by a finite system of differential polynomials is strongly related to the study of the analytic solutions of this system. Indeed, in algebraic geometry, it is well known that the set of the polynomials which vanish over the solutions of a given polynomial system form an ideal and even a radical ideal [78, Section VII.3, Theorem 14]. For differential equations, the set of the differential polynomials which vanish over the analytic[8] solutions of a given differential polynomial system form a differential ideal and even a radical differential ideal [61, Sections II.4 and II.7].

Ritt's example (continued). The analytic solutions of the differential equation $\dot{u}^2 - 4\,u = 0$ are the function $u(t) = 0$ and the family of functions $u(t) = (t + c)^2$. These solutions are also solutions of all the derivatives of the differential equation:

$$2\,\dot{u}\,(\ddot{u} - 2) = 0, \quad 2\,\dot{u}\,\dddot{u} + 2\,\ddot{u}\,(\ddot{u} - 2) = 0, \quad \ldots$$

More generally, they are solutions of every differential polynomial, a power of which is a finite linear combination of the derivatives of $\dot{u}^2 - 4\,u$ with arbitrary differential polynomials as coefficients i.e. every element of the radical of the differential ideal generated by $\dot{u}^2 - 4\,u$.

The problem of computing a representation of the radical of the differential ideal generated by a finite set of differential polynomials is thus an important problem, related to the study of the analytic solutions of this system. So is the membership problem to radical differential ideals which is solved by *Rosenfeld-Gröbner*. To present it, one needs to define the concept of *ranking* and Ritt's reduction.

Definition 2.2 If U is a finite set of dependent variables, a *ranking* over U is a total ordering over the set ΘU of all the derivatives of the elements of U which satisfies: $a < \dot{a}$ and $a < b \Rightarrow \dot{a} < \dot{b}$ for all $a,\,b \in \Theta U$.

---

[6] In the differential algebra theory, the terminology *differential indeterminates* is preferred to *dependent variables* for derivations are abstract and differential indeterminates are not even assumed to correspond to functions. In order not to mix different expressions in this paper, the second expression, which seems to be more widely known, was chosen.

[7] An ideal $\mathfrak{A}$ is said to be *radical* if $a \in \mathfrak{A}$ whenever there exists some nonnegative integer p such that $a^p \in \mathfrak{A}$. The radical of an ideal $\mathfrak{A}$ is the set of all the ring elements a power of which belongs to $\mathfrak{A}$. The radical of a (differential) ideal is a radical (differential) ideal [65, Section 4].

[8] Over some unspecified domain.

Let U be a finite set of dependent variables. A ranking such that, for every u, $v \in U$, the ith derivative of u is greater than the jth derivative of v whenever $i > j$ is said to be *orderly* [43, Section I.8]. If U and V are two finite sets of differential variables, one denotes $U \gg V$ every ranking such that any derivative of any element of U is greater than any derivative of any element of V. Such rankings are said to *eliminate* U w.r.t. V.

Definition 2.3 Assume that some ranking is fixed. Then one may associate with any differential polynomial $f \in K\{U\} \setminus K$ the greatest (w.r.t. the given ranking) derivative $v \in \Theta U$ such that $\deg(f, v) > 0$. This derivative is called the *leading derivative* or the *leader* of f.

Ritt's reduction. It is a generalization of the Euclidean division. It is well known that, if f and g are two polynomials, in one variable v, with coefficients in a field, the Euclidean division of f by g (g nonzero) is possible. It yields a unique pair (q, r) of polynomials such that $f = g q + r$ and $\deg r < \deg g$. If f and g have coefficients in a ring, the Euclidean division is no more possible in general for the leading coefficient of g may not be invertible. The closest available algorithm is the *pseudodivision* which consists in multiplying f by the leading coefficient c of g, raised at the power $p = \deg f - \deg g + 1$ before performing the Euclidean division [73, Section 6.12]. It yields a unique pair (q, r) of polynomials such that $c^p f = g q + r$ and $\deg r < \deg g$. The polynomial r is called the *pseudoremainder* of f by g and is denoted $\mathrm{prem}(f, g)$ or $\mathrm{prem}(f, g, v)$ when the variable is not clear from the context (case of polynomials depending on many different variables). The pseudodivision generalizes to the differential setting, providing Ritt's reduction algorithm [43, Section I.9], described below. Observe that only the "remainder" is computed.

Let f be a differential polynomial, to be reduced by a finite set $C = \{g_1, \ldots, g_n\}$ of differential polynomials. Denote v_i the leader of g_i for $1 \leq i \leq n$ (assuming that none of the g_i lies in the base field). Ritt's reduction builds a sequence $f_0, \ldots, f_r$ of differential polynomials starting at $f_0 = f$. The result is the polynomial

$$f_r = \mathrm{Ritt\_reduction}(f, C).$$

To compute $f_{\ell+1}$ from f_ℓ, three cases may occur. First case: if, for each $1 \leq i \leq n$, the differential polynomial f_ℓ does not depend on any proper derivative[9] $v_i^{(k)}$ of v_i and $\deg(f_\ell, v_i) < \deg(g_i, v_i)$ then the computation stops and $f_\ell = f_r$ is returned. Second case: if there exists some index $1 \leq i \leq n$ such that $\deg(f_\ell, v_i) \geq \deg(g_i, v_i)$ then $f_{\ell+1} = \mathrm{prem}(f_\ell, g_i, v_i)$. Third case: if there exists some index $1 \leq i \leq n$ such that f_ℓ depends on some proper derivative $v_i^{(k)}$ of v_i then $f_{\ell+1} = \mathrm{prem}(f_\ell, g_i^{(k)}, v_i^{(k)})$.

Remarks. The second rule could actually be viewed as a particular case of the third one. The sequence $f_0, \ldots, f_r$ described above is not uniquely defined. One could define a precise algorithm by specifying that the sequence of the reduced derivatives

---

[9] One denotes $v_i^{(k)}$ the kth derivative of v. When $k \geq 1$, $v_i^{(k)}$ is said to be a *proper* derivative of v_i. When $k = 0$, one defines $v_i^{(k)} = v_i$.

$v_i^{(k)}$ must be decreasing. This is the usual strategy but any other strategy could be applied. Last, observe that whenever $k \geq 1$, the differential polynomial $g_i^{(k)}$ has degree one in $v_i^{(k)}$ and admits the *separant* $s_i = \partial g_i / \partial v_i$ for leading coefficient. In this case, writing $g_i^{(k)} = s_i \, v_i^{(k)} + t_{i,k}$, one sees that the pseudodivision of f_ℓ by $g_i^{(k)}$ amounts to the following: first perform the following substitution in f_ℓ

$$v_i^{(k)} \longrightarrow -\frac{t_{i,k}}{s_i}$$

then clear the denominator of the obtained rational fraction. The resulting polynomial is free of $v_i^{(k)}$.

Example. Let us apply Ritt's reduction over $f_0 = \ddot{u} - v \, \dot{u}$ and $C = \{\dot{u}^2 + v\}$. The ranking is $u \gg v$ so that the leader of $g = \dot{u}^2 + v$ is $\dot{u}$. The polynomial f_0 gets pseudoreduced by the first derivative of g i.e. $2 \, \dot{u} \, \ddot{u} + \dot{v}$. First one substitutes $\ddot{u} \longrightarrow -\dot{v}/(2 \, \dot{u})$ over f_0, giving the rational fraction

$$-\frac{\dot{v}}{2 \, \dot{u}} - v \, \dot{u}.$$

Second, the denominator is cleared, giving $f_1 = -\dot{v} - 2 \, v \, \dot{u}^2$. This polynomial f_1 gets pseudoreduced by g: one substitutes $\dot{u}^2 \longrightarrow -v$ over f_1, giving the differential polynomial f_2 (there is no denominator to clear).

$$f_2 = -\dot{v} + 2 \, v^2.$$

Ritt's reduction stops at this step and $f_2 = f_r$ is returned.

Normal forms. Observe that in general, the set of all the differential polynomials which are reduced to zero by Ritt's reduction has no clear structure. It does not even need to be an ideal. Observe also that the returned polynomial f_r is not equivalent to f modulo the differential ideal generated by C because of the denominator clearing step. A more careful version was designed in [12]. It returns a rational fraction instead of a polynomial. When C is a *characteristic set* of the ideal $\mathfrak{A}$ that it defines, the rational fraction is guaranteed to be a *normal form* of the residue class of f modulo $\mathfrak{A}$. Such a normal form algorithm may be used to detect linear dependencies between residue classes modulo $\mathfrak{A}$, following the idea of [29]. See [8] or [5, Section 6.1].

Rosenfeld-Gröbner. The *Rosenfeld-Gröbner* algorithm gathers as input a finite system F of differential polynomials and a ranking. It returns a finite family (possibly empty) $C_1, \ldots, C_r$ of finite subsets of $K\{U\} \setminus K$. Each system C_i defines a differential ideal $\mathfrak{C}_i$ in the sense that, for any $f \in K\{U\}$, we have

$$f \in \mathfrak{C}_i \quad \text{iff} \quad \text{Ritt\_reduction}(f, C_i) = 0.$$

The relationship with the radical $\mathfrak{A}$ of the differential ideal generated by F is the following:

$$\mathfrak{A} = \mathfrak{C}_1 \cap \cdots \cap \mathfrak{C}_r.$$

When $r = 0$ we have $\mathfrak{A} = K\{U\}$. Combining both relations, one gets an algorithm to decide membership in $\mathfrak{A}$. Indeed, given any $f \in K\{U\}$ we have:

$$f \in \mathfrak{A} \quad \text{iff} \quad \text{Ritt\_reduction}(f, C_i) = 0, \quad 1 \leq i \leq r.$$

The systems C_i are often called *(differential) characteristic sets* or *differential regular chains*[10] in the literature. The differential ideals $\mathfrak{C}_i$ do not need to be prime. They are however necessarily radical, thanks to Lazard's lemma. Observe that it is possible to refine further the intersection in order to get prime differential ideals. It is sufficient for this to apply a usual primary decomposition algorithm. However, no algorithm is known to decide inclusion between differential ideals presented by characteristic sets, even when they are prime [43, Section IV.9, Problem 3]. Thus the computed representation can by no means be guaranteed to be minimal though this latter theoretically exists.

Ritt's example (continued). When $U = \{u\}$ there exists only one ranking:

$$\cdots > \ddot{u} > \dot{u} > u.$$

Take $F = \{\dot{u}^2 - 4u\}$ and denote $\mathfrak{A}$ the radical differential ideal generated by F. If one applies the *Rosenfeld-Gröbner* to F and this ranking, one gets an intersection $\mathfrak{A} = \mathfrak{C}_1 \cap \mathfrak{C}_2$ with

$$C_1 = \{\dot{u}^2 - 4u\}, \qquad C_2 = \{u\}.$$

The differential polynomial u is reduced to zero by C_2, not by C_1. Thus $u \notin \mathfrak{A}$. The differential polynomial $\ddot{u} - 2$ is reduced to zero by C_1, not by[11] C_2. Thus $\ddot{u} - 2 \notin \mathfrak{A}$. The product $\dot{u}(\ddot{u}-2)$ is reduced to zero by C_1 and C_2. Thus it lies in $\mathfrak{A}$ (it is one-half of the first derivative of $\dot{u}^2 - 4u$). This proves that the ideal $\mathfrak{A}$ is not prime. The ideal $\mathfrak{C}_1$ corresponds to the family of parabolas $u(t) = (t+c)^2$. The ideal $\mathfrak{C}_2$ corresponds to the solution $u(t) = 0$.

Complexity. From a theoretical point of view, differential elimination is a very powerful tool. It permits to decide if a system of differential equations admits analytic solutions over some unspecified domain[12]. See [67, Embedding theorem] and [60, 47]. Moreover, non differential polynomial elimination can be reduced to differential elimination in two different ways. First any non differential polynomial system can be viewed as a differential system of order zero (one seeks constant functions solutions instead of numbers) and the differential characteristic sets computed by *Rosenfeld-Gröbner* are exactly those that non differential algorithms [45, 42, 53] would compute. Second, any non differential polynomial system can be encoded as a system of linear partial differential equations in one dependent variable and constant coefficients ; the differential characteristic set computed by *Rosenfeld-Gröbner* over this linear system is (up to the inverse encoding) the reduced Gröbner basis of the non differential system w.r.t. the admissible ordering induced by the ranking. This last reduction proves that

---

[10] There is a slight difference between these two notions but it does not matter in this paper.

[11] Proving that $\mathfrak{C}_1 \not\subset \mathfrak{C}_2$ though C_1 is reduced to zero by C_2.

[12] One encounters undecidability results when the domain is precised. See [21, Theorem 4.11].

the membership problem to radical differential ideal is exspace hard [44]. See also [5, Section 9.7].

2.3 Computational Introduction

There are many different ways to tackle systems of ordinary differential equations in a computer algebra software. *Differential elimination* is one of them. It is presented here by comparison with numerical integration and closed form integration and illustrated over the *differential index reduction* problem. Most computations are performed using the *diffalg* package of MAPLE 9. A short presentation of the *BLAD* libraries is provided too.

Numerical integration.   Here is an example of an ordinary differential equation with an initial condition. The dependent variable x represents an unknown time varying function (one denotes $\dot{x}$ the first derivative of x).

$$\dot{x} = x\,(3 - x), \qquad x(0) = 1.$$

Numerical integration of an ordinary differential equation with an initial condition consists in computing a discrete approximation of the graph of the integral curve of the equation as a finite number of points. In principle, it is always possible to carry it out. The simplest method is Euler's explicit method [36, page 132]. Numerical integration is not considered as a method of computer algebra. The commands below show how to numerically integrate the above example using MAPLE 9 (the method is not the one of Euler but an adaptative stepsize Runge-Kutta scheme). The output of the numerical integrator is a function which evaluates the solution.

```
ode := diff(x(t),t) = x(t)*(3-x(t));
```

$$ode := \frac{d}{dt}x\,(t) = x\,(t)\,(3 - x\,(t))$$

```
sol := dsolve ({ode, x(0)=1}, x(t), numeric):
sol (0.5);
```

$$[t = 0.5, \ x(t) = 2.07431460567341386]$$

Closed form integration.   *Closed form integration* of an ordinary differential equation consists in computing its solutions as finite formulae. See [15] for an introductory text. Over the example, it is possible and yields the formula below. Observe that the formula involves an arbitrary constant _C1 for no initial condition is specified. Closed form integration is part of computer algebra. It is however not possible in general. It is different from differential elimination.

```
dsolve (ode, x(t));
```

$$x(t) = \frac{3}{(1 + 3\,e^{-3\,t}\_C1)}$$

Differential elimination.   To explain what *differential elimination* is, one needs to consider a system of at least two ordinary differential equations. The following example is borrowed from [37, Chapter VII, page 454]. Since it mixes ordinary differential equations and non differential equations, this type of system is sometimes called a *differential algebraic system*[13]. There are three unknown time varying functions (three dependent variables) x, y and z :

$$\dot{x} = 0.7\,y + \sin(2.5\,z),$$
$$\dot{y} = 1.4\,x + \cos(2.5\,z),$$
$$1 = x^2 + y^2.$$

Even readers not familiar with differential algebraic systems may see that such systems raise problems. Assume that some initial conditions $x(0)$, $y(0)$ and $z(0)$ are given and let us try to numerically integrate the system with Euler's method for some stepsize h. Evaluating the right-hand sides of the two first equations at $t = 0$ one gets $\dot{x}(0)$ and $\dot{y}(0)$. Using these numbers, Euler's method permits us to compute the estimations $x(h) \simeq x(0) + h\,\dot{x}(0)$ and $y(h) \simeq y(0) + h\,\dot{y}(0)$. However, one cannot estimate the value of $z(h)$ since no ordinary differential equation of the form (2.1) is available. Thus Euler's method cannot perform the next step.

$$\dot{z} = \quad something. \tag{2.1}$$

The point here is that the ordinary differential equation (2.1) which seems to be missing is actually not missing but hidden in some differential ideal[14]. It can be automatically extracted from the initial system by means of *differential elimination*. Before showing how to proceed with the help of the *diffalg* package of MAPLE, one needs to convert the system as a *polynomial differential system*. For this, one denotes s the sine, c the cosine and one introduces a few more equations. The following differential polynomial system is equivalent to the above one.

$$\dot{x} = 0.7\,y + s, \qquad \dot{s} = 2.5\,\dot{z}\,c,$$
$$\dot{y} = 1.4\,x + c, \qquad \dot{c} = -2.5\,\dot{z}\,s,$$
$$1 = x^2 + y^2. \qquad 1 = s^2 + c^2.$$

Let's now compute the hidden equation using *diffalg*. One first stores the differential polynomial system in the variable *syst*, converting floating point numbers as rational numbers.

```
with (diffalg):
syst := [diff(x(t),t) - 7/10*y(t) - s(t),
         diff(y(t),t) - 14/10*x(t) - c(t),
         x(t)^2 + y(t)^2 - 1,
         diff(s(t),t) - 25/10*diff(z(t),t)*c(t),
         diff(c(t),t) + 25/10*diff(z(t),t)*s(t),
         s(t)^2 + c(t)^2 - 1]:
```

[13] For readers familiar with this notion, it has *differentiation index* 2 [37, Section VII.1, Definition 1.2].

[14] All the differential algebra terminology used in this section is precisely defined in Section 2.2.

Then one assigns to the variable R the context of the computation: one indicates that the only derivation is taken with respect to the time, that the notation is the standard *diff* notation of MAPLE and one provides the *ranking*. For short[15], let us just say that the fact that z stands on the rightmost place of the list indicates that we are looking for an ordinary differential equation of the form (2.1).

```
R := differential_ring (derivations = [t], notation = diff,
                        ranking = [[s, c, x, y, z]]):
```

Next the *Rosenfeld-Gröbner* function is applied to *syst* and R. It returns a list of MAPLE tables. Each table provides a *characteristic set*. The list should be understood as an intersection. Over the example, the list only involves one characteristic set so that the characteristic set does represent the radical differential ideal generated by the input system. The desired equation stands on the second place of the characteristic set (only the two first equations are displayed). Enlarging the input system with this equation, it is now easy to perform any numerical integration method and our problem is solved. Technically speaking, differential elimination has permitted the reduction to zero of the *differentiation index* of the input system: it was 2 ; it is now 0. See [31, 58] for related works.

```
ideal := Rosenfeld_Groebner (syst, R):
rewrite_rules (ideal [1]);
```

$$\left[\frac{d}{dt} y\,(t) = \frac{7}{5}\,x\,(t) + c\,(t)\, , \right.$$

$$\frac{d}{dt} z\,(t) = \frac{1}{25}\, \frac{3500 - 12348\,(y\,(t))^6 + 13230\,c\,(t)\,x\,(t)\,(y\,(t))^4 + 25809\,(y\,(t))^4}{441\,(y\,(t))^6 - 882\,(y\,(t))^4 + 541\,(y\,(t))^2 - 100}$$

$$\left. + \frac{1}{25}\, \frac{-14700\,x\,(t)\,(y\,(t))^2\,c\,(t) - 16961\,(y\,(t))^2 + 3940\,x\,(t)\,c\,(t)}{441\,(y\,(t))^6 - 882\,(y\,(t))^4 + 541\,(y\,(t))^2 - 100}, \quad \ldots \right]$$

Let us now perform some slight change on the chosen ranking. Strictly speaking, the ranking below is different from the above one[16] but it also indicates that we are looking for an ordinary differential equation of the form (2.1). However, if one applies *Rosenfeld-Gröbner* over *syst* for this ranking, one never gets any result because of the size of the equations the algorithm tries to compute.

```
R := differential_ring (derivations = [t], notation = diff,
                        ranking = [[s, c, x, y], z]):
ideal := Rosenfeld_Groebner (syst, R):
```

Warning, computation interrupted

To summarize, differential elimination is a process which takes as input a system of differential equations (ordinary or with partial derivatives) and a ranking. It rewrites the

---

[15] With the terminology inroduced in Section 2.2, this is the *orderly* ranking such that $s > c > x > y > z$.
[16] It is the ranking $(s, c, x, y) \gg z$ which eliminates s, c, x and y and such that $s > c > x > y$.

input system into another equivalent system (or an equivalent finite family of systems when case splittings are necessary). The ranking permits to control the elimination process, indicating what should be eliminated. Differential elimination methods are considered as computer algebra. In principle, differential elimination is always possible. However, in practice, it is restricted by its terrifying worst case complexity and the related problem of choosing rankings.

A few packages are available for differential elimination: the *diffgrob* package of Mansfield [50], the *rif* package of Reid, Wittkopf and Boulton [59], the *epsilon* package of Wang [76] and the *diffalg* package which was illustrated just above. The first version of the *diffalg* package was written by the author in 1995 for MAPLE 5 [7, 9]. However, the version involved in MAPLE 9 is not the original one since it was much improved by Évelyne Hubert [40] and, more recently, by François Lemaire [12].

The BLAD libraries. In order to overcome (at least partially) the difficulties stated above, the author has developed a C library, called *BLAD*, from the model of the *GMP* library. This library aims at providing differential elimination methods to scientific software which are not necessarily computer algebra systems. It is available on [4]. One of the important functionnalities it provides consists in bounding in advance the time and the memory allocated to a given differential elimination request. In the case of a failure, the calling program gets back a clean working environment. The following C program performs the first elimination provided above. It reads the data in characters strings and prints the result of the differential elimination on the standard output. Of course, this is not a natural way to use the *BLAD* libraries.

```
#include "bad.h"

int main ()
{   struct bad_intersectof_regchain ideal;
    struct bap_tableof_polynom_mpz eqns, ineqns;
    bav_Iordering r;

    bad_restart (0, 0);
    ba0_sscanf2
      ("ordering (derivations = [t], blocks = [[s, y, c, x, z]])",
        "%ordering", &r);
    bav_R_push_ordering (r);
    bad_init_intersectof_regchain (&ideal);
    ba0_sscanf2
      ("intersectof_regchain ([], \
           [differential, primitive, autoreduced, normalized])",
         "%intersectof_regchain", &ideal);
    ba0_init_table ((ba0_table)&eqns);
    ba0_init_table ((ba0_table)&ineqns);
    ba0_sscanf2 ("[10*x[t] - 7*y - 10*s, 10*y[t] - 14*x - 10*c, \
                   10*s[t] - 25*z[t]*c, 10*c[t] + 25*z[t]*s, \
                   x^2 + y^2 - 1, c^2 + s^2 - 1]",
                "%t[%Az]", &eqns);
    bad_Rosenfeld_Groebner (&ideal, &eqns, &ineqns, 0);
```

```
    ba0_printf ("%intersectof_regchain\n", &ideal);
    bad_terminate (ba0_init_level);
    return (0);
}
```

There are four stacked *BLAD* libraries. From top down: *bad* (differential elimination), *bap* (differential polynomials), *bav* (rankings) and *ba0* (kernel). Functions identifiers are prefixed by the library they belong to. The *main* function starts by defining some variables: *ideal* which is going to contain the result, *eqns* and *ineqns* which will serve to store the input system and *r* which will contain the ranking. The first instruction (*bad_restart*) starts a *sequence of calls* to the library. This sequence terminates with the call to *bad_terminate*. The two parameters provided to *bad_restart* give the limits, in time and in memory, allocated to the sequence of calls. A zero parameter means that there is no limit. Then the ranking is read from a string and stored in *r* (the *ba0_sscanf2* function provides a generalization of the *sscanf* function of the standard C library). The variable *ideal* is initialized to an empty intersection of regular differential chains (characteristic sets) endowed with some attributes which will serve to parametrize the elimination: *"differential"* indicates that the ideal represented by the variable is differential, the other attributes set some technical properties that the regular differential chains will have to satisfy. Then the array *eqns* is initialized with the system to process (*x[t]* denotes $\dot{x}$). We do not need to bother with *ineqns* which is not used here. Last *Rosenfeld-Gröbner* is called and the content of *ideal* is printed on the screen. Here is the result of the execution. The desired equation starts on the third line.

```
intersectof_regchain ([regchain ([[100*c^2 - 420*c*x^3 + 420*c*x -
441*x^4 + 341*x^2, y^2 + x^2 - 1, 10*s*x + 10*y*c + 21*y*x,
11025*z[t]*x^5 - 11025*z[t]*x^3 + 2500*z[t]*x + 13230*c*x^4 -
11760*c*x^2 + 2470*c + 12348*x^5 - 11235*x^3 + 2387*x, 5*x[t]*x
+ 5*y*c + 7*y*x], [differential, autoreduced, primitive,
squarefree, coherent, normalized])], [differential, autoreduced,
primitive, squarefree, coherent, normalized])
```

3 Parameters Estimation

This section describes an application of differential elimination and, more precisely, an application of algorithms which perform changes of rankings over characteristic sets. The principle of this application was designed by Ghislaine Joly-Blanchard, Lilianne Denis-Vidal and Céline Noiret [23] and presented in [56]. The addressed problem is this one: estimate parameters values of parametric ordinary differential systems the dependent variables of which are not all *observed*. When all the dependent variables of the system are observed, the method still works but differential elimination is no more necessary. The work of Joly-Blanchard, Denis-Vidal and Noiret is strongly related to the problem of the *identifiability* study of differential systems, for which a huge literature is available. See e.g. [74, 30, 57, 24, 26, 25, 49, 2, 64]. The method of Joly-Blanchard, Denis-Vidal and Noiret is original for two reasons: it relies on rigorous differential elimination methods and it carries out the study of real examples up to the final numerical treatment. It mixes symbolics and numerics.

It assumes that the phenomenon under study is quite accurately modelled and that quite precise measures are available for the observed variables. Thus, though it was applied with quite some success in pharmacokinetics [20, 71], biological modelling may not be the most suitable field of application of the method. The method is described over an example coming from biology anyway, but it is more presented as an academic challenge than as a real application.

Here is a summary of the rest of this section. The addressed problem is stated over an example. The classical numerical solution is recalled. It relies on the use of a numerical nonlinear least squares solver, i.e. a Newton method. Differential elimination gets involved in the process to help solving the most difficult part of the Newton method: guessing the starting point. Last the difficulties of the overall method are discussed.

3.1 Statement of the Problem over an Example

Figure 3.1 represents a *compartmental model*. The two *compartments* represent the blood and some organ. A medical product is injected in the blood at $t = 0$. It can go from the blood to the organ and conversely. It may also get degraded and exit from the system. In order to write the corresponding differential system, some hypotheses must be made on the nature of the exchanges: exchanges between the two compartments are assumed to be linear, i.e. that, over every small enough interval of time, the amount of product going from compartment i to compartment j is proportional to the concentration of product in compartment i. The proportionality constant is denoted k_{ij}. The degradation is assumed to follow a *Michaelis-Menten* law. This law is a bit more difficult to explain. It can be derived from the modelling of an enzyme-catalyzed reaction by means of some *model reduction*. Two parameters are associated to this degradation: a maximal speed V_e and another constant k_e.

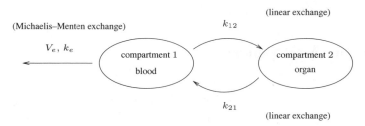

Figure 3.1 Compartmental model

From Figure 3.1, it is possible to derive a system of parametric ordinary differential equations. One associates to compartments 1 and 2, dependent variables x_1 and x_2 which represent the concentrations of product present in these compartments. Differential equations are built by considering exchanges the ones after the other ones. Each exchange appends one term to the right-hand side of the differential equation of the source compartment (with a minus sign) and a term to the right-hand side of the differential equation of the target compartment (with a plus sign). Beware to the trap: quantities are conserved by exchanges while exchanges are defined from the concentrations, which depend on the volumes of the compartments. For simplicity, it is assumed here that both compartments have a unitary volume. Applying the above process, one gets the following differential system. The second one is either linear or polynomial (it depends the way parameters are viewed). The first one is a rational fraction but it is equivalent to a polynomial since its denominator cannot vanish: parameters and dependent variables are positive real numbers.

$$
\begin{aligned}
\dot{x}_1 &= -k_{12}\, x_1 + k_{21}\, x_2 - \frac{V_e\, x_1}{k_e + x_1}, \\
\dot{x}_2 &= k_{12}\, x_1 - k_{21}\, x_2.
\end{aligned}
$$

Let us consider now some instance of the above model and assume that some extra information is available: parameters k_{12} and k_{21} are completely unknown, an interval of possible values $70 \leq V_e \leq 110$ is known for V_e and that $k_e = 7$ is known[1]. Some information is available on compartments also: compartment 1 is assumed to be *observed* i.e. a file of measures is assumed to be available for x_1. Compartment 2 is assumed to be non observed. One just knows that[2] $x_2(0) = 0$, i.e. that no product is initially present in the organ. To fix ideas and help the reader to reproduce the example studied in this section, here is a part of a file of 31 measures[3] for x_1.

```
     t              x1
0.00000e-01 5.00000e+01
5.00000e-02 4.45078e+01
      ...          ...
1.50000e+00 4.95270e-02
```

We are now ready to state the problem over this example: *given the system of parametric ordinary differential equations, the file of measures and the extra information, estimate the values of the three unknown parameters: V_e, k_{12} and k_{21}.*

3.2   The Numerical Method

There exists a purely numerical method to solve this problem. It is a *nonlinear* least squares solving method, i.e. a Newton method. Precisely, a Levenberg-Marquardt solver is called. The idea is simple: pick *random* values for the three unknown parameters. Integrate numerically the differential system w.r.t. these values and compare the curve obtained by simulation with the file of measures. The *error* is defined as the sum, for all abscissas, of the squares of the ordinates differences between the two curves. The Levenberg-Marquardt method updates the values of the three unknown

---

[1] It is realistic to assume that one of the parameters is known since equations can often be normalized by dividing some of the system parameters by one of them or, more generally, by studying their Lie symmetries.

[2] Unknown initial conditions do not raise any problem for they can be handled as plain parameters. See e.g. [36, Section I.14].

[3] The file was produced by numerical integration with $x_1(0) = 50$, $x_2(0) = 0$, $V_e = 101$, $k_{12} = 0.5$ and $k_{21} = 3$. The time ranges from $t = 0$ to $t = 1.5$ by steps of length 0.05.

parameters if the error is considered as too large. It stops either if the error is small enough of if a stationary point is reached.

Let us try and take the following values: $V_e = 70$, $k_{12} = 4.5$ and $k_{21} = 1.5$. One gets the two curves on the left-hand side picture of Figure 3.2. After a few loops, the Levenberg-Marquardt yields the two curves on the right-hand side picture with $V_e = 82.8$, $k_{12} = .76$ and $k_{21} = .16$. Numerical computations (numerical integration of ODE, Levenberg-Marquardt method) were performed by the *Gnu Scientific Library* (*GSL*). The picture was produced by *gnuplot*. According to the pictures, the purely numerical method seems to work perfectly. However, the obtained parameters values are wrong: the Levenberg-Marquardt ended in a local minimum.

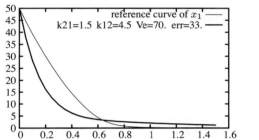 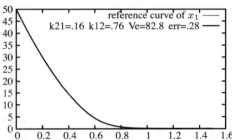

Figure 3.2: The reference (thin) and the simulated (thick) curves, before (left) and after (right) running the Levenberg-Marquardt method

3.3 The Symbolic-Numeric Method

The previous method has a drawback: it relies on *nonlinear* least squares which require the *a priori* knowledge of a good approximation of the parameters values. Thanks to differential elimination and to *linear* least squares, it is possible to estimate a first approximation of the parameters values. This first approximation may be used afterwards by the purely numerical method as a starting point.

The idea here consists in *eliminating the non observed variables* of the model. In other words, the idea consists in computing a differential polynomial which lies in the differential ideal generated by the model equations and which only involves the observed variable x_1, its derivatives up to any order and the model parameters. Let us show how to do this with the help of *diffalg*.

To compute this polynomial, the *Rosenfeld-Gröbner* algorithm is applied over the model equations. The ranking eliminates x_2 w.r.t. x_1:

$$x_2 \gg x_1.$$

In other words, the ranking indicates that we are looking for a polynomial free of x_2. The right-hand side of the first model equation is a rational fraction. It is decomposed as a numerator and a denominator. The numerator is stored in the list of the equations (first parameter to *Rosenfeld-Gröbner*). The denominator is stored in the list of the *inequations*[4] (second parameter to *Rosenfeld-Gröbner*). To avoid splitting cases on parameters values, one views them as (transcendental) elements of the base field of the differential polynomials.

```
K := field_extension
            (transcendental_elements = [k21, k12, ke, Ve]):
R := differential_ring
            (derivations = [t], notation = diff,
             field_of_constants = K, ranking = [x2, x1]):
ideal := Rosenfeld_Groebner
            ([numer (eq1), eq2], [denom (eq1)], R);
```

$$ideal := [characterizable]$$

The characteristic set *ideal* involves two polynomials. The one which does not involve x_2 is the second one, which is displayed below, slightly pretty printed. The expressions enclosed between square brackets are called *parameters blocks*.

$$\ddot{x}_1 \left(x_1 + k_e\right)^2 + [k_{12} + k_{21}] \, \dot{x}_1 \left(x_1 + k_e\right)^2 + [V_e] \, \dot{x}_1 \, k_e + [k_{21} \, V_e] \, x_1 \left(x_1 + k_e\right) = 0.$$

This equation tells us that the model is *globally identifiable* i.e. that, given a function x_1 and a parameter value k_e, the three unknown parameters are uniquely defined. Indeed, assume that the function x_1 is known. Then so are its derivatives $\dot{x}_1$ and $\ddot{x}_1$. These three functions can therefore be evaluated for three different values of the time t. The known parameter k_e can be replaced by its value. One thereby gets an exactly determined system of three linear equations whose unknowns are the parameters blocks. This system admits a unique solution. The values of the parameters blocks being fixed, it is obvious (over this example !) that the values of k_{12}, k_{21} and V_e also are uniquely defined. QED.

In practice, the function x_1 is known from a file of measures and one can try to numerically estimate the values of its first and its second derivative. If the measures are free of noise, the first derivative can be quite accurately estimated but this is usually not the case for the second derivative. To overcome these difficulties due to numerical approximations, one builds an overdetermined linear system that one solves by means of *linear* least squares. Over the example, one gets the following values:

$$[k_{12} + k_{21}] = 2.1, \quad [V_e] = 87.29, \quad [k_{21} \, V_e] = 144.01.$$

The values of the blocks of parameters being known, one still has to recover the values of the parameters by solving the above algebraic system. Over this example, it is very easy and one gets:

$$V_e = 87.29, \quad k_{12} = 0.45, \quad k_{21} = 1.65.$$

The above values can now be used as a starting point for the purely numerical method. Still over the example, one gets the correct parameters values:

$$V_e = 101, \quad k_{12} = 0.5, \quad k_{21} = 3.$$

---

[4] Inequations are polynomials which are considered as invertible. Indeed, if h is an inequation and some polynomial $h \, p$ lies in the ideal then p lies in the ideal. The ideal theoretic corresponding operation is the *saturation*.

3.4   Issues and Implementation

In general, there is no guarantee that the first estimation provided by the symbolic-numeric method leads the purely numerical method in the global minimum. Estimating parameters only makes sense for models at least *locally identifiable*. However, testing this property does not raise any difficulty. Some seminumerical algorithms are available [64]. These are probabilistic algorithms for which the failure probability is known and can be decreased up to any value. Numerically estimating the derivatives raises an important difficulty. To overcome it, a good method consists in converting the differential equations as integral equations as suggested in [22]. Under some conditions, integral equations are less sensitive to the noise than differential equations. There exists another important difficulty: there may exist algebraic relations between the parameters blocks. There is no such relation over the example. But assume, for the sake of the explanation, that the computed differential polynomial involves the three following blocks of parameters so that the third block is the product of the two first ones:

$$[V_e], \quad [k_{21}], \quad [V_e \, k_{21}].$$

There is no doubt that the numerical values produced during the resolution of the linear overdetermined system would not satisfy this relation. This would imply that the final algebraic system to solve in order to get the values of the parameters would be inconsistent. A way to overcome this problem consists in applying a nonlinear least squares method to solve the algebraic system. But then one needs to provide a first estimation of the parameters values: the problem to be overcome ! A symbolic method, based on algebraic elimination would be much more interesting. Indeed, it would provide the desired solution and could also compute the number of solutions of the algebraic system. It would solve in the same time the problem of estimating the parameters values and the problem of the identifiability of the model. Is it reasonable to try to apply algebraic elimination here ? One may think so, provided that many model variables are observed (at least one half). In this case, the differential elimination is fast and the parameters blocks are small: the algebraic elimination should be cheap.

A first draft of the above method was implemented in the *LEPISME* project [6]. The *Gnu Scientific Library* was used to perform the numerical methods. The *BLAD* libraries were used to perform the differential elimination. The method is difficult to implement in a satisfactory way: it involves many different steps. Each of these steps can be performed using a few different methods. When any method fails, it is difficult to provide synthetic informations on the failure to the user. In *BLAD*, instead of *Rosenfeld-Gröbner*, the more specialized and more efficient *PARDI* algorithm is used [11]. It takes advantage of the fact that the model equations generate a differential prime ideal and already form a characteristic set of this ideal w.r.t. some orderly ranking. It avoids all the discussions that *Rosenfeld-Gröbner* would perform and always computes only one characteristic set.

3.5 Prospects

In spite of all the difficulties, the project is being continued[5]: even in the case the symbolic-numeric method fails, the purely numerical method is still available. The existing method is thus improved. The use of the *BLAD* libraries is particularly interesting here for they permit to bound in advance the time and the memory allocated to the symbolic part of the symbolic-numeric method. Observe that the limitations are often due to the numerical part of the computations.

4 Model Reduction

The green alga *ostreococcus tauri* (Figure 4.1) was discovered in 1994 in the *Étang de Thau*, in the south of France. It is the minimal non parasitic known organism. Its genom, constituted of 11 millions of pairs of bases was published in 2006. Though very simple, this unicellular organism is endowed by a *circadian clock*[1]. See [51] for an historical perspective essay on circadian clocks and [28, Chapter 9] or [32] for more general texts about oscillations in biology. This clock permits the alga to raise itself at the top of the water before the sunrise. The alga is one of the main objects of study of the the *Observatoire Océanologique de Banyuls*.

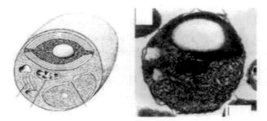

Figure 4.1: Ostreococcus tauri: a nucleus with a hole (bottom right), a chloroplast (top) with an amide ball (white spot), a Golgi apparatus (bottom left) and a mitochondry (center). The size is about one micrometer.

The author has been involved for two years in a pluridisciplinary working group (including computer scientists, physicists and biologists), led by François-Yves Bouget of the *Observatoire océanologique de Banyuls* for the biological part and Marc Lefranc of the *nonlinear dynamics* team for the physics and computer science part. This working group aims at modelling the cell division cycle of *ostreococcus tauri*. Our first goal has been to try to model the circadian clock of *ostreococcus tauri* which controls[2] the

---

[5] The author is getting involved in a project which aims at applying this method for modelling the biosynthesis of fatty acids and oil in oilseed embryos.

[1] A circadian clock is a clock the period of which is about 24 hours. The qualifier is built from *circa* (around) and *dies* (day).

[2] This is our simplifying working assumption. The clock itself might actually very well be regulated by the division cycle.

division cycle. In the genom of the green alga, two genes (named *TOC* and *CCA1*) were identified. They are known to be central components of clocks. We have thus been seeking a model under the form of a system of parametric ordinary differential equations, describing a two genes regulatory network and producing oscillating trajectories. We have very quickly met the following difficulty: many systems of parametric ordinary differential equations have integral curves which do not oscillate at all and, even the ones which have oscillating integral curves, only have such curves for very restricted ranges of parameters. Our problem can thus be reformulated as follows: *given a system of parametric ordinary differential equations, does there exist ranges of parameters w.r.t. which integral curves oscillate ?*

This problem can be addressed by looking for conditions on parameters which produce *Hopf bifurcations* [38, Chapter 11]. This approach was recently studied in the computer algebra community [27, 77, 35, 34]. It applies the Routh-Hurwitz criterion [36, Section I.13] and involves non differential elimination. It is not discussed in this paper. Another approach consists in applying the Poincaré-Bendixson theorem [38, Chapter 12] together with *differential elimination*. It was applied by members of the biology community in [72] over an abstract two genes regulatory network. This is the approach described in this chapter[3].

What does the Poincaré-Bendixson theorem state and how can it be applied in this context ? Roughly speaking, the theorem states that, if the integral curves of an autonomous ordinary differential system in *two dependent variables* stay in a bounded area and if this area does not involve any stable steady point then this area involves limit cycles. Limit cycles correspond to oscillating trajectories. Where is differential elimination involved ? The initial model (Section 4.1) involves seven dependent variables. The idea consists in approximating it by a *reduced model* of two ordinary differential equations in two variables by means of *model reduction*. Differential elimination permits to simplify[4] the reduced model (Section 4.2). The application of the Poincaré-Bendixson theorem is afterwards pretty straightforward. Indeed, in biology, trajectories of variables are always bounded. So are the ones of the reduced model, at least for parameters values which are biologically consistent (positivity is the least requirement). The steady points of the reduced model can be computed by algebraic elimination (e.g. Gröbner bases methods). There are three steady points but only one of them correspond to positive values of the variables (the other ones are discarded). Its stability can be studied by linearizing the model in the neighborhood of the steady point: the point is unstable if and only if at least one of the eigenvalues of the coefficients matrix J of the linearized system has a positive real part [36, Section I.13]. The conditions on parameters values which make the reduced system oscillate correspond thus to conditions on parameters values which make the trace and the determinant of the matrix J (which is 2×2) both positive. These parameters ranges make the reduced system oscillate. Do they make the initial model oscillate ? Yes ... provided that the model reduction is a good reduction ! This theoretically very difficult question can

---

[3] The author would like to thank Natacha Skrzypczak: an important part of the following analysis was initiated by her in [69].

[4] The author of [72] did not actually use any differential elimination method: they simplified their system interactively with MATHEMATICA. As shown later, the use of a differential elimination method permits to improve their result.

actually be checked, as in [72], by numerically integrating the initial model for many different parameters values picked in the estimated parameters ranges.

4.1 The Initial Model

This section describes the initial abstract model of [72]. The model involves two genes: an activator $\mathscr{A}$ and a repressor $\mathscr{R}$. These genes get transcribed into two mRNA M_A and M_R. The mRNAs then get translated into proteins A and R. Protein A can fix itself on the promotors of both genes $\mathscr{A}$ and $\mathscr{R}$, speeding up both transcription rates. The two proteins A and R can react together and form a complex C. Intuitively, one sees that the action of gene $\mathscr{A}$ consists in speeding up the reaction by producing protein A while the action of gene $\mathscr{R}$ consists in slowing down it by producing protein R which catches A to form the complex.

The seven model variables. The variables M_A and M_R denote the concentrations of mRNA transcribed from genes $\mathscr{A}$ and $\mathscr{R}$. The variables A, R and C denote the concentrations of the corresponding proteins. For each gene, one needs to introduce a variable to distinguish the case where protein A is bound to its promotor from the case where protein A is not bound to its promotor[5]. This variable is not a concentration. It should rather be considered as a probability or a mean value: the variable D_A corresponds to the gene $\mathscr{A}$. The value 1 indicates that protein A is bound to the promotor of $\mathscr{A}$. The value 0 indicates that protein A is not bound to the promotor. A similar variable D_R is introduced for gene $\mathscr{R}$. There are 15 parameters, denoted by Greek letters.

The model equations. They are derived from a picture. Since the complete picture might be a bit difficult to interpret for casual readers, it is explained and built piece by piece. Picture 4.2 describes the possible binding of protein A on the promotors of genes $\mathscr{A}$ and $\mathscr{R}$.

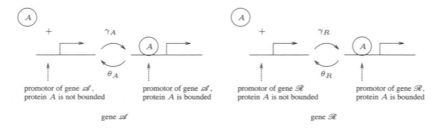

Figure 4.2 The two possible states of genes $\mathscr{A}$ and $\mathscr{R}$.

---

[5] The *promotor* of a gene is an area located in front of the gene. For a gene to be transcribed into mRNA, it is necessary that some protein binds itself to the gene promotor. Many different proteins may play this role. In this model, it is implicitly assumed that some unspecified proteins different from A may bind themselves to the promotors of the two genes, but with more difficulty than A so that the transcription rates of the two genes are higher when A is bound than when A is not bound.

The corresponding model equations[6] are given below. The "plus signs" in the diagram indicate that the binding rate of protein A is proportional to the product of the concentration of A by the variables D_A and D_R. It is a variant of the *mass action law*, variables D_A and D_R being handled as concentrations[7]. Observe that one temporarily omits the differential equation which describes the evolution of A because it would be incomplete at this step.

$$\dot{D}_A = -\gamma_A\,A\,D_A + \theta_A\,(1 - D_A), \quad \dot{D}_R = -\gamma_R\,A\,D_R + \theta_R\,(1 - D_R).$$

The leftmost part of Figure 4.3 shows that gene $\mathscr{A}$ gets transcribed into mRNA M_A at different rates depending on whether protein A is bound or not to its promotor. The mRNA M_A can be degraded at rate δ_{M_A}. The rightmost part of the figure shows a symmetric phenomenon for gene $\mathscr{R}$. The corresponding model equations are given

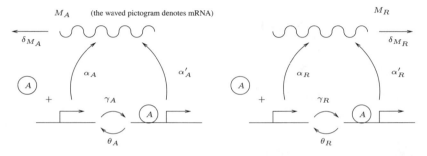

Figure 4.3 Transcriptions of the two genes into mRNA.

below. They enlarge the above set of two differential equations. Observe that the terms $\alpha_A\,D_A + \alpha'_A\,(1 - D_A)$ and $\alpha_R\,D_R + \alpha'_R\,(1 - D_R)$ are not subtracted to the right-hand sides of the differential equations which describe the evolutions of D_A and D_R (contrarily to what is usually done when translating chemical reactions into differential equations) since genes are not consumed by transcriptions.

$$\dot{M}_A = \alpha'_A\,(1 - D_A) + \alpha_A\,D_A - \delta_{M_A}\,M_A, \quad \dot{M}_R = \alpha'_R\,(1 - D_R) + \alpha_R\,D_R - \delta_{M_R}\,M_R.$$

The complete diagram is given in Figure 4.4. It indicates that mRNAs M_A and M_B get translated into proteins A and R. Since translations do not consume mRNA, the terms $\beta_A\,M_A$ and $\beta_R\,M_R$ are not subtracted to the right-hand sides of the two differential equations above. Figure 4.4 also shows that proteins A and R can react together to form[8] a complex C. The complex C may break, producing back protein R. There are degradations rates for proteins A and R. The new model equations are given below.

---

[6] The model given in [72] involves nine variables instead of seven: two extra variables were introduced to avoid the $(1 - D)$ terms.

[7] One may wonder why differential equations are used to model such phenomenons while stochastic equations might better correspond to the reality. An answer is that the qualitative analysis of the model is much easier with the rich theory of systems of ordinary differential equations than with stochastic equations. Of course, the conclusions derived from the differential model should be validated afterwards by stochastic simulations as the authors of [72] actually do.

[8] This *dimerization* of the two proteins does not seem to occur in the context of *ostreococcus tauri*. This causes a difficulty to apply the model of [72] to the green alga for the oscillating behaviour of the model seems to be strongly related to the dimerization.

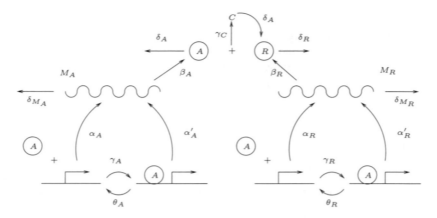

Figure 4.4 The complete diagram.

They enlarge the set of four equations previously built.

$$\dot{C} = \gamma_C \, A \, R - \delta_A \, C, \quad \dot{R} = \beta_R \, M_R - \gamma_C \, A \, R + \delta_A \, C - \delta_R \, R,$$
$$\dot{A} = \theta_A \, (1 - D_A) + \theta_R \, (1 - D_R) + \beta_A \, M_A - (\gamma_A \, D_A + \gamma_R \, D_R + \gamma_C \, R + \delta_A) \, A.$$

4.2   Reduction of the Model

One tries to approximate the model built in Section 4.1, which involves seven parametric ordinary differential equations, in seven variables, by a system of two parametric ordinary differential equations, in two variables. To eliminate five variables, the idea consists in separating the seven variables into a set of two *slow* variables, a set of five *fast* variables and to proceed to a *steady state approximation* [36, Section I.16]. Roughly speaking, here is the idea: consider a differential system of the following form, where ε denotes a *small* positive constant:

$$\dot{x} = f(x, y), \quad \varepsilon \, \dot{y} = g(x, y).$$

Over a generic point $(x, y) \in \mathbb{R}^2$ and, in particular, in the neighborhood of the initial conditions, the speed of y is high and thus rapidly approaches an area where $g(x, y) \simeq 0$. It is thus reasonable to approximate such a system by the following one:

$$\dot{x} = f(x, y), \quad 0 = g(x, y)$$

which mixes differential and algebraic equations. The study of such systems is not easy, in particular when there are many different algebraic equations $g_i = 0$. Numerical integrators cannot usually guarantee that the computed integral curves stay on the algebraic variety defined by the algebraic equations. For such systems, there may also exist hidden algebraic equations, consequences of the $g_i = 0$, which must be satisfied. Differential elimination is a tool which may simplify such systems and uncover these hidden equations. The authors of [72] decided that D_A, D_R, M_A, M_R and A are fast and R and C are slow. They were thus led to study the following differential algebraic system:

$$
\begin{aligned}
/\dot{D}_A\,0 &= \theta_A(1-D_A) - \gamma_A\,D_A\,A, \\
/\dot{D}_R\,0 &= \theta_R\,(1-D_R) - \gamma_R\,D_R\,A, \\
/\dot{M}_A\,0 &= \alpha'_A\,(1-D_A) + \alpha_A\,D_A - \delta_{M_A}\,M_A, \\
/\dot{M}_R\,0 &= \alpha'_R\,(1-D_R) + \alpha_R\,D_R - \delta_{M_R}\,M_R, \\
/\dot{A}\,0 &= \theta_A\,(1-D_A) + \theta_R\,(1-D_R) + \beta_A\,M_A \\
&\quad -(\gamma_A\,D_A + \gamma_R\,D_R + \gamma_C\,R + \delta_A)\,A, \\
\dot{R} &= \beta_R\,M_R - \gamma_C\,A\,R + \delta_A\,C - \delta_R\,R, \\
\dot{C} &= \gamma_C\,A\,R - \delta_A\,C.
\end{aligned}
$$

The authors of [72] did actually perform a differential elimination process over the above example, without stating the words *differential elimination*. They did it interactively, using MATHEMATICA. The *diffalg* package of MAPLE can indeed perform the same task. We are somehow looking for a differential system involving only R and C. A natural ranking to choose is the following one, which eliminates the fast variables:

$$\text{(fast variables)} \gg \text{(slow variables)}.$$

However, to avoid a pointless expression swell and to obtain exactly the same result as [72], it is better to keep the fast variable A in the set of the slow variables. Here are the coresponding MAPLE commands:

```
syst := [thetaA*(1 - DA) - gammaA*DA*A,
        thetaR*(1 - DR) - gammaR*DR*A,
        alphaAp*(1 - DA) + alphaA*DA - deltaMA*MA,
        alphaRp*(1 - DR) + alphaR*DR - deltaMR*MR,
        thetaA*(1 - DA) + thetaR*(1 - DR) + betaA*MA
           - (gammaA*DA + gammaR*DR + gammaC*R + deltaA)*A,
        R[t] - (betaR*MR - gammaC*A*R + deltaA*C - deltaR*R),
        C[t] - (gammaC*A*R - deltaA*C)]:

K := field_extension (transcendental_elements = [thetaA, thetaR,
    gammaA, gammaR, gammaC, alphaA, alphaAp, alphaR, alphaRp,
    betaA, betaR, deltaA, deltaR, deltaMA, deltaMR]):
Ring := differential_ring (derivations = [t],
        field_of_constants = K,
        ranking = [[DA, DR, MA, MR], [A, R, C]]);

ideal := Rosenfeld_Groebner (syst, Ring):
```

The list *ideal* only involves one characteristic set, involving seven equations. The three last equations only depend on R, C, A, $\dot{R}$ and $\dot{C}$. They have the following form:

$$
\begin{aligned}
\dot{R} &= \text{a rational fraction,} \\
\dot{C} &= \gamma_C\,A\,R - \delta_A\,C, \\
0 &= (\gamma_A\,\delta_{M_A}\,\delta_A + \gamma_A\,\delta_{M_A}\,\gamma_C\,R)\,A^2 + \\
&\quad (\delta_A\,\theta_A\,\delta_{M_A} + \theta_A\,\delta_{M_A}\,\gamma_C\,R - \alpha'_A\,\gamma_A\,\beta_A)\,A - \beta_A\,\theta_A\,\alpha_A.
\end{aligned}
$$

Observe that the above system is not so easy to integrate numerically: solving the third equation implies to choose a root of a degree two polynomial. Here, the choice is straightforward for A, being a concentration, needs to be positive and the equation always has only one positive root[9]. However, this argument needs some understanding of the system: the user has to manipulate the equation and solve it explicitly in order to select the positive root. Having a differential elimination algorithm at hand permits us however to try to compute many different representations of the same system. In particular, if one tries the following ranking, obtained by permuting A, R and C in the second block, one gets a simpler representation[10]:

```
Ring := differential_ring (derivations = [t],
         field_of_constants = K,
         ranking = [[DA, DR, MA, MR], [R, C, A]]):
ideal := Rosenfeld_Groebner (syst, Ring);
```

The list *ideal* only involves one characteristic set. The last three equations provide another presentation of the reduced model with two ordinary differential equations and a degree one algebraic equation:

$$\dot{C} = a \ rational \ fraction, \ \dot{A} = a \ rational \ fraction, \ R = a \ rational \ fraction.$$

Moreover, the variable R does not appear anywhere in the two differential equations since any occurence of R would have been replaced by the right-hand side of the last equation. One can thus just omit the third, algebraic, equation.

The above system might be surprising for readers not familiar with steady state approximations. Indeed, the reduced model was obtained by letting the speeds of the fast variables (including A) equal to zero. How is it then possible to end up with a differential equation defining a nonzero speed for A ? The answer comes from the fact that the above sentence is wrong: the speed of A was not set[11] to zero ! Indeed, the differential equations describing the evolutions of the fast variables were removed. The resulting system of two ordinary differential equations was just specialized on the algebraic variety defined by the right-hand sides of the removed equations.

4.3 Prospects

Ranges of parameters values which make the reduced and the initial model oscillate are given in [72] but the authors do not describe the method they applied to compute these ranges. Clearly, one now needs a method able to automatically derive ranges of parameters from the sign conditions on the trace and the determinant of the computed matrix. Observe that even heuristic methods would be helpful and that the derived

---

[9] The first coefficient is positive and the last coefficient is negative: the number of positive real roots is at least one. Now, whatever the sign of the central coefficient, the number of sign changes is one. By Descartes rule of sign [3], the number of positive real roots is at most one. The polynomial thus always has exactly one positive real root.

[10] Observe that there are other possible permutations over the second block of variables. Most of them lead to untractable computations. This example illustrates the need of software able to try many different reasonable rankings with a time limit. The *BLAD* libraries are designed to offer such a functionnality.

[11] The interested reader may try to apply *Rosenfeld-Gröbner* over the initial system enlarged with the five ordinary differential equations setting to zero the speeds of the five fast variables. One gets an inconsistent system.

ranges of parameters do not need to be complete in any sense. Though such methods can certainly be designed in theory (e.g. based on interval arithmetic [39]), the choice is not so easy in practice. It must still be done in order to get an automatic method which would help modelling the circadian clock of *ostreococcus tauri*. Last, observe that it would be very interesting to compare the approach based on the Poincaré-Bendixson theorem, and the one based on the direct application of the Routh-Hurwitz criterion over the initial model. This study also is still in progress.

5 Conclusion

Differential elimination is a tool which may play a real role to improve some applied mathematics methods. As illustrated in Section 2.3 and 4, it permits to reduce the differentiation index of differential-algebraic systems. It permits also to compute different representations of the same system. Both features may help designing better numerical integrators. Differential elimination may help guessing good starting points for Newton methods (Section 3). It may also be involved in the qualitative analysis of dynamical system for it permits to simplify these systems after model reduction (Section 4). These examples show that differential elimination is complementary to numerical methods. It is interesting here to compare the non differential and the differential elimination. From an algorithmic point of view, both theories are very close to each other. From the applications standpoint, the situations are very different. In the non differential setting, one may hope to bypass all numerical methods. For instance, in the zerodimensional case, there exists symbolic algorithms [1, 63] able to isolate the real roots of large polynomials. It thus makes sense to compute large Gröbner bases or characteristic sets. In the differential setting however, no such algorithms are known. Cooperating with numerical methods is thus mandatory. Now, differential systems which are considered as difficult from the numerical point of view are actually very small and very easy from the symbolic one (see [37, Section IV.1]). It thus may not really make sense to compute large differential characteristic sets. Note that this observation is an argument which minimizes the importance of the terrible worst case complexity of differential methods ! The examples considered in this paper also show that differential elimination only play very local roles in the different processes: it helps but may quite often be bypassed. Since moreover, the theory is rather difficult and usually not taught in traditional university courses, it seems very important to develop easy to use software components. The *BLAD* libraries are an attempt in that direction.

Bibliography

[1] A. G. Akritas, *There is no Uspensky's method*, Proceedings of ISSAC'86, 1986.

[2] S. Audoly, G. Bellu, L. D'Angio, M. P. Saccomani, and C. Cobelli, *Global Identifiability of Nonlinear Models of Biological Systems*, IEEE Transactions on Biomedical Engineering 48 (2001), pp. 55–65.

[3] S. Basu, R. Pollack, and M.-F. Roy, *Algorithms in Real Algebraic Geometry*, Algorithms and Computation in Mathematics, vol. 10, Springer Verlag, 2003.

[4] F. Boulier, *The BLAD libraries*, `http://www.lifl.fr/~boulier/BLAD`, 2004.

[5] ———, *Réécriture algébrique dans les systèmes d'équations différentielles polynomiales en vue d'applications dans les Sciences du Vivant*, May 2006, Mémoire d'habilitation à diriger des recherches, Université Lille I, LIFL, 59655 Villeneuve d'Ascq, France.

[6] F. Boulier, L. Denis-Vidal, T. Henin, and F. Lemaire, *LÉPISME*, presented at the ICPSS conference, 2004, submitted to the Journal of Symbolic Computation.

[7] F. Boulier, *Étude et implantation de quelques algorithmes en algèbre différentielle*, Ph.D. thesis, Université Lille I, 59655, Villeneuve d'Ascq, France, 1994.

[8] ———, *Efficient computation of regular differential systems by change of rankings using Kähler differentials*, Tech. report, Université Lille I, 59655, Villeneuve d'Ascq, France, November 1999, (ref. LIFL 1999–14, presented at the MEGA2000 conference).

[9] F. Boulier, D. Lazard, F. Ollivier, and M. Petitot, *Representation for the radical of a finitely generated differential ideal*, Proceedings of ISSAC'95 (Montréal, Canada), 1995, pp. 158–166.

[10] ———, *Computing representations for radicals of finitely generated differential ideals*, Tech. report, Université Lille I, LIFL, 59655, Villeneuve d'Ascq, France, 1997, (ref. IT306, december 1998 version published in the habilitation thesis of Michel Petitot).

[11] F. Boulier, F. Lemaire, and M. Moreno Maza, *PARDI !*, Proceedings of ISSAC'01 (London, Ontario, Canada), 2001, pp. 38–47.

[12] F. Boulier and F. Lemaire, *Computing canonical representatives of regular differential ideals*, Proceedings of ISSAC 2000 (St Andrews, Scotland), 2000, pp. 37–46.

[13] F. Boulier, F. Lemaire, and M. Moreno Maza, *Well known theorems on triangular systems and the D^5 principle*, Proceedings of Transgressive Computing 2006 (Granada, Spain), 2006, pp. 79–91.

[14] D. Bouziane, A. Kandri Rody, and H. Maârouf, *Unmixed–Dimensional Decomposition of a Finitely Generated Perfect Differential Ideal*, Journal of Symbolic Computation 31 (2001), pp. 631–649.

[15] M. Bronstein, *Integration and Differential Equations in Computer Algebra*, Programmirovanie 5 (1992), pp. 26–44.

[16] B. Buchberger, *Ein Algorithmus zum Auffinden der Basiselemente des Restklassenringes nach einem nulldimensionalen Polynomideals*, Ph.D. thesis, University of Innsbruck, Austria, Math. Institute, Austria, 1966, English translation in [17].

[17] ———, *An Algorithm for Finding the Basis Elements in the Residue Class Ring Modulo a Zero Dimensional Polynomial Ideal*, Journal of Symbolic Computation (Special Issue on Logic, Mathematics, and Computer Science: Interactions) 41 (2006), pp. 475–511.

[18] A. Buium and P. Cassidy, *Differential Algebraic Geometry and Differential Algebraic Groups: From Algebraic Differential Equations To Diophantine Geometry*, pp. 567–636, Amer. Math. Soc., Providence, RI, 1998.

[19] G. Carra-Ferro, *Gröbner bases and differential ideals*, Notes of AAECC 5 (Menorca, Spain), Springer Verlag, 1987, pp. 129–140.

[20] S. Demignot and D. Domurado, *Effect of prosthetic sugar groups on the pharmacokinetics of glucose-oxidase*, Drug Design Deliv. 1 (1987), pp. 333–348.

[21] J. Denef and L. Lipshitz, *Power Series Solutions of Algebraic Differential Equations*, Mathematische Annalen 267 (1984), pp. 213–238.

[22] L. Denis-Vidal, *Identifiabilité de modèles non linéaires paramétriques de dynamiques classiques et à retard. Planification d'expériences et estimation de paramètres*, Mémoire d'Habilitation à Diriger des Recherches, Université de Technologie de Compiègne, may 2004.

[23] L. Denis-Vidal, G. Joly-Blanchard, and C. Noiret, *System identifiability (symbolic computation) and parameter estimation (numerical computation)*, Numerical Algorithms, vol. 34, 2003, pp. 282–292.

[24] S. Diop, *Elimination in Control Theory*, Mathematics of Control, Signal and Systems 4 (1991), pp. 17–42.

[25] ———, *Differential algebraic decision methods and some application to system theory*, Theoretical Computer Science 98 (1992), pp. 137–161.

[26] S. Diop and M. Fliess, *Nonlinear observability, identifiability, and persistent trajectories*, Proc. 30th CDC (Brighton), 1991, pp. 714–719.

[27] M'H. El Kahoui and A. Weber, *Deciding Hopf bifurcations by quantifier elimination in a software–component architecture*, Journal of Symbolic Computation 30 (2000), pp. 161–179.

[28] C. P. Fall, E. S. Marland, J. M. Wagner, and John J. Tyson, *Computational Cell Biology*, Interdisciplinary Applied Mathematics, vol. 20, Springer Verlag, 2002.

[29] J.-C. Faugère, P. Gianni, D. Lazard, and T. Mora, *Efficient computation of Gröbner bases by change of orderings*, Journal of Symbolic Computation 16 (1993), pp. 329–344.

[30] M. Fliess, *Automatique et corps différentiels*, Forum Math. 1 (1989), pp. 227–238.

[31] M. Fliess, J. Lévine, P. Martin, and P. Rouchon, *Index and Decomposition of Nonlinear Implicit Differential Equations*, Proceedings of IFAC, 1995, pp. 43–48.

[32] J.-P. Françoise, *Oscillations en biologie*, Mathématiques et Applications, vol. 46, Springer Verlag, 2005.

[33] G. Gallo, B. Mishra, and F. Ollivier, *Some constructions in rings of differential polynomials*, Lecture Notes in Computer Science, vol. 539, pp. 171–182, Montréal, Canada, 1991.

[34] K. Gatermann, M. Eiswirth, and A. Sensse, *Toric Ideals and graph theory to analyze Hopf bifurcations in mass action systems*, Journal of Symbolic Computation 40 (2005), pp. 1361–1382.

[35] K. Gatermann and S. Hosten, *Computational algebra for bifurcation theory*, Journal of Symbolic Computation 40 (2005), pp. 1180–1207.

[36] E. Hairer, S. P. Norsett, and G. Wanner, *Solving ordinary differential equations I. Nonstiff problems*, 2. ed., Springer Series in Computational Mathematics, vol. 8, Springer–Verlag, New York, 1993.

[37] E. Hairer and G. Wanner, *Solving ordinary differential equations II. Stiff and Differential–Algebraic Problems*, 2. ed., Springer Series in Computational Mathematics, vol. 14, Springer–Verlag, New York, 1996.

[38] J. K. Hale and H. Koçak, *Dynamics and Bifurcations*, Texts in Applied Mathematics, vol. 3, Springer–Verlag, New York, 1991.

[39] E. Hansen, *Global Optimization Using Interval Analysis*, Marcel Dekker Inc., New York, 1992.

[40] É. Hubert, *Factorization free decomposition algorithms in differential algebra*, Journal of Symbolic Computation 29 (2000), pp. 641–662.

[41] ———, *Notes on triangular sets and triangulation–decomposition algorithm II: Differential Systems*, Symbolic and Numerical Scientific Computing 2001 (2003), pp. 40–87.

[42] M. Kalkbrener, *A Generalized Euclidean Algorithm for Computing Triangular Representations of Algebraic Varieties*, Journal of Symbolic Computation 15 (1993), pp. 143–167.

[43] E. R. Kolchin, *Differential Algebra and Algebraic Groups*, Academic Press, New York, 1973.

[44] K. Kühnle and E. W. Mayr, *Exponential Space Computation of Gröbner Bases*, Proceedings of ISSAC'96 (Zürich, Switzerland), 1996, pp. 63–71.

[45] D. Lazard, *A new method for solving algebraic systems of positive dimension*, Discrete Applied Mathematics 33 (1991), pp. 147–160.

[46] ———, *Solving Zero–dimensional Algebraic Systems*, Journal of Symbolic Computation 13 (1992), pp. 117–131.

[47] F. Lemaire, *An orderly linear PDE system with analytic initial conditions with a non analytic solution*, Special Issue on Computer Algebra and Computer Analysis, Journal of Symbolic Computation 35 (2003), pp. 487–498.

[48] Z. Li and D. Wang, *Coherent, regular and simple systems in zero decompositions of partial differential systems*, Systems Science and Mathematical Sciences 12 (1999), pp. 43–60.

[49] L. Ljung and S. T. Glad, *On global identifiability for arbitrary model parametrisations*, Automatica 30 (1994), pp. 265–276.

[50] E. L. Mansfield, *Differential Gröbner Bases*, Ph.D. thesis, University of Sydney, Australia, 1991.

[51] C. R. McClung, *Plant Circadian Rhythms*, The Plant Cell 18 (2006), pp. 792–803.

[52] M. Moreno Maza, *Calculs de Pgcd au–dessus des Tours d'Extensions Simples et Résolution des Systèmes d'Équations Algébriques*, Ph.D. thesis, Université Paris VI, France, 1997.

[53] ———, *On Triangular Decompositions of Algebraic Varieties*, Tech. report, NAG, 2000, (presented at the MEGA2000 conference, submitted to the JSC).

[54] S. Morrison, *Yet another proof of Lazard's lemma*, private communication, december 1995.

[55] ———, *The Differential Ideal* $[P] : M^\infty$, Journal of Symbolic Computation 28 (1999), pp. 631–656.

[56] C. Noiret, *Utilisation du calcul formel pour l'identifiabilité de modèles paramétriques et nouveaux algorithmes en estimation de paramètres*, Ph.D. thesis, Université de Technologie de Compiègne, 2000.

[57] F. Ollivier, *Le problème de l'identifiabilité structurelle globale : approche théorique, méthodes effectives et bornes de complexité*, Ph.D. thesis, École Polytechnique, Palaiseau, France, 1990.

[58] G. J. Reid, Ping Lin, and A. D. Wittkopf, *Differential Elimination–Completion Algorithms for DAE and PDAE*, Studies in Applied Mathematics 106 (2001), pp. 1–45.

[59] G. J. Reid, A. D. Wittkopf, and A. Boulton, *Reduction of systems of nonlinear partial differential equations to simplified involutive forms*, European Journal of Applied Math. (1996), pp. 604–635.

[60] C. Riquier, *Les systèmes d'équations aux dérivées partielles*, Gauthier–Villars, Paris, 1910.

[61] J. F. Ritt, *Differential Algebra*, Dover Publications Inc., New York, 1950, Available at `http://www.ams.org/online_bks/coll33`.

[62] A. Rosenfeld, *Specializations in differential algebra*, Trans. Amer. Math. Soc. 90 (1959), pp. 394–407.

[63] F. Rouillier and P. Zimmermann, *Efficient isolation of a polynomial real roots*, Journal of Computational and Applied Mathematics 162 (2004), pp. 33–50.

[64] A. Sedoglavic, *A Probabilistic Algorithm to Test Local Algebraic Observability in Polynomial Time*, Journal of Symb. Comp. 33 (2002), pp. 735–755.

[65] A. Seidenberg, *Some basic theorems in differential algebra (characteristic p arbitrary)*, Trans. Amer. Math. Soc. 73 (1952), pp. 174–190.

[66] _____, *An elimination theory for differential algebra*, Univ. California Publ. Math. (New Series) 3 (1956), pp. 31–65.

[67] _____, *Abstract differential algebra and the analytic case*, Proc. Amer. Math. Soc. 9 (1958), pp. 159–164.

[68] W. M. Seiler, *Computer Algebra and Differential Equations. An Overview*, mathPAD 7 (1997), pp. 34–49, available at `http://www.iwr.uni-heidelberg.de/groups/compalg/seiler`.

[69] N. Skrzypczak, *Modélisation des réseaux de gènes : formalisme SBML2. Réduction des modèles biologiques*, Tech. report, Mémoire de Master 2. Université Lille I, France, 2005.

[70] W. W. Tsün, *On the foundation of algebraic differential geometry*, Mechanization of Mathematics, research preprints 3 (1989), pp. 2–27.

[71] N. Verdière, *Identifiabilité de systèmes d'équations aux dérivées partielles semi–discrétisées et applications à l'identifiabilité paramétrique de modèles en pharmacocinétique ou en pollution*, Ph.D. thesis, Université de Technologie de Compiègne, France, 2005.

[72] J. M. G. Vilar, H. Y. Kueh, N. Barkai, and S. Leibler, *Mechanisms of noise-resistance in genetic oscillators*, Proceedings of the National Academy of Science of the USA 99 (2002), pp. 5988–5992.

[73] J. von zur Gathen and J. Gerhard, *Modern Computer Algebra*, Cambridge University Press, United Kingdom, 1999.

[74] É. Walter, *Identifiability of State Space Models*, Lecture Notes in Biomathematics, vol. 46, Springer Verlag, 1982.

[75] D. Wang, *An elimination method for differential polynomial systems I*, Tech. report, LIFIA–IMAG, Grenoble, France, 1994.

[76] _____, *Elimination Practice: Software Tools and Applications*, Imperial College Press, London, 2003.

[77] D. Wang and B. Xia, *Stability Analysis of Biological Systems with Real Solution Classification*, Proceedings of ISSAC 05 (Beijing, China), 2005, pp. 354–361.

[78] O. Zariski and P. Samuel, *Commutative Algebra*, Van Nostrand, New York, 1958.

Author information

François Boulier, LIFL, University Lille I, 59655 Villeneuve d'Ascq, France.
Email: `boulier@lifl.fr`

Radon Series Comp. Appl. Math **2**, 139–168

Janet Bases and Applications

Daniel Robertz

Key words. Janet basis, involutive basis, implementations in Maple and C++, Hilbert series.

AMS classification. 13N10, 13P10, 13D02, 13D40.

1 Introduction

This paper surveys a work in the area of Janet's algorithm and involutive bases at Lehrstuhl B für Mathematik, RWTH Aachen. The emphasis is on the Maple and C++ implementations of the involutive basis algorithm [16] by V. P. Gerdt and Y. A. Blinkov that are continuously developed and improved. However, for theoretical issues we stick to the original language of Janet's algorithm, in which we develop for instance the notion of generalized Hilbert series. The outline is as follows: Section 2 reviews Janet's algorithm without proofs, the main combinatorial point being the notion of a decomposition into disjoint cones of a multiple-closed set of monomials. The presentation of Janet's algorithm is along the lines of [37]. In Section 2.3, Janet's algorithm is generalized to a certain class of Ore algebras. Section 3 discusses Maple and C++ implementations of the involutive basis algorithm. In Section 4, the notion of generalized Hilbert series is introduced and illustrated for the case of partial differential equations. Finally, applications of the software described in Section 3 are given in Section 5.

We would like to thank two anonymous referees for their useful suggestions.

2 Janet Bases

2.1 Multiple-closed Sets of Monomials

We start by fixing some notation. Let k be a field and $R := k[x_1, \ldots, x_n]$ the (commutative) polynomial algebra over k. The set of monomials of R resp. in $\{x_1, \ldots, x_n\}$ is denoted by

$$\mathrm{Mon}(R) := \mathrm{Mon}(\{x_1, \ldots, x_n\}) := \{x^a \mid a \in (\mathbb{Z}_{\geq 0})^n\}, \qquad x^a := x_1^{a_1} \cdots x_n^{a_n}.$$

For each subset $\mu \subseteq \{x_1, \ldots, x_n\}$ we set

$$\mathrm{Mon}(\mu) := \{x^a \mid a \in (\mathbb{Z}_{\geq 0})^n; \ a_i = 0 \text{ for all } 1 \leq i \leq n \text{ such that } x_i \notin \mu\}.$$

Let $q \in \mathbb{N}$. In what follows, the standard basis vectors of the free R-module R^q are denoted by $e_1, \ldots, e_q$. Moreover, we define the set of monomials of R^q by

$$\mathrm{Mon}(R^q) := \bigcup_{i=1}^{q} \mathrm{Mon}(R)e_i.$$

Finally, for any set M we write $\mathcal{P}(M)$ for the power set of M.

Definition 2.1 A set $S \subseteq \mathrm{Mon}(R^q)$ is said to be $\mathrm{Mon}(R)$-*multiple closed*, if

$$ms \in S \quad \text{for all } m \in \mathrm{Mon}(R), \quad s \in S.$$

Every set $G \subseteq \mathrm{Mon}(R^q)$ satisfying

$$\mathrm{Mon}(R)G = \{mg \mid m \in \mathrm{Mon}(R), \ g \in G\} = S$$

is called a *generating set* for S. By $[G]$ we denote the $\mathrm{Mon}(R)$-multiple closed set generated by G in $\mathrm{Mon}(R^q)$.

Example 2.2 Let $R = k[x_1, x_2]$ and $G := \{x_1 x_2^2, x_1^3 x_2, x_1^4\}$. We consider the $\mathrm{Mon}(R)$-multiple closed set $S = [G]$ generated by G. If we visualize the monomial $x_1^i x_2^j$ as the point (i, j) in the positive quadrant of a x_1-x_2-coordinate system, then the set S of monomials can be viewed as the discrete set of points in the upper-right region in Figure 2.1.

The corollary to the following lemma by M. Janet ensures that Janet's algorithm, which is discussed in the next section, terminates. This lemma is a special case of Hilbert's basis theorem, which in turn is proved by Janet's theory, too. For a given finite generating set L for a submodule M of R^q and a fixed monomial ordering Janet's algorithm possibly removes elements from L and inserts new elements of M into L repeatedly in order to achieve finally that the $\mathrm{Mon}(R)$-multiple closed set generated by the leading monomials of the elements in L equals the set of all leading monomials of elements in M.

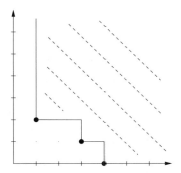

Figure 2.1 $\mathrm{Mon}(R)$-multiple closed set $S = [G]$

Lemma 2.3 *[28, 29] Every* $\mathrm{Mon}(R)$-*multiple closed set has a finite generating set.*

Moreover, every $\mathrm{Mon}(R)$-multiple closed set has a unique minimal generating set.

Corollary 2.4 *Every ascending sequence of* $\mathrm{Mon}(R)$-*multiple closed sets becomes stationary.*

Definition 2.5 A pair $(C, \mu) \in \mathcal{P}(\mathrm{Mon}(R^q)) \times \mathcal{P}(\{x_1, \dots, x_n\})$ is called a *cone* if there exists $v \in C$ such that

$$\mathrm{Mon}(\mu)v = \{mv \mid m \in \mathrm{Mon}(\mu)\} = C.$$

Then $v(C) := v$ is called the *vertex* of the cone (C, μ) and the elements of μ are the *multiplicative variables* for (C, μ) (or simply for C). Finally, $\overline{\mu} := \{x_1, \dots, x_n\} - \mu$ denotes the set of *non-multiplicative variables* for (C, μ) (or simply for C).

Definition 2.6 Let S be any set of monomials in $\mathrm{Mon}(R^q)$. A *decomposition of S into disjoint cones* is given by a finite set

$$\{(C_1, \mu_1), \dots, (C_l, \mu_l)\} \subset \mathcal{P}(\mathrm{Mon}(R^q)) \times \mathcal{P}(\{x_1, \dots, x_n\})$$

such that each pair (C_i, μ_i) is a cone,

$$\bigcup_{i=1}^{l} C_i = S \quad \text{and} \quad C_i \cap C_j = \emptyset \quad \text{for all} \quad i \neq j.$$

Passing over to cone vertices, we will also call a finite set

$$T = \{(m_1, \mu_1), \dots, (m_l, \mu_l)\} \subset \mathrm{Mon}(R^q) \times \mathcal{P}(\{x_1, \dots, x_n\})$$

a *decomposition of S into disjoint cones* if

$$\{(\mathrm{Mon}(\mu_1)m_1, \mu_1), \dots, (\mathrm{Mon}(\mu_l)m_l, \mu_l)\} \tag{2.1}$$

satisfies the above conditions. Then, T is also said to be *complete*.

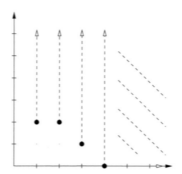

Figure 2.2 Decomposition of S into disjoint cones

Given a finite set $\{m_1, \ldots, m_l\}$ of monomials, there are many possible ways of how to arrange sets of multiplicative variables $\mu_1, \ldots, \mu_l$ such that (2.1) is a set of cones such that $\mathrm{Mon}(\mu_i)m_i \cap \mathrm{Mon}(\mu_j)m_j = \emptyset$ for $i \neq j$. These possibilities are addressed by the notion of *involutive division* which was introduced by V. P. Gerdt and Y. A. Blinkov [17, 18]. Important for this paper is only the Janet division:

Definition 2.7 [17] Let $M \subset \mathrm{Mon}(R)$ be finite. For each $m \in M$, the *Janet division* defines the set μ of multiplicative variables for the cone generated by m as follows. Let $m = x^a \in M$. For $1 \leq i \leq n$, we have:

$$x_i \in \mu \quad \Longleftrightarrow \quad \max\{b_i \mid x^b \in M;\ b_j = a_j \text{ for all } j < i\} = a_i,$$

i.e. x_i is a multiplicative variable for the cone generated by m if and only if its exponent in m is maximal among all exponents of monomials in M which have the same sequence of exponents of $x_1, \ldots, x_{i-1}$ as m.

There are also other common involutive divisions, e.g. J. Thomas [43] proposed another way of defining the multiplicative variables of cones.

Example 2.8 A decomposition of the $\mathrm{Mon}(R)$-multiple closed set S defined in Example 2.2 is

$$\{\, (x_1^4, \{x_1, x_2\}), \quad (x_1^3 x_2, \{x_2\}), \quad (x_1^2 x_2^2, \{x_2\}), \quad (x_1 x_2^2, \{x_2\}) \,\},$$

which is visualized in Figure 2.2.

2.2   Janet's Algorithm

In this section we describe Janet's algorithm. The combinatorial part of this algorithm is driven by the set of leading monomials of a given generating set for a module of tuples over a polynomial ring. In Janet's algorithm the generating set from the input is possibly enlarged or reduced in many steps in order to finally achieve one whose set of leading monomials generates the multiple-closed set of all leading monomials of the module under consideration. Decompositions of multiple-closed sets of monomials as described in the previous section are applied to the set of leading monomials. It indicates by the non-multiplicative variables how new leading monomials can be found. Janet-reduction is defined which produces normal forms of representatives for residue classes modulo the given module.

We continue to use the notation defined in the previous section and fix a monomial ordering $<$ on $\mathrm{Mon}(R^q)$. Then for any $p \in R^q - \{0\}$, the leading monomial of p (with respect to $<$) is denoted by $\mathrm{lm}(p)$ and its coefficient by $\mathrm{lc}(p)$. Moreover, for any set $L \subseteq R^q$ we define $\mathrm{lm}(L) := \{\, \mathrm{lm}(p) \mid p \in L - \{0\} \,\}$.

Definition 2.9 Let $T = \{\, (b_1, \mu_1), \ldots, (b_l, \mu_l) \,\} \subset (R^q - \{0\}) \times \mathcal{P}(\{x_1, \ldots, x_n\})$.

1. T is said to be *complete*, if $\{\, (\mathrm{lm}(b_1), \mu_1), \ldots, (\mathrm{lm}(b_l), \mu_l) \,\}$ is a decomposition into disjoint cones of the $\mathrm{Mon}(R)$-multiple closed set generated by $\{\, \mathrm{lm}(b_1), \ldots, \mathrm{lm}(b_l) \,\}$.

2. Let $p = \sum_{i=1}^{q} \sum_{a \in (\mathbb{Z}_{\geq 0})^n} c_{i,a}\, x^a\, e_i \in R^q$. Then the *support of p* is defined to be $\mathrm{supp}(p) := \{x^a\, e_i \mid c_{i,a} \neq 0\}$. The element p is *Janet-reducible modulo T* if there is an element $(b, \mu) \in T$ such that $\mathrm{Mon}(\mu)\, \mathrm{lm}(b)$ contains an element of $\mathrm{supp}(p)$. In this case, (b, μ) is called a *Janet-divisor* of p. If p is not Janet-reducible modulo T, then p is also said to be *Janet-reduced modulo T*.

The following algorithm subtracts suitable multiples of Janet-divisors from a given $p \in R^q$ as long as a monomial in p is Janet-reducible.

Algorithm 2.10 (Janet-reduce)

Input: $(p, T, <)$, where $p \in R^q$,
$\quad T = \{\, (b_1, \mu_1), \ldots, (b_l, \mu_l) \,\} \subset (R^q - \{0\}) \times \mathcal{P}(\{x_1, \ldots, x_n\})$ is complete,
$\quad$ and $<$ is a monomial ordering on $\mathrm{Mon}(R^q)$

Output: $r \in R^q$ such that $r + \langle b_1, \ldots, b_l \rangle = p + \langle b_1, \ldots, b_l \rangle$ and r is Janet-reduced modulo T

Algorithm:

1: $r \leftarrow 0$

2: **while** $p \neq 0$ **do**

3: $\quad$ **if** $\exists\, (b, \mu) \in T : \mathrm{lm}(p) \in \mathrm{Mon}(\mu)\, \mathrm{lm}(b)$ **then**

4: $\quad\quad p \leftarrow p - \frac{\mathrm{lc}(p)}{\mathrm{lc}(b)} \frac{\mathrm{lm}(p)}{\mathrm{lm}(b)}\, b$

5: $\quad$ **else**

6: $\quad\quad r \leftarrow r + \mathrm{lc}(p)\, \mathrm{lm}(p)$

7: $\quad\quad p \leftarrow p - \mathrm{lc}(p)\, \mathrm{lm}(p)$

8: $\quad$ **fi**

9: **od**

10: **return** r

Remark 2.11   1. As long as $p \neq 0$, the leading monomial of p gets properly smaller with respect to the monomial ordering $<$. Since $<$ is a well-ordering, Algorithm 2.10 terminates. The result r of Algorithm 2.10 is uniquely defined for the given input because T is complete. As opposed to reduction procedures for (tuples of) polynomials disregarding multiplicative variables, the course of Algorithm 2.10 is uniquely determined.

2. Let $p_1, p_2 \in R^q$ and T be as in the input of Algorithm 2.10. In general, the equality $p_1 + \langle b_1, \ldots, b_l \rangle = p_2 + \langle b_1, \ldots, b_l \rangle$ does not imply that the results of applying Janet-reduce to p_1 and p_2 are equal. But if T is a Janet basis (cf. the following definition), then the result of Janet-reduce constitutes a unique representative for every coset in $R^q / \langle b_1, \ldots, b_l \rangle$. It is called the *Janet-normal form of p modulo T* and is denoted by $\mathrm{NF}(p, T)$.

Definition 2.12 Let $T = \{ (b_1, \mu_1), \ldots, (b_l, \mu_l) \} \subset (R^q - \{0\}) \times \mathcal{P}(\{x_1, \ldots, x_n\})$ and assume that T is complete. Then T is said to be *passive*, if

$$\mathrm{NF}(x \cdot b_i, T) = 0$$

holds for all $x \in \overline{\mu_i}$, $1 \leq i \leq l$. In this case T is also called a *Janet basis for* $\langle b_1, \ldots, b_l \rangle \leq R^q$.

Now Janet's algorithm is presented which computes a Janet basis for a given submodule of R^q.

Algorithm 2.13 (JanetBasis)

Input: $(L, <)$, where $L \subset R^q$ is a finite set and $<$ is a monomial ordering on $\mathrm{Mon}(R^q)$

Output: $J \subset R^q \times \mathcal{P}(\{x_1, \ldots, x_n\})$, a Janet basis satisfying $\langle p \mid (p, \mu) \in J \rangle = \langle L \rangle$

Algorithm:

1: $G \leftarrow L$
2: **do**
3:    $G \leftarrow$ **Auto-reduce**(G)
4:    $J \leftarrow$ **Decompose**(G)
5:    $P \leftarrow \{ \mathrm{NF}(x \cdot p, J) \mid (p, \mu) \in J, x \notin \mu \}$
6:    $G \leftarrow \{ p \mid (p, \mu) \in J \} \cup P$
7: **od while** $P \neq \{0\}$
8: **return** J

In Algorithm 2.13, the sub-algorithm Auto-reduce subtracts suitable multiples of elements of G from other elements of G in order to arrive at a set $G' \subset R^q$ such that no monomial occurring with non-zero coefficient in an element of G' is divisible by a leading monomial of any other element of G'. The sub-algorithm Decompose returns a decomposition into disjoint cones of the $\mathrm{Mon}(R)$-multiple closed set generated by G. Finally, $\mathrm{NF}(x \cdot p, J)$ is computed by means of Algorithm 2.10. We have the following main theorem on Janet bases.

Theorem 2.14 *[37] Let $L \subset R^q$ be a finite set, $<$ a monomial ordering on $\mathrm{Mon}(R^q)$, and $J \subset R^q \times \mathcal{P}(\{x_1, \ldots, x_n\})$ the result of Algorithm 2.13 applied to $(L, <)$.*

1. *A k-basis of $\langle L \rangle$ is given by $\bigcup_{(g,\mu) \in J} \mathrm{Mon}(\mu) g$.*

2. *A k-basis of $R^q / \langle L \rangle$ is given by the cosets represented by $\bigcup_{(m,\mu) \in T} \mathrm{Mon}(\mu) m$, where T is a decomposition into disjoint cones of the complement of the $\mathrm{Mon}(R)$-multiple closed set generated by $\{ \mathrm{lm}(p) \mid (p, \mu) \in J \}$ in $\mathrm{Mon}(R^q)$.*

3. *Given $p_1, p_2 \in R^q$, we have*

$$p_1 + \langle L \rangle = p_2 + \langle L \rangle \quad \Longleftrightarrow \quad \mathrm{NF}(p_1, J) = \mathrm{NF}(p_2, J).$$

4. *Let $J = \{(g_1, \mu_1), \ldots, (g_r, \mu_r)\}$ and define the homomorphism of R-modules $\pi : R^r \to R$ by $e_i \mapsto g_i$, $i = 1, \ldots, r$, where $e_1, \ldots, e_r$ are the standard basis vectors of R^r. Then the*

$$x_j e_i - \sum_l \alpha_{i,j,l} e_l, \qquad x_j \notin \mu_i, \quad i = 1, \ldots, r,$$

form a Janet basis of $\ker \pi$ for a suitable monomial ordering, where $x_j g_i = \sum_l \alpha_{i,j,l} g_l$.

Statement 4 in the previous theorem allows to write down a free resolution of R^q / J by performing one Janet basis computation only.

Remark 2.15 Janet's algorithm can also be performed over the integers $\mathbb{Z}$ to obtain Janet bases for submodules of $\mathbb{Z}[x_1, \ldots, x_n]^q$. To this end, the monomial ordering $<$ is extended to terms $a \cdot m$, where $a \in \mathbb{Z}$ and $m \in \mathrm{Mon}(R^q)$, such that the absolute values of the coefficients of two terms with equal monomials are compared to break ties. Essentially the only other necessary modification of Janet's algorithm is to replace Algorithm 2.10 by a corresponding method which uses Euclidean division instead of exact division for the coefficients $\mathrm{lc}(p)$, $\mathrm{lc}(b)$.

An application of Janet bases over the integers can be found in Section 5.1.

2.3   Ore Algebras

In this section, Janet's algorithm is adapted to a certain class of Ore algebras, which are non-commutative polynomial rings. Previous results on Gröbner bases for non-commutative polynomial rings have been obtained by many researchers: Gröbner bases for algebras of solvable type were investigated by A. Kandri-Rody and V. Weispfenning in [30]. Buchberger's algorithm to compute Gröbner bases has been generalized to Ore algebras by F. Chyzak (see [7], [11], where it is applied to the study of special functions and combinatorial sequences). Involutive divisions were studied for the Weyl algebra case by W. M. Seiler [27]. Gröbner bases were also addressed in the framework of G-algebras by V. Levandovskyy [31]. Very recently, the concept of involutive division was extended to non-commutative rings in the thesis [13]. However, an adaptation of Janet's algorithm to Ore algebras does not need this generality, so that we give a presentation of it using only the necessary concepts. Restricted to Ore algebras, the exposition of Janet's algorithm is as efficient as for commutative polynomial rings.

Ore algebras are certain algebras of linear operators. For instance, for a field k, the Weyl algebra $A_1(k)$ consists of the polynomials in the differential operator $\frac{d}{dt}$ whose coefficients are polynomials in t with coefficients in k. In systems theory, many types of linear systems can be analyzed structurally by viewing them as modules over appropriate Ore algebras. The Ore algebra is chosen to contain all polynomials in the operators occurring in the equations that describe the system. For more details we refer to [9].

The construction of an Ore algebra can be thought of as an iteration of "Ore extensions" of a field or a polynomial ring or an Ore algebra itself. An "Ore extension" forms a skew polynomial ring in one indeterminate where the given ground field or ring provides the coefficients. After giving the definition of skew polynomial rings and Ore algebras following [7, 11], an adaptation of Janet's algorithm to a certain class of Ore algebras is discussed.

In what follows, let k be a field and A a domain[1] which is also a k-algebra.

Definition 2.16 [32] The *skew polynomial ring* $A[\partial; \sigma, \delta]$ is the (not necessarily commutative) ring consisting of all polynomials in ∂ with coefficients in A obeying the commutation rule

$$\partial a = \sigma(a) \partial + \delta(a), \qquad a \in A,$$

where $\sigma : A \to A$ is a k-algebra endomorphism (i.e. σ is multiplicative, k-linear and satisfies $\sigma(1) = 1$) and $\delta : A \to A$ is a σ-derivation, i.e. δ is k-linear and satisfies:

$$\delta(a\,b) = \sigma(a)\,\delta(b) + \delta(a)\,b, \qquad a, b \in A.$$

Remark 2.17 If σ is injective, then $A[\partial; \sigma, \delta]$ is a domain because the degree in ∂ of a product of two non-zero elements of $A[\partial; \sigma, \delta]$ equals the sum of the degrees in ∂ of the factors. Then the construction of a skew polynomial ring can be iterated.

We recall the notion of Ore algebra as defined in [7], [11].

Definition 2.18 Let $A = k$ or $A = k[x_1, \ldots, x_n]$, the commutative polynomial algebra over k. The *Ore algebra*

$$R = A[\partial_1; \sigma_1, \delta_1] \ldots [\partial_m; \sigma_m, \delta_m]$$

is the (not necessarily commutative) ring consisting of all polynomials in $\partial_1, \ldots, \partial_m$ with coefficients in A, where

$$\partial_i \partial_j = \partial_j \partial_i \qquad \text{for all} \quad 1 \leq i, j \leq m,$$

and all other commutation rules in R are defined by

$$\partial_i a = \sigma_i(a) \partial_i + \delta_i(a), \qquad a \in R, \qquad i = 1, \ldots, m, \tag{2.2}$$

---

[1] A *domain* is a (not necessarily commutative) ring A with 1 which satisfies for all $a_1, a_2 \in A$ that $a_1 \neq 0$, $a_2 \neq 0$ implies $a_1 a_2 \neq 0$.

where for $i = 1, \ldots, m$ the maps $\sigma_i : R \to R$ are k-algebra endomorphisms and $\delta_i : R \to R$ are σ_i-derivations (cf. Definition 2.16) satisfying

$$\begin{cases} \sigma_i \circ \sigma_j &= \sigma_j \circ \sigma_i, \\ \delta_i \circ \delta_j &= \delta_j \circ \delta_i, \\ \sigma_i \circ \delta_j &= \delta_j \circ \sigma_i, \qquad \text{for all} \quad 1 \leq i, j \leq m. \\ \sigma_i(\partial_j) &= \partial_j, \\ \delta_i(\partial_j) &= 0 \end{cases} \qquad (2.3)$$

Remark 2.19 Let $R = A[\partial_1; \sigma_1, \delta_1] \ldots [\partial_m; \sigma_m, \delta_m]$ be an Ore algebra.

1. A straightforward calculation shows that the conditions (2.3) ensure that the commutation rules (2.2) are compatible with the postulation that ∂_i and ∂_j commute in R.

2. If $\sigma_i = \mathrm{id}_R$ and $\delta_i = 0$ for all $i = 1, \ldots, m$, then R is the commutative polynomial algebra over k in $n + m$ indeterminates.

3. Although the Ore algebra R in Definition 2.18 is defined "in one go", it can be thought of as an iterated skew polynomial ring. However, formally it is more elaborate then to specify the maps $\sigma_i, \delta_i, i = 1, \ldots, m$.

When the commutation rules for an Ore algebra are clear from the context (e.g. for the Weyl algebras), then we also write

$$k[x_1, \ldots, x_n][\partial_1, \ldots, \partial_m]$$

instead of

$$k[x_1, \ldots, x_n][\partial_1; \sigma_1, \delta_1] \ldots [\partial_m; \sigma_m, \delta_m],$$

which is, of course, a non-commutative ring in general.

Let k be a field, $q \in \mathbb{N}$, and $R = k[x_1, \ldots, x_n][\partial_1; \sigma_1, \delta_1] \ldots [\partial_m; \sigma_m, \delta_m]$ an Ore algebra, where σ_i is an automorphism of $k[x_1, \ldots, x_n]$ for all $i = 1, \ldots, m$. The ring R is left Noetherian [32]. In order to be able to base Janet's algorithm for Ore algebras on the notion of multiple-closed sets of monomials as treated before in the commutative case, we restrict to the class of Ore algebras R for which the commutation rules have the form

$$\begin{cases} \sigma_i(x_j) &= c_{ij} x_j + d_{ij}, \qquad c_{ij} \in k - \{0\}, \quad d_{ij} \in k, \\ \deg(\delta_i(x_j)) &\leq 1, \qquad\qquad 1 \leq i \leq m, \quad 1 \leq j \leq n. \end{cases} \qquad (2.4)$$

Definition 2.20 We define the set of *monomials* of R by

$$\mathrm{Mon}(R) := \{x^a \partial^b \mid a \in (\mathbb{Z}_{\geq 0})^n, b \in (\mathbb{Z}_{\geq 0})^m\}, \ x^a := x_1^{a_1} \cdots x_n^{a_n}, \ \partial^b := \partial_1^{b_1} \cdots \partial_m^{b_m}.$$

Moreover, we set $\mathrm{Mon}(R^q) := \bigcup_{i=1}^q \mathrm{Mon}(R) e_i$.

Remark 2.21 $\mathrm{Mon}(R^q)$ is a k-basis of R^q, i.e. every element $p \in R^q$ has a unique representation

$$p = \sum_{i=1}^{q} \sum_{\substack{a \in (\mathbb{Z}_{\geq 0})^n \\ b \in (\mathbb{Z}_{\geq 0})^m}} c_{i,a,b}\, x^a\, \partial^b\, e_i \tag{2.5}$$

as linear combination of the elements of $\mathrm{Mon}(R^q)$ with coefficients $c_{i,a,b} \in k$, where only finitely many $c_{i,a,b}$ are non-zero. Since x_i and ∂_j do not commute in general, by the previous definition of monomials we distinguish a *normal form* (2.5) for the elements $p \in R^q$.

If R is non-commutative, it is not reasonable to generalize the notion of $\mathrm{Mon}(R)$-multiple closed set of monomials from Section 2.1 to "left $\mathrm{Mon}(R)$-multiple closed set" of monomials because left multiples of elements in $\mathrm{Mon}(R^q)$ are not elements in $\mathrm{Mon}(R^q)$ in general: for instance, in the Weyl algebra $A_1(k)$ with commutation rule $\partial t = t \partial + 1$, the left multiple ∂t of t is not a monomial. However, the type of commutation rules implied by (2.4) allows to use a bijection between the non-commutative monomials and commutative monomials in order to reduce the combinatorial part of Janet's algorithm in the present situation to the concepts of Section 2.1.

Remark 2.22 Let $\tilde{R} = k[\tilde{x}_1, \ldots, \tilde{x}_n, \tilde{\partial}_1, \ldots, \tilde{\partial}_m]$ be the commutative polynomial algebra in $n + m$ indeterminates over the field k. By using the bijection

$$\tilde{} : \mathrm{Mon}(R^q) \to \mathrm{Mon}(\tilde{R}^q)\ :\ x^a\, \partial^b\, e_i \mapsto \tilde{x}^a\, \tilde{\partial}^b\, e_i,$$

$a \in (\mathbb{Z}_{\geq 0})^n$, $b \in (\mathbb{Z}_{\geq 0})^m$, $1 \leq i \leq q$, the sets $\mathrm{Mon}(R^q)$ and $\mathrm{Mon}(\tilde{R}^q)$ can be identified.

Definition 2.23 Let $\tilde{R}$ and $\tilde{}$ be defined as in Remark 2.22.

1. A *monomial ordering* on $\mathrm{Mon}(R^q)$ is a total ordering $<$ on $\mathrm{Mon}(R^q)$ for which there exists a monomial ordering $\tilde{<}$ on $\mathrm{Mon}(\tilde{R}^q)$ such that

$$m_1 < m_2 \quad \Longleftrightarrow \quad \widetilde{m_1} \, \tilde{<} \, \widetilde{m_2} \qquad \text{for all } m_1, m_2 \in \mathrm{Mon}(R^q).$$

2. Fix a monomial ordering $<$ on $\mathrm{Mon}(R^q)$. Let $p \in R^q - \{0\}$ and consider its unique representation (2.5) with respect to the k-basis $\mathrm{Mon}(R^q)$. The *leading monomial* $\mathrm{lm}(p)$ of p is defined as the $<$-greatest element $x^a\, \partial^b\, e_i$ in $\mathrm{Mon}(R^q)$ for which $c_{i,a,b} \neq 0$. Then the *leading coefficient* $\mathrm{lc}(p)$ of p is defined by $c_{i,a,b}$. Hence, we have maps $\mathrm{lm} : R^q - \{0\} \to \mathrm{Mon}(R^q)$ and $\mathrm{lc} : R^q - \{0\} \to k$.

In what follows, we fix a monomial ordering $<$ on $\mathrm{Mon}(R^q)$ such that $\mathrm{Im}(\delta_i(x_j)) < x_j \delta_i$ for all $1 \leq i \leq m$, $1 \leq j \leq n$.

We are going to discuss now the adaptation of Janet's algorithm to Ore algebras. We succeed in doing so by keeping Algorithm 2.13 literally and explaining the differences caused by the non-commutativity of R.

Remark 2.24 Let M be a left submodule of R^q and let G be a finite generating set for M. The set $\mathrm{lm}(G)$ is utilized again to steer the course of Janet's algorithm. In Algorithm 2.13 it generated a $\mathrm{Mon}(R)$-multiple closed set. Here we apply the bijection $\sim$ of Remark 2.22 in order to consider $\mathrm{lm}(G)$ as a set of monomials in a commutative polynomial ring in $n + m$ variables. A procedure to decompose $[\mathrm{lm}(G)]$ into disjoint cones must respect that the module generators are multiplied by monomials only from the left. Accordingly, the notion of Janet-reducibility and the process of Janet-reduction are adopted to the present situation by restricting all multiplications of monomials in $\mathrm{Mon}(R)$ by arbitrary elements in R^q to left multiplications as was already indicated in Definition 2.9 and Algorithm 2.10. Because of the form of the commutation rules of R implied by (2.4) cancellation of the leading term is achieved similarly to the commutative case. Since $<$ is a well-ordering, Janet-reduction terminates.

Again the non-multiplicative variables of the cones guide Janet's algorithm to new module generators. Since Janet's algorithm for R again produces an ascending sequence of sets of leading monomials, Corollary 2.4 shows the termination. The correctness is proved in the same way as the correctness of Janet's algorithm is proved in the commutative case.

3 Implementations of the Involutive Basis Algorithm

In this section, implementations of the involutive basis algorithm [16] for commutative polynomial algebras respectively for certain differential rings are described. The programs have been developed in Maple and in C++ at Lehrstuhl B für Mathematik, RWTH Aachen, in cooperation with V. P. Gerdt and Y. A. Blinkov. The development of the Maple packages Involutive and Janet was started by C. F. Cid in 2000 and has been continued by the author from 2001 on. Meanwhile, these implementations have been adjusted to more recent versions of the involutive basis algorithm by Gerdt and Blinkov. In particular, many unnecessary involutive reductions are avoided by remembering multiplications by non-multiplicative variables which have already been considered in previous passivity checks and by involutive analogues of Buchberger's criteria for Gröbner basis computations [5, 6, 1, 26]. Strategies for selecting the next non-multiplicative multiple which is necessary to consider are currently adapted to new heuristics [21]. Involutive and Janet also contain some combinatorial tools, e.g. the computation of generalized Hilbert series, Hilbert polynomials, Cartan characters etc., and several valuable procedures for commutative algebra and differential algebra. A first description of Involutive and Janet was published in [3, 4]. These packages (including help pages) are available online at

http://wwwb.math.rwth-aachen.de/Janet.

More recently, the development of a new C++ module for Python called ginv has been initiated by V. P. Gerdt and Y. A. Blinkov. This open source software

http://invo.jinr.ru

Involutive			
Basic commands:			
InvolutiveBasis	GroebnerBasis	PolTabVar	PolZeroSets
PolHilbertSeries	FactorModuleBasis	PolInvReduce	Syzygies
Further commands for the computation of involutive bases:			
AddRhs	AssertInvBasis	InvolutivePreprocess	Substitute
Commands for special applications:			
PolMinPoly	PolRepres	PolResolution	
Various invariants derivable from PolHilbertSeries:			
PolCartanCharacter	PolHilbertPolynomial	PolHilbertFunction	
PolHP	PolHF	PolIndexRegularity	
Commands for module theory:			
PolIntersection	PolSubFactor	PolHom	PolDefect
PolHomHom	PolKernel	PolCokernel	PolExt1
PolExtn	PolParametrization	PolTorsion	
Auxiliary commands:			
InvolutiveOptions	LeadingMonomial	JanetGraph	Stats

Figure 3.1 Main commands of Involutive

is dedicated to the computation of involutive bases for systems of polynomials, differential equations and difference equations. A Maple interface to ginv is also available in the Maple package Involutive.

The Maple package JanetOre, which is a variant of Involutive, is only alluded to briefly because of its similarity to Involutive.

In general, we want to point to the Maple help pages of these packages for further information and details.

Finally, we would like to thank an anonymous referee for the reference [46] to another Maple implementation of the involutive basis technique for linear partial differential equations.

3.1 Involutive

The Maple package Involutive implements the involutive basis algorithm specialized for Janet and Janet-like division [23, 24, 20, 19] for commutative polynomial algebras over certain fields and the rational integers. The most important commands of Involutive are listed in Table 3.1.

Let k be a field for which there are constructive methods in Maple to compute in k, i.e. $k = \mathbb{Q}$ or $\mathbb{Z}/p\mathbb{Z}$, where p is a prime number, or a finitely generated field extension of $\mathbb{Q}$ or $\mathbb{Z}/p\mathbb{Z}$. We consider the polynomial algebra $R = k[x_1, \ldots, x_n]$ over k and the free R-module R^q for some $q \in \mathbb{N}$. A submodule M of R^q is given by a list of generators in R^q. (In case $q = 1$, M is an ideal of R.) Then InvolutiveBasis computes the minimal Janet basis (or the minimal Janet-like Gröbner basis) of M with respect to a certain monomial ordering. Among the supported monomial orderings are the lexicographical one, the degree reverse lexicographical one, block orderings and their extensions to the module case which either select the first non-zero entry of a tuple ("position over term") or the greatest monomial in all entries of the tuple ("term over position"). InvolutiveBasis computes $J \subset R^q \times \mathcal{P}(\{x_1, \ldots, x_n\})$ with the same meaning as in Algorithm 2.13, but returns only the list of $p \in R^q$ for which $(p, \mu) \in J$. The complete information about the Janet basis is displayed by PolTabVar. A k-vector space basis of R^q/M is enumerated by the generalized Hilbert series (cf. Section 4) which is returned by FactorModuleBasis. The classical Hilbert series is obtained by means of PolHilbertSeries. Moreover, the command GroebnerBasis extracts the minimal Gröbner basis of M from the involutive basis without further computation.

After a Janet basis of M has been obtained by InvolutiveBasis, PolInvReduce provides normal forms of elements of R^q modulo M and therefore allows computations in the residue class module R^q/M. The command Syzygies provides a generating set of relations between the given module generators. A free resolution of R^q/M can be obtained by means of PolResolution which works along the lines of Theorem 2.14 4. For the case $q = 1$, PolMinPoly determines the minimal polynomial of an element of R/M if it exists. If moreover, R/M is a finite-dimensional k-vector space, PolRepres returns the matrix which represents the multiplication by a given polynomial with respect to a given vector space basis of R/M.

In contrast to the monomial ordering which is chosen separately for each involutive basis computation, more general options for these computations can be set globally by means of InvolutiveOptions. The characteristic of k can be specified, or k can be chosen to be the ring $\mathbb{Z}$ so that subsequent Janet basis computations are performed over the rational integers. (For an application of Janet bases over the integers cf. Section 5.1.) Janet or Janet-like division can be switched on and the involutive criteria [26] for avoiding unnecessary reductions can be selected almost independently with InvolutiveOptions.

The computation of an involutive basis can be passed to a C++ program whose result is then read into Maple again. Apart from the command which does this explicitly, this much faster way of computing involutive bases can be turned on globally by means of InvolutiveOptions which then affects many of the higher level commands.

Several procedures provide combinatorial information about the residue class module R^q/M under consideration. Filtered and graded Hilbert polynomials and functions are returned by PolHP, PolHilbertPolynomial, PolHF, PolHilbertFunction, and the Cartan characters [39] are computed by PolCartanCharacter. Finally, JanetGraph returns the Janet graph [37] as a list.

It happens quite often that problems which seem to be intractable using the current Maple implementations of the involutive basis algorithms can be handled successfully by eliminating some of the variables which are present in the given equations. In case $q = 1$, the commands InvolutivePreprocess and Substitute search for an equation in the input in which a variable occurs only linearly with constant coefficient. Such an equation can be used to eliminate the variable in all other equations. The involutive basis computation is then performed with a smaller number of variables.

InvolutivePreprocess gives many suggestions of how to eliminate variables, Substitute also performs the substitution and iterates this strategy.

The package Involutive also includes a couple of procedures for module-theoretic applications. For instance, the intersection of two submodules of R^q can be computed; a presentation of the module of homomorphisms between two finitely presented R-modules is returned by PolHom; and the Ext-functor is implemented by means of PolExt1 and PolExtn. Since these module-theoretic methods had to be implemented in several Maple packages which are dedicated to computations with modules over different kinds of rings, the motivation for a Maple package which implements techniques from homological algebra arose. Such a package has been initiated by M. Barakat and the author under the name homalg. For more details we refer to [2].

3.2   Janet

Let k be a differential field of characteristic zero with n commuting derivations ∂_1, ..., ∂_n whose elements can be dealt with in Maple. The Maple package Janet implements the involutive basis algorithm over the iterated skew polynomial ring $R :=$ $k[\partial_1, \ldots, \partial_n]$. We consider the free (left) R-module R^q, where the rank $q \in \mathbb{N}$ usually is the number of dependent variables of a system of linear partial differential equations. In this context, multiplication of module generators by a variable in $\{\partial_1, \ldots, \partial_n\}$ corresponds to partial differentiation in Janet's algorithm. For each Janet basis element these variables are partitioned into variables which may be applied to this element and those which may not be used for differentiation.In case k is a field of constants, R is a commutative ring and the package Involutive could be seen as a special case of the package Janet.

The main procedures of Janet (cf. Table 3.2) construct Janet bases (or Janet-like Gröbner bases) for linear partial differential equations (JanetBasis), reduce linear differential expressions w.r.t. a Janet basis (InvReduce), give combinatorial information about the processed system (PrincDeriv, ParamDeriv, HilbertSeries, HilbertPolynomial, etc.), and construct polynomial or (truncated) power series solutions (PolySol, SolSeries, cf. Example 4.6). Moreover, a generating set of compatibility conditions for a given set of affine equations is returned by CompCond, and Resolution automatizes the iterated computation of compatibility conditions as indicated in Theorem 2.14 4.

The supported monomial orderings are the same as in Involutive. Again JanetOptions allows to choose Janet or Janet-like division and to select the involutive criteria.

During an involutive basis computation, the differential expressions in k by which the algorithm divides are collected in a list. After the involutive basis computation is completed, the command ZeroSets returns the list of these expressions. This functionality for instance allows to find configurations of system parameters which are in some sense singular, cf. Section 5.2.

Given a non-linear system of partial differential equations, the command Linearize returns a generic linearization [42] with non-constant coefficients of the system, for which no solution of the given set of equations is needed. In the case of non-linear ordinary differential equations, the linearization along a certain trajectory is obtained from the generic linearization by substituting the trajectory into the coefficients of the

Janet			
Basic commands:			
JanetBasis	InvReduce	PrincDeriv	ParamDeriv
HilbertSeries	ZeroSets	SolSeries	PolySol
Commands for special applications:			
CompCond	Resolution	SyzOp	SubFactor
Various invariants derivable from HilbertSeries:			
CartanCharacter	HilbertPolynomial	HilbertFunction	
HP	HF	IndexRegularity	
Translation between polynomials and differential expressions:			
Diff2Pol	Pol2Diff		
Commands dealing with differential operators:			
Diff2Op	Pol2Op	JAdjoint	CmpOp
AppOp	ElementaryDivisors		
Commands involving jet notation:			
Diff2Ind	Ind2Diff	Pol2Ind	AppOpInd
Useful commands for control theory:			
Linearize	FlatOutput	Parametrization	Torsion
Auxiliary commands:			
JanetOptions	LeadingDeriv	AssertJanetBasis	

Figure 3.2 Main commands of Janet

linearized equations. This is a particular setting in which ZeroSets is often applied.

The package Janet includes several commands which translates a differential expression into jet notation, or into a matrix representing a differential operator, or, if possible, into a polynomial. Matrices representing differential operators can be composed (CmpOp), applied (AppOp), and their adjoint operators can be computed (JAdjoint).

Furthermore, Janet contains more specific procedures for the case $R = \mathbb{Q}(t)[\partial]$. The elementary divisor theory can be adapted from commutative principal ideal domains to the Weyl algebra $R = \mathbb{Q}(t)[\partial]$: for every matrix $A \in R^{l \times m}$ there exist $r \in \mathbb{N}$ and matrices $P \in \mathrm{GL}(l, R)$ and $Q \in \mathrm{GL}(m, R)$ such that

$$PAQ = \mathrm{diag}(\underbrace{1, \ldots, 1}_{r}, \lambda, 0, \ldots, 0)$$

with a monic element $\lambda \in R$. This *Jacobson normal form* [12] uses the simplicity of the ring R. For rather small matrices A it can be computed by means of ElementaryDivisors.

Finally, the package Janet contains some commands for control-theoretic applications. In the algebraic analysis approach to linear control systems, cf. Section 5.2, the torsion submodule of a finitely presented (left) R-module is of special interest. By means of Torsion a presentation for the torsion submodule can be computed; more generally, the command Ext implements the Ext-functor with values in the ring R, which is of interest in the system-theoretic context, too.

3.3 JanetOre

The Maple package JanetOre is a variant of the package Involutive which was generalized to Ore algebras as described in Section 2.3. Since its commands are very similar to the ones of Involutive, we omit a longer discussion here.

JanetOre can be used in combination with the Maple package OreModules [8, 10] which was designed for system-theoretic applications.

For examples of applying JanetOre we refer to [42].

4 The Generalized Hilbert Series

In this section the generalized Hilbert series is presented as a very useful tool in the context of Janet bases. We adopt the notation from Section 2.

Definition 4.1 For any set $S \subseteq \mathrm{Mon}(R^q)$ of monomials, the *generalized Hilbert series* of S is the formal power series

$$H_S(x_1, \ldots, x_n) := \sum_{m \in S} m \in \bigoplus_{i=1}^q \mathbb{Z}[[x_1, \ldots, x_n]] \, e_i.$$

Remark 4.2 The Hilbert series usually encountered in commutative algebra is obtained from the generalized Hilbert series as $H_S(\lambda, \ldots, \lambda)$ for an indeterminate λ.

The importance of the complement in $\mathrm{Mon}(R^q)$ of the $\mathrm{Mon}(R)$-multiple closed set generated by the leading monomials of the elements of a Janet basis is explained in the following remark.

Remark 4.3 Let M be a submodule of R^q and let J be a Janet basis for M with respect to some monomial ordering on $\mathrm{Mon}(R^q)$. We denote by S the $\mathrm{Mon}(R)$-multiple closed set generated by $\{\mathrm{lm}(p) \mid (p, \mu) \in J\}$. According to Theorem 2.14 3., a k-basis of R^q/M is formed by the cosets in R^q/M represented by the monomials in $\bigcup_{(m,\mu) \in T} \mathrm{Mon}(\mu)m$, where T is a decomposition of $C := \mathrm{Mon}(R^q) - S$ into disjoint cones. Therefore, the generalized Hilbert series $H_C(x_1, \ldots, x_n)$ enumerates a k-basis of R^q/M.

The next remark shows that the computation of the generalized Hilbert series of a set S of monomials is trivial if a decomposition of S into disjoint cones (cf. Definition 2.6) is available.

Remark 4.4 Let $(C, \mu) \in \mathcal{P}(\mathrm{Mon}(R)) \times \mathcal{P}(\{x_1, \ldots, x_n\})$ be a cone. We use the geometric series

$$\frac{1}{1-y} = \sum_{i \geq 0} y^i$$

to write down the generalized Hilbert series $H_C(x_1, \ldots, x_n)$ as follows:

$$H_C(x_1, \ldots, x_n) = \frac{v(C)}{\prod_{y \in \mu} (1 - y)}.$$

More generally, every decomposition of a $\mathrm{Mon}(R)$-multiple closed set S into disjoint cones allows to compute the generalized Hilbert series of S by adding the generalized Hilbert series of the cones. In an analogous way this remark applies to the complements of $\mathrm{Mon}(R)$-multiple closed sets.

We only mention the following theorem which is proved in [42] by using the generalized Hilbert series.

Theorem 4.5 *Let $D = k[x_1, \ldots, x_n][\partial_1; \sigma_1, \delta_1] \ldots [\partial_m; \sigma_m, \delta_m]$ be an Ore algebra as in Section 2.3 and let $\mathcal{F} := \hom_k(D, k)$. Then $\mathcal{F}$ is an injective cogenerator for the category of left D-modules.*

Among the algebras which are in the scope of Theorem 4.5 are e.g. commutative polynomial rings and the Weyl algebras. Both are used for an algebraic treatment of systems of linear partial differential equations. In what follows, we exemplify the generalized Hilbert series for the case of partial differential equations with constant coefficients.

Let $R = k[\partial_1, \ldots, \partial_n]$ be the commutative polynomial algebra over a field k of characteristic zero, and let $\mathcal{F} := k[[x_1, \ldots, x_n]] \cong \hom_k(R, k)$ be the k-algebra of formal power series. Then every $p \in \mathcal{F}$ can be represented uniquely as

$$\sum_{\alpha \in (\mathbb{Z}_{\geq 0})^n} c_\alpha \frac{x^\alpha}{\alpha!}, \qquad c_\alpha \in k, \qquad \alpha! := \alpha_1! \cdots \alpha_n!. \tag{4.1}$$

We let R act on $\mathcal{F}$ from the left by partial differentiation of the formal power series (4.1) and consider $(x^\alpha/\alpha! \mid \alpha \in (\mathbb{Z}_{\geq 0})^n)$ as the dual basis of $(\partial^\alpha \mid \alpha \in (\mathbb{Z}_{\geq 0})^n)$ with respect to the pairing

$$(\ ,\ ) : R \times \mathcal{F} \to k : (d, \lambda) \mapsto \lambda(d),$$

i.e.

$$(\partial^\beta, \textstyle\sum_{\alpha \in (\mathbb{Z}_{\geq 0})^n} c_\alpha \frac{x^\alpha}{\alpha!}) = c_\beta.$$

For a given system of linear partial differential equations with constant coefficients for one unknown function we compute a Janet basis J for the ideal of R which is generated by the left hand sides of these equations. By considering the differential equations as linear equations for $(\partial^\beta, \lambda)$, $\beta \in (\mathbb{Z}_{\geq 0})^n$, where $\lambda \in \mathcal{F}$ is a formal power series solution, Janet's algorithm partitions $\mathrm{Mon}(R)$ into a set of monomials m for which $(m, \lambda) \in k$ can be chosen arbitrarily and the set S of monomials m for which $(m, \lambda) \in k$ is uniquely determined by these choices. The set S is the $\mathrm{Mon}(R)$-multiple closed set generated by the leading monomials of the polynomials in the Janet basis J. In particular, the k-dimension of the space of formal power series solutions can be computed as the number of monomials in the complement C of S in $\mathrm{Mon}(R)$. In fact, the generalized Hilbert series $H_C(\partial_1, \ldots, \partial_n)$ of C enumerates a basis for the Taylor coefficients $(\partial^\beta, \lambda)$ of λ whose values can be assigned freely.

In this context, M. Janet calls the monomials ∂^β in $\mathrm{Mon}(R) - S$ *parametric derivatives* because the corresponding Taylor coefficients $(\partial^\beta, \lambda)$ of a formal power series solution λ can be chosen arbitrarily. The monomials in S are called *principal derivatives*. The Taylor coefficients $(\partial^\beta, \lambda)$ which correspond to principal derivatives ∂^β are uniquely determined by k-linear equations in terms of the Taylor coefficients of parametric derivatives. Of course, the extension of this method to determine the formal power series solutions of a linear system of partial differential equations is extended to the case of more than one unknown function in a straightforward way.

The previous remark also applies to linear systems of partial differential equations whose coefficients are rational functions in the independent variables $x_1, \ldots, x_n$, i.e. $R = k[\partial_1, \ldots, \partial_n]$ is replaced by the Weyl algebra $B_n(k) := k(x_1, \ldots, x_n)[\partial_1, \ldots, \partial_n]$. Of course, in this case a formal power series solution is only well-defined if the left submodule M of $B_n(k)^q$ which represents the left hand sides of the equations is also a left submodule of $A[\partial_1, \ldots, \partial_n]^q$, where A is a k-subalgebra of $B_n(k)$ whose elements do not have a pole in $0 \in k^n$ and the Janet basis for M is computed within $A[\partial_1, \ldots, \partial_n]^q$. In other words, a formal power series solution is only well-defined if $0 \in k^n$ is not a zero of any denominator occurring in the course of Janet's algorithm.

Example 4.6 [29] Let us consider the following system of linear partial differential equations in one unknown function u of three variables x, y, t:

$$\begin{cases} \frac{\partial^2 u}{\partial x^2} - y \frac{\partial^2 u}{\partial t^2} & = & 0, \\ \frac{\partial^2 u}{\partial y^2} & = & 0. \end{cases} \qquad (4.2)$$

We want to determine all formal power series solutions of (4.2).

```
> with(Janet):
```

The independent variables are x, y, t, the dependent variable is u:

```
> ivar:=[x,y,t]:   dvar:=[u]:
```

Next we enter the left hand sides of the partial differential equations:

```
> L:=[diff(u(x,y,t),x$2) - y*diff(u(x,y,t),t$2),
> diff(u(x,y,t),y$2)];
```

$$L := [(\tfrac{\partial^2}{\partial x^2}\, u(x,\,y,\,t)) - y\,(\tfrac{\partial^2}{\partial t^2}\, u(x,\,y,\,t)),\ \tfrac{\partial^2}{\partial y^2}\, u(x,\,y,\,t)]$$

We compute the Janet basis of L with respect to the degree reverse lexicographical ordering on the partial derivatives.

```
>    JB:=JanetBasis(L,ivar,dvar);
```

$JB := [[\frac{\partial^2}{\partial y^2} \mathrm{u}(x, y, t), (\frac{\partial^2}{\partial x^2} \mathrm{u}(x, y, t)) - y\,(\frac{\partial^2}{\partial t^2} \mathrm{u}(x, y, t)), \frac{\partial^3}{\partial y\,\partial t^2} \mathrm{u}(x, y, t),$

$\frac{\partial^3}{\partial y^2\,\partial x} \mathrm{u}(x, y, t), \frac{\partial^4}{\partial t^4} \mathrm{u}(x, y, t), \frac{\partial^4}{\partial y\,\partial x\,\partial t^2} \mathrm{u}(x, y, t), \frac{\partial^5}{\partial x\,\partial t^4} \mathrm{u}(x, y, t)], [x, y, t], [u]]$

The result of JanetBasis is a list with three entries. The first entry gives the Janet basis, the second one the list of independent variables, the third one the list of dependent variables. Now the parametric derivatives are obtained as follows:

```
>    ParamDeriv(ivar,dvar);
```

$[\mathrm{u}(x, y, t), \frac{\partial}{\partial t} \mathrm{u}(x, y, t), \frac{\partial}{\partial y} \mathrm{u}(x, y, t), \frac{\partial}{\partial x} \mathrm{u}(x, y, t), \frac{\partial^2}{\partial t^2} \mathrm{u}(x, y, t), \frac{\partial^2}{\partial y\,\partial t} \mathrm{u}(x, y, t),$

$\frac{\partial^2}{\partial x\,\partial t} \mathrm{u}(x, y, t), \frac{\partial^2}{\partial y\,\partial x} \mathrm{u}(x, y, t), \frac{\partial^3}{\partial t^3} \mathrm{u}(x, y, t), \frac{\partial^3}{\partial x\,\partial t^2} \mathrm{u}(x, y, t), \frac{\partial^3}{\partial y\,\partial x\,\partial t} \mathrm{u}(x, y, t),$

$\frac{\partial^4}{\partial x\,\partial t^3} \mathrm{u}(x, y, t)]$

Of course, the generalized Hilbert series for the parametric derivatives is just the sum of the above twelve elements.

In this example, all formal power series solutions are in fact polynomials. The command PolySol of the Janet package returns all polynomial solutions of a system of linear partial differential equations up to a given degree. The first list in the result gives the polynomial solutions up to the given degree for each dependent variable; the second list consists of all parameters which occur in these polynomials.

```
>    PolySol(JB,5);
```

$[\mathrm{u}(x, y, t) = C1_{0,0,0} + C1_{1,0,0}\,x + C1_{0,1,0}\,y + C1_{0,0,1}\,t + \frac{1}{2}\,C1_{0,0,2}\,t^2$

$+ C1_{1,0,1}\,x\,t + C1_{0,1,1}\,y\,t + C1_{1,1,0}\,x\,y + \frac{1}{6}\,C1_{0,0,3}\,t^3 + \frac{1}{2}\,C1_{1,0,2}\,x\,t^2$

$+ C1_{1,1,1}\,x\,y\,t + \frac{1}{2}\,C1_{0,0,2}\,x^2\,y + \frac{1}{6}\,C1_{1,0,3}\,x\,t^3 + \frac{1}{2}\,C1_{0,0,3}\,x^2\,y\,t$

$+ \frac{1}{6}\,C1_{1,0,2}\,x^3\,y + \frac{1}{6}\,C1_{1,0,3}\,x^3\,y\,t], [C1_{0,0,0},\ C1_{1,0,0},\ C1_{0,1,0},\ C1_{0,0,1},$

$C1_{0,0,2},\ C1_{1,0,1},\ C1_{0,1,1},\ C1_{1,1,0},\ C1_{0,0,3},\ C1_{1,0,2},\ C1_{1,1,1},\ C1_{1,0,3}]$

5   Applications

The implementations of the involutive basis algorithm described in Section 3 have been applied already for the construction of polynomial invariants of finite groups [36], for the construction of matrix representations of Hurwitz groups [38] and for elimination problems [35]. In Section 5.1 we give an example from the context of [38].

The package Janet can be applied easily to control-theoretic problems. In Section 5.2 we give two examples. More applications of Janet and JanetOre can be found in [42].

Finally, a variant of the Janet package which was adapted to systems of linear difference equations was discussed in [25]. This implementation has been applied for the generation of finite difference schemes for partial differential equations in [22].

5.1 Constructing Matrix Representations of Finitely Presented Groups

Example 5.1 In order to find matrix representations of degree 3 over various fields K of the group
$$G_{2,3,13;4} := \langle\, a, b \mid a^2,\, b^3,\, (ab)^{13},\, [a,b]^4 \,\rangle,$$

the images of a and b under such a representation are first written down as matrices A and B with indeterminate entries. The relators a^2, b^3, $(ab)^{13}$, $[a,b]^4$ of the above presentation are translated into relations for commutative polynomials obtained from the entries of the matrix equations

$$A^2 = I_3, \quad B^3 = I_3, \quad (AB)^{13} = I_3, \quad [A,B]^4 = I_3,$$

where I_3 is the (3×3)-identity matrix. For more details on this approach we refer to [38].

```
>   with(Involutive):
>   with(LinearAlgebra):
```

We may choose a K-basis (v_1, v_2, v_3) of K^3 with respect to which the K-linear action of the images of a, b in $G_{2,3,13;4}$ on K^3 is represented by A, B. We let v_1 be an eigenvector of $A \cdot B$ with eigenvalue λ, possibly in an algebraic field extension of K, and let $v_2 = Bv_1$, $v_3 = Bv_2$. We confine ourselves to finding irreducible representations, which implies that (v_1, v_2, v_3) is K-linearly independent. Then, without loss of generality, A, B have the following form (where $c_1 = \lambda^{-1}$, $c_2 = \lambda$):

```
>   A:=Matrix(3,3,[[0,c[2],c[3]],[c[1],0,c[4]],
>   [0,0,c[5]]]);
>   B:=SubMatrix(Matrix(3,3)+1,1..3,[2,3,1]);
```

$$A := \begin{bmatrix} 0 & c_2 & c_3 \\ c_1 & 0 & c_4 \\ 0 & 0 & c_5 \end{bmatrix}$$

$$B := \begin{bmatrix} 0 & 0 & 1 \\ 1 & 0 & 0 \\ 0 & 1 & 0 \end{bmatrix}$$

By construction, we have $B^3 = I_3$.

We are going to derive a system of algebraic equations for the unknowns $c_1, \ldots, c_5$.

```
>   A^2;
```

$$\begin{bmatrix} c_1 c_2 & 0 & c_2 c_4 + c_3 c_5 \\ 0 & c_1 c_2 & c_1 c_3 + c_4 c_5 \\ 0 & 0 & c_5{}^2 \end{bmatrix}$$

```
>   Determinant(A);
```

$$-c_1 c_2 c_5$$

Now $\det A^2 = 1$, $\det B = 1$, and $\det(AB)^{13} = 1$ imply $\det A = 1$. Hence, $c_5 = -1$ because $c_1 c_2 = 1$.

```
>   A:=subs(c[5]=-1, A);
```

$$A := \begin{bmatrix} 0 & c_2 & c_3 \\ c_1 & 0 & c_4 \\ 0 & 0 & -1 \end{bmatrix}$$

So we are left with four unknowns and we are going to collect the equations which arise from equating all entries of $A^2 - I_3$, $(AB)^{13} - I_3$, $[A, B]^4 - I_3$ to zero.

```
>   var:=[c[1],c[2],c[3],c[4]];
```

$$var := [c_1, c_2, c_3, c_4]$$

```
>   L1:=map(op, convert(A^2-1, listlist));
```

$$L1 := [c_2 c_1 - 1, 0, c_2 c_4 - c_3, 0, c_2 c_1 - 1, c_1 c_3 - c_4, 0, 0, 0]$$

```
>   L2:=map(op, convert((A.B)^13-1, listlist)):
>   L3:=map(op, convert((A.B.A.B^2)^2-(B.A.(B^2).A)^2,
>   listlist)):
```

We let Maple delegate the involutive basis computation to the C++ implementation.

```
>   InvolutiveBasisFast(map(op, [L1,L2,L3]), var);
```

$$[1]$$

The previous involutive basis computation (over the rational numbers $\mathbb{Q}$) revealed that there is no non-trivial irreducible matrix representation $G_{2,3,13;4} \rightarrow \mathrm{GL}(3, \mathbb{C})$. Next we check whether there are such matrix representations of $G_{2,3,13;4}$ in positive characteristic. To this end we switch to involutive basis computations over the integers $\mathbb{Z}$.

```
>   InvolutiveOptions("rational", false);
```

true

```
>   InvolutiveBasisFast(map(op, [L1,L2,L3]),var);
```

$$[15,\ 15\,c_4,\ c_1 + 4\,c_2 + c_3 + 4\,c_4,\ 15\,c_4{}^2,\ 15\,c_3\,c_4,\ c_2\,c_4 - c_3,$$
$$c_3{}^2 + 4\,c_3\,c_4 + c_4{}^2 + c_3 + c_4 + 4,\ c_2\,c_3 - 4\,c_4{}^2 - 4\,c_3 - 1,\ c_2{}^2 + c_4{}^2 + 2\,c_3 - 7,$$
$$c_4{}^3 + 2\,c_3\,c_4 + 4\,c_4{}^2 + 4\,c_3 - 7\,c_4 + 1,\ c_3\,c_4{}^2 - 4\,c_3\,c_4 - c_4{}^2 + c_2 + 7\,c_3 - 2\,c_4 - 4]$$

We find that solutions of the system of algebraic equations exist only if $15 = 0$, i.e. possibly in characteristic 3 or 5. We are going to check both possibilities and start by changing the ground field for involutive basis computations to $\mathbb{F}_3 := \mathbb{Z}/3\mathbb{Z}$.

```
>   InvolutiveOptions("char", 3);
```

$$0$$

```
>   IB3:=InvolutiveBasisFast(map(op,[L1,L2,L3]),var);
```

$$IB3 := [c_1 + c_2 + c_3 + c_4,\ c_2\,c_4 + 2\,c_3,\ c_3{}^2 + c_3\,c_4 + c_4{}^2 + c_3 + c_4 + 1,$$
$$c_2\,c_3 + 2\,c_4{}^2 + 2\,c_3 + 2,\ c_2{}^2 + c_4{}^2 + 2\,c_3 + 2,\ c_4{}^3 + 2\,c_3\,c_4 + c_4{}^2 + c_3 + 2\,c_4 + 1,$$
$$c_3\,c_4{}^2 + 2\,c_3\,c_4 + 2\,c_4{}^2 + c_2 + c_3 + c_4 + 2]$$

Next we compute the Hilbert series of the factor module $\mathbb{F}_3[c_1, c_2, c_3, c_4]/J$, where J is the ideal of $\mathbb{F}_3[c_1, c_2, c_3, c_4]$ which is generated by $L1$, $L2$, $L3$.

```
>   AssertInvBasis(IB3,var):   PolHilbertSeries(t);
```

$$1 + 3\,t + 2\,t^2$$

Hence, $\mathbb{F}_3[c_1, c_2, c_3, c_4]/J$ is 6-dimensional as an $\mathbb{F}_3$-vector space. We compute the involutive basis for the ideal J with respect to the lexicographical ordering (indicated by the option "1"):

```
>   InvolutiveBasis(IB3,var,1);
```

$$[c_4{}^4 + c_4 + 1 + c_4{}^6,\ c_3 + 2\,c_4{}^5 + 2\,c_4{}^4 + c_4{}^3 + 2\,c_4 + 2,\ c_2 + c_4{}^5 + 2\,c_4{}^4 + c_4{}^2,$$
$$c_1 + 2\,c_4{}^4 + 2\,c_4{}^3 + 2\,c_4{}^2 + 2\,c_4 + 1]$$

```
>   Factor(c[4]^4+c[4]+1+c[4]^6) mod 3;
```

$$(c_4{}^3 + c_4{}^2 + 2)\,(c_4{}^3 + 2\,c_4{}^2 + 2\,c_4 + 2)$$

Hence, we have found matrix representations of $G_{2,3,13;4}$ of degree 3 over the fields $\mathbb{F}_3[\zeta]/(\zeta^3 + \zeta^2 + 2)$ and $\mathbb{F}_3[\zeta]/(\zeta^3 + 2\zeta^2 + 2\zeta + 2)$. For the sake of brevity we restrict to the first case. We substitute the expressions for c_1, c_2, and c_3 in $\zeta = c_4$ into the matrix A and check the imposed relations.

```
>   A3:=subs([c[1]=zeta^4+zeta^3+zeta^2+zeta+2,
>   c[2]=2*zeta^5+zeta^4+2*zeta^2,
>   c[3]=zeta^5+zeta^4+2*zeta^3+zeta+1, c[4]=zeta], A);
```

$$A3 := \begin{bmatrix} 0 & 2\zeta^5 + \zeta^4 + 2\zeta^2 & \zeta^4 + 2\zeta^3 + \zeta + \zeta^5 + 1 \\ \zeta^4 + \zeta^3 + \zeta^2 + \zeta + 2 & 0 & \zeta \\ 0 & 0 & -1 \end{bmatrix}$$

```
> map(i->Rem(i, zeta^3+zeta^2+2, zeta) mod 3, A3^2);
```

$$\begin{bmatrix} 1 & 0 & 0 \\ 0 & 1 & 0 \\ 0 & 0 & 1 \end{bmatrix}$$

```
> map(i->Rem(i, zeta^3+zeta^2+2, zeta) mod 3,
> (A3.B)^13);
```

$$\begin{bmatrix} 1 & 0 & 0 \\ 0 & 1 & 0 \\ 0 & 0 & 1 \end{bmatrix}$$

```
> map(i->Rem(i, zeta^3+zeta^2+2, zeta) mod 3,
> (A3.B.A3.B^2)^4);
```

$$\begin{bmatrix} 1 & 0 & 0 \\ 0 & 1 & 0 \\ 0 & 0 & 1 \end{bmatrix}$$

Finally, we repeat the above procedure for characteristic 5.

```
> InvolutiveOptions("char", 5);
```
$$3$$
```
> IB5:=InvolutiveBasisFast(map(op,[L1,L2,L3]),var);
```

$IB5 := [c_1 + 4\,c_2 + c_3 + 4\,c_4,\ c_2\,c_4 + 4\,c_3,\ c_3^2 + 4\,c_3\,c_4 + c_4^2 + c_3 + c_4 + 4,$
$c_2\,c_3 + c_4^2 + c_3 + 4,\ c_2^2 + c_4^2 + 2\,c_3 + 3,\ c_2^3 + 2\,c_3\,c_4 + 4\,c_4^2 + 4\,c_3 + 3\,c_4 + 1,$
$c_3\,c_4^2 + c_3\,c_4 + 4\,c_4^2 + c_2 + 2\,c_3 + 3\,c_4 + 1]$

Again, $\mathbb{F}_5[c_1, c_2, c_3, c_4]/J$ is 6-dimensional as a vector space over $\mathbb{F}_5 := \mathbb{Z}/5\mathbb{Z}$:

```
> AssertInvBasis(IB5,var):  PolHilbertSeries(t);
```
$$1 + 3\,t + 2\,t^2$$
```
> InvolutiveBasis(IB5,var,1);
```

$[2\,c_4^2 + c_4 + 1 + c_4^4 + c_4^3 + c_4^6,\ c_3 + 4\,c_4^5 + 2\,c_4^4 + 2\,c_4^2 + c_4 + 1,$
$c_2 + 4\,c_4^5 + 4\,c_4^4 + c_4^3 + 4\,c_4^2,\ c_1 + 2\,c_4^4 + c_4^3 + 2\,c_4^2 + 3\,c_4 + 4]$

```
> Factor(2*c[4]^2+c[4]+1+c[4]^4+c[4]^3+c[4]^6) mod 5;
```
$$(c_4^2 + 2\,c_4 + 4)\,(c_4^4 + 3\,c_4^3 + c_4^2 + 2\,c_4 + 4)$$

Hence, the matrix representations of $G_{2,3,13;4}$ in this case are defined over the fields $\mathbb{F}_5[\zeta]/(\zeta^2 + 2\zeta + 4)$ and $\mathbb{F}_5[\zeta]/(\zeta^4 + 3\zeta^3 + \zeta^2 + 2\zeta + 4)$. Again, we confine ourselves to the first case.

```
> A5:=subs([c[1]=3*zeta^4+4*zeta^3+3*zeta^2+2*zeta+1,
> c[2]=zeta^5+zeta^4+4*zeta^3+zeta^2,
> c[3]=zeta^5+3*zeta^4+3*zeta^2+4*zeta+4, c[4]=zeta],
> A): A5;
```

$$
\begin{bmatrix}
0 & \zeta^5 + \zeta^4 + 4\zeta^3 + \zeta^2 & 3\zeta^2 + 4\zeta + 4 + 3\zeta^4 + \zeta^5 \\
3\zeta^4 + 4\zeta^3 + 3\zeta^2 + 2\zeta + 1 & 0 & \zeta \\
0 & 0 & -1
\end{bmatrix}
$$

```
> map(i->Rem(i, zeta^2+2*zeta+4, zeta) mod 5, A5^2);
```

$$
\begin{bmatrix}
1 & 0 & 0 \\
0 & 1 & 0 \\
0 & 0 & 1
\end{bmatrix}
$$

```
> map(i->Rem(i, zeta^2+2*zeta+4, zeta) mod 5,
> (A5.B)^13);
```

$$
\begin{bmatrix}
1 & 0 & 0 \\
0 & 1 & 0 \\
0 & 0 & 1
\end{bmatrix}
$$

```
> map(i->Rem(i, zeta^2+2*zeta+4, zeta) mod 5,
> (A5.B.A5.B^2)^4);
```

$$
\begin{bmatrix}
1 & 0 & 0 \\
0 & 1 & 0 \\
0 & 0 & 1
\end{bmatrix}
$$

5.2   Control Theory

Example 5.2 [40] Let us consider a bipendulum, more precisely a bar on which two pendula of certain lengths $l1$ resp. $l2$ are fixed. The bar is movable horizontally. Its horizontal position is denoted by u. The horizontal positions of the end points of the two pendula are $x1$ resp. $x2$. Then u, $x1$, $x2$ fulfill the ordinary differential equations

$$
\begin{cases}
\dfrac{d^2 x1}{dt^2} + \dfrac{g}{l1}\, x1 - \dfrac{g}{l1}\, u & = 0, \\[2mm]
\dfrac{d^2 x2}{dt^2} + \dfrac{g}{l2}\, x2 - \dfrac{g}{l2}\, u & = 0,
\end{cases}
\tag{5.1}
$$

where g is the gravitational constant[1].

---

[1] The deduction of the given linear ordinary differential equations which describe the bipendulum relies on the approximation $\sin\theta \approx \theta$ for small angles θ of the pendula to the vertical.

```
>  with(Janet):
```

For the equations which describe the bipendulum, the independent variable is t, the dependent variables are $x1$, $x2$, and u, i.e. we investigate (5.1) for smooth functions $x1$, $x2$, and u of t.

```
>  ivar:=[t]:  dvar:=[x1,x2,u]:
```

We enter the left hand sides of the equations. The variables g, $l1$, and $l2$ are parameters for the system.

```
>  L1:=[g*x1(t)+l1*diff(x1(t),t,t)+g*u(t),
>  g*x2(t)+l2*diff(x2(t),t,t)+g*u(t)];
```

$$L1 := [g\,\mathrm{x}1(t) + l1\,(\tfrac{d^2}{dt^2}\,\mathrm{x}1(t)) + g\,\mathrm{u}(t),\ g\,\mathrm{x}2(t) + l2\,(\tfrac{d^2}{dt^2}\,\mathrm{x}2(t)) + g\,\mathrm{u}(t)]$$

Since we deal with ordinary differential equations with constant coefficients, it is appropriate to consider the system over the commutative polynomial algebra

$$R = \mathbb{Q}(g, l1, l2)[\partial],$$

where ∂ represents the differential operator $\frac{d}{dt}$. Then (5.1) can be written as

$$A\,(x1 \quad x2 \quad u)^T = 0$$

with

$$A := \begin{pmatrix} l1\,\partial^2 + g & 0 & g \\ 0 & l2\,\partial^2 - g & g \end{pmatrix} \in R^{2\times 3}.$$

These equations define the R-module

$$M := R^{1\times 3}/R^{1\times 2}\,A.$$

The algebraic analysis approach to linear control systems [33, 34, 40, 41, 44, 45, 9] translates structural properties of the control system into properties of the R-module M. In particular, controllability of the bipendulum is equivalent to $t(M) = \{0\}$, where $t(M)$ is the torsion submodule of M. More precisely, there is a one-to-one correspondence of the elements of $t(M)$ to *autonomous elements* of the control system, i.e. system variables whose behaviour is prescribed by an autonomous (differential) equation so that these variables cannot be controlled. A presentation of $t(M)$ is computed by the command Torsion of the package Janet:

```
>  Torsion(L1,ivar,dvar);
```

$$[[\_T1(t) = 0],\ [\_T1(t)],\ 0]$$

The result means that the torsion submodule $t(M)$ of M is trivial. In general, the first list of the result consists of equations whose right hand sides are generators for the presentation of the torsion submodule and whose left hand sides define names for these generators. The second entry of the result is a list of generating relations satisfied by the generators for the presentation of $t(M)$ (more details are given below). The result includes the Hilbert series of $t(M)$ as last entry. In the present case, we conclude that $t(M) = \{0\}$. However, this consequence only holds for the generic configurations of the parameters for the bipendulum, i.e. for a generic choice of g, $l1$, and $l2$. We check by which expressions the computation of $t(M)$ has divided:

> `ZeroSets(ivar);`

$$[g,\ l1,\ l2,\ l1 - l2]$$

Of course, g, $l1$, and $l2$ are assumed to be strictly positive. But the algorithm divided by the difference $l1 - l2$ which is zero if and only if the lengths of the two pendula are equal. Therefore, we study this particular case separately:

> `L2:=subs(l2=l1,L1);`

$$L2 := [g\,x1(t) + l1\,(\tfrac{d^2}{dt^2}\,x1(t)) + g\,u(t),\ g\,x2(t) + l1\,(\tfrac{d^2}{dt^2}\,x2(t)) + g\,u(t)]$$

By M' we denote the R-module which is defined by this particular system ($l1 = l2$). We compute a presentation of the torsion submodule $t(M')$ of M':

> `T:=Torsion(L2,ivar,dvar);`

$$T := [[\_T1(t) = x1(t) - x2(t)],\ [\_T1(t)\,g + (\tfrac{d^2}{dt^2}\,\_T1(t))\,l1],\ 1 + s]$$

The list of expressions by which the algorithm divided assures that this result is valid in any case:

> `ZeroSets(ivar);`

$$[g,\ l1]$$

The first list in T gives one generator $x1 - x2$ for $t(M')$. For the presentation of $t(M')$ this generator is called $\_T1$. The second list in T is a generating set of the differential relations satisfied by $\_T1$. It serves as the list of relations for the presentation of $t(M')$. Finally, the Hilbert series of $t(M')$ given in the last entry of T states that $t(M')$ is two-dimensional as a $\mathbb{Q}(g, l1)$-vector space. More precisely, $t(M')$ has the $\mathbb{Q}(g, l1)$-basis $(\_T1, \tfrac{d\,\_T1}{dt})$.

We conclude that the bipendulum is controllable if and only if the lengths of the two pendula are different. This coincides with the intuition that for the case $l1 = l2$ we consider two copies of a system consisting of one pendulum alone which are joined by the bar. If the initial configurations of the pendula of equal lengths are the same (positions and velocities), then the link of these two systems enforces the motions of the two pendula to be the same.

The next example is of academic interest only, but demonstrates the use of the Janet-command ZeroSets for non-constant parameters of the system under consideration.

Example 5.3 [15] Let

$$A = \begin{pmatrix} 1 & 0 & 0 \\ a(t) & 1 & 0 \\ 0 & 0 & 1 \end{pmatrix}, \qquad B = \begin{pmatrix} 1 & 1 \\ 0 & 1 \\ 0 & 1 \end{pmatrix},$$

where $a(t)$ is a smooth function of t, and consider the linear system

$$\begin{pmatrix} \dot{x}_1 \\ \dot{x}_2 \\ \dot{x}_3 \end{pmatrix} = A \begin{pmatrix} x_1 \\ x_2 \\ x_3 \end{pmatrix} + B \begin{pmatrix} u_1 \\ u_2 \end{pmatrix}.$$

Similarly to the previous example, we want to check controllability of this linear system.

> `with(Janet):`

The independent variable is t, the dependent variables are $x1$, $x2$, $x3$, $u1$, $u2$.

> `ivar:=[t]; dvar:=[x1,x2,x3,u1,u2];`
$$ivar := [t]$$
$$dvar := [x1,\ x2,\ x3,\ u1,\ u2]$$

Next we enter the system equations:

> `L:=[diff(x1(t),t)-x1(t)-u1(t)-u2(t),`
> `diff(x2(t),t)-a(t)*x1(t)-x2(t)-u2(t),`
> `diff(x3(t),t)-x3(t)-u2(t)];`

$L := [(\frac{d}{dt} x1(t)) - x1(t) - u1(t) - u2(t),\ (\frac{d}{dt} x2(t)) - a(t) x1(t) - x2(t) - u2(t),$
$(\frac{d}{dt} x3(t)) - x3(t) - u2(t)]$

Similarly to the previous example, we check whether the linear system under consideration has non-trivial autonomous elements.

> `Torsion(L,ivar,dvar);`
$$[[\_T1(t) = 0],\ [\_T1(t)],\ 0]$$

For generic $a(t)$, no autonomous elements exist. Next we check the list of differential expressions by which Torsion has divided.

> `ZeroSets(ivar);`
$$[a(t)]$$

Hence, for the case $a(t) = 0$, controllability has to be checked separately. However, for the case $a(t) \neq 0$, a flat output [14] of the control system can be obtained as follows:

> `FlatOutput(L,ivar,dvar);`

$$[-\frac{x3(t)}{a(t)^2}, -\frac{x2(t)}{a(t)^2}]$$

Finally, we check whether the system has autonomous elements in case $a(t) = 0$.

> `L0:=subs(a(t)=0,L);`

$L0 := [(\frac{d}{dt} x1(t)) - x1(t) - u1(t) - u2(t), (\frac{d}{dt} x2(t)) - x2(t) - u2(t),$
$(\frac{d}{dt} x3(t)) - x3(t) - u2(t)]$

> `Torsion(L0,ivar,dvar);`

$$[[\_T1(t) = x2(t) - x3(t)], [-\_T1(t) + (\frac{d}{dt} \_T1(t))], 1]$$

In fact, $y := x2 - x3$ is an autonomous element in this case, which satisfies the differential equation $\dot{y} - y = 0$. Therefore, the system is not completely controllable in this case.

Bibliography

[1] J. Apel and R. Hemmecke, *Detecting unnecessary reductions in an involutive basis computation*, J. Symbolic Computation 40 (2005), pp. 1131–1149.

[2] M. Barakat and D. Robertz, *homalg: An abstract package for homological algebra*, submitted for publication, see also arXiv: math.AC/0701146 v2 23 Jul 2007.

[3] Y. A. Blinkov, C. F. Cid, V. P. Gerdt, W. Plesken, and D. Robertz, *The MAPLE Package "Janet": I. Polynomial Systems*, Proceedings of the 6th International Workshop on Computer Algebra in Scientific Computing, Passau (Germany), 2003, `http://wwwb.math.rwth-aachen.de/Janet`, pp. 31–40.

[4] ———, *The MAPLE Package "Janet": II. Linear Partial Differential Equations*, Proceedings of the 6th International Workshop on Computer Algebra in Scientific Computing, Passau (Germany), 2003, `http://wwwb.math.rwth-aachen.de/Janet`, pp. 41–54.

[5] B. Buchberger, *Ein Algorithmus zum Auffinden der Basiselemente des Restklassenringes nach einem nulldimensionalen Polynomideal*, Ph.D. thesis, Univ. Innsbruck, Austria, 1965.

[6] ———, *A Criterion for detecting unnecessary reductions in the construction of Gröbner bases*, Symbolic and Algebraic Computation, Proc. EUROSAM '79, Lecture Notes in Computer Science, vol. 72, Springer, 1979, pp. 3–21.

[7] F. Chyzak, *Fonctions holonomes en calcul formel*, Ph.D. thesis, Ecole Polytechnique, France, 1998.

[8] F. Chyzak, A. Quadrat, and D. Robertz, OREMODULES *project*, 2003-2007, `http://wwwb.math.rwth-aachen.de/OreModules`.

[9] ———, *Effective algorithms for parametrizing linear control systems over Ore algebras*, Applicable Algebra in Engineering, Communication and Computing 16 (2005), pp. 319–376.

[10] _____, OREMODULES: *A symbolic package for the study of multidimensional linear systems*, Applications of Time-Delay Systems (J. Chiasson and J.-J. Loiseau, eds.), Lecture Notes in Control and Information Sciences 352, Springer, 2007, pp. 233–264.

[11] F. Chyzak and B. Salvy, *Non-commutative elimination in Ore algebras proves multivariate identities*, Journal of Symbolic Computation 26 (1998), pp. 187–227.

[12] P. M. Cohn, *Free Rings and their Relations*, 2. ed., Academic Press, 1985.

[13] G. A. Evans, *Noncommutative Involutive Bases*, Ph.D. thesis, University of Wales, Bangor, 2006.

[14] M. Fliess, J. Lévine, P. Martin, and P. Rouchon, *Flatness and defect of nonlinear systems: introductory theory and examples*, International Journal of Control 61 (1995), pp. 1327–1361.

[15] E. Freund, *Zeitvariable Mehrgrößensysteme*, Springer, 1971.

[16] V. P. Gerdt, *Involutive Algorithms for Computing Gröbner Bases*, Computational Commutative and Non-Commutative Algebraic Geometry (S. Cojocaru, G. Pfister, and V. Ufnarovski, eds.), NATO Science Series, IOS Press, 2005, pp. 199–225.

[17] V. P. Gerdt and Y. A. Blinkov, *Involutive bases of polynomial ideals*, Mathematics and Computers in Simulation 45 (1998), pp. 519–541.

[18] _____, *Minimal involutive bases*, Mathematics and Computers in Simulation 45 (1998), pp. 543–560.

[19] _____, *Janet-like Gröbner Bases*, Computer Algebra in Scientific Computing CASC 2005 (V. G. Ganzha, E. W. Mayr, and E. V. Vorozhtsov, eds.), Springer, 2005, pp. 184–195.

[20] _____, *Janet-like Monomial Division*, Computer Algebra in Scientific Computing CASC 2005 (V. G. Ganzha, E. W. Mayr, and E. V. Vorozhtsov, eds.), Springer, 2005, pp. 174–183.

[21] _____, *On Computing Janet Bases for Degree-Compatible Orderings*, Proceedings of the 10-th Rhine Workshop on Computer Algebra, Basel (Switzerland), 2006, see also arXiv: math.AC/0603161 v2 10 Apr 2006, pp. 107–117.

[22] V. P. Gerdt, Y. A. Blinkov, and V. V. Mozzhilkin, *Gröbner Bases and Generation of Difference Schemes for Partial Differential Equations*, Symmetry, Integrability and Geometry: Methods and Applications 2 (2006).

[23] V. P. Gerdt, Y. A. Blinkov, and D. A. Yanovich, *Construction of Janet Bases, I. Monomial Bases*, Computer Algebra in Scientific Computing CASC 2001 (V. G. Ganzha, E. W. Mayr, and E. V. Vorozhtsov, eds.), Springer, 2001, pp. 233–247.

[24] _____, *Construction of Janet Bases, II. Polynomial Bases*, Computer Algebra in Scientific Computing CASC 2001 (V. G. Ganzha, E. W. Mayr, and E. V. Vorozhtsov, eds.), Springer, 2001, pp. 249–263.

[25] V. P. Gerdt and D. Robertz, *A Maple Package for Computing Gröbner Bases for Linear Recurrence Relations*, Nuclear Instruments and Methods in Physics Research, A: Accelerators, Spectrometers, Detectors and Associated Equipment 559 (2006), pp. 215–219.

[26] V. P. Gerdt and D. A. Yanovich, *Experimental Analysis of Involutive Criteria*, Algorithmic Algebra and Logic (A. Dolzmann, A. Seidl, and T. Sturm, eds.), BOD Norderstedt, 2005, pp. 105–109.

[27] M. Hausdorf, W. M. Seiler, and R. Steinwandt, *Involutive Bases in the Weyl Algebra*, Journal of Symbolic Computation 34 (2002), pp. 181–198.

[28] D. Hilbert, *Über die Theorie der algebraischen Formen*, Mathematische Annalen 36 (1890), pp. 473–534.

[29] M. Janet, *Leçons sur les systèmes des équationes aux dérivées partielles*, Cahiers Scientifiques IV, Gauthiers-Villars, Paris, 1929.

[30] A. Kandri-Rody and V. Weispfenning, *Non-commutative Gröbner Bases in Algebras of Solvable Type*, Journal of Symbolic Computation 9 (1990), pp. 1–26.

[31] V. Levandovskyy, *Non-commutative Computer Algebra for polynomial algebras: Gröbner bases, applications and implementation*, Ph.D. thesis, Universität Kaiserslautern, Germany, 2005.

[32] J. C. McConnell and J. C. Robson, *Noncommutative Noetherian Rings*, American Mathematical Society, 2000.

[33] U. Oberst, *Multidimensional constant linear systems*, Acta Applicandae Mathematicae 20 (1990), pp. 1–175.

[34] U. Oberst and F. Pauer, *The Constructive Solution of Linear Systems of Partial Difference and Differential Equations with Constant Coefficients*, Multidimensional Systems and Signal Processing 12 (2001), pp. 253–308.

[35] W. Plesken and D. Robertz, *Elimination for coefficients of special characteristic polynomials*, in preparation.

[36] ———, *Constructing Invariants for Finite Groups*, Experimental Mathematics 14 (2005), pp. 175–188.

[37] ———, *Janet's approach to presentations and resolutions for polynomials and linear pdes*, Archiv der Mathematik 84 (2005), pp. 22–37.

[38] ———, *Representations, commutative algebra, and Hurwitz groups*, Journal of Algebra 300 (2006), pp. 223–247.

[39] J.-F. Pommaret, *Partial Differential Equations and Group Theory*, Kluwer Academic Publishers, 1994.

[40] ———, *Partial Differential Control Theory*, Kluwer Academic Publishers, 2001.

[41] A. Quadrat, *Analyse algébrique des systèmes de contrôle linéaires multidimensionnels*, Ph.D. thesis, Ecole Nationale des Ponts et Chaussées, France, 1999.

[42] D. Robertz, *Formal Computational Methods for Control Theory*, Ph.D. thesis, RWTH Aachen, Germany, 2006, available online at `http://darwin.bth.rwth-aachen. de/opus/volltexte/2006/1586`.

[43] J. Thomas, *Differential Systems*, American Mathematical Society, 1937.

[44] J. Wood, *Modules and behaviours in nD systems theory*, Multidimensional Systems and Signal Processing 11 (2000), pp. 11–48.

[45] E. Zerz, *Topics in Multidimensional Linear Systems Theory*, Lecture Notes in Control and Information Sciences 256, Springer, 2000.

[46] Shan-qing Zhang and Zhi-bin Li, *An Implementation for the Algorithm of Janet bases of Linear Differential Ideals in the Maple System*, Acta Mathematicae Applicatae Sinica 20 (2004), pp. 605–616.

Author information

Daniel Robertz, Lehrstuhl B für Mathematik, RWTH Aachen, Templergraben 64, D–52062 Aachen, Germany.

Email: `daniel@momo.math.rwth-aachen.de`

Radon Series Comp. Appl. Math **2**, 169–216

Spencer Cohomology, Differential Equations, and Pommaret Bases

Werner M. Seiler

Key words. Spencer cohomology, Koszul homology, quasi-regularity, partial differential equation, formal integrability, involution, Pommaret basis, δ-regularity, Castelnuovo–Mumford regularity.

AMS classification. 13P10, 35G20, 35N99, 58H10.

1 Introduction

A key notion in the theory of general (i. e. including under- or overdetermined) systems of differential equations is involution. As we will see it may be understood as a simultaneous abstraction and generalisation of Gröbner bases for polynomial ideals to differential equations (without any restriction to linear or polynomial systems). Without the concept of involution (or some variation of it like passivity in Janet–Riquier theory [27, 48] or differential Gröbner bases [37]), one cannot prove general existence and uniqueness theorems like the Cartan–Kähler theorem.

The terminology "involutive" appeared probably first in the 19th century in the analysis of overdetermined systems of first-order linear differential equations in one unknown function. Nowadays these works are subsumed by the Frobenius theorem which is usually treated in differential geometry (where one still has the notion of an

Work supported by EU NEST-Adventure grant 5006 (GIFT).

involutive distribution) and no longer in differential equations theory. The first complete theories of arbitrary differential equations were reached in the early 20th century with the Janet–Riquier and the Cartan–Kähler theory [6, 9, 24, 29], the latter one formulated in the language of exterior differential systems (Cartan also provided a differential equations version for linear first-order systems in [8]).

The Janet–Riquier theory is completely based on local coordinate computations and requires the introduction of a ranking in analogy to the term orders used to define Gröbner bases. By contrast, the Cartan–Kähler theory is in principle intrinsic, but in the classical approach the decision whether or not a given exterior differential system is involutive requires for the so-called Cartan test at least the introduction of a local basis on the tangent bundle which is often done via coordinates.

Only much later it was realised that Cartan's test is actually of a homological nature. The homological approach to involution was mainly pioneered by Spencer [57] and collaborators [18, 19, 46]; later discussions can be found e. g. in [6, 10, 12, 28, 32, 30, 34, 36, 38, 42]. However, one should mention that the Spencer cohomology appeared first not in the context of differential equations but in deformation theory [56].

This contribution is largely a review; most results are well known to specialists. However, these specialists are divided into two classes: many experts in the formal theory of differential equations are familiar with Spencer cohomology but much less with commutative algebra; conversely, few experts in commutative algebra know the formal theory. This clear division into two communities is the main reason why even elementary facts like that the degree of involution and the Castelnuovo–Mumford regularity coincide have remained unnoticed for a long time. It is our hope that this article may help to bridge this gap.

Some novel aspects are contained in the use of Pommaret bases; this concerns in particular Chapter 5 (parts of this material is also contained in [22]). While we do not discuss here any algorithmic aspects (this is done in [22]), it should be mentioned that by relating concepts like involution or the Castelnuovo–Mumford regularity to Pommaret bases, we make them immediately accessible for effective computations.

The article is organised as follows. The next chapter introduces axiomatically the polynomial de Rham complex and its dualisation, the Koszul complex. The Spencer cohomology and the Koszul homology of a (co)module arise then by (co)tensoring with the (co)module. Since the symmetric algebra is Noetherian by Hilbert's basis theorem, it is straightforward to prove a number of finiteness statements for the Spencer cohomology via dualisation to the Koszul side which are otherwise quite hard to obtain. It seems that the duality between the Spencer cohomology and the Koszul homology was first noted by Singer and Sternberg [55] (but see also [46, Lemma 5.5]) who attributed it to private discussions with Grothendieck and Mumford. An independent proof was later given by Ruiz [49]. The chapter closes by defining involution as the vanishing of the Spencer cohomology (or the dual Koszul homology, resp.).

Chapter 2 is concerned with the Cartan test for deciding whether a symbolic system is involutive. It represents a homological reformulation of the classical Cartan test in the theory of exterior differential systems and is due to Matsushima [39, 40]. We then discuss the dual version of the Cartan test developed by Serre in a letter appended to [21]. While the notion of involution is intrinsically defined, any form of Cartan's test requires the introduction of coordinates and it turns out that in certain "bad" coordinate systems the test fails. This problem is known under the name δ- or quasi-regularity and appears in all versions of the Cartan test.

Chapter 4 recalls briefly the notion of an involutive basis for a polynomial ideal with particular emphasis on Pommaret bases. Involutive bases represent a special kind of Gröbner bases with additional combinatorial properties; they were introduced by Gerdt and Blinkov [15] combining ideas from the Janet–Riquier theory of differential equations with the classical theory of Gröbner bases. It is shown that the Pommaret basis with respect to the degree reverse lexicographic order contains many structural information. This chapter essentially summarises some of the results of [51].

Most invariants that can be read off from a Pommaret basis are of a homological nature. Therefore we study in Chapter 5 the relation between the Pommaret basis (for the degree reverse lexicographic order) of an ideal $\mathcal{I}$ and the Koszul homology of the factor algebra $\mathcal{P}/\mathcal{I}$ in more detail. The presented results only scratch at the surface of this question. It is a conjecture of us that for Pommaret bases the Schreyer Theorem can be significantly generalised so that it yields explicit bases for the whole Koszul homology and not only for the degree-1-part. This entails that in contrast to general Gröbner bases this special kind of bases is to a large extent determined by the structure of the ideal.

The last two chapters demonstrate how the algebraic theory developed in the previous chapters can be applied to general differential equations. For this purpose, a differential equation is defined geometrically as submanifold of a jet bundle. The fundamental identification leads to a natural polynomial structure in the hierarchy of jet bundles. It allows us to associate with each differential equation a symbolic system (or dually a polynomial module) so that involution can be effectively decided with any form of Cartan's test. We also discuss why 2-acyclicity implies formal integrability; going to the Koszul side this becomes an elementary statements about syzygies.

Finally, some conclusions are given. Two small appendices fix the used notations concerning multi indices and term orders, respectively. A slightly larger appendix gives an introduction to coalgebras and comodules.

2   Spencer Cohomology and Koszul Homology

Let $\mathcal{V}$ be an n-dimensional vector space over a field $\Bbbk$;[1] over $\mathcal{V}$ one has the symmetric algebra $S\mathcal{V}$ and the exterior algebra $\Lambda\mathcal{V}$. We introduce two natural complexes based on the product spaces $S_q\mathcal{V} \otimes \Lambda_p\mathcal{V}$. Any element of such a space may be written as a $\Bbbk$-linear sum of separable elements, i. e. elements of the form $w_1 \cdots w_q \otimes v_1 \wedge \cdots \wedge v_p$ with $w_i, v_j \in \mathcal{V}$. By convention, we set $S_j\mathcal{V} = 0$ for $j < 0$.

Definition 2.1 For any integer $r \geq 0$ the complex

$$0 \longrightarrow S_r\mathcal{V} \xrightarrow{\delta} S_{r-1}\mathcal{V} \otimes \mathcal{V} \xrightarrow{\delta} S_{r-2}\mathcal{V} \otimes \Lambda_2\mathcal{V} \xrightarrow{\delta} \cdots$$

$$\cdots \xrightarrow{\delta} S_{r-n}\mathcal{V} \otimes \Lambda_n\mathcal{V} \longrightarrow 0$$

(2.1)

---

[1] For simplicity, we assume throughout that char $\Bbbk = 0$.

where the differential δ is defined by[2]

$$\delta(w_1 \cdots w_q \otimes v_1 \wedge \cdots \wedge v_p) = \sum_{i=1}^{q} w_1 \cdots \widehat{w_i} \cdots w_q \otimes w_i \wedge v_1 \wedge \cdots \wedge v_p \qquad (2.2)$$

is called the *polynomial de Rham complex* $R_r(S\mathcal{V})$ at degree r over the vector space $\mathcal{V}$. The *Koszul complex* $K_r(S\mathcal{V})$ at degree r over $\mathcal{V}$ is given by

$$0 \longrightarrow S_{r-n}\mathcal{V} \otimes \Lambda_n\mathcal{V} \xrightarrow{\partial} S_{r-n+1}\mathcal{V} \otimes \Lambda_{n-1}\mathcal{V} \xrightarrow{\partial} \cdots$$

$$\cdots \xrightarrow{\partial} S_r\mathcal{V} \longrightarrow 0 \qquad (2.3)$$

where now the differential ∂ is defined as

$$\partial(w_1 \cdots w_q \otimes v_1 \wedge \cdots \wedge v_p) = \sum_{i=1}^{p} (-1)^{i+1} w_1 \cdots w_q v_i \otimes v_1 \wedge \cdots \wedge \widehat{v_i} \wedge \cdots \wedge v_p . \quad (2.4)$$

It is trivial to verify that, due to the skew-symmetry of the wedge product, the differentials satisfy $\delta^2 = 0$ and $\partial^2 = 0$, so that we are indeed dealing with complexes.

Let $\{x^1, \ldots, x^n\}$ be a basis of $\mathcal{V}$. Then a basis of the vector space $S_q\mathcal{V}$ is given by all terms x^μ with μ a multi index[3] of length q. For a basis of the vector space $\Lambda_p\mathcal{V}$ we use the following convention: let I be a *sorted* repeated index of length p, i. e. $I = (i_1, \ldots, i_p)$ with $1 \le i_1 < i_2 < \cdots < i_p \le n$; then we write x^I for $x^{i_1} \wedge \cdots \wedge x^{i_p}$ and the set of all such "terms" provides a basis of $\Lambda_p\mathcal{V}$. With respect to these bases, we obtain the following expressions for the above differentials:

$$\delta(x^\mu \otimes x^I) = \sum_{i=1}^{n} \text{sgn}\left(\{i\} \cup I\right) \mu_i x^{\mu-1_i} \otimes x^{\{i\}\cup I} \qquad (2.5)$$

and

$$\partial(x^\mu \otimes x^I) = \sum_{j=1}^{p} (-1)^{j+1} x^{\mu+1_{i_j}} \otimes x^{I\setminus\{i_j\}} . \qquad (2.6)$$

Formally, (2.5) looks like the exterior derivative applied to a differential p-form with polynomial coefficients. This observation explains the name "polynomial de Rham complex" for (2.1) and in principle one should use the usual symbol d for the differential but the notation δ has become standard.

Remark 2.2 While the de Rham differential δ indeed depends on the algebra structure of the exterior algebra $\Lambda\mathcal{V}$, it exploits only the vector space structure of the symmetric algebra $S\mathcal{V}$. Thus we may substitute the symmetric algebra $S\mathcal{V}$ by the symmetric *coalgebra*[4] $\mathfrak{S}\mathcal{V}$ and define δ on the components of the free $\mathfrak{S}\mathcal{V}$-comodule $\mathfrak{S}\mathcal{V} \otimes \Lambda\mathcal{V}$,

---

[2] The hat signals that the corresponding factor is omitted.
[3] See Appendix A for the used conventions on multi indices.
[4] See Appendix C for some information about coalgebras and comodules.

since both are identical as vector spaces. It is not difficult to verify that with this interpretation the differential δ is a comodule morphism. In fact, we will see later that in our context this comodule interpretation is even more natural. It is somewhat surprising that this point of view was introduced only very recently in [33]. For the Koszul differential ∂ we have the opposite situation: we need the algebra $S\mathcal{V}$ but only the vector space $\Lambda\mathcal{V}$. Thus one could similarly $\Lambda\mathcal{V}$ replace by the exterior coalgebra, however, this will not become relevant for us.

Lemma 2.3 *We have* $(\delta \circ \partial + \partial \circ \delta)(\omega) = (p+q)\omega$ *for all* $\omega \in S_q\mathcal{V} \otimes \Lambda_p\mathcal{V}$.

Proof. For $\omega = w_1 \cdots w_q \otimes v_1 \wedge \cdots \wedge v_p$ one readily computes that

$$(\partial \circ \delta)(\omega) = q\omega + \sum_{i=1}^{q}\sum_{j=1}^{p}(-1)^j w_1 \cdots \widehat{w_i} \cdots w_q v_j \otimes w_i \wedge v_1 \wedge \cdots \wedge \widehat{v_j} \wedge \cdots \wedge v_p \quad (2.7)$$

and similarly

$$(\delta \circ \partial)(\omega) = p\omega + \sum_{j=1}^{p}\sum_{i=1}^{q}(-1)^{j+1} w_1 \cdots \widehat{w_i} \cdots w_q v_j \otimes w_i \wedge v_1 \wedge \cdots \wedge \widehat{v_j} \wedge \cdots \wedge v_p \quad (2.8)$$

which immediately implies our claim. □

Proposition 2.4 *The complexes* $R_q(S\mathcal{V})$ *and* $K_q(S\mathcal{V})$ *are exact for all values* $q > 0$. *For* $q = 0$ *both complexes are of the form* $0 \to \Bbbk \to 0$.

Proof. This is an immediate consequence of Lemma 2.3. It implies that for $q > 0$ the map ∂ induces a contracting homotopy for $R_q(S\mathcal{V})$ and conversely δ for $K_q(S\mathcal{V})$ connecting the respective identity and zero maps. It is well known that the existence of such a map entails exactness. □

For the polynomial de Rham complex, this result is also known as the *formal Poincaré Lemma*, as one may interpret it as a special case of the Poincaré Lemma for general differential forms. We consider the complexes $R_q(S\mathcal{V})$ and $K_q(S\mathcal{V})$ as homogeneous components of complexes $R(S\mathcal{V})$ and $K(S\mathcal{V})$ over the $S\mathcal{V}$-modules $S\mathcal{V} \otimes \Lambda_i\mathcal{V}$. Since $S_0\mathcal{V} = \Bbbk$, we find that the Koszul complex $K(S\mathcal{V})$ defines a free resolution of the ground field $\Bbbk$. Similarly, the polynomial de Rham complex $R(S\mathcal{V})$ may be considered as a free coresolution of $\Bbbk$.

The polynomial de Rham and the Koszul complex are related by duality [46, 49, 55]. Recall that we may introduce for any complex of $\mathcal{R}$-modules its dual complex obtained by applying the functor $\mathrm{Hom}_{\mathcal{R}}(\cdot, \mathcal{R})$. In the case of finite-dimensional vector spaces, it is well known that the homology of the dual complex is the dual space of the cohomology of the original complex.

Remark 2.5 There exists a canonical isomorphism $S_q(\mathcal{V}^*) \cong (S_q\mathcal{V})^*$: any separable element $\phi_1 \cdots \phi_q \in S_q(\mathcal{V}^*)$ is interpreted as the linear map on $S_q\mathcal{V}$ obtained by setting

$$(\phi_1 \cdots \phi_q)(v_1 \cdots v_q) = \sum_{\pi \in \mathcal{S}_q}\prod_{i=1}^{q}\phi_i(v_{\pi(i)}) \quad (2.9)$$

where $\mathcal{S}_q$ denotes the symmetric group of all permutations of $1, \ldots, q$. The same construction can be applied to exterior products and thus we can extend to a canonical isomorphism $S_q(\mathcal{V}^*) \otimes \Lambda_p(\mathcal{V}^*) \cong (S_q\mathcal{V} \otimes \Lambda_p\mathcal{V})^*$.

At the level of bases, this isomorphism takes the following form. We denote again by $\{x^1, \ldots, x^n\}$ a basis of $\mathcal{V}$ and by $\{y_1, \ldots, y_n\}$ the corresponding dual basis of $\mathcal{V}^*$. Then the monomials x^μ with $|\mu| = q$ form a basis of $S_q\mathcal{V}$ and similarly the monomials $y_\mu = y_1^{\mu_1} \cdots y_n^{\mu_n}$ with $|\mu| = q$ form a basis of $S_q(\mathcal{V}^*)$. However, these two bases are *not* dual to each other, since according to (2.9) $y_\mu(x^\nu) = \mu! \delta_\mu^\nu$. Thus the dual basis consists of the *divided powers* $\frac{y_\mu}{\mu!}$. For the exterior algebra no such combinatorial factor arises, as the evaluation of the expression corresponding to the right hand side of (2.9) on basis vectors yields only one non-vanishing summand.

Another way to see that the dualisation leads to the divided powers is based on the coalgebra approach of Remark 2.2. If we substitute in the definition of the polynomial de Rham complex the symmetric algebra $S\mathcal{V}$ by the symmetric coalgebra $\mathfrak{S}\mathcal{V}$, then the dual algebra is $S(\mathcal{V}^*)$ and evaluation of the convolution product (C.3) leads to (2.9).

Proposition 2.6 $\bigl(R(S\mathcal{V})^*, \delta^*\bigr)$ *is isomorphic to* $\bigl(K(S(\mathcal{V}^*)), \partial\bigr)$.

Proof. There only remains to show that ∂ is indeed the pull-back of δ. Choosing the above described dual bases, this is a straightforward computation. By definition of the pull-back,

$$\delta^* \left(\frac{y^\mu}{\mu!} \otimes y^I \right)(x^\nu \otimes x^J) = \begin{cases} v_j \, \mathrm{sgn}\left(\{j\} \cup J\right) & \text{if } \exists j : \begin{cases} \mu = \nu - 1_j \\ I = \{j\} \cup J \end{cases}, \\ 0 & \text{otherwise .} \end{cases} \qquad (2.10)$$

Note that $\nu_j = \frac{\nu!}{\mu!}$ if $\mu = \nu - 1_j$; hence we find that

$$\delta^*(y^\mu \otimes y^I) = \sum_{j=1}^p (-1)^{j+1} y^{\mu + 1_{i_j}} \otimes y^{I \setminus \{i_j\}} . \qquad (2.11)$$

Comparison with (2.6) yields the desired result.   □

For reasons that will become apparent in Chapter 7 when we apply the here developed algebraic theory to differential equations, we prefer to consider the Koszul complex over the vector space $\mathcal{V}$ and the polynomial de Rham complex over its dual space $\mathcal{V}^*$. Thus we will always use $R(S(\mathcal{V}^*))$ and $K(S\mathcal{V})$. If $\mathcal{U}$ is a further finite-dimensional vector space over $\Bbbk$ with $\dim \mathcal{U} = m$, then we may extend to the tensor product complex $R(S(\mathcal{V}^*) \otimes \mathcal{U}) = R(S(\mathcal{V}^*)) \otimes \mathcal{U}$ and dually to $K(S\mathcal{V} \otimes \mathcal{U}^*) = K(S\mathcal{V}) \otimes \mathcal{U}^*$. Everything we have done so far remains valid with trivial modifications, as the differentials of the complexes are essentially unaffected by this operation. Basically, one must only add a factor $u \in \mathcal{U}$ (or $\nu \in \mathcal{U}^*$, respectively) to each equation and consider all our computations above as componentwise.

Definition 2.7 Let $\mathcal{N}_q \subseteq S_q(\mathcal{V}^*) \otimes \mathcal{U}$ be an arbitrary vector subspace. Its *(first) prolongation* is the subspace

$$\mathcal{N}_{q,1} = \{f \in S_{q+1}(\mathcal{V}^*) \otimes \mathcal{U} \mid \delta(f) \in \mathcal{N}_q \otimes \mathcal{V}^*\} . \qquad (2.12)$$

A sequence of vector subspaces $\left(\mathcal{N}_q \subseteq S_q(\mathcal{V}^*) \otimes \mathcal{U}\right)_{q \in \mathbb{N}_0}$ is called a *symbolic system* over $\mathcal{V}^*$, if $\mathcal{N}_{q+1} \subseteq \mathcal{N}_{q,1}$ for all $q \in \mathbb{N}_0$.

We may equivalently introduce the prolongation as

$$\mathcal{N}_{q,1} = (\mathcal{V}^* \otimes \mathcal{N}_q) \cap \left(S_{q+1}(\mathcal{V}^*) \otimes \mathcal{U}\right) \tag{2.13}$$

with the intersection understood to take place in $\mathcal{V}^* \otimes \left(S_q(\mathcal{V}^*) \otimes \mathcal{U}\right)$. This follows immediately from the definition of the differential δ. The extension to higher prolongations $\mathcal{N}_{q,r} \subseteq S_{q+r}(\mathcal{V}^*) \otimes \mathcal{U}$ proceeds either by induction, $\mathcal{N}_{q,r+1} = (\mathcal{N}_{q,r})_{,1}$ for all $r \in \mathbb{N}$, or alternatively by generalising (2.13) to $\mathcal{N}_{q,r} = \left(\bigotimes_{i=1}^{r} \mathcal{V}^* \otimes \mathcal{N}_q\right) \cap \left(S_{q+r}(\mathcal{V}^*) \otimes \mathcal{U}\right)$ where the intersection is now understood to take place in $\bigotimes_{i=1}^{r} \mathcal{V}^* \otimes \left(S_q(\mathcal{V}^*) \otimes \mathcal{U}\right)$.

The notion of a symbolic system is fairly classical in the formal theory of differential equations (see Proposition 7.6). The next result shows, however, that if we take the coalgebra point of view of the polynomial de Rham complex mentioned in Remark 2.2, then a symbolic system is equivalent to a simple algebraic structure.

Lemma 2.8 *Let* $(\mathcal{N}_q)_{q \in \mathbb{N}_0}$ *be a symbolic system. Then* $\mathcal{N} = \bigoplus_{q=0}^{\infty} \mathcal{N}_q$ *is a graded (right) subcomodule of the free* $\mathfrak{S}(\mathcal{V}^*)$-*comodule* $\mathfrak{S}(\mathcal{V}^*) \otimes \mathcal{U}$. *Conversely, the sequence* $(\mathcal{N}_q)_{q \in \mathbb{N}_0}$ *of the components of any graded (right) subcomodule* $\mathcal{N} \subseteq \mathfrak{S}(\mathcal{V}^*) \otimes \mathcal{U}$ *defines a symbolic system.*

Proof. Let $(\mathcal{N}_q)_{q \in \mathbb{N}_0}$ be a symbolic system and $f \in \mathcal{N}_q$. Then $\delta f \in \mathcal{N}_{q-1} \otimes \mathcal{V}$ and hence $\partial f / \partial x^i \in \mathcal{N}_{q-1}$ for all $1 \leq i \leq n$, since our differential δ is just the exterior derivative. Using induction we thus find that $\partial^{|\mu|} f / \partial x^\mu \in \mathcal{N}_{q-r}$ for all μ with $|\mu| = r$. By the definition of the polynomial coproduct, this is equivalent to $\Delta(f) \in \mathcal{N} \otimes \mathfrak{S}(\mathcal{V}^*)$ and hence $\mathcal{N}$ is a subcomodule. For the converse, we simply revert every step of this argument to find that $\Delta(f) \in \mathcal{N} \otimes \mathfrak{S}(\mathcal{V}^*)$ implies that $\mathcal{N}_q \subseteq \mathcal{N}_{q-1,1}$ for all $q > 0$.   □

Example 2.9 Let $\mathcal{V}$ be a two-dimensional space. The subspaces $\mathcal{N}_0 = \Bbbk$, $\mathcal{N}_1 = \mathcal{V}^*$ and $\mathcal{N}_q = \langle y_1^q \rangle \subset S_q(\mathcal{V}^*)$ for $q \geq 2$ define a symbolic system where $\mathcal{N}_{q,1} = \mathcal{N}_{q+1}$ for all $q \geq 2$. Indeed, if $k + \ell = q$, then $\delta(y_1^k y_2^\ell) = y_1^{k-1} y_2^\ell \otimes y_1 + y_1^k y_2^{\ell-1} \otimes y_2$ so that the result lies in $\mathcal{N}_{q-1} \otimes \mathcal{V}^*$ only for $\ell = 0$. We will see later that this symbolic system is associated with partial differential equations of the form $u_{22} = F(\mathbf{x}, u^{(1)})$, $u_{12} = G(\mathbf{x}, u^{(1)})$ where the shorthand $u^{(q)}$ denotes the unknown function u depending on $\mathbf{x} = (x^1, x^2)$ and all its derivatives up to order q.

Another simple symbolic system over the same dual space $\mathcal{V}^*$ is given by $\mathcal{N}_0 = \Bbbk$, $\mathcal{N}_1 = \mathcal{V}^*$, $\mathcal{N}_2 = S_2(\mathcal{V}^*)$, $\mathcal{N}_3 = \langle y_1^2 y_2, y_1 y_2^2 \rangle$, $\mathcal{N}_4 = \langle y_1^2 y_2^2 \rangle$ and $\mathcal{N}_q = 0$ for all $q \geq 5$. This system is related to partial differential equations of the form $u_{222} = F(\mathbf{x}, u^{(2)})$, $u_{111} = G(\mathbf{x}, u^{(2)})$. One can show that any such equation has a finite-dimensional solution space and this fact is reflected by the vanishing of the associated symbolic system beyond a certain degree.

From now on, we will not distinguish between a symbolic system $(\mathcal{N}_q)_{q \in \mathbb{N}_0}$ and the corresponding subcomodule $\mathcal{N} \subseteq \mathfrak{S}(\mathcal{V}^*) \otimes \mathcal{U}$. We are particularly interested in subcomodules $\mathcal{N}$ where almost all components $\mathcal{N}_q$ are different from zero (i.e. as a vector space $\mathcal{N}$ is infinite-dimensional). Recall that it follows immediately from the definition of the polynomial coproduct that cogeneration in $\mathfrak{S}(\mathcal{V}^*)$ always leads to elements of at most the same degree as the cogenerator; hence a finitely cogenerated comodule is necessarily finite-dimensional as vector space. However, the duality between $\mathfrak{S}(\mathcal{V}^*)$ and $S\mathcal{V}$ yields easily the following result.

Corollary 2.10 *Let $(\mathcal{N}_q)_{q\in\mathbb{N}_0}$ be an arbitrary symbolic system. There exists an integer $r_0 \geq 0$ such that $\mathcal{N}_{r+1} = \mathcal{N}_{r,1}$ for all $r \geq r_0$.*

Proof. It is well known that the annihilator $\mathcal{N}^0 \subseteq S\mathcal{V} \otimes \mathcal{U}^*$ is an $S\mathcal{V}$-submodule. Now $\mathcal{N}_{r+1} \subsetneq \mathcal{N}_{r,1}$ implies that any minimal basis of $\mathcal{N}^0$ contains at least one generator of degree r. Since, by Hilbert's Basis Theorem, any polynomial ring in a finite number of variables and hence also the symmetric algebra $S\mathcal{V}$ is Noetherian, an upper bound r_0 for such values r exists. $\square$

By this corollary, we may consider symbolic systems as a kind of finitely cogenerated "differential comodules": since the truncated comodule $\mathcal{N}_{\leq r_0}$ is a finite-dimensional vector space, it is obviously finitely cogenerated and by repeated prolongations of the component $\mathcal{N}_{r_0}$ we obtain the remainder of the comodule $\mathcal{N}$. Thus we conclude that every symbolic system is uniquely determined by a finite number of elements.

Definition 2.11 Let $\mathcal{N}$ be a graded comodule over the coalgebra $\mathcal{C} = \mathfrak{S}(\mathcal{V}^*)$. Its *Spencer complex* $(R(\mathcal{N}),\delta)$ is the cotensor product[5] complex $\mathcal{N} \boxtimes_\mathcal{C} R(\mathfrak{S}(\mathcal{V}^*))$. The *Spencer cohomology* of $\mathcal{N}$ is the corresponding bigraded cohomology; the cohomology group at $\mathcal{N}_q \otimes \Lambda_p(\mathcal{V}^*)$ in $(R_{q+p}(\mathcal{N}),\delta)$ is denoted by $H^{q,p}(\mathcal{N})$.

Since $\mathcal{N} \boxtimes_\mathcal{C} \mathcal{C} \cong \mathcal{N}$ for any $\mathcal{C}$-comodule $\mathcal{N}$, the components of the cotensored complex $\mathcal{N} \boxtimes_\mathcal{C} R(\mathfrak{S}(\mathcal{V}^*))$ are indeed just the vector spaces $\mathcal{N}_q \otimes \Lambda_p(\mathcal{V}^*)$. We are mainly interested in the special case that $\mathcal{N}$ is a subcomodule of a free comodule $\mathcal{C} \otimes \mathcal{U}$ and then the differential in the Spencer complex $R(\mathcal{N})$ is simply given by the restriction of the differential δ in the polynomial de Rham complex $R(\mathfrak{S}(\mathcal{V}^*))$ to the subspaces $\mathcal{N}_q \otimes \Lambda_p(\mathcal{V}^*) \subseteq \mathfrak{S}_q(\mathcal{V}^*) \otimes \Lambda_p(\mathcal{V}^*) \otimes \mathcal{U}$; this observation explains why we keep the notation δ for the differential. One can also verify by direct computation that this restriction makes sense whenever $(\mathcal{N}_q)_{q\in\mathbb{N}_0}$ defines a symbolic system (this is basically the same computation as the one showing the equivalence of the two definitions (2.12) and (2.13) of the prolongation); in fact, this restriction is the classical approach to define the Spencer complex.

Remark 2.12 It is important to note here that the Spencer cohomology is bigraded. If we ignore the polynomial degree and consider only the form degree, we obtain the modules $H^p(\mathcal{N}) = \bigoplus_{q=0}^{\infty} H^{q,p}(\mathcal{N})$. For these, another point of view is possible. Since any free comodule is injective, we have exactly the situation of the definition of *cotorsion*: we are given an injective coresolution (of $\Bbbk$) and cotensor it with a comodule. Thus we may consider the Spencer cohomology as the right derived functor of $\mathcal{N} \boxtimes_\mathcal{C} \cdot$ and identify $H^p(\mathcal{N}) = \mathrm{Cotor}_\mathcal{C}^p(\mathcal{N},\Bbbk)$.

---

[5] The definition of the cotensor product $\boxtimes_\mathcal{C}$ over a coalgebra $\mathcal{C}$ is dual to the one of the usual tensor product; it was introduced by Eilenberg and Moore [13].

As for arbitrary derived functors, the definition of $\mathrm{Cotor}_C^p(\mathcal{N}, \Bbbk)$ is independent of the coresolution used for its computation or, more precisely, the results obtained with different coresolutions are isomorphic. However, given some other way to explicitly determine $\mathrm{Cotor}_C^p(\mathcal{N}, \Bbbk)$, say via a coresolution of $\mathcal{N}$, it may be a non-trivial task to recover the bigrading of the Spencer cohomology.

Lemma 2.13 *Let* $\mathcal{N} \subseteq \mathfrak{S}(\mathcal{V}^*) \otimes \mathcal{U}$ *be a symbolic system. Then* $H^{q,0}(\mathcal{N}) = 0$ *and* $\dim H^{q-1,1}(\mathcal{N}) = \dim \left(\mathcal{N}_{q-1,1} / \mathcal{N}_q \right)$ *for all* $q > 0$.

Proof. The first claim follows immediately from the formal Poincaré Lemma (Proposition 2.4). For the second claim consider a non-vanishing element $f \in \mathcal{N}_{q-1,1} \setminus \mathcal{N}_q$. Then $g = \delta f \in \ker \delta|_{\mathcal{N}_{q-1} \otimes \mathcal{V}^*}$ and, because of the formal Poincaré Lemma, $g \neq 0$. However, by construction, $g \notin \mathrm{im}\, \delta|_{\mathcal{N}_q}$ and hence we find $0 \neq [g] \in H^{q-1,1}(\mathcal{N})$. This implies immediately the inequality $\dim H^{q-1,1}(\mathcal{N}) \geq \dim \left(\mathcal{N}_{q-1,1} / \mathcal{N}_q \right)$. Conversely, consider an arbitrary non-vanishing cohomology class $[g] \in H^{q-1,1}(\mathcal{N})$. Again by the formal Poincaré Lemma, an element $f \in \mathfrak{S}_q(\mathcal{V}^*) \otimes \mathcal{U}$ exists such that $g = \delta f$ and, by definition of the prolongation, $f \in \mathcal{N}_{q-1,1} \setminus \mathcal{N}_q$. Thus we also have the opposite inequality $\dim H^{q-1,1}(\mathcal{N}) \leq \dim \left(\mathcal{N}_{q-1,1} / \mathcal{N}_q \right)$. $\square$

Note that Corollary 2.10 implies that $H^{q,1}(\mathcal{N}) = 0$ for a sufficiently high degree q. Dualisation of Definition 2.11 leads to the following classical construction in commutative algebra with a polynomial module.

Definition 2.14 Let $\mathcal{M}$ be a graded module over the symmetric algebra $\mathcal{P} = S\mathcal{V}$. Its *Koszul complex* $\left(K(\mathcal{M}), \partial \right)$ is the tensor product complex $\mathcal{M} \otimes_{\mathcal{P}} K(S\mathcal{V})$. The *Koszul homology* of $\mathcal{M}$ is the corresponding bigraded homology; the homology group at $\mathcal{M}_q \otimes \Lambda_p \mathcal{V}$ is denoted by $H_{q,p}(\mathcal{M})$.

Remark 2.15 We observed already above that the Koszul complex defines a free resolution of the field $\Bbbk$. Hence, as for the Spencer cohomology, we may take another point of view and consider the Koszul homology as the right derived functor of $\mathcal{M} \otimes_{\mathcal{P}} \cdot$. According to the definition of the torsion modules, this leads to the identification $H_p(\mathcal{M}) = \bigoplus_{q=0}^{\infty} H_{q,p}(\mathcal{M}) = \mathrm{Tor}_p^{\mathcal{P}}(\mathcal{M}, \Bbbk)$ where we consider $\Bbbk$ as a $\mathcal{P}$-module. But again this interpretation ignores the natural bigrading of the Koszul complex $K(\mathcal{M})$.

An alternative way to compute $\mathrm{Tor}_p^{\mathcal{P}}(\mathcal{M}, \Bbbk)$ consists of using a free resolution of the module $\mathcal{M}$. If $\mathcal{F} \to \mathcal{M} \to 0$ is such a resolution, then the Koszul homology $H_\bullet(\mathcal{M})$ is isomorphic to the homology of the tensor product complex $\mathcal{F} \otimes_{\mathcal{P}} \Bbbk$. Each component in $\mathcal{F}$ is of the form $\mathcal{P}^m$ and therefore $\mathcal{P}^m \otimes_{\mathcal{P}} \Bbbk = \Bbbk^m$. Now assume that we actually have a *minimal* resolution. In this case all differentials in $\mathcal{F}$ possess a positive degree and it follows from the $\mathcal{P}$-action on $\Bbbk$ that the induced differential on the complex $\mathcal{F} \otimes_{\mathcal{P}} \Bbbk$ is the zero map. Hence we find that $H_\bullet(\mathcal{M}) \cong \mathcal{F} \otimes_{\mathcal{P}} \Bbbk$ and $\dim H_p(\mathcal{M})$ is just the pth Betti number of $\mathcal{M}$. In this sense we may say that the Koszul homology corresponds to a minimal free resolution.

Lemma 2.16 *Let $\mathcal{M}$ be a graded $\mathcal{P}$-module. Then $H_{q,0}(\mathcal{M}) = \mathcal{M}_q/\mathcal{V}\mathcal{M}_{q-1}$ and thus* $\dim H_{q,0}(\mathcal{M})$ *gives the numbers of generators of degree q in any minimal basis of $\mathcal{M}$.* *Furthermore,*

$$H_{q,n}(\mathcal{M}) \cong \left\{ m \in \mathcal{M}_q \mid \mathrm{Ann}\,(m) = \mathcal{S}_+\mathcal{V} \right\} . \qquad (2.14)$$

Proof. The first assertion follows trivially from the definition of the Koszul homology. Elements of $H_{q,n}(\mathcal{M})$ are represented by cycles in $\mathcal{M}_q \otimes \Lambda_n \mathcal{V}$. If $\{x^1,\dots,x^n\}$ is a basis of $\mathcal{V}$, these are forms $\omega = m \otimes x^1 \wedge \cdots \wedge x^n$ and the condition $\partial\omega = 0$ is equivalent to $x^i m = 0$ for $1 \leq i \leq n$. $\qquad\square$

Lemma 2.17 *Let $\mathcal{M}$ be a graded $\mathcal{P}$-module. Multiplication by an arbitrary element of $\mathcal{S}_+\mathcal{V}$ induces the zero map on the Koszul homology $H_\bullet(\mathcal{M})$.*

Proof. We first observe that if $\omega \in \mathcal{M}_q \otimes \Lambda_p \mathcal{V}$ is a cycle, i.e. $\partial\omega = 0$, then for any $v \in \mathcal{V}$ the form $v\omega$ is a boundary, i.e. $v\omega \in \mathrm{im}\,\partial$. Indeed,

$$\partial\bigl(v \wedge \omega\bigr) = -v \wedge (\partial\omega) + v\omega = v\omega . \qquad (2.15)$$

Since ∂ is $\mathcal{S}\mathcal{V}$-linear, this observation remains true, if we take for v an arbitrary element of $\mathcal{S}_+\mathcal{V}$, i.e. a polynomial without constant term. $\qquad\square$

Each subcomodule $\mathcal{N} \subseteq \mathfrak{S}(\mathcal{V}^*) \otimes \mathcal{U}$ induces a submodule $\mathcal{M} = \mathcal{N}^0 \subseteq \mathcal{S}\mathcal{V} \otimes \mathcal{U}^*$, its annihilator. Conversely, the annihilator of any submodule $\mathcal{M} \subseteq \mathcal{S}\mathcal{V} \otimes \mathcal{U}^*$ defines a subcomodule $\mathcal{N} = \mathcal{M}^0 \in \mathfrak{S}(\mathcal{V}^*) \otimes \mathcal{U}$. In view of the duality between the polynomial de Rham and the Koszul complex, we expect a simple relation between the Spencer cohomology $H^\bullet(\mathcal{N})$ of the comodule $\mathcal{N}$ and the Koszul homology $H_\bullet(\mathcal{N}^0)$ of its annihilator $\mathcal{N}^0$.

Such a relation is easily obtained with the help of the $\mathcal{S}\mathcal{V}$-module $\mathcal{N}^*$ dual to $\mathcal{N}$. If we take the dual $\pi^* : \bigl((\mathfrak{S}(\mathcal{V}^*) \otimes \mathcal{U})/\mathcal{N}\bigr)^* \to \bigl(\mathfrak{S}(\mathcal{V}^*) \otimes \mathcal{U}\bigr)^* = \mathcal{S}\mathcal{V} \otimes \mathcal{U}^*$ of the canonical projection $\pi : \mathfrak{S}(\mathcal{V}^*) \otimes \mathcal{U} \to \bigl(\mathfrak{S}(\mathcal{V}^*) \otimes \mathcal{U}\bigr)/\mathcal{N}$, then $\mathrm{im}\,\pi^* = \mathcal{N}^0$. Like for any map, we have for π the canonical isomorphism $\mathrm{coker}\,(\pi^*) \cong (\ker \pi)^* = \mathcal{N}^*$ and hence may identify $\mathcal{N}^*$ with the factor module $(\mathcal{S}\mathcal{V} \otimes \mathcal{U}^*)/\mathcal{N}^0$.

Proposition 2.18 *Let $\mathcal{N} \subseteq \mathfrak{S}(\mathcal{V}^*) \otimes \mathcal{U}$ be a symbolic system. Then for all $q \geq 0$ and $1 \leq p \leq n$*

$$\bigl(H^{q,p}(\mathcal{N})\bigr)^* \cong H_{q,p}(\mathcal{N}^*) \cong H_{q+1,p-1}(\mathcal{N}^0) \qquad (2.16)$$

where the second isomorphism is induced by the Koszul differential ∂.

Proof. The first isomorphism follows from Proposition 2.6. For the second one we note that the considerations above lead to the short exact sequence

$$0 \longrightarrow \mathcal{N}^0 \overset{\iota}{\hookrightarrow} \mathcal{S}\mathcal{V} \otimes \mathcal{U}^* \overset{\pi}{\longrightarrow} \mathcal{N}^* \longrightarrow 0 \qquad (2.17)$$

where the first map is the natural inclusion and the second one the canonical projection. Tensoring with the vector space $\Lambda_p \mathcal{V}$ is a flat functor and hence does not affect the exactness so that we obtain a short exact sequence of Koszul complexes:

$$0 \longrightarrow K(\mathcal{N}^0) \overset{}{\hookrightarrow} K(\mathcal{S}\mathcal{V} \otimes \mathcal{U}^*) \longrightarrow K(\mathcal{N}^*) \longrightarrow 0 . \qquad (2.18)$$

Since $K(S\mathcal{V} \otimes \mathcal{U}^*)$ is exact in positive exterior degree, the long exact homological sequence for (2.18) yields an isomorphism $H_p(\mathcal{N}^*) \to H_{p-1}(\mathcal{N}^0)$. Furthermore, as the maps in the exact sequence (2.17) are so simple, it follows straightforwardly from the construction of the connecting homomorphism that this isomorphism is induced by the Koszul differential ∂. Hence, taking the bigrading into account, we obtain an isomorphism $H_{q,p}(\mathcal{N}^*) \to H_{q+1,p-1}(\mathcal{N}^0)$. □

Remark 2.19 From a computational point of view, it is often more convenient to work with the annihilator $\mathcal{N}^0$ instead of the dual module $\mathcal{N}^*$. The way we proved the lemma gave the isomorphism only in one direction. However, Lemma 2.3 allows us to derive easily an explicit expression for the inverse.

Let $\omega \in \mathcal{N}_q^* \otimes \Lambda_p \mathcal{V}$ be a cycle and $\tilde{\omega} \in S_q \mathcal{V} \otimes \Lambda_p \mathcal{V} \otimes \mathcal{U}^*$ an arbitrary form such that $\pi(\tilde{\omega}) = \omega$. Then $\partial \omega = 0$ implies that $\bar{\omega} = \partial \tilde{\omega} \in \mathcal{N}_{q+1}^0 \otimes \Lambda_{p-1} \mathcal{V}$. Now the isomorphism used in the proof above simply maps $[\omega] \mapsto [\bar{\omega}]$. For the inverse we note that, by Lemma 2.3, $\delta \bar{\omega} = (p+q)\tilde{\omega} - \partial(\delta \tilde{\omega})$ and hence $\left[\frac{1}{p+q}\delta\bar{\omega}\right] = [\tilde{\omega}]$. But this implies that the inverse of our isomorphism is given by the map $[\bar{\omega}] \mapsto \left[\frac{1}{p+q}\pi(\delta\bar{\omega})\right]$.

For our purposes, the most important property of the Spencer cohomology is the following finiteness result obviously requiring the bigrading. A direct proof would probably be not easy, but the duality to the Koszul homology (Proposition 2.18) allows us to restrict to the dual situation where the finiteness is a corollary to Lemma 2.17.

Theorem 2.20 Let $\mathcal{N} \subseteq \mathfrak{S}(\mathcal{V}^*) \otimes \mathcal{U}$ be a symbolic system. Then there exists an integer $q_0 \geq 0$ such that $H^{q,p}(\mathcal{N}) = 0$ for all $q \geq q_0$ and $0 \leq p \leq n$. Dually, let $\mathcal{M}$ be a finitely generated graded polynomial module. Then there exists an integer $q_0 \geq 0$ such that $H_{q,p}(\mathcal{M}) = 0$ for all $q \geq q_0$ and $0 \leq p \leq n$.

Proof. As mentioned above, it suffices to consider the case of a polynomial module $\mathcal{M}$. The cycles in $\mathcal{M} \otimes \Lambda_p \mathcal{V}$ form a finitely generated $S\mathcal{V}$-module. Thus there exists an integer $q_0 \geq 0$ such that the polynomial degree of all elements in a finite generating set of it is less than q_0. All cycles of higher polynomial degree are then linear combinations of these generators with polynomial coefficients without constant terms. By Lemma 2.17, they are therefore boundaries. Hence $H_{q,p}(\mathcal{M}) = 0$ for all $q \geq q_0$. □

Definition 2.21 The *degree of involution* of the $\mathfrak{S}(\mathcal{V}^*)$-comodule $\mathcal{N}$ is the smallest value q_0 such that $H^{q,p}(\mathcal{N}) = 0$ for all $q \geq q_0$ and $0 \leq p \leq n = \dim \mathcal{V}$. More generally, we say that $\mathcal{N}$ is *s-acyclic* at degree q_0 for an integer $0 \leq s \leq n$, if $H^{q,p}(\mathcal{N}) = 0$ for all $q \geq q_0$ and $0 \leq p \leq s$. A comodule that is n-acyclic at degree q_0 is called *involutive* at degree q_0. Dually, we call an $S\mathcal{V}$-module $\mathcal{M}$ *involutive* at degree q_0, if its Koszul homology vanishes beyond degree q_0: $H_{q,p}(\mathcal{M}) = 0$ for all $q \geq q_0$ and $0 \leq p \leq n$.

With this terminology we may formulate Lemma 2.13 as follows: if the symbolic system $\mathcal{N}$ is such that its annihilator $\mathcal{N}^0$ is generated in degree less than or equal to r_0, then $\mathcal{N}$ is 1-acyclic at degree r_0, and if conversely r_0 is the smallest degree at which $\mathcal{N}$ is 1-acyclic, then any generating set of $\mathcal{N}^0$ contains an element of degree r_0 or higher. We will see later in Theorem 7.15 that 2-acyclicity is very important for checking formal integrability. It follows trivially from the definition that if $\mathcal{N}$ is involutive at some degree q_0, then it is also involutive at any higher degree $q \geq q_0$.

For complexity considerations, it is of great interest to bound for a given comodule $\mathcal{N}$ or module $\mathcal{M}$, respectively, its degree of involution. In our applications to differential equations we will be mainly concerned with the special case that $\mathcal{M}$ is a submodule of a free $S\mathcal{V}$-module of rank m generated by homogeneous elements of degree q. Sweeney [61, Corollary 7.7] derived for this situation a bound $\bar{q}$ depending only on the values of n, m and q. It may be expressed as a nested recursion relation:

$$\bar{q}(n, m, q) = \bar{q}\left(n, m\binom{q+n-1}{n}, 1\right) ,$$

$$\bar{q}(n, m, 1) = m\left(\frac{\bar{q}(n-1, m, 1) + n}{n-1}\right) + \bar{q}(n-1, m, 1) + 1 , \qquad (2.19)$$

$$\bar{q}(0, m, 1) = 0 .$$

Table 2.1 shows $\bar{q}(n, m, 1)$ for different values of m and n. One sees that the values rapidly explode, if n increases. The situation is still worse for modules generated in higher order. It seems to be an open question whether this bound is sharp, i. e. whether for some modules the degree of involution is really that high. Fortunately, $\bar{q}(n, m, q)$ yields usually a coarse over-estimate of the actual degree of involution.

$n \backslash m$	1	2	3	4
1	2	3	4	5
2	7	14	23	34
3	53	287	999	2 699
4	29 314	8 129 858	503 006 503	13 151 182 504

Table 2.1 $\bar{q}(n, m, 1)$ for different values of m and n.

Example 2.22 Let us consider the homogeneous ideal $\mathcal{I}$ (i. e. $m = 1$) generated by the two monomials $(x^1)^q$ and $(x^2)^q$ for some value $q > 0$ in the polynomial ring $\Bbbk[x^1, x^2]$ in $n = 2$ variables. For the value $q = 3$ this ideal is just the annihilator $\mathcal{N}^0$ of the second symbolic system $\mathcal{N}$ considered in Example 2.9. A trivial computation yields that the only non-vanishing Koszul homology modules are $H_{q,0}(\mathcal{I}) = \langle [(x^1)^q], [(x^2)^q] \rangle$ and $H_{2q-1,1}(\mathcal{I}) = \langle [(x^1)^q(x^2)^{q-1} \otimes x^2 - (x^1)^{q-1}(x^2)^q \otimes x^1] \rangle$. Hence the degree of involution of $\mathcal{I}$ is $2q - 1$. By contrast, evaluation of Sweeney's bound (2.19) yields

$$\bar{q}(2, 1, q) = \frac{1}{4}q^4 + \frac{1}{2}q^3 + \frac{9}{4}q^2 + 2q + 2 , \qquad (2.20)$$

i. e. a polynomial in q of degree 4.

3 Cartan's Test

We study now some explicit criteria for a (co)module to be involutive. We start with a symbolic system $\mathcal{N} \subseteq \mathfrak{S}(\mathcal{V}^*) \otimes \mathcal{U}$. As before, let $\{x^1, \ldots, x^n\}$ be an ordered basis of $\mathcal{V}$ and $\{y_1, \ldots, y_n\}$ the corresponding dual basis of $\mathcal{V}^*$. Then we introduce for any $0 \leq k \leq n$ the following subspaces of the homogeneous component $\mathcal{N}_q$:

$$
\mathcal{N}_q^{(k)} = \Big\{ f \in \mathcal{N}_q \mid f(x^i, v_1, \ldots, v_{q-1}) = 0,
$$

$$
\forall 1 \leq i \leq k,\ \forall v_1, \ldots, v_{q-1} \in \mathcal{V} \Big\} \tag{3.1}
$$

$$
= \Big\{ f \in \mathcal{N}_q \mid \frac{\partial f}{\partial y_i} = 0 \ \forall 1 \leq i \leq k \Big\} .
$$

In the first line we interpreted elements of $\mathcal{N}_q$ as multilinear maps on $\mathcal{V}$ and in the last line we considered them as polynomials in the "variables" $y_1, \ldots, y_n$.

Obviously, these subspaces define a filtration

$$
0 = \mathcal{N}_q^{(n)} \subseteq \mathcal{N}_q^{(n-1)} \subseteq \cdots \subseteq \mathcal{N}_q^{(1)} \subseteq \mathcal{N}_q^{(0)} = \mathcal{N}_q . \tag{3.2}
$$

It is clear that this filtration (and in particular the dimensions of the involved subspaces) depend on the chosen basis for $\mathcal{V}^*$. Thus it distinguishes certain bases. This effect is known as the problem of δ-regularity. In the next chapter we will see it reappear in a different form for Pommaret bases.

Definition 3.1 Let $\mathcal{N} \subseteq \mathfrak{S}(\mathcal{V}^*) \otimes \mathcal{U}$ be a symbolic system. With respect to a given basis $\{y_1, \ldots, y_n\}$ of $\mathcal{V}^*$, we define the *Cartan characters* of the component $\mathcal{N}_q$ as

$$
\alpha_q^{(k)} = \dim \mathcal{N}_q^{(k-1)} - \dim \mathcal{N}_q^{(k)} , \qquad 1 \leq k \leq n . \tag{3.3}
$$

A basis $\{y_1, \ldots, y_n\}$ of $\mathcal{V}^*$ is δ-*regular* for the component $\mathcal{N}_q$, if the sum $\sum_{k=1}^n k \alpha_q^{(k)}$ attains a minimal value for it.

One can show that generic bases are always δ-regular. Hence conceptually trivial solutions of the problem of δ-regularity are to use either a random basis (which is δ-regular with probability 1) or to work with a general (i. e. parametrised) basis. However, from a computational point of view both approaches are extremely expensive and useless in larger calculations. In the context of Pommaret bases much more efficient solutions have been developed (see Remark 4.6 below for references).

We know from the proof of Lemma 2.8 that differentiation with respect to a variable y_k maps $\mathcal{N}_{q+1}$ into $\mathcal{N}_q$. It follows trivially from the definition of the subspaces $\mathcal{N}_q^{(k)}$ that we may consider the restrictions $\partial_{y_k} : \mathcal{N}_{q+1}^{(k-1)} \to \mathcal{N}_q^{(k-1)}$.

Proposition 3.2 *Let $\mathcal{N} \subseteq \mathfrak{S}(\mathcal{V}^*) \otimes \mathcal{U}$ be a symbolic system and $\{y_1, \ldots, y_n\}$ a basis of $\mathcal{V}^*$. Then we have for any $q \geq 0$ the inequality*

$$
\dim \mathcal{N}_{q+1} \leq \sum_{k=0}^{n-1} \dim \mathcal{N}_q^{(k)} = \sum_{k=1}^n k \alpha_q^{(k)} . \tag{3.4}
$$

Equality holds, if and only if the restricted maps $\partial_{y_k} : \mathcal{N}_{q+1}^{(k-1)} \to \mathcal{N}_q^{(k-1)}$ *are surjective for all* $1 \le k \le n$.

Proof. By definition of the subspaces $\mathcal{N}_q^{(k)}$, we have the exact sequences

$$0 \longrightarrow \mathcal{N}_{q+1}^{(k)} \overset{\subset}{\longrightarrow} \mathcal{N}_{q+1}^{(k-1)} \overset{\partial_{y_k}}{\longrightarrow} \mathcal{N}_q^{(k-1)} \qquad (3.5)$$

implying the inequalities $\dim \mathcal{N}_{q+1}^{(k-1)} - \dim \mathcal{N}_{q+1}^{(k)} \le \dim \mathcal{N}_q^{(k-1)}$. Summing over all $0 \le k \le n$ yields immediately the inequality (3.4). Equality in (3.4) is obtained, if and only if in all these dimension relations equality holds. But this is the case, if and only if all the maps ∂_{y_k} are surjective. □

Proposition 3.3 *The symbolic system* $\mathcal{N} \subseteq \mathfrak{S}(\mathcal{V}^*) \otimes \mathcal{U}$ *is involutive at degree* q_0, *if and only if a basis* $\{y_1, \ldots, y_n\}$ *of* $\mathcal{V}^*$ *exists such that the maps* $\partial_{y_k} : \mathcal{N}_{q+1}^{(k-1)} \to \mathcal{N}_q^{(k-1)}$ *are surjective for all degrees* $q \ge q_0$ *and all values* $1 \le k \le n$.

Proof. We prove only one direction; the converse will follow from our subsequent considerations for the dual Koszul homology of $\mathcal{N}^*$ (see Remark 3.13). Let us take an arbitrary cycle $\omega \in \mathcal{N}_q \otimes \Lambda_p(\mathcal{V}^*)$ with $1 \le p \le n$ and $q \ge q_0$; we will show that, if all maps ∂_{y_k} are surjective, then a form $\eta \in \mathcal{N}_{q+1} \otimes \Lambda_{p-1}(\mathcal{V}^*)$ exists with $\omega = \delta\eta$. This implies that $H^{q,p}(\mathcal{N}) = 0$.

We do this demonstration in an iterative process, assuming first that the exterior part of ω depends only on $y_k, y_{k+1}, \ldots, y_n$. Then we may decompose $\omega = \omega_1 + y_k \wedge \omega_2$ where the exterior parts of both ω_1 and ω_2 depend only on $y_{k+1}, \ldots, y_n$. Since ω is a cycle, we have $\delta\omega = \delta\omega_1 - y_k \wedge \delta\omega_2 = 0$. Consider now in this equation those terms where the exterior part is of the form $y_\ell \wedge y_k \wedge \cdots$ with $\ell \le k$. Such terms occur only in the second summand and hence we must have $\partial\omega_2 / \partial y_l = 0$ for all $1 \le \ell < k$. This implies $\omega_2 \in \mathcal{N}_q^{(k-1)} \otimes \Lambda_{p-1}(\mathcal{V}^*)$.

By assumption, the map $\partial_{y_k} : \mathcal{N}_{q+1}^{(k-1)} \to \mathcal{N}_q^{(k-1)}$ is surjective so that there exists a form $\eta^{(k)} \in \mathcal{N}_{q+1}^{(k-1)} \otimes \Lambda_{p-1}(\mathcal{V}^*)$ such that $\partial_{y_k} \eta^{(k)} = \omega_2$. Hence the exterior part of the form $\omega^{(k)} = \omega - \delta\eta^{(k)}$ depends only on $y_{k+1}, \ldots, y_n$ and we can iterate. Thus starting with $k = 1$ we finally obtain $\omega = \delta\big(\eta^{(1)} + \cdots + \eta^{(n-1)}\big)$. □

While Proposition 3.3 is nice from a theoretical point of view, it is not very useful computationally, as we must check infinitely many conditions, namely one for each degree $q \ge q_0$. Under modest assumptions it suffices to consider only the one degree q_0 and then we obtain an effective criterion for involution representing an algebraic reformulation of the classical Cartan test in the theory of exterior differential systems. It uses only linear algebra with the two finite-dimensional components $\mathcal{N}_q$ and $\mathcal{N}_{q+1}$ (note, however, that the test can only be applied in δ-regular bases). In particular, it is not necessary to determine explicitly any Spencer cohomology module. In the context of differential equations, this observation will later translate into the fact that it is easier to check involution than formal integrability.

Theorem 3.4 (Cartan Test) *Let $\mathcal{N} \subseteq \mathfrak{S}(\mathcal{V}^*) \otimes \mathcal{U}$ be a symbolic system such that $\mathcal{N}_{q,1} = \mathcal{N}_{q+1}$ for all $q \geq q_0$. Then $\mathcal{N}$ is involutive at degree q_0, if and only if a basis $\{y_1, \ldots, y_n\}$ of $\mathcal{V}^*$ exists such that we have equality in (3.4) for $q = q_0$.*

Implicitly, a proof of this result was already given by Janet [26]. Later the theorem was explicitly demonstrated by Matsushima [39, 40]. We do not give here a proof, as it will follow automatically from later results on Pommaret bases (see Remark 4.15); in spirit this corresponds to the proof of Janet.

Example 3.5 Let us consider over a three-dimensional vector space $\mathcal{V}$ the symbolic system $\mathcal{N} \subset \mathfrak{S}(\mathcal{V}^*)$ defined by $\mathcal{N}_0 = \Bbbk$, $\mathcal{N}_1 = \mathcal{V}^*$, $\mathcal{N}_2 = \langle y_1^2, y_1 y_2, y_1 y_3, y_2^2 \rangle$ and $\mathcal{N}_q = \mathcal{N}_{q-1,1}$ for $q \geq 3$. One easily verifies that here $\mathcal{N}_2^{(1)} = \langle y_2^2 \rangle$ and $\mathcal{N}_2^{(2)} = \mathcal{N}_2^{(3)} = 0$ and therefore the only non-vanishing Cartan characters of $\mathcal{N}$ are $\alpha_2^{(1)} = 3$ and $\alpha_2^{(2)} = 1$. Furthermore, $\mathcal{N}_3 = \langle y_1^3, y_1^2 y_2, y_1^2 y_3, y_1 y_2^2, y_2^3 \rangle$. Since $\alpha_2^{(1)} + 2\alpha_2^{(2)} = 5 = \dim \mathcal{N}_3$, the symbolic system $\mathcal{N}$ passes the Cartan test and is involutive at degree $q = 2$. One also immediately sees that the map $\partial_{y_1} : \mathcal{N}_3 \to \mathcal{N}_2$ is indeed surjective and that the map $\partial_{y_2} : \mathcal{N}_3^{(1)} = \langle y_2^3 \rangle \to \mathcal{N}_2^{(1)}$ is even bijective (there is no need to consider also ∂_{y_3}, since both $\mathcal{N}_2^{(2)}$ and $\mathcal{N}_3^{(2)}$ vanish).

Example 3.6 For an instance where the Cartan test is not passed, we return to the second symbolic system in Example 2.9. Since $\mathcal{N}$ vanishes beyond degree 5, it is trivially involutive at degree 5. We verify now that it is not involutive at a lower degree. It is clear that $\partial_{y_1} : \mathcal{N}_5 \to \mathcal{N}_4$ cannot be surjective and also $\alpha_4^{(1)} = 1 > \dim \mathcal{N}_5 = 0$. Hence $\mathcal{N}$ is not involutive at degree 4.

Given the duality between the polynomial de Rham and the Koszul complex, we expect that a similar criterion for involution exists for polynomial modules. The essence of the proof of Proposition 2.6 is that differentiation with respect to y_k is dual to multiplication with x^k. Hence when we now study, following the letter of Serre appended to [21], the dualisation of the considerations above, it is not surprising that the multiplication with elements $v \in \mathcal{V}$ is central.

Lemma 3.7 *Let $\mathcal{M}$ be a finitely generated graded $\mathcal{SV}$-module and $q > 0$ an integer. Then the following statements are equivalent.*

  (i) $H_{r,n}(\mathcal{M}) = 0$ for all $r \geq q$.

  (ii) If $\mathrm{Ann}\,(m) = \mathcal{S}_+\mathcal{V}$ for an $m \in \mathcal{M}$, then $m \in \mathcal{M}_{<q}$.

  (iii) The existence of an element $v \in \mathcal{V}$ with $v \cdot m = 0$ entails $m \in \mathcal{M}_{<q}$.

  (iv) For all $v \in \mathcal{V}$ except the elements of a finite number of proper subspaces of $\mathcal{V}$ the equation $v \cdot m = 0$ entails $m \in \mathcal{M}_{<q}$.

Proof. The equivalence of (i) and (ii) follows immediately from Lemma 2.16. Furthermore, it is trivial that (iv) implies (iii) implies (ii). Hence there only remains to show that (iv) is a consequence of (ii).

Assume that (ii) holds and let $\mathcal{A} = \{m \in \mathcal{M}_{<q} \mid \mathrm{Ann}\,(m) = \mathcal{S}_+\mathcal{V}\}$. We choose a complement $\mathcal{K}$ such that $\mathcal{M}_{<q} = \mathcal{A} \oplus \mathcal{K}$ and set $\overline{\mathcal{M}} = \mathcal{K} \oplus \bigoplus_{r \geq q} \mathcal{M}_r$. Because of

(ii) no element of $\overline{\mathcal{M}} \setminus \{0\}$ is annihilated by $S_+ \mathcal{V}$ and hence $S_+ \mathcal{V}$ is not an associated prime ideal of the module $\overline{\mathcal{M}}$. By a standard result in commutative algebra, the set $\operatorname{Ass} \overline{\mathcal{M}}$ of all associated prime ideals of $\overline{\mathcal{M}}$ contains only finitely many elements. The intersection of any of these with $\mathcal{V}$ is a proper subspace. If we choose $v \in \mathcal{V}$ such that it is not contained in any of these subspaces, then $v \cdot m = 0$ entails $m \in \mathcal{M}_{<q}$.   □

The property of v in Part (iii) will become so important in the sequel that we provide a special name for it. It is closely related to the classical notion of a regular sequence in commutative algebra except that for the latter it is not permitted that the multiplication with v has a non-trivial kernel whereas here we only restrict the degree of the kernel.

Definition 3.8 A vector $v \in \mathcal{V}$ is called *quasi-regular* at degree q for the module $\mathcal{M}$, if $v \cdot m = 0$ entails $m \in \mathcal{M}_{<q}$. A finite sequence $(v_1, \ldots, v_k)$ of elements of $\mathcal{V}$ is *quasi-regular* at degree q for the module $\mathcal{M}$, if each v_i is quasi-regular at degree q for the factor module $\mathcal{M}/\langle v_1, \ldots, v_{i-1}\rangle\mathcal{M}$.

If a vector v is quasi-regular at degree q, it is also quasi-regular at any degree $r > q$. Furthermore, the vectors in a quasi-regular sequence are linearly independent. Thus such a sequence of length $n = \dim \mathcal{V}$ defines a basis of the vector space $\mathcal{V}$.

Lemma 3.9 *Let $v \in \mathcal{V}$ be quasi-regular at degree q. For each $r \geq q$ and $1 \leq p \leq n$ there is a short exact sequence*

$$0 \longrightarrow H_{r,p}(\mathcal{M}) \overset{\alpha}{\longrightarrow} H_{r,p}(\mathcal{M}/v\mathcal{M}) \overset{\beta}{\longrightarrow} H_{r,p-1}(\mathcal{M}) \longrightarrow 0 \qquad (3.6)$$

and the multiplication with v is injective on $\mathcal{M}_{\geq q}$.

Proof. As above we decompose $\mathcal{M} = \mathcal{A} \oplus \overline{\mathcal{M}}$. Since $\mathcal{A} \subseteq \mathcal{M}_{<q}$, we have the equality $H_{r,p}(\mathcal{M}) = H_{r,p}(\overline{\mathcal{M}})$ for all $r \geq q$ and, because of $v\mathcal{A} = 0$, similarly $H_{r,p}(\mathcal{M}/v\mathcal{M}) = H_{r,p}(\overline{\mathcal{M}}/v\overline{\mathcal{M}})$ for all $r \geq q$.

It follows trivially from the definition of quasi-regularity that multiplication with v is injective on $\mathcal{M}_{\geq q}$. In fact, it is injective on $\overline{\mathcal{M}}$. Indeed, suppose that $v \cdot m = 0$ for some homogeneous element $m \in \overline{\mathcal{M}}$. Let us assume first that $m \in \mathcal{M}_{q-1}$. Then $v \cdot (w \cdot m) = 0$ for all $w \in \mathcal{V}$ and since $w \cdot m \in \mathcal{M}_q$, this is only possible, if $w \cdot m = 0$ and thus $\operatorname{Ann}(m) = S_+ \mathcal{V}$ implying $m \in \mathcal{A}$. Iterating this argument, we conclude that m cannot be contained in $\overline{\mathcal{M}}_{q-2}$ either and so on. Hence $m \in \mathcal{A}$.

Because of the injectivity, the sequence

$$0 \longrightarrow \mathcal{M} \overset{v}{\longrightarrow} \mathcal{M} \overset{\pi}{\longrightarrow} \mathcal{M}/v\mathcal{M} \longrightarrow 0 \qquad (3.7)$$

of graded modules is exact at all degrees $r \geq q$. Tensoring with the vector space $\Lambda\mathcal{V}$ yields a similar sequence for the corresponding Koszul complexes $K(\mathcal{M})$ and $K(\mathcal{M}/v\mathcal{M})$, respectively, with the same exactness properties. Now we consider the associated long exact homological sequence

$$\cdots \longrightarrow H_{r-1,p}(\mathcal{M}) \xrightarrow{H(v)} H_{r,p}(\mathcal{M}) \xrightarrow{H(\pi)} \tag{3.8}$$

$$\longrightarrow H_{r,p}(\mathcal{M}/v\mathcal{M}) \xrightarrow{\beta} H_{r,p-1}(\mathcal{M}) \xrightarrow{H(v)} \cdots .$$

Since, by Lemma 2.17, $H(v)$ is the zero map, it decomposes into the desired short exact sequences with $\alpha = H(\pi)$ and $\beta([\omega]) = [\frac{1}{v} \circ \partial(\omega)]$.   □

Proposition 3.10 *Let $\mathcal{M}$ be a finitely generated graded $\mathcal{P}$-module and the sequence $(v_1, \ldots, v_k)$ quasi-regular at degree q. Then $H_{r,p}(\mathcal{M}) = 0$ for all values $r \geq q$ and $n - k < p \leq n$. If we set $\mathcal{M}^{(i)} = \mathcal{M}/\langle v_1, \ldots, v_i \rangle \mathcal{M}$, then*

$$H_{r,n-k}(\mathcal{M}) \cong H_{r,n-k+1}\big(\mathcal{M}^{(1)}\big) \cong \cdots \cong H_{r,n}\big(\mathcal{M}^{(k)}\big) \tag{3.9}$$

for all $r \geq q$.

Proof. We proceed by induction over the length k of the quasi-regular sequence. For $k = 1$, it follows from Lemma 3.7 that $H_{r,n}(\mathcal{M}) = 0$ for all $r \geq q$. Entering this result into the short exact sequence (3.6) of Lemma 3.9 gives immediately an isomorphism $H_{r,n-1}(\mathcal{M}) \cong H_{r,n}(\mathcal{M}^{(1)})$.

Assume now that the proposition holds for any quasi-regular sequence of length less than k. Then we know already that $H_{r,p}(\mathcal{M}) = 0$ for all $r \geq q$ and $n - k + 1 < p \leq n$ and that $H_{r,n-k+1}(\mathcal{M}) \cong H_{r,n}(\mathcal{M}^{(k-1)})$. Since v_k is quasi-regular at degree q for $\mathcal{M}^{(k-1)}$, the latter homology group vanishes by Lemma 3.7 proving the first assertion.

Applying the induction hypothesis to the module $\mathcal{M}^{(i-1)}$ and the quasi-regular sequence $(v_i, \ldots, v_k)$ shows that $H_{r,n-k+i}(\mathcal{M}^{(i-1)}) = 0$. Now we may use again the exact sequence of Lemma 3.9 to conclude that $H_{r,n-k+i}(\mathcal{M}^{(i)}) \cong H_{r,n-k+i-1}(\mathcal{M}^{(i-1)})$. This proves the second assertion.   □

Proposition 3.11 *Let $\mathcal{M}$ be a graded SV-module finitely generated in degree less than $q > 0$. The module $\mathcal{M}$ is involutive at degree q, if and only if a basis $\{x^1, \ldots, x^n\}$ of $\mathcal{V}$ exists such that the maps*

$$\mu_k : \mathcal{M}_r/\langle x^1, \ldots, x^{k-1}\rangle \mathcal{M}_{r-1} \longrightarrow \mathcal{M}_{r+1}/\langle x^1, \ldots, x^{k-1}\rangle \mathcal{M}_r \tag{3.10}$$

induced by the multiplication with x^k are injective for all $r \geq q$ and $1 \leq k \leq n$.

Proof. We first note that the statement that $\mathcal{M}$ is generated in degree less than q is equivalent to $H_{r,0}(\mathcal{M}) = 0$ for all $r \geq q$ by Lemma 2.16.

If $\mathcal{M}$ is involutive at degree q, then $H_{r,n}(\mathcal{M}) = 0$ for all $r \geq q$ and Lemma 3.7 implies that a generic vector $x^1 \in \mathcal{V}$ is quasi-regular. Now we proceed by iteration. Setting $\mathcal{M}^{(k)} = \mathcal{M}/\langle x^1, \ldots, x^k \rangle \mathcal{M}$, we find that $H_{r,n}(\mathcal{M}^{(k)}) = H_{r,n-k}(\mathcal{M}) = 0$ for all $r \geq q$ by Lemma 3.9. Thus we may again apply Lemma 3.7 in order to show that for any $1 \leq k < n$ the quasi-regular sequence $(x^1, \ldots, x^k)$ can be extended by a generic vector $x^{k+1} \in \mathcal{V}$. As already remarked above, an quasi-regular sequence of length n defines a basis of $\mathcal{V}$. Now the injectivity of the maps μ_k follows from Lemma 3.9.

Conversely, if all maps μ_k are injective, then obviously $(x^1, \ldots, x^n)$ defines an quasi-regular sequence of length n. Now the vanishing of all homology groups $H_{r,p}(\mathcal{M})$ with $r \geq q$ and $1 \leq p \leq n$ follows from Proposition 3.10 and $\mathcal{M}$ is involutive.   □

Again we face the problem that this proposition requires an infinite number of checks and thus cannot be applied effectively. Quillen [46, App., Prop. 8] was the first to show that for a certain class of modules, it suffices to consider only the components $\mathcal{M}_q$ and $\mathcal{M}_{q+1}$. This leads to the following dual formulation of the Cartan test (Theorem 3.4); again we refer to Remark 4.15 for a proof (or alternatively to [35]).

Theorem 3.12 (Dual Cartan Test) *Let $\mathcal{N}^0 \subseteq S\mathcal{V} \otimes \mathcal{U}^*$ be a homogeneous submodule of the free $S\mathcal{V}$-module $S\mathcal{V} \otimes \mathcal{U}^*$ finitely generated in degree less than $q > 0$. Then the factor module $\mathcal{M} = (S\mathcal{V} \otimes \mathcal{U}^*)/\mathcal{N}^0$ is involutive at degree q, if and only if a basis $\{x^1, \ldots, x^n\}$ of $\mathcal{V}$ exists such that the maps*

$$\mu_k : \mathcal{M}_q/\langle x^1, \ldots, x^{k-1}\rangle \mathcal{M}_{q-1} \longrightarrow \mathcal{M}_{q+1}/\langle x^1, \ldots, x^{k-1}\rangle \mathcal{M}_q \qquad (3.11)$$

induced by the multiplication with x^k are injective for all $1 \leq k \leq n$.

Remark 3.13 Let $\mathcal{N} \subseteq \mathfrak{S}(\mathcal{V}^*) \otimes \mathcal{U}$ be a symbolic system and consider the dual $S\mathcal{V}$-module $\mathcal{M} = \mathcal{N}^* \cong (S\mathcal{V} \otimes \mathcal{U}^*)/\mathcal{N}^0$. Let furthermore $\{x^1, \ldots, x^n\}$ be a basis of $\mathcal{V}$ and $\{y_1, \ldots, y_n\}$ the dual basis of $\mathcal{V}^*$. Then we find that $\mu_1^* = \partial_{y_1}$ and hence that $\left(\mathcal{N}^{(1)}\right)^* = (\ker \partial_{y_1})^* \cong \operatorname{coker} \mu_1 = \mathcal{M}^{(1)}$. Iteration of this argument yields $\left(\mathcal{N}^{(k)}\right)^* \cong \mathcal{M}^{(k)}$ for all $1 \leq k \leq n$ (considering always ∂_{y_k} as a map on $\mathcal{N}^{(k-1)}$ so that $\mathcal{N}^{(k)} = \ker \partial_{y_k}$ and μ_k as a map on $\mathcal{M}^{(k-1)}$ so that $\mathcal{M}^{(k)} = \operatorname{coker} \mu_k$). We also have $\mu_k = \partial_{y_k}^*$ and hence obtain the isomorphisms $(\ker \mu_k)^* \cong \operatorname{coker} \partial_{y_k}^*$ (again considering the maps on the appropriate domains of definition). Thus injectivity of all the maps μ_k is equivalent to surjectivity of all the maps ∂_{y_k}. Hence applying Proposition 3.11 to $\mathcal{M}$ proves dually Proposition 3.3 for $\mathcal{N}$ and similarly for the Theorems 3.4 and 3.12. Furthermore, it is obvious that if the basis $\{x^1, \ldots, x^n\}$ is quasi-regular at degree q, then the dual basis $\{y_1, \ldots, y_n\}$ is δ-regular for $\mathcal{N}_q$. The converse does not necessarily hold, as δ-regularity is a much weaker condition than quasi-regularity (the latter implies involution via the dual Cartan test; the former is only a necessary condition for applying the Cartan test).

Example 3.14 Recall the symbolic system $\mathcal{N}$ of Example 3.5. Its annihilator $\mathcal{N}^0$ is the ideal $\mathcal{I} \subset \mathcal{P} = S\mathcal{V}$ generated by the monomials $x^2 x^3$ and $(x^3)^2$. We apply now the dual Cartan test to the the factor module $\mathcal{M} = \mathcal{P}/\mathcal{I}$. For the two relevant module components we obtain after a trivial computation that $\mathcal{M}_2 \cong \langle (x^1)^2, x^1 x^2, x^1 x^3, (x^2)^2 \rangle \cong \mathcal{N}_2$ and $\mathcal{M}_3 \cong \langle (x^1)^3, (x^1)^2 x^2, (x^1)^2 x^3, x^1 (x^2)^2, (x^2)^3 \rangle \cong \mathcal{N}_3$. Similarly, we find that the non-vanishing factor modules required for the dual Cartan test are given by $\mathcal{M}_2^{(1)} \cong \langle (x^2)^2 \rangle \cong \mathcal{N}_2^{(1)}$ and $\mathcal{M}_3^{(1)} \cong \langle (x^2)^3 \rangle \cong \mathcal{N}_3^{(1)}$. It is now trivial to see that the map $\mu_1 : \mathcal{M}_2 \to \mathcal{M}_3$ induced by the multiplication with x^1 is injective and the map $\mu_2 : \mathcal{M}_2^{(1)} \to \mathcal{M}_3^{(1)}$ induced by the multiplication with x^2 is even bijective. Hence by the dual Cartan test the module $\mathcal{M}$ is involutive at degree 2.

Remark 3.15 Another way to formulate the assumptions of Theorem 3.12 is to require that $\mathcal{M}$ is a finitely generated graded $S\mathcal{V}$-module such that $H_{r,0}(\mathcal{M}) = H_{r,1}(\mathcal{M}) = 0$ for all $r \geq q$. Indeed, any such module can be finitely presented and thus is isomorphic to a factor module $(S\mathcal{V} \otimes \mathcal{U}^*)/\mathcal{N}^0$ for an appropriately chosen $\mathcal{U}$. By the same argument as in the proof of Proposition 2.18, $H_{r,1}(\mathcal{M}) \cong H_{r+1,0}(\mathcal{N}^0)$. Since the latter homology is determined by the minimal generators of the submodule $\mathcal{N}^0$, the two sets of assumptions are equivalent.

Consider the monomial ideal $\mathcal{I} = \langle (x^1)^3, (x^2)^3 \rangle \subset \Bbbk[x^1, x^2]$ generated in degree 3; it is the annihilator of the second symbolic system $\mathcal{N}$ in Example 2.9. It is trivial that here $\mu_1 : \mathcal{I}_4 \to \mathcal{I}_5$ is injective. For the map μ_2 we note that $\mathcal{I}_4/x^1\mathcal{I}_3 \cong \langle (x^1)^3 x^2, (x^2)^4 \rangle$ and thus it is again easy to see that μ_2 is injective.

If we consider the map $\mu_2 : \mathcal{I}_5/x^1\mathcal{I}_4 \to \mathcal{I}_6/x^1\mathcal{I}_5$, then we find (using the identification $\mathcal{I}_5/x^1\mathcal{I}_4 \cong \langle (x^1)^3(x^2)^2, (x^2)^5 \rangle$) that $\mu_2\big([(x^1)^3(x^2)^2]\big) = [(x^1)^3(x^2)^3] = 0$ so that μ_2 is not injective and the Theorem 3.12 is not valid here. The observation that at some lower degree the maps μ_1 and μ_2 are injective may be understood from the syzygies. $\mathrm{Syz}(\mathcal{I}) \cong H_1(\mathcal{I})$ is generated by the single element $(x^2)^3\mathbf{e}_1 - (x^1)^3\mathbf{e}_2$ of degree 6. As its coefficients are of degree 3, nothing happens with the maps μ_i before we encounter $\mathcal{I}_6$ and then the equation $\mu_2\big([(x^1)^3(x^2)^2]\big) = 0$ is a trivial consequence of this syzygy.

4 Involutive Bases

Involutive bases are a special form of Gröbner bases with additional combinatorial properties. They were introduced by Gerdt and Blinkov [15, 16] generalising earlier ideas by Janet [25] in the theory of partial differential equations (a special case was slightly earlier discovered by Wu [63]); an introduction into their basic theory can be found in [50]. We assume in the sequel that the reader is familiar with the basic concepts in the theory of Gröbner bases; classical introductory texts are [1, 4, 11]. A Gröbner basis is defined with respect to a term order; for an involutive basis we need one further ingredient, namely a so-called involutive division.

While the precise definition of an involutive division is somewhat technical, the underlying idea is simple. We consider first the monomial case. Let $\mathcal{T} = \{t_1, \dots, t_s\}$ be a set of terms in the ring $\mathcal{P} = \Bbbk[X]$ where $X = \{x^1, \dots, x^n\}$. Then the ideal $\mathcal{I} \subseteq \mathcal{P}$ generated by $\mathcal{T}$ consists of all polynomials $f = \sum_{i=1}^{s} P_i t_i$ where the coefficients P_i are arbitrary polynomials, i. e. $\mathcal{I}$ is the linear span $\langle \mathcal{T} \rangle$. An involutive division L assigns to each generator t_i a set of multiplicative variables $X_{L,\mathcal{T}}(t_i) \subseteq X$ (the remaining variables are denoted by $\overline{X}_{L,\mathcal{T}}(t_i)$). The involutive span $\langle \mathcal{T} \rangle_L$ consists of all linear combinations $f = \sum_{i=1}^{s} P_i t_i$ where now the coefficients must satisfy $P_i \in \Bbbk[X_{L,\mathcal{T}}(t_i)]$. Thus in general it contains only a subset of the ideal $\mathcal{I}$.

We call the set $\mathcal{T}$ a *weak involutive basis* of $\mathcal{I}$ for the involutive division L, if $\langle \mathcal{T} \rangle_L = \mathcal{I}$. For a term $x^\mu \in \mathcal{I}$, we call any generator $t_i \in \mathcal{T}$ such that $x^\mu \in \Bbbk[X_{L,\mathcal{T}}(t_i)] \cdot t_i$ an *involutive divisor*. Thus $\mathcal{T}$ is a weak involutive basis, if every term in $\mathcal{I}$ has at least one involutive divisor. For a *(strong) involutive basis* we require additionally that this involutive divisor is unique (in other words, for any two generators $t_i \neq t_j \in \mathcal{T}$ we have $\Bbbk[X_{L,\mathcal{T}}(t_i)] \cdot t_i \cap \Bbbk[X_{L,\mathcal{T}}(t_j)] \cdot t_j = \{0\}$).

The assignment of the multiplicative variables by the involutive division cannot be arbitrary but must satisfy certain conditions which we omit here (they can be found in the above references), as we are here only interested in one particularly simple division, namely the *Pommaret*[1] *division* P. While for general involutive divisions the assignment of multiplicative variables depends on the set $\mathcal{T}$ (i. e. the same term t may be assigned different variables if considered as element of different sets $\mathcal{T}$), the Pommaret division is a so-called global division where the assignment is independent of $\mathcal{T}$. If the term t is of the form $t = x^\mu$ with a multi index of class $\mathrm{cls}\,\mu = k$, then we simply assign as multiplicative variables $X_P(t) = \{x^1, \ldots, x^k\}$.

Example 4.1 Consider again the second symbolic system $\mathcal{N}$ in Example 2.9. Its annihilator is the ideal $\mathcal{I}$ generated by $\mathcal{T} = \{(x^1)^3, (x^2)^3\}$. Since the first generator is of class 1, x^2 is not multiplicative for it with respect to the Pommaret division. As a consequence the monomial $(x^1)^3 x^2 \in \mathcal{I}$ is not contained in $\langle \mathcal{T} \rangle_P$ and thus the minimal basis $\mathcal{T}$ of $\mathcal{I}$ is not an involutive basis. An involutive basis is obtained, if we add the monomials $(x^1)^3 x^2$ and $(x^1)^3 (x^2)^2$, as one easily verifies.

The extension to general ideals is now straightforward. Let $\mathcal{F} = \{f_1, \ldots, f_s\}$ be an arbitrary set of polynomials, $\prec$ a term order and L an involutive division. $\mathcal{F}$ is a *weak involutive basis* of $\mathcal{I} = \langle \mathcal{F} \rangle$ for $\prec$ and L, if the monomial set $\mathrm{lt}\,\mathcal{F} = \{\mathrm{lt}\,f_1, \ldots, \mathrm{lt}\,f_s\}$ is a weak involutive basis of the leading ideal $\mathrm{lt}\,\mathcal{I}$. Note that this definition trivially implies that any weak involutive basis is a Gröbner basis, too. We call $\mathcal{F}$ a *(strong) involutive basis*, if no two elements of $\mathcal{F}$ have the same leading term and $\mathrm{lt}\,\mathcal{F}$ is a strong involutive basis of $\mathrm{lt}\,\mathcal{I}$. An involutive basis with respect to the Pommaret division is briefly called *Pommaret basis*. We also introduce multiplicative variables for generators $f \in \mathcal{F}$ by setting $X_{L,\mathcal{F},\prec}(f) = X_{L,\mathrm{lt}\,\mathcal{F}}(\mathrm{lt}\,f)$.

Remark 4.2 One easily shows that by a simple elimination process any weak involutive basis can be reduced to a strong one and thus we will exclusively work with strong bases (this is no longer possible in more general situations with e. g. local term orders or coefficient rings) [50]. A particular property of the Pommaret division (in fact of any global division) is that the Pommaret basis of any monomial ideal is unique.

The above definition does not provide us with an effective criterion for recognising an involutive basis. For arbitrary involutive divisions no such criterion has been discovered so far. However, the Pommaret division belongs to the so-called *continuous* divisions for which the situation is more favourable. A finite set $\mathcal{F} \subset \mathcal{P}$ is *locally involutive* for the division L, if for every polynomial $f \in \mathcal{F}$ and for every non-multiplicative variable $x^j \in \overline{X}_{L,\mathcal{F},\prec}(f)$ the product $x^j f$ can be involutively reduced[2] to zero with respect to $\mathcal{F}$. Obviously, this property can be checked effectively.

Theorem 4.3 *If the finite set $\mathcal{F} \subset \mathcal{P}$ is involutively (head) autoreduced and locally involutive for a continuous division L, then $\mathcal{F}$ is an involutive basis of $\langle \mathcal{F} \rangle$ for L.*

---

[1] Historically the terminology "Pommaret division" is a misnomer, as this division was already introduced by Janet. But the name has become generally accepted and therefore we stick to it.

[2] Involutive reducibility is defined as in the standard Gröbner theory; the sole difference is that a reduction is permitted only, if the reducing element $f \in \mathcal{F}$ is multiplied with a polynomial in $\mathrm{k}[X_{L,\mathcal{F},\prec}(f)]$.

As the proof of this theorem is rather technical and requires some concepts not introduced here, we refer to [15, 50]. We show here only a simpler special case. However, it will turn out later (Remarks 4.5 and 4.15) that this special case entails the Cartan test (in fact, this approach is almost identical with Janet's proof [26]).

Proposition 4.4 *Let* $\mathcal{H}_q \subset \mathcal{P}_q$ *be a finite triangular set of homogeneous polynomials of degree* q *which is locally involutive for the Pommaret division and a term order* $\prec$. *Then the set*

$$\mathcal{H}_{q+1} = \{x^i h \mid h \in \mathcal{H}_q, \ x^i \in X_P(h)\} \subset \mathcal{P}_{q+1} \tag{4.1}$$

is also triangular and locally involutive (by induction this implies that $\mathcal{H}_q$ *is involutive).*

Proof. It is trivial to see that $\mathcal{H}_{q+1}$ is again triangular (all leading terms are different). For showing that it is also locally involutive, we consider an element $x^i h \in \mathcal{H}_{q+1}$. By construction, $\mathrm{cls}\,(x^i h) = i \leq \mathrm{cls}\,h$. We must show that for any non-multiplicative index $i < j \leq n$ the polynomial $x^j(x^i h)$ is expressible as a linear combination of polynomials $x^k \bar{h}$ where $\bar{h} \in \mathcal{H}_{q+1}$ and $x^k \in X_{P,\prec}(\bar{h})$. In the case that $j \leq \mathrm{cls}\,h$, this is trivial, as we may choose $\bar{h} = x^j h$ and $k = i$.

Otherwise x^j is non-multiplicative for h and since $\mathcal{H}_q$ is assumed to be locally involutive, the polynomial $x^j h$ can be written as a $\Bbbk$-linear combination of elements of $\mathcal{H}_{q+1}$. For exactly one summand $\bar{h}$ in this linear combination we have $\mathrm{lt}\,\bar{h} = \mathrm{lt}\,(x^j h)$ and hence $x^i \in X_{P,\prec}(\bar{h})$. If x^i is also multiplicative for all other summands, we are done. If the variable x^i is non-multiplicative for some summand $\bar{h}' \in \mathcal{H}_{q+1}$, then we analyse the product $x^i \bar{h}'$ in the same manner writing $\bar{h}' = x^k h'$ for some $h' \in \mathcal{H}_q$. Since $\mathrm{lt}\,\bar{h}' \prec \mathrm{lt}\,(x^j h)$, this process terminates after a finite number of steps leading to an involutive standard representation of $x^j(x^i h)$. $\square$

Remark 4.5 Proposition 4.4 may be considered as an involutive basis version of the Cartan test. Let $\mathcal{I} \subseteq \mathcal{P}$ be a homogeneous ideal and $\mathcal{H}_q$ a triangular vector space basis of $\mathcal{I}_q$ for some degree q. The set $\mathcal{H}_q$ is locally involutive (and thus a Pommaret basis of $\mathcal{I}_{\geq q}$), if and only if the set $\mathcal{H}_{q+1}$ defined by (4.1) is a vector space basis of $\mathcal{I}_{q+1}$. Denoting by $\beta_q^{(k)}$ the number of elements of $\mathcal{F}$ where the leading term is of class k, these considerations lead to the inequality

$$\dim \mathcal{I}_{q+1} \geq \sum_{k=1}^{n} k \beta_q^{(k)} \tag{4.2}$$

and equality holds, if and only if $\mathcal{F}$ is a Pommaret basis. Remark 4.15 below shows that (4.2) does not only formally looks like (3.4) but that it is actually equivalent.

While it is almost trivial to prove that any ideal $\mathcal{I} \subseteq \mathcal{P}$ possesses a Gröbner basis for any term order, the existence of involutive bases is a more difficult question and depends on the precise form of the chosen division. For some divisions the existence is always guaranteed; one speaks of *Noetherian* divisions.

For Pommaret bases the situation is more complicated. It is easy to find ideals without a Pommaret basis—consider for example $\langle x^1 x^2 \rangle \subset \Bbbk[x^1, x^2]$ where a Pommaret basis would have to include all terms $x^1(x^2)^k$ with $k \geq 1$. A closer look reveals that this is actually only a problem of the chosen coordinates. If we begin as above with the symmetric algebra $S\mathcal{V}$ and consider $\mathcal{I}$ as an ideal in it, then for a generic basis of $\mathcal{V}$ the corresponding polynomial ideal $\mathcal{I}$ has a Pommaret basis. We call a basis (or co-ordinates) $\{x^1, \ldots, x^n\}$ such that $\mathcal{I}$ possesses a Pommaret basis δ-*regular* for the ideal $\mathcal{I}$.[3] The use of the same terminology as in the Cartan test is no coincidence, as we will show in the next chapter. In the example above the transformation $x^1 \rightarrow x^1 + x^2$ leads to the ideal $\langle (x^2)^2 + x^1 x^2 \rangle$ which has a Pommaret basis for any term order where $x^2 \succ x^1$ (if $x^2 \prec x^1$, we can use the transformation $x^2 \rightarrow x^1 + x^2$).

Remark 4.6 For general information about the algorithmic determination of involutive bases we refer to [15, 16, 17]. Effective criteria for recognising δ-singular and effective methods for the construction of δ-regular coordinates for a given ideal $\mathcal{I}$ are discussed in detail in [22]. From a strictly algorithmic point of view, it is unpleasant that the Pommaret division is not Noetherian. But we will see in the remainder of this chapter that this seeming disadvantage has a number of benefits, as for many applications in algebraic geometry it is of considerable interest to know "good" coordinates.

We turn now to properties of involutive bases, in particular to those not shared by ordinary Gröbner bases. For simplicity, we always assume that we are dealing with a homogeneous ideal $\mathcal{I}$ and that also all considered bases of $\mathcal{I}$ are homogeneous.

If $\mathcal{H} = \{h_1, \ldots, h_s\}$ is a Gröbner basis of the ideal $\mathcal{I}$, then it is well known that any polynomial $f \in \mathcal{I}$ possesses a standard representation $f = \sum_{i=1}^{s} P_i h_i$ where the coefficients $P_i \in \mathcal{P}$ satisfy $\mathrm{lt}\,(P_i h_i) \preceq \mathrm{lt}\, f$ whenever $P_i \neq 0$. However, even with this constraint this representation is in general not unique. This changes, if we assume that $\mathcal{H}$ is an involutive basis for an involutive division L. Imposing now the additional constraint that $P_i \in \Bbbk \big[X_{L, \mathrm{lt}\,\mathcal{H}}(\mathrm{lt}\, h_i) \big]$, we obtain the unique *involutive standard representation* of f with respect to $\mathcal{H}$. This uniqueness is the key to most applications of involutive bases.[4]

Another way to express the uniqueness of the involutive standard representation is to say that $\mathcal{H}$ induces a *Stanley decomposition* of $\mathcal{I}$. Because of the assumed homogeneity, $\mathcal{I}$ may be considered as a graded vector space with respect to the natural grading given by the total degree. A Stanley decomposition is then an isomorphism of graded vector spaces $\mathcal{I} \cong \bigoplus_{t \in \mathcal{T}} \Bbbk[X_t] \cdot t$ where $\mathcal{T}$ is some finite set of generators and $X_t \subseteq X$ is some subset of variables. In our case, we obtain the decomposition

$$\mathcal{I} \cong \bigoplus_{i=1}^{s} \Bbbk \big[X_{L, \mathrm{lt}\,\mathcal{H}}(\mathrm{lt}\, h_i) \big] \cdot h_i \,. \tag{4.3}$$

Pommaret bases lead to a special kind of Stanley decompositions, so-called *Rees decompositions* [47], where the subsets X_t are always of the form $\{x^1, \ldots, x^{k_t}\}$ for some value $0 \leq k_t \leq n$.

---

[3] Of course the used term order is here of great importance: it follows immediately from the definition of an involutive basis that the δ-regularity of a coordinate system is completely determined by $\mathrm{lt}\,\mathcal{I}$.

[4] One easily shows that with respect to a *weak* involutive basis also every ideal member has an involutive standard representation. However, it will be unique, if and only if one is dealing with a strong involutive basis. For this reason, for most advanced applications of involutive bases only the strong ones are of real interest.

A simple application of a Stanley decomposition (in fact, the one which motivated its introduction by Stanley [58][5]) is that one can trivially read off the *Hilbert series*:

$$\mathcal{H}_{\mathcal{I}}(\lambda) = \sum_{t \in \mathcal{T}} \frac{\lambda^{q_t}}{(1 - \lambda)^{k_t}} \tag{4.4}$$

where we introduced $q_t = \deg t$ and $k_t = |X_t|$. In particular, the *(Krull) dimension* of the ideal $\mathcal{I}$ is given by $D = \max_{t \in \mathcal{T}} k_t$ and the *multiplicity* (or *degree*) by the number of generators $t \in \mathcal{T}$ with $k_t = D$.

For most purposes, it is of greater interest to obtain a *complementary decomposition*, i. e. a Stanley decomposition of the factor algebra $\mathcal{A} = \mathcal{P}/\mathcal{I}$. Sturmfels and White [60] presented a recursive algorithm for computing such a decomposition given a Gröbner basis of $\mathcal{I}$. Somewhat surprising, the knowledge of an arbitrary involutive basis does not seem to give an advantage here. The situation changes, if one considers special involutive divisions. In the context of determining formally well-posed initial conditions for overdetermined systems of partial differential equations, Janet [27, §15] presented an algorithmic solution to this problem already in the 1920s. For Pommaret bases the solution is almost trivial; in fact, one only needs the degree of the basis (the reason will become evident below when we discuss the Castelnuovo–Mumford regularity).

Proposition 4.7 *The homogeneous ideal $\mathcal{I} \subseteq \mathcal{P}$ possesses a Pommaret basis $\mathcal{H}$ with* $\deg \mathcal{H} = q$, *if and only if the two sets* $\bar{\mathcal{T}}_0 = \{x^\mu \in \mathbb{T} \setminus \operatorname{lt} \mathcal{I} \mid \deg x^\mu < q\}$ *and* $\bar{\mathcal{T}}_1 = \{x^\mu \in \mathbb{T} \setminus \operatorname{lt} \mathcal{I} \mid \deg x^\mu = q\}$ *yield the complementary decomposition*

$$\mathcal{A} \cong \bigoplus_{t \in \bar{\mathcal{T}}_0} \Bbbk \cdot t \oplus \bigoplus_{t \in \bar{\mathcal{T}}_1} \Bbbk[X_P(t)] \cdot t . \tag{4.5}$$

Remark 4.8 Stanley decompositions are not unique. The complementary decomposition (4.5) is generally rather redundant. One can show that any Pommaret basis is simultaneously a Janet basis. Applying Janet's algorithm to it almost always yields a more compact decomposition with less generators. However, from a theoretical point of view, Proposition 4.7 is very useful, as it provides a closed formula and not only an algorithm. Note that (4.5) is again a Rees decomposition.

Proposition 4.9 *Let $\mathcal{H}$ be a homogeneous Pommaret basis of the homogeneous ideal $\mathcal{I} \subseteq \mathcal{P}$ with $\deg \mathcal{H} = q$. Then $D = \dim \mathcal{A}$ is given by*

$$D = \min \left\{ i \mid \langle \mathcal{H}, x^1, \dots, x^i \rangle_q = \mathcal{P}_q \right\} . \tag{4.6}$$

Remark 4.10 As a corollary to this result, one can easily show that $\{x^1, \dots, x^D\}$ is a maximal strongly independent set modulo $\mathcal{I}$ (see [20, 31] for the notion of an independent set modulo an ideal and its relation to the dimension). Here we see for the first time that the knowledge of δ-regular coordinates is of some interest, as generally no maximal independent set of this particularly simple form exists.

---

[5] One should note that in the context of partial differential equations Janet [27] derived already much earlier a similar expression for the Hilbert function.

In fact, combining this observation with Proposition 4.7 yields that the restriction of the canonical projection $\pi : \mathcal{P} \to \mathcal{A} = \mathcal{P}/\mathcal{I}$ to the subring $\Bbbk[x^1, \ldots, x^D]$ is a *Noether normalisation* of $\mathcal{A}$. Thus computing δ-regular coordinates determines automatically a Noether normalisation. One can show that δ-regularity is equivalent to simultaneous Noether normalisations of $\operatorname{lt} \mathcal{I}$ and all its primary components [5, 51].

Another measure for the size of $\mathcal{A}$ is its *depth*. It can also be immediately read off from a Pommaret basis. The proof of this fact provided by [51] relies on a direct verification that the given sequence is regular. Later in this article (Theorem 5.6) we will provide a homological proof of the following statement about the depth.

Proposition 4.11 *Let $\mathcal{H}$ be a homogeneous Pommaret basis of the homogeneous ideal $\mathcal{I} \subseteq \mathcal{P}$ for a class respecting term order $\prec$ and $d = \min_{h \in \mathcal{H}} \operatorname{cls} h$. Then $(x^1, \ldots, x^{d-1})$ is a maximal regular sequence for $\mathcal{A}$ and hence* $\operatorname{depth} \mathcal{A} = d - 1$.

Remark 4.12 Combining Propositions 4.9 and 4.11 leads immediately to the so-called *Hironaka criterion* for Cohen–Macaulay algebras: the factor algebra $\mathcal{A} = P/\mathcal{I}$ is Cohen–Macaulay, if and only if it possesses a Rees decomposition where all generators are of the same class.

Definition 4.13 A homogeneous ideal $\mathcal{I} \subseteq \mathcal{P}$ is called *q-regular*, if its ith syzygy module can be generated by elements of degree less than or equal to $q+i$; the *Castelnuovo–Mumford regularity* $\operatorname{reg} \mathcal{I}$ is the least value q for which $\mathcal{I}$ is q-regular.

Among other applications, $\operatorname{reg} \mathcal{I}$ represents an important measure for the complexity of Gröbner basis computations [2]. According to Bayer and Stillman [3], *generically* the reduced Gröbner basis with respect to the degree reverse lexicographic order has the degree $\operatorname{reg} \mathcal{I}$ and no other term order yields a lower degree. However, one rarely knows whether or not one is in the generic case so that this result is only of limited use for concrete computations. For Pommaret bases we rediscover here again simply the question of δ-regularity.

Theorem 4.14 *Let $\mathcal{I} \subseteq \mathcal{P}$ be a homogeneous ideal. Then $\operatorname{reg} \mathcal{I} = q$, if and only if $\mathcal{I}$ has in some coordinates a homogeneous Pommaret basis $\mathcal{H}$ with respect to the degree reverse lexicographic order such that* $\deg \mathcal{H} = q$.

This result implies that in δ-regular coordinates the equality $\operatorname{reg} \mathcal{I} = \operatorname{reg}(\operatorname{lt} \mathcal{I})$ holds whereas in general we only have the inequality $\operatorname{reg} \mathcal{I} \leq \operatorname{reg}(\operatorname{lt} \mathcal{I})$. Another remarkable implication is that in arbitrary coordinates $x^1, \ldots, x^n$ the ideal $\mathcal{I}$ either does not possess a finite Pommaret basis or the basis is of the fixed degree $\operatorname{reg} \mathcal{I}$.

Remark 4.15 According to Remark 2.15, the Koszul homology of a module $\mathcal{M}$ is equivalent to its minimal free resolution. Thus if $\operatorname{reg} \mathcal{I} = q$, then all homology modules $H_{r,p}(\mathcal{I})$ with $r > q$ vanish. Taking into account the degree shift in (2.16), for the factor module $\mathcal{M} = \mathcal{P}/\mathcal{I}$ thus all homology modules $H_{r,p}(\mathcal{M})$ with $r \geq q$ vanish. Hence the Castelnuovo–Mumford regularity of $\mathcal{I}$ is the same as the degree of involution of $\mathcal{M}$. It is very surprising that this elementary fact remained unobserved until very recently; it is implicitly contained in [53] and explicitly mentioned by Malgrange [35].

Combining this observation with Theorem 4.14 and Remark 4.5, we finally see that Proposition 4.4 may indeed be considered as an involutive basis version of the Cartan test. Let—as in Remark 4.5—the set $\mathcal{H}_q$ be a basis of the vector space $\mathcal{I}_q$ where all generators have different leading terms. Then we may choose as representatives of a basis of $\mathcal{M}_q$ polynomials which have as leading terms exactly those terms which do not appear in $\mathrm{lt}\,\mathcal{H}_q$. Elementary combinatorics shows that if $\mathcal{H}_q$ contains $\beta_q^{(k)}$ elements with a leading term of class k, then our basis of $\mathcal{M}_q$ contains

$$\alpha_q^{(k)} = m\binom{q+n-k-1}{q-1} - \beta_q^{(k)} \tag{4.7}$$

representatives with a leading term of class k.

It is no coincidence that we use here the same notation as for the Cartan characters. As a vector space the symbolic system $\mathcal{N} = \mathcal{I}^0$ is isomorphic to $\mathcal{P}/\mathcal{I}$; a concrete isomorphism is given by replacing in the above representatives x^i by y_i. If we use a class respecting term order, then it follows from Lemma B.1 that $\dim \mathcal{N}_q^{(k)} = \sum_{j=k+1}^{n} \alpha_q^{(j)}$ so that the numbers $\alpha_q^{(k)}$ are indeed the Cartan characters. A well known identity for binomial coefficients proves now that (3.4) and (4.2) are equivalent inequalities.

Theorem 4.14 represents probably the simplest method for computing $\mathrm{reg}\,\mathcal{I}$. It requires the knowledge of δ-regular coordinates, but as mentioned in Remark 4.6 these can be constructed effectively. In recent years, a number of methods for the determination of $\mathrm{reg}\,\mathcal{I}$ have been developed [3, 5, 62]. However, they all also require the use of generic coordinates (in [51] their relation to Pommaret bases is studied in detail).

Example 4.16 Consider the homogeneous ideal

$$\mathcal{I} = \langle z^8 - wxy^6,\ y^7 - x^6 z,\ yz^7 - wx^7 \rangle \subset \mathbb{Q}[w, x, y, z]\ . \tag{4.8}$$

The given basis of degree 8 is already a Gröbner basis for the degree reverse lexicographic term order. If we perform a permutation of the variables and consider $\mathcal{I}$ as an ideal in $\mathbb{Q}[w, y, x, z]$, then we obtain for the degree reverse lexicographic term order (in the new variables!) the following Gröbner basis of degree 50:

$$\{y^7 - x^6 z,\ yz^7 - wx^7,\ z^8 - wxy^6,\ y^8 z^6 - wx^{13},$$
$$y^{15} z^5 - wx^{19},\ y^{22} z^4 - wx^{25},\ y^{29} z^3 - wx^{31},$$
$$y^{36} z^2 - wx^{37},\ y^{43} z - wx^{43},\ y^{50} - wx^{49}\}\ . \tag{4.9}$$

Unfortunately, neither coordinate system is generic: as $\mathrm{reg}\,\mathcal{I} = 13$, one yields a basis of too low degree and the other one one of too high degree.

With a Pommaret basis it is no problem to determine the Castelnuovo-Mumford regularity, as the first coordinate system is δ-regular. A Pommaret basis of $\mathcal{I}$ for the degree reverse lexicographic term order is obtained by adding the polynomials $z^k(y^7 - x^6 z)$ for $1 \leq k \leq 6$ and thus the degree of the basis is indeed 13.

Remark 4.17 In order to obtain their above mentioned result, Bayer and Stillman first proved the following characterisation of a q-regular ideal (which may be considered as a variant of δ-regularity): if $\mathcal{I}$ is a homogeneous ideal which can be generated by elements of degree less than or equal to q, then it is q-regular, if and only if for some value $0 \leq d \leq n$ linear forms $y_1, \ldots, y_d \in \mathcal{P}_1$ exist such that

$$\left(\langle \mathcal{I}, y_1, \ldots, y_{j-1}\rangle : y_j\right)_q = \langle \mathcal{I}, y_1, \ldots, y_{j-1}\rangle_q , \quad 1 \leq j \leq d , \tag{4.10a}$$

$$\langle \mathcal{I}, y_1, \ldots, y_d\rangle_q = \mathcal{P}_q . \tag{4.10b}$$

We will discuss later in Remark 5.5 that this characterisation of q-regularity is equivalent to the dual Cartan test.

5   Pommaret Bases and Homology

Now we study the relationship between Pommaret bases and the homological constructions introduced in Chapters 2 and 3. We assume throughout that a fixed basis $\{x^1, \ldots, x^n\}$ of $\mathcal{V}$ has been chosen so that we may identify $S\mathcal{V} = \Bbbk[x^1, \ldots, x^n] = \mathcal{P}$. For simplicity, we restrict to homogeneous ideals $\mathcal{I} \subseteq \mathcal{P}$. We only consider Pommaret bases for the degree reverse lexicographic order $\prec_{\text{degrevlex}}$, as for any other term order the corresponding Pommaret basis (if it exists) cannot be of lower degree by the inequality $\operatorname{reg} \mathcal{I} \leq \operatorname{reg}(\operatorname{lt} \mathcal{I})$ and Theorem 4.14.

It turns out that this relationship takes its simplest form, if we compare the Pommaret basis of the ideal $\mathcal{I}$ and the Koszul homology of its factor algebra $\mathcal{P}/\mathcal{I}$ which we consider here as a $\mathcal{P}$-module in order to be consistent with the terminology introduced in Chapters 2 and 3. Like for general Gröbner bases, essentially everything relevant for involutive bases can be read off the leading ideal. Therefore, we show first that at least for our chosen term order quasi-regularity is also already decided by the leading ideal.

Lemma 5.1 *Let $\mathcal{I} \subseteq \mathcal{P}$ be a homogeneous ideal and $\prec$ the degree reverse lexicographic order. The sequence $(x^1, \ldots, x^n)$ is quasi-regular at degree q for the module $\mathcal{M} = \mathcal{P}/\mathcal{I}$, if and only if it is quasi-regular at degree q for $\mathcal{M}' = \mathcal{P}/\operatorname{lt} \mathcal{I}$.*

Proof. Let $\mathcal{G}$ be a Gröbner basis of $\mathcal{I}$ for $\prec$. Then the normal form with respect to the basis $\mathcal{G}$ defines an isomorphism between the vector spaces $\mathcal{M}$ and $\mathcal{M}'$. One direction is now trivial, as an obvious necessary condition for $m = [f] \in \mathcal{M}$ to satisfy $x^1 \cdot m = 0$ is that $x^1 \cdot [\operatorname{lt} f] = 0$ in $\mathcal{M}'$. Hence quasi-regularity of x^1 for $\mathcal{M}'$ implies quasi-regularity of x^1 for $\mathcal{M}$ and by iteration the same holds true for the whole sequence (note that here we could have used any term order).

For the converse let $r \geq q$ be an arbitrary degree. Because of the mentioned isomorphism, we may choose for the vector space $\mathcal{M}_r$ a basis where each member is represented by a monomial, i.e. the representatives simultaneously induce a basis of $\mathcal{M}'_r$. Let x^μ be one of these monomials. As we assume that x^1 is quasi-regular for $\mathcal{M}$, we must have $x^1 \cdot [x^\mu] \neq 0$ in $\mathcal{M}$. Suppose now that $x^1 \cdot [x^\mu] = 0$ in $\mathcal{M}'$ so that x^1 is not quasi-regular for $\mathcal{M}'$.

Thus $x^{\mu+1_1} \in \operatorname{lt} \mathcal{I}$. Since $\operatorname{lt} \mathcal{I} = \langle \operatorname{lt} \mathcal{G} \rangle$ by the definition of a Gröbner basis, $\mathcal{G}$ must contain a polynomial g with $\operatorname{lt} g \mid x^{\mu+1_1}$. Because of the assumption $x^\mu \notin \operatorname{lt} \mathcal{I}$, we must have $\operatorname{cls}(\operatorname{lt} g) = 1$. By Lemma B.1, this implies that every term in g is of class 1. Iteration of this argument shows that the normal form of $x^{\mu+1_1}$ with respect to $\mathcal{G}$ is divisible by x^1, i. e. it can be written as $x^1 f$ with $f \in \mathcal{P}_r$ and $\operatorname{lt} f \prec x^\mu$. Consider now the polynomial $\bar{f} = x^\mu - f \in \mathcal{P}_r \setminus \{0\}$. As it consists entirely of terms not contained in $\operatorname{lt} \mathcal{I}$, we have $[\bar{f}] \neq 0$ in $\mathcal{M}_r$. However, by construction $x^1 \cdot [\bar{f}] = 0$ contradicting the injectivity of multiplication by x^1 on $\mathcal{M}_r$.

For the remaining elements of the sequence $(x^1, \ldots, x^n)$ we note the isomorphism $\mathcal{M}^{(k)} = \mathcal{M}/\langle x^1, \ldots, x^k \rangle \mathcal{M} \cong \mathcal{P}^{(k)}/\mathcal{I}^{(k)}$ for each $1 \leq k < n$ where we introduced the abbreviations $\mathcal{P}^{(k)} = \Bbbk[x^{k+1}, \ldots, x^n]$ and $\mathcal{I}^{(k)} = \mathcal{I} \cap \mathcal{P}^{(k)}$. It implies that we may iterate the arguments above so that indeed quasi-regularity of $(x^1, \ldots, x^n)$ for $\mathcal{M}'$ is equivalent to quasi-regularity of the sequence for $\mathcal{M}'$. $\qquad\square$

Note that restriction to the degree reverse lexicographic order is here essential, as in general we have only the inequality $\operatorname{reg} \mathcal{M} \leq \operatorname{reg}(\operatorname{lt} \mathcal{M})$ and if it is strict, then a sequence may be quasi-regular for $\mathcal{M}$ at any degree $\operatorname{reg} \mathcal{M} \leq q < \operatorname{reg}(\operatorname{lt} \mathcal{M})$, but it cannot be quasi-regular for $\mathcal{M}'$ at such a degree by the results below.

Theorem 5.2 *The basis $\{x^1, \ldots, x^n\}$ is δ-regular for the homogeneous ideal $\mathcal{I} \subseteq \mathcal{P}$ in the sense that $\mathcal{I}$ possesses a Pommaret basis $\mathcal{H}$ for the degree reverse lexicographic term order with $\deg \mathcal{H} = q$, if and only if the sequence $(x^1, \ldots, x^n)$ is quasi-regular for the factor algebra $\mathcal{P}/\mathcal{I}$ at degree q but not at any lower degree.*

Proof. It suffices to consider monomial ideals $\mathcal{I}$: for Pommaret bases it is obvious from their definition that a basis is δ-regular for $\mathcal{I}$, if and only if it is so for $\operatorname{lt} \mathcal{I}$; a similar statement holds for quasi-regularity by Lemma 5.1.

Let us first assume that the basis is $\{x^1, \ldots, x^n\}$ is δ-regular in the described sense. By Proposition 4.7, the leading terms $\operatorname{lt} \mathcal{H}$ induce a complementary decomposition of the form (4.5) of $\mathcal{M} = \mathcal{P}/\mathcal{I}$ where all generators are of degree $q = \deg \mathcal{H}$ or less. Thus, if $\mathcal{M}_q \neq 0$ (otherwise there is nothing to show), then we can choose a vector space basis of it as part of the complementary decomposition and the variable x^1 is multiplicative for all its members. But this observation immediately implies that multiplication with x^1 is injective from degree q on, so that x^1 is quasi-regular for $\mathcal{M}$ at degree q.

For the remaining elements of $\{x^1, \ldots, x^n\}$ we proceed as in the proof of Lemma 5.1 and use the isomorphism $\mathcal{M}^{(k)} \cong \mathcal{P}^{(k)}/\mathcal{I}^{(k)}$. One easily verifies that a Pommaret basis of $\mathcal{I}^{(k)}$ is obtained by setting $x^1 = \cdots = x^k = 0$ in the partial basis $\mathcal{H}^{(k)} = \{h \in \mathcal{H} \mid \operatorname{cls} h > k\}$. Thus we can again iterate for each $1 < k \leq n$ the argument above so that indeed $(x^1, \ldots, x^n)$ is a quasi-regular sequence for $\mathcal{M}$ at degree q.

For the converse, we first show that quasi-regularity of the sequence $(x^1, \ldots, x^n)$ implies the existence of a Rees decomposition for $\mathcal{P}/\mathcal{I}$. Exploiting again the isomorphism $\mathcal{M}^{(k)} \cong \mathcal{P}^{(k)}/\mathcal{I}^{(k)}$, one easily sees that a vector space basis of $\mathcal{M}_q^{(k)}$ is induced by all terms $x^\mu \notin \mathcal{I}$ with $|\mu| = q$ and $\operatorname{cls} \mu \geq k$. By the definition of quasi-regularity, multiplication with x^k is injective on $\mathcal{M}^{(k)}$, hence we take $\{x^1, \ldots, x^k\}$ as multiplicative variables for such a term (which is exactly the assignment used in the Rees decomposition induced by a Pommaret basis according to Proposition 4.7).

We claim now that this assignment yields a Rees decomposition of $\mathcal{M}_{\geq q}$ (and hence induces one of $\mathcal{P}/\mathcal{I}$, since we only have to add all terms $x^\mu \notin \mathcal{I}$ such that $|\mu| < q$ without any multiplicative variables). The only thing to prove is that our decomposition indeed covers all of $(\mathcal{P}/\mathcal{I})_{\geq q}$. But this is trivial. If $x^\mu \notin \mathcal{I}$ is an arbitrary term with $|\mu| = q + 1$ and $\mathrm{cls}\,\mu = k$, then we can write $x^\mu = x^k \cdot x^{\mu - 1_k}$. Obviously, $x^\mu \notin \mathcal{I}$ implies $x^{\mu - 1_k} \notin \mathcal{I}$ and $\mathrm{cls}\,(\mu - 1_k) \geq k$ so that x^k is multiplicative for it. Hence all of $\mathcal{M}_{q+1}$ is covered and an easy induction shows that we have a decomposition of $\mathcal{M}_{\geq q}$.

Proposition 4.7 entails now that $\mathcal{I}$ possesses a *weak* Pommaret basis of degree q. Since the reduction to a strong basis as mentioned in Remark 4.2 can only decrease the degree, we conclude that $\mathcal{I}$ has a strong Pommaret basis of degree at most q. However, if the degree of the basis actually decreased, then, by the converse statement already proven, $(x^1, \ldots, x^n)$ would be a quasi-regular sequence for $\mathcal{M}$ at a lower degree than q contradicting our assumptions.

The same "reverse" argument shows that if $\mathcal{I}$ has a Pommaret basis of degree q, then the sequence $(x^1, \ldots, x^n)$ cannot be quasi-regular for $\mathcal{M}$ at any degree less than q, as otherwise a Pommaret basis of lower degree would exist which is not possible by the discussion following Theorem 4.14.   □

For *monomial* ideals $\mathcal{I} \subseteq \mathcal{P}$ a much stronger statement is possible. Using again the isomorphism $\mathcal{M}^{(k)} \cong \mathcal{P}^{(k)}/\mathcal{I}^{(k)}$, we may identify elements of $\mathcal{M}^{(k)}$ with linear combinations of the terms $x^\nu \notin \mathcal{I}$ satisfying $\mathrm{cls}\,x^\nu > k$. Finally, if we denote as before by $\mu_k : \mathcal{M}^{(k-1)} \to \mathcal{M}^{(k-1)}$ the map induced by multiplication with x^k, then we obtain a simple relationship between the (unique!) Pommaret basis of the monomial ideal $\mathcal{I}$ and the kernels of the maps μ_k.

Proposition 5.3 *Let the basis $\{x^1, \ldots, x^n\}$ of V be δ-regular for the monomial ideal $\mathcal{I} \subseteq \mathcal{P}$. Furthermore, let $\mathcal{H}$ be the Pommaret basis of $\mathcal{I}$ and set $\mathcal{H}_k = \{x^\nu \in \mathcal{H} \mid \mathrm{cls}\,\nu = k\}$ for any $1 \leq k \leq n$. Then the set $\{x^{\nu - 1_k} \mid x^\nu \in \mathcal{H}_k\}$ is a basis of $\ker \mu_k$.*

Proof. Assume that $x^\nu \in \mathcal{H}_k$. Then $x^{\nu - 1_k} \notin \mathcal{I}$, as otherwise the Pommaret basis $\mathcal{H}$ was not involutively autoreduced, and hence we find $x^{\nu - 1_k} \in \ker \mu_k$.

Conversely, suppose that $x^\nu \in \ker \mu_k$. Obviously, this implies $x^{\nu + 1_k} \in \mathcal{I}$ and the Pommaret basis $\mathcal{H}$ must contain an involutive divisor of $x^{\nu + 1_k}$. If this divisor was not $x^{\nu + 1_k}$ itself, the term x^ν would have to be an element of $\mathcal{I}$ which is obviously not possible. Since $x^\nu \in \ker \mu_k$ entails $\mathrm{cls}\,(\nu + 1_k) = k$, we thus find $x^{\nu + 1_k} \in \mathcal{H}_k$.   □

We noted already in Remark 4.15 that the degree of involution is nothing but the Castelnuovo–Mumford regularity. There we used the equivalence of the Koszul homology to the minimal free resolution. With the help of Theorem 5.2, we can also give a simple direct proof.

Corollary 5.4 *Let $\mathcal{I} \subseteq \mathcal{P}$ be a homogeneous ideal. Then the factor module $\mathcal{M} = \mathcal{P}/\mathcal{I}$ is involutive at degree q but not at any lower degree, if and only if the Castelnuovo–Mumford regularity takes the value $\mathrm{reg}\,\mathcal{I} = q$.*

Proof. By Theorem 4.14, $\mathrm{reg}\,\mathcal{I} = q$, if and only if $\mathcal{I}$ possesses in suitable variables $x^1, \ldots, x^n$ a Pommaret basis $\mathcal{H}$ with $\deg \mathcal{H} = q$. According to Theorem 5.2, the sequence $(x^1, \ldots, x^n)$ is then quasi-regular for $\mathcal{M}$ at degree q but not any lower degree, so that by the dual Cartan test (Theorem 3.12) the module $\mathcal{M}$ is involutive at degree q but not any lower degree. $\qquad\square$

Remark 5.5 Given this result, it is not so surprising to see that the characterisation of the Castelnuovo–Mumford regularity mentioned in Remark 4.17 and the dual Cartan test in Theorem 3.12 are equivalent. Consider a homogeneous ideal $\mathcal{I} \subseteq \mathcal{P}$ for which the basis $\{x^1, \ldots, x^n\}$ of $\mathcal{V}$ is δ-regular and assume that for some degree $q \geq 0$ the condition (4.10a) is violated for some $1 \leq j \leq D = \dim(\mathcal{P}/\mathcal{I})$. Thus there exists a polynomial $f \in \mathcal{P}_{q-1}$ such that $f \notin \langle \mathcal{I}, x^1, \ldots, x^{j-1} \rangle$ but $x^j f$ is contained in this ideal. If we set $\mathcal{M}_j = \mathcal{P}/\langle \mathcal{I}, x^1, \ldots, x^j \rangle$, then obviously the equivalence class $[f]$ lies in the kernel of the map $\mu_j : \mathcal{M}_{j-1} \to \mathcal{M}_{j-1}$ induced by multiplication with x^j. Since trivially for $\mathcal{M} = \mathcal{P}/\mathcal{I}$ the module $\mathcal{M}^{(j)} = \mathcal{M}/\langle x^1, \ldots, x^j \rangle \mathcal{M}$ is isomorphic to $\mathcal{M}_j$, the conditions of Theorem 3.12 are not satisfied for $\mathcal{M}$ either. Conversely, any representative of a non-trivial element of $\ker \mu_j$ of degree q provides us at once with such a polynomial f. There is no need to consider a value $j > D$, since we know from Proposition 4.9 that $(\mathcal{M}_D)_{\geq \operatorname{reg} \mathcal{I}} = 0$.

As an application we consider the following theorem providing a classical characterisation of the depth via Koszul homology which in fact is often even used as definition of $\operatorname{depth} \mathcal{M}$ (see e.g. [54, Sect. IV.A.4]). Note that, taking into account the relation between the minimal free resolution of a module and its Koszul homology discussed in Remark 2.15, it also trivially implies the Auslander–Buchsbaum formula relating depth and projective dimension.

Theorem 5.6 *Let $\mathcal{M}$ be a $\mathcal{P}$-module. Then* $\operatorname{depth} \mathcal{M} = d$, *if and only if* $H_{n-d}(\mathcal{M}) \neq 0$ *and* $H_{n-d+1}(\mathcal{M}) = \cdots = H_n(\mathcal{M}) = 0$.

Proof. For simplicity, we give the proof only for the case of an ideal $\mathcal{I} \subseteq \mathcal{P}$. The extension to modules is straightforward. Let $\mathcal{H}$ be a Pommaret basis of $\mathcal{I}$ with respect to the degree reverse lexicographic order, $d = \min_{h \in \mathcal{H}} \operatorname{cls} h$ (and thus $\operatorname{depth} \mathcal{I} = d$) and $\mathcal{H}_d = \{h \in \mathcal{H} \mid \operatorname{cls} h = d\}$. We choose a polynomial $\bar{h} \in \mathcal{H}_d$ of maximal degree and show now that it induces a non-zero element of $H_{n-d}(\mathcal{I})$.

By Lemma B.1, $\bar{h} \in \langle x^1, \ldots, x^d \rangle$ and thus it possesses a unique representation $\bar{h} = x^1 \bar{h}^{(1)} + \cdots + x^d \bar{h}^{(d)}$ with $\bar{h}^{(i)} \in \Bbbk[x^i, \ldots, x^n]$. The polynomial $\bar{h}^{(d)}$ cannot lie in $\mathcal{I}$, as otherwise there would exist an $h \in \mathcal{H}$ with $\operatorname{lt} h \mid_P \operatorname{lt} \bar{h}^{(d)} \mid_P \operatorname{lt} \bar{h}$ contradicting the fact that any Pommaret basis is involutively head autoreduced. We claim now that for any $d < k \leq n$ polynomials $P_h \in \langle x^1, \ldots, x^d \rangle$ exist such that $x^k \bar{h} = \sum_{h \in \mathcal{H}} P_h h$.

Obviously, the variable x^k is non-multiplicative for $\bar{h}$. By definition of a Pommaret basis, for each generator $h \in \mathcal{H}$ a polynomial $P_h \in \Bbbk[x^1, \ldots, x^{\operatorname{cls} h}]$ exists such that $x^k \bar{h} = \sum_{h \in \mathcal{H}} P_h h$. No polynomial h with $\operatorname{cls} h > d$ lies in $\langle x^1, \ldots, x^d \rangle$ (obviously $\operatorname{lt} h \notin \langle x^1, \ldots, x^d \rangle$). As the leading terms cannot cancel in the sum, this implies already that $P_h \in \langle x^1, \ldots, x^d \rangle$ for all $h \in \mathcal{H} \setminus \mathcal{H}_d$. For all $h \in \mathcal{H}_d$ we know that $P_h \in \Bbbk[x^1, \ldots, x^d]$ and thus the only possibility for $P_h \notin \langle x^1, \ldots, x^d \rangle$ is that P_h contains a constant term. However, as $\mathcal{I}$ is a homogeneous ideal and as the degree of $\bar{h}$ is maximal in $\mathcal{H}_d$, this is not possible for degree reason. As above, each of the coefficients may thus be uniquely decomposed $P_h = x^1 P_h^{(1)} + \cdots + x^d P_h^{(d)}$ with $P_h^{(i)} \in \Bbbk[x^i, \ldots, x^n]$.

Because of the uniqueness of these decompositions we find that $x^k \bar{h}^{(i)} = \sum_{h \in \mathcal{H}} P_h^{(i)} h$ and therefore we conclude that $x^k \bar{h}^{(i)} \in \mathcal{I}$ for any $d < k \leq n$.

Let $I = (i_1, \ldots, i_{d-1})$ be a repeated index with $i_1 < i_2 < \cdots < i_{d-1}$. Then its complement $\bar{I} = \{1, \ldots, n\} \setminus I$ is a repeated index of length $n - d + 1$ and we may represent any element $\bar{\omega} \in \mathcal{P} \otimes \Lambda_{n-d+1}\mathcal{V}$ in the form $\bar{\omega} = \sum_{|I|=d-1} \bar{f}_I \mathrm{d}x^{\bar{I}}$. We consider now in particular all repeated indices with $i_{d-1} \leq d$. For each of them a unique value $i \in \{1, \ldots, d\}$ exists such that $i \notin I$ and we set $\bar{f}_I = (-1)^{d+i}\bar{h}^{(i)}$. For all remaining coefficients we only assume that $\bar{f}_I \in \mathcal{I}$. Then, by our considerations above, the so constructed form $\bar{\omega}$ is *not* contained in $\mathcal{I} \otimes \Lambda_{n-d+1}\mathcal{V}$.

We claim that $\omega = \partial\bar{\omega} \in \mathcal{I} \otimes \Lambda_{n-d}\mathcal{V}$. If we write $\omega = \sum_{|I|=d} f_I \mathrm{d}x^{\bar{I}}$, then by definition of the Koszul differential $f_I = \sum_{j=1}^{d} (-1)^j x^{i_j} \bar{f}_{I \setminus \{i_j\}}$. Let us first assume that $i_d > d$. Then it follows from our choice of $\bar{\omega}$ that $f_{I \setminus \{i_j\}} \in \mathcal{I}$ for all $j < d$ and that always $x^{i_d} f_{I \setminus \{i_d\}} \in \mathcal{I}$ implying trivially that $f_I \in \mathcal{I}$. If $i_d = d$, then one easily verifies that we have chosen $\bar{\omega}$ precisely such that $f_I = \bar{h} \in \mathcal{I}$. Hence our claim is proven.

If we can now show that it is not possible to choose a form $\tilde{\omega} \in \mathcal{P} \otimes \Lambda_{n-d+2}\mathcal{V}$ such that $\bar{\omega} + \partial\tilde{\omega} \in \mathcal{I} \otimes \Lambda_{n-d+1}\mathcal{V}$, then we have constructed a non-zero element $[\omega] \in H_{n-d}(\mathcal{I})$. But this is easy to achieve by considering in particular the coefficient $\bar{f}_{(1,2,\ldots,d-1)} = \bar{h}^{(d)} \notin \mathcal{I}$. The corresponding coefficient of the form $\partial\tilde{\omega}$ is given by $\sum_{j=1}^{d-1} (-1)^j x^j \tilde{f}_{(1,2,\ldots,d-1)\setminus\{j\}} \in \langle x^1, \ldots, x^{d-1} \rangle$. As noted above, we have $\bar{h}^{(d)} \in \mathrm{k}[x^d, \ldots, x^n]$ so that it is not possible to eliminate it in this manner and hence no form $\bar{\omega} + \partial\tilde{\omega}$ can be contained in $\mathcal{I} \otimes \Lambda_{n-d+1}\mathcal{V}$.

There remains to show that $H_{n-d+1}(\mathcal{I}) = \cdots = H_n(\mathcal{I}) = 0$ under our assumptions. $H_n(\mathcal{I}) = 0$ follows immediately from Lemma 2.16. Consider now a cycle $\omega \in \mathcal{I} \otimes \Lambda_{n-k}\mathcal{V}$ with $0 < k < d$. Since the Koszul complex $K(\mathcal{P})$ is exact, a form $\bar{\omega} \in \mathcal{P} \otimes \Lambda_{n-k+1}\mathcal{V}$ exists with $\partial\bar{\omega} = \omega$. For all I we have by assumption $f_I = \sum_{j=1}^{d} (-1)^j x^{i_j} \bar{f}_{I \setminus \{i_j\}} \in \mathcal{I}$; our goal is to show that (modulo im ∂) we can always choose $\bar{\omega}$ such that all coefficients $\bar{f}_J \in \mathcal{I}$, too.

Without loss of generality, we may assume that all coefficients $\bar{f}_J$ are in normal form with respect to the Gröbner basis $\mathcal{H}$, as the difference is trivially contained in $\mathcal{I}$. In addition, we may assume that $\mathrm{lt}\, f_I = \mathrm{lt}\, (x^{i_j}\bar{f}_{I\setminus\{i_j\}})$ for some value j. Indeed, it is easy to see that cancellations between such leading terms can always be eliminated by subtracting a suitable form $\partial\tilde{\omega}$ from $\bar{\omega}$.

We begin with those repeated indices $I = (i_1, \ldots, i_k)$ for which all indices satisfy $i_j < d = \min_{h \in \mathcal{H}} \mathrm{cls}\, h$. In this case $\mathrm{lt}\, f_I \in \langle \mathrm{lt}\, \mathcal{H} \rangle_P = \mathrm{lt}\, \mathcal{I}$ implies that already $\mathrm{lt}\, \bar{f}_{I\setminus\{i_j\}} \in \mathrm{lt}\, \mathcal{I}$ for the above j. But unless $\bar{f}_{I\setminus\{i_j\}} = 0$, this observation contradicts our assumption that all f_J are in normal form and thus do not contain any terms from $\mathrm{lt}\, \mathcal{I}$. Therefore all $\bar{f}_J$ where all entries of J are less than d must vanish.

We continue with those repeated indices $I = (i_1, \ldots, i_k)$ where only one index $i_\ell > d$. Then, by our considerations above, $\bar{f}_{I\setminus\{i_\ell\}} = 0$ and hence $\mathrm{lt}\, f_I = \mathrm{lt}\, (x^{i_j}\bar{f}_{I\setminus\{i_j\}})$ for some value $j \neq \ell$. Thus $i_j < d$ and the same argument as above implies that all such $\bar{f}_{I\setminus\{i_j\}} = 0$. A trivial induction proves now that in fact all $\bar{f}_J = 0$ and therefore we find $\bar{\omega} \in \mathcal{I} \otimes \Lambda_{n-k+1}\mathcal{V}$. $\qquad\square$

6 Formal Geometry of Differential Equations

In the next chapter we will demonstrate how the algebraic and homological theory presented so far naturally appears in the analysis of differential equations. Perhaps somewhat paradoxically, the key for applying algebraic methods lies in first providing a differential geometric framework. For this purpose, we must briefly recall some basic notions from the formal geometry of differential equations [30, 32, 42, 53].

Let $\pi : \mathcal{E} \to \mathcal{X}$ be a fibred manifold with an n-dimensional base space $\mathcal{X}$ and an $(m + n)$-dimensional total space $\mathcal{E}$ (in the simplest case $\mathcal{X} = \mathbb{R}^n$ and $\mathcal{E} = \mathbb{R}^{n+m}$ with π being the projection on the first n components). Local coordinates on $\mathcal{X}$ are $\mathbf{x} = (x^1, \dots, x^n)$ and fibre coordinates on $\mathcal{E}$ are $\mathbf{u} = (u^1, \dots, u^m)$. A section is then a map[1] $\sigma : \mathcal{X} \to \mathcal{E}$ satisfying $\pi \circ \sigma = \mathrm{id}_{\mathcal{X}}$. In local coordinates, such a section σ corresponds to a smooth function $\mathbf{u} = \mathbf{s}(\mathbf{x})$, as $\sigma(\mathbf{x}) = (\mathbf{x}, \mathbf{s}(\mathbf{x}))$.

A *q-jet* is an equivalence class $[\sigma]_{\mathbf{x}_0}^{(q)}$ of sections where two sections σ_1, σ_2 are considered as equivalent, if their graphs have at the point $\sigma_i(\mathbf{x}_0)$ a contact of order q, in other words if their Taylor expansions at $\mathbf{x}_0$ coincides up to order q (thus we may consider a q-jet as a truncated Taylor series). The *qth order jet bundle* $J_q\pi$ is then defined to be the set of all such q-jets. One easily verifies that $J_q\pi$ is an $(n+m\binom{n+q}{q})$-dimensional manifold. Projection on the expansion point $\mathbf{x}_0$ defines a fibration $\pi^q : J_q\pi \to \mathcal{X}$. As fibre coordinates for the point $[\sigma]_{\mathbf{x}_0}^{(q)}$ we may use $\mathbf{u}^{(q)} = (u_\mu^\alpha)$ with $1 \leq \alpha \leq n$ and a multi index μ where $0 \leq |\mu| \leq q$ and the interpretation that u_μ^α is the value of $\partial^{|\mu|} s^\alpha / \partial x^\mu$ at the expansion point $\mathbf{x}_0 \in \mathcal{X}$.

A *differential equation* is now a fibred submanifold $\mathcal{R}_q \subseteq J_q\pi$. We will always assume that it may be locally described as the zero set of a function $\Phi : J_q\pi \to \mathbb{R}^t$; thus we recover the usual picture of a differential equation $\Phi(\mathbf{x}, \mathbf{u}^{(q)}) = 0$ (note that we do not distinguish between a scalar equation and a system). The *prolongation* of a section $\sigma : \mathcal{X} \to \mathcal{E}$ is the section $j_q\sigma : \mathcal{X} \to J_q\pi$ locally defined by $u_\mu^\alpha = \partial^{|\mu|} s^\alpha / \partial x^\mu(\mathbf{x})$. We call σ a *solution* of the differential equation $\mathcal{R}_q$, if $\mathrm{im}\,(j_q\sigma) \subseteq \mathcal{R}_q$. Expressed in coordinates, this is equivalent to the usual definition.

If $q > r$, then we also have the canonical fibrations $\pi_r^q : J_q\pi \to J_r\pi$ defined by simply "forgetting" the higher-order derivatives. This leads to two natural operations with a differential equation $\mathcal{R}_q$. The first one is the *projection* to lower order: given a differential equation $\mathcal{R}_q \subseteq J_q\pi$ its r-fold projection is $\mathcal{R}_{q-r}^{(r)} = \pi_{q-r}^q(\mathcal{R}_q) \subseteq J_{q-r}\pi$. While the projection is easy to describe geometrically, it is, in particular for nonlinear equations, hard to perform effectively, as it requires the elimination of variables.

In the second basic operation, the *prolongation* to higher order, we encounter the opposite situation: while it is easy to perform effectively in local coordinates, it is somewhat cumbersome to provide a rigorous intrinsic definition. Given a differential equation $\mathcal{R}_q \subseteq J_q\pi$, we may consider the restriction $\hat{\pi}^q : \mathcal{R}_q \to \mathcal{X}$ of the projection π^q which provides $\mathcal{R}_q$ with the structure of a fibred manifold over which we may again construct jet bundles. If we consider now both $J_r\hat{\pi}^q$ and $J_{q+r}\pi$ as submanifolds of $J_r\pi^q$ (which is possible with certain straightforward identifications), then the r-fold

---

[1] For notational simplicity, we do not explicitly mention local charts and use a global notation. Nevertheless all construction are to be understood purely locally.

prolongation of $\mathcal{R}_q$ is the differential equation $\mathcal{R}_{q+r} = J_r \hat{\pi}^q \cap J_{q+r}\pi \subseteq J_{q+r}\pi$.

In local coordinates, prolongation requires only the *formal derivative*. If Φ is an arbitrary smooth function $J_q\pi \to \mathbb{R}$, then its formal derivative $D_i\Phi$ with respect to the variable x^i is a smooth function $J_{q+1}\pi \to \mathbb{R}$ given by the chain rule:

$$D_i\Phi = \frac{\partial\Phi}{\partial x^i} + \sum_{\alpha=1}^{m} \sum_{0 \leq |\mu| \leq q} \frac{\partial\Phi}{\partial u_\mu^\alpha} u_{\mu+1_i}^\alpha . \tag{6.1}$$

If now the differential equation $\mathcal{R}_q$ is locally described as the zero set of the functions $\Phi^\tau : J_q\pi \to \mathbb{R}$ for $1 \leq \tau \leq t$, then its first prolongation $\mathcal{R}_{q+1}$ is the common zero set of the functions Φ^τ and their formal derivatives $D_i\Phi^\tau$ for $1 \leq i \leq n$ and $1 \leq \tau \leq t$. Higher prolongations are obtained by iteration.

It should be noted that in general neither a projection $\mathcal{R}_{q-r}^{(r)}$ nor a prolongation $\mathcal{R}_{q+r}$ is again a manifold, as we must expect that singularities appear. For simplicity, we will ignore this problem and always assume that we are dealing with a *regular* differential equation where all operations yield manifolds.

One could think that prolongation and projection are some kind of "inverse" operations: if one first prolongs an equation $\mathcal{R}_q \subseteq J_q\pi$ to $\mathcal{R}_{q+r} \subseteq J_{q+r}\pi$ for some $r > 0$ and subsequently projects back to $J_q\pi$ with π_q^{q+r}, one might naively expect that the obtained equation $\mathcal{R}_q^{(r)}$ coincides with the original equation $\mathcal{R}_q$. However, this is in general not correct, as *integrability conditions* may arise: we only get that always $\mathcal{R}_q^{(r)} \subseteq \mathcal{R}_q$.

Example 6.1 From a computational point of view, one may distinguish two different mechanisms for the generation of integrability conditions during prolongations and projections (for ordinary differential equations only the first one occurs): (i) the local representation of $\mathcal{R}_q$ comprises equations of different orders and formal differentiation of the lower-order equations leads to new (i. e. *algebraically* independent equations), (ii) generalised cross-derivatives.

As a concrete example for the first mechanism consider the trivial ordinary differential equation $\mathcal{R}_1$ in two dependent variables u^1, u^2 and one independent variable x defined by $(u^1)' = u^2$ and $u^1 = x$. The local representation of $\mathcal{R}_2$ contains in addition the equations $(u^1)'' = (u^2)'$ and $(u^1)' = 1$. As the second one is of first order, it survives a subsequent projection back to second order and (after an obvious simplification) the projected system $\mathcal{R}_2^{(1)}$ is given by $(u^1)' = 1$, $u^1 = x$ and $u^2 = 1$.

For demonstrating of the second mechanism, we use a classical example due to Janet, namely the partial differential equation $\mathcal{R}_2$ in one dependent variable u and three independent variables x^1, x^2, x^3 locally described by $u_{33} + x^2 u_{11} = 0$ and $u_{22} = 0$. Among others, the second prolongation $\mathcal{R}_4$ contains the equations $u_{2233} + x^2 u_{1122} + u_{112} = 0$, $u_{1122} = 0$ and $u_{2233} = 0$. An obvious linear combination of these equations yields the integrability condition $u_{112} = 0$ and hence $\mathcal{R}_3^{(1)}$ is a proper subset of $\mathcal{R}_3$. Note that here the integrability condition is of higher order than the original system; this is a typical phenomenon for partial differential equations.

It is important to note that integrability conditions are not additional restrictions on the solution space of the considered equation $\mathcal{R}_q$; any solution of $\mathcal{R}_q$ *automatically* satisfies them. They represent conditions implicitly contained or hidden in $\mathcal{R}_q$ and which can be made visible by performing a suitable sequence of prolongations and projections. They may be considered as obstructions for the order by order construction of formal power series solutions. In practice, it often considerable simplifies the integration of the equation, if at least some integrability conditions are added.

These considerations motivate the following definition where the term "integrable" is used in its most basic meaning: existence of solutions. As we discuss here only formal solutions, we speak of formal integrability.[2] One should not confuse this concept with other notions like complete integrability where properties like the existence of first integrals or symmetries are considered.

Definition 6.2 The differential equation $\mathcal{R}_q \subseteq J_q\pi$ is called *formally integrable*, if for all $r \geq 0$ the equality $\mathcal{R}_{q+r}^{(1)} = \mathcal{R}_{q+r}$ holds.

While this geometric definition of formal integrability is very natural, it has an obvious and serious drawback: it requires the satisfaction of an infinite number of conditions (surjectivity of the projections $\hat{\pi}_{q+r}^{q+r+1} : \mathcal{R}_{q+r+1} \to \mathcal{R}_{q+r}$ for all $r \geq 0$). Thus in the given form formal integrability cannot be verified effectively. We will see in the next chapter that algebraic and homological methods lead to a finite criterion for formal integrability. The key for the application of these methods lies in a natural polynomial structure hidden in the jet bundle hierarchy, the so-called fundamental identification. As this topic is often ignored in the literature, we discuss it here in some detail. It is based the following crucial observation.

Proposition 6.3 *The jet bundle $J_q\pi$ of order q is affine over the jet bundle $J_{q-1}\pi$ of order $q - 1$.*

Proof. The simplest approach to proving this proposition consists of studying the effect of fibred changes of coordinates $\bar{\mathbf{x}} = \bar{\mathbf{x}}(\mathbf{x})$ and $\bar{\mathbf{u}} = \bar{\mathbf{u}}(\mathbf{x}, \mathbf{u})$ in the total space $\mathcal{E}$ on the derivatives which are fibre coordinates in $J_q\pi$. Using the chain rule one easily computes that in repeated index notation the result for the highest-order derivatives is

$$\bar{u}_{j_1 \cdots j_q}^\alpha = \left(\frac{\partial \bar{u}^\alpha}{\partial u^\beta} \frac{\partial x^{i_1}}{\partial \bar{x}^{j_1}} \cdots \frac{\partial x^{i_q}}{\partial \bar{x}^{j_q}} \right) u_{i_1 \cdots i_q}^\beta + \cdots \qquad (6.2)$$

where the dots represent a complicated expression in the derivatives of lower order and where $\partial\mathbf{x}/\partial\bar{\mathbf{x}}$ represents the inverse of the Jacobian $\partial\bar{\mathbf{x}}/\partial\mathbf{x}$. But this implies that (6.2) is indeed affine in the derivatives of order q as claimed. $\qquad\square$

An affine space is always modelled on a vector space: the difference between two points may be interpreted as a vector. In our case it is easy to identify this vector space. Let $[\sigma]_\mathbf{x}^{(q)}$ and $[\sigma']_\mathbf{x}^{(q)}$ be two points in $J_q\pi$ such that $[\sigma]_\mathbf{x}^{(q-1)} = [\sigma']_\mathbf{x}^{(q-1)}$, i. e. the two points belong to the same fibre with respect to the fibration π_{q-1}^q. Thus $[\sigma]_\mathbf{x}^{(q)}$ and $[\sigma']_\mathbf{x}^{(q)}$ correspond to two Taylor series truncated at degree q which coincide up to degree $q - 1$.

---

[2] In some applications like Lie symmetry theory *local solvability* is very important [41]. A differential equation $\mathcal{R}_q$ is locally solvable, if for every point $\rho \in \mathcal{R}_q$ a solution σ exists such that $\rho \in \mathrm{im}\, j_q\sigma$. Again in the sense of existence of the formal solutions, formal integrability trivially implies local solvability.

Obviously, this observation implies that their difference consists of one homogeneous polynomial of degree q for each dependent variable u^α.

In a more intrinsic language, we may formulate this result as follows. Let $\rho = [\sigma]_{\mathbf{x}}^{(q)}$ be a point in $J_q\pi$ and $\bar\rho = [\sigma]_{\mathbf{x}}^{(q-1)} = \pi_{q-1}^q(\rho)$ its projection to $J_{q-1}\pi$; we furthermore set $\xi = \sigma(\mathbf{x}) = \pi_0^q(\rho) \in \mathcal{E}$. Then according to Proposition 6.3, the fibre $(\pi_{q-1}^q)^{-1}(\bar\rho)$ is an affine space modelled on the vector space $S_q(T_{\mathbf{x}}^*\mathcal{X}) \otimes V_\xi\pi$ where S_q denotes again the q-fold symmetric product and $V_\xi\pi \subset T_\xi\mathcal{E}$ is the *vertical bundle* defined as the kernel of the tangent map $T_\xi\pi$. Indeed, this follows immediately from our discussion so far: the symmetric algebra $S(T_{\mathbf{x}}^*\mathcal{X})$ is a coordinate-free form of the polynomial ring and one easily verifies that the homogeneous part of (6.2) obtained by dropping the terms represented by the dots describes the transformation behaviour of vectors in $S_q(T_{\mathbf{x}}^*\mathcal{X}) \otimes V_\xi\mathcal{E}$ (note that we must use the *cotangent* space $T_{\mathbf{x}}^*\mathcal{X}$, as tangent vectors would transform with the inverse matrix).

By Proposition 6.3, the jet bundle $J_q\pi$ is an affine bundle over $J_{q-1}\pi$. This fact implies that the tangent space to the affine space $(\pi_{q-1}^q)^{-1}(\bar\rho)$ at the point $\rho \in J_q\pi$ is canonically isomorphic to the corresponding vector space, i. e. to $S_q(T_{\mathbf{x}}^*\mathcal{X}) \otimes V_\xi\mathcal{E}$. This isomorphism is called the *fundamental identification*. We derive now a local coordinate expression for it. On one side we have the tangent space to the fibre $(\pi_{q-1}^q)^{-1}(\bar\rho)$ at the point ρ, i. e. the vertical space $V_\rho\pi_{q-1}^q$ defined as the kernel of the tangent map $T_\rho\pi_{q-1}^q$. Obviously, it is spanned by all the vectors $\partial_{u_\mu^\alpha}$ with $|\mu| = q$. Let us consider one of these vectors; it is tangent to the curve $\gamma : t \mapsto \rho(t)$ where $\rho(0) = \rho$ and all coordinates of a point $\rho(t)$ coincide with those of ρ except for the one coordinate u_μ^α corresponding to the chosen vector which is increased by t.

On the other side, we may compute the difference quotient $(\rho(t) - \rho)/t$ interpreting the points as above as truncated Taylor series. The u^α-component of the result is obviously the polynomial $(u_\mu^\alpha(t) - u_\mu^\alpha)x^\mu/\mu!$. Hence the fundamental identification is just the map $\epsilon_q : V_\rho\pi_{q-1}^q \to S_q(T_{\mathbf{x}}^*\mathcal{X}) \otimes V_\xi\mathcal{E}$ given by

$$\epsilon_q(\partial_{u_\mu^\alpha}) = \frac{1}{\mu!}\mathrm{d}x^\mu \otimes \partial_{u^\alpha} . \tag{6.3}$$

Note the combinatorial factor $\frac{1}{\mu!}$ having its origin in Taylor's formula!

7   Algebraic Analysis of Differential Equations

Definition 7.1 Let $\mathcal{R}_q \subseteq J_q\pi$ be a differential equation. The *(geometric) symbol* $\mathcal{N}_q$ of $\mathcal{R}_q$ is a family of vector spaces over $\mathcal{R}_q$ where the value at $\rho \in \mathcal{R}_q$ is given by

$$(\mathcal{N}_q)_\rho = T_\rho\mathcal{R}_q \cap V_\rho\pi_{q-1}^q = V_\rho\big(\pi_{q-1}^q|_{\mathcal{R}_q}\big) . \tag{7.1}$$

Thus the symbol is the vertical part of the tangent space of the submanifold $\mathcal{R}_q$ with respect to the fibration π_{q-1}^q. If $\mathcal{R}_q$ is globally described by a map $\Phi : J_q\pi \to \mathcal{E}'$ with a vector bundle $\pi' : \mathcal{E}' \to \mathcal{X}$, then we introduce the *symbol map* $\sigma : V\pi_{q-1}^q \to T\mathcal{E}'$ given by $\sigma = T\Phi|_{V\pi_{q-1}^q}$ and define $\mathcal{N}_q = \ker\sigma$. Locally, this leads to the following picture.

Let $(\mathbf{x}, \mathbf{u}^{(q)})$ be coordinates on $J_q\pi$ in a neighbourhood of ρ. We first determine $T_\rho\mathcal{R}_q$ as a subspace of $T_\rho(J_q\pi)$. Let $(\mathbf{x}, \mathbf{u}^{(q)}; \dot{\mathbf{x}}, \dot{\mathbf{u}}^{(q)})$ be the induced coordinates on $T_\rho(J_q\pi)$; every vector $X \in T_\rho(J_q\pi)$ has the form $X = \dot{x}^i\partial_{x^i} + \dot{u}^\alpha_\mu\partial_{u^\alpha_\mu}$. Assuming that $\mathcal{R}_q$ is locally defined by $\Phi^\tau(\mathbf{x}, \mathbf{u}^{(q)}) = 0$ with $\tau = 1, \ldots, t$, its tangent space $T_\rho\mathcal{R}_q$ consists of all vectors X such that $\mathrm{d}\Phi^\tau(X) = X\Phi^\tau = 0$. The symbol $\mathcal{N}_q$ is by definition the vertical part of this tangent space. Hence we are only interested in those solutions of the above conditions where $\dot{\mathbf{x}} = \dot{u}^{(q-1)} = 0$ and locally $\mathcal{N}_q$ can be described as the solution space of the following system of linear equations:

$$(\mathcal{N}_q)_\rho : \left\{ \sum_{\substack{1 \le \alpha \le m \\ |\mu|=q}} \frac{\partial\Phi^\tau}{\partial u^\alpha_\mu}(\rho)\, \dot{u}^\alpha_\mu = 0\,, \qquad \tau = 1, \ldots, t\,. \right. \tag{7.2}$$

This is a system with real coefficients, as the derivatives $\partial\Phi^\tau/\partial u^\alpha_\mu$ are evaluated at the point $\rho \in \mathcal{R}_q$. We call its matrix the *symbol matrix* and denote it by $M_q(\rho)$. It is also the matrix of the symbol map σ in local coordinates.

The symbol is most easily understood for linear differential equations. Loosely speaking, the geometric symbol is then simply the highest-order or principal part of the system (considered as algebraic equations). For nonlinear systems we perform a brute force linearisation at the point ρ in order to obtain $(\mathcal{N}_q)_\rho$. Obviously, $\dim(\mathcal{N}_q)_\rho$ might vary with ρ. For this reason, we speak in Definition 7.1 only of a family of vector spaces and not of a vector bundle. Only if the dimension remains constant over $\mathcal{R}_q$, the symbol $\mathcal{N}_q$ is a vector subbundle of $V\pi^q_{q-1}$. For simplicity, we will assume that all considered symbols are vector bundles.

Example 7.2 For Janet's partial differential equation $\mathcal{R}_2$ considered in Example 6.1 the symbol equations are $\dot{u}_{33}+x^2\dot{u}_{11} = 0$ and $\dot{u}_{22} = 0$ (as $\mathcal{R}_2$ is a linear system without lower-order terms, the symbol equations look formally like the differential equation itself; however, the symbol equations are *algebraic* and not differential equations). Hence the symbol $\mathcal{N}_2$ is here the one-dimensional distribution spanned by the vector field $\partial_{u_{11}} - x^2\partial_{u_{33}}$.

Of course, not only the original equation $\mathcal{R}_q$ has a symbol $\mathcal{N}_q$, but also every prolongation $\mathcal{R}_{q+r} \subseteq J_{q+r}\pi$ of it possesses a symbol $\mathcal{N}_{q+r} \subseteq T(J_{q+r}\pi)$. It follows easily from the coordinate expression (6.1) of the formal derivative that for obtaining a local representation of the prolonged symbol $\mathcal{N}_{q+r}$, there is no need to explicitly compute a local representation of the prolonged differential equation $\mathcal{R}_{q+r}$. We can directly derive it from a local representation of $\mathcal{N}_q$, as we need only the partial derivatives $\partial D_i\Phi^\tau/\partial u^\alpha_\nu$ with $1 \le i \le n$ and $|\nu| = q + 1$, i.e. the highest-order part of the formal derivative $D_i\Phi^\tau$, for determining the symbol $\mathcal{N}_{q+1}$. It is given by $\partial D_i\Phi^\tau/\partial u^\alpha_\nu = \partial\Phi^\tau/\partial u^\alpha_{\nu-1_i}$ (if $\nu_i = 0$, the derivative vanishes) and thus a local representation of $\mathcal{N}_{q+1}$ is

$$(\mathcal{N}_{q+1})_\rho : \left\{ \sum_{\substack{1 \le \alpha \le m \\ |\mu|=q}} \frac{\partial\Phi^\tau}{\partial u^\alpha_\mu}\, \dot{u}^\alpha_{\mu+1_i} = 0\,, \qquad \begin{array}{l} \tau = 1, \ldots, p\,, \\ i = 1, \ldots, n\,. \end{array} \right. \tag{7.3}$$

In our geometric approach to integrability conditions, their existence is signalled by a dimension inequality: $\dim \mathcal{R}_q^{(1)} < \dim \mathcal{R}_q$. By the following result, which follows from a straightforward analysis of the Jacobians of the involved differential equations, the dimension of $\mathcal{R}_q^{(1)}$ is related to $\dim \mathcal{N}_{q+1}$, i.e. analysing the prolonged symbol matrix M_{q+1} gives information about possible integrability conditions.

Proposition 7.3 *If $\mathcal{N}_{q+1}$ is a vector bundle, then* $\dim \mathcal{R}_q^{(1)} = \dim \mathcal{R}_{q+1} - \dim \mathcal{N}_{q+1}$.

In the classical theory of partial differential equations a different notion of symbol is used which should not be confused with the geometric symbol introduced above: the classical symbol is *not* an intrinsic object. Our notion of symbol is closely related to what is traditionally called the principal symbol which is intrinsically defined.

Assume we are given a one-form $\chi \in T^*\mathcal{X}$. It induces for every $q > 0$ a map $\iota_{\chi,q} : V\pi \to V\pi_{q-1}^q$ defined by $\iota_{\chi,q}(v) = \epsilon_q(\chi^q \otimes v)$ where ϵ_q is the fundamental identification and χ^q denotes the q-fold symmetric product of χ. In local coordinates, we write $\chi = \chi_i dx^i$ and obtain $\iota_{\chi,q} : v^\alpha \partial_{u^\alpha} \mapsto \chi_\mu v^\alpha \partial_{u_\mu^\alpha}$ where μ runs over all multi indices of length q and $\chi_\mu = \chi_1^{\mu_1} \cdots \chi_n^{\mu_n}$.

Let σ be the symbol map of the differential equation $\mathcal{R}_q$ globally described by the map $\Phi : J_q\pi \to \mathcal{E}'$. Then the *principal symbol* of $\mathcal{R}_q$ is the linear map $\tau_\chi : V\pi \to T\mathcal{E}'$ defined by $\tau_\chi = \sigma \circ \iota_{\chi,q}$. Locally, we can associate a matrix $T[\chi]$ with τ_χ:

$$T_\alpha^\tau[\chi] = \sum_{|\mu|=q} \frac{\partial \Phi^\tau}{\partial u_\mu^\alpha} \chi^\mu . \tag{7.4}$$

If $\dim \mathcal{E} = m$ and $\dim \mathcal{E}' = p$, it has p rows and m columns. Its entries are homogeneous polynomials of degree q in the coefficients of χ. We may think of $T[\chi]$ as a kind of contraction of the symbol matrix M_q. Both matrices have the same number of rows. The column with index α of $T[\chi]$ is a linear combination of all columns in M_q corresponding to a variable $\dot{u}_\mu^\alpha$ with the coefficients given by χ_μ.

Remark 7.4 Using the matrix $T[\chi]$ of the principal symbol, we may relate the construction of integrability conditions with syzygy computations. Assume that the functions Φ^τ with $1 \leq \tau \leq t$ locally representing the differential equation $\mathcal{R}_q$ lie in a differential field $\mathbb{F}$. Then the entries of $T[\chi]$ are polynomials in $\mathcal{P} = \mathbb{F}[\chi_1, \ldots, \chi_n]$. The rows of $T[\chi]$ may be considered as elements of $\mathcal{P}^m$ and generate a submodule $\mathcal{M} \subseteq \mathcal{P}^m$. Let $\mathbf{S} \in \mathrm{Syz}(\mathcal{M}) \subseteq \mathcal{P}^t$ be a syzygy of the rows of $T[\chi]$. The substitution $\chi_i \to D_i$ transforms each component S_τ of $\mathbf{S}$ into a differential operator $\hat{S}_\tau$. By construction, $\Psi = \sum_{\tau=1}^t \hat{S}_\tau \Phi^\tau$ is a linear combination of differential consequences of $\mathcal{R}_q$ in which the highest-order terms cancel. In fact, this represents nothing but the rigorous mathematical formulation of "taking a cross-derivative."

Example 7.5 For Janet's differential equation the module generated by the rows of $T[\chi]$ is the ideal $\mathcal{I}_1 = \langle \chi_3^2 + x^2 \chi_1^2, \chi_2^2 \rangle$. Obviously, its syzygy module is spanned by $\mathbf{S}_1 = \chi_2^2 \mathbf{e}_1 - (\chi_3^2 + x^2 \chi_1^2) \mathbf{e}_2$ and applying the corresponding differential operator to Janet's equation yields the above mentioned integrability condition $u_{112} = 0$.

The fundamental identification ϵ_q allows us to identify the symbol $(\mathcal{N}_q)_\rho$ with a subspace of $S_q(T_x^* \mathcal{X}) \otimes V_\xi \pi$ where $\xi = \pi_0^q(\rho)$ and $x = \pi(\xi)$. In local coordinates, ϵ_q is given by (6.3); hence its main effect is the introduction of some combinatorial factors which can be absorbed in the choice of an appropriate basis. More precisely, we recover here the discussion in Remark 2.5. If we take $\{\partial_{x^1}, \ldots, \partial_{x^n}\}$ as basis of the tangent space $T_x \mathcal{X}$ and the dual basis $\{\mathrm{d}x^1, \ldots, \mathrm{d}x^n\}$ for the cotangent space $T_x^* \mathcal{X}$, then the "terms" $\partial_{x^\mu} = \partial_{x^1}^{\mu_1} \cdots \partial_{x^n}^{\mu_n}$ with $|\mu| = q$ form a basis of $S_q(T_x \mathcal{X})$ whereas the dual basis of $S_q(T_x^* \mathcal{X})$ is given by the "divided powers" $\frac{1}{\mu!} \mathrm{d}x^\mu$. If we express an element $f \in S_q(T_x^* \mathcal{X}) \otimes V_\xi \pi$ in this basis as $f = \frac{1}{\mu!} f_\mu^\alpha \mathrm{d}x^\mu \otimes \partial_{u^\alpha}$ where μ runs over all multi indices with $|\mu| = q$, then the symbol $\mathcal{N}_q$ consists of all such f satisfying the linear system of equations $\sum_{1 \leq \alpha \leq m, |\mu|=q} \frac{\partial \Phi^\tau}{\partial u_\mu^\alpha} f_\mu^\alpha = 0$ with $\tau = 1, \ldots, t$. Obviously, this is the same linear system as (7.2) defining the symbol as a subspace of $V_\rho \pi_{q-1}^q$.

Proposition 7.6 *Let $\mathcal{R}_q \subseteq J_q \pi$ be a differential equation and $\left(\rho_r \in \mathcal{R}_r\right)_{r \geq q}$ be a sequence of points such that $\pi_q^r(\rho_r) = \rho_q$ and set $\xi = \pi_0^q(\rho_q)$ and $x = \pi^q(\rho_q)$. If we set $\mathcal{N}_r = S_r(T_x^* \mathcal{X}) \otimes V_\xi \pi$ for $0 \leq r < q$, then the sequence $\left((\mathcal{N}_r)_{\rho_r}\right)_{r \in \mathbb{N}_0}$ defines a symbolic system in $S(T_x^* \mathcal{X}) \otimes V_\xi \pi$ which satisfies $\mathcal{N}_{r+1} = \mathcal{N}_{r,1}$ for all $r \geq q$.*

Proof. For notational simplicity, we consider only $r = q$. Let $f = \frac{1}{\nu!} f_\nu^\alpha \mathrm{d}x^\nu \otimes \partial_{u^\alpha}$ where ν runs over all multi indices with $|\nu| = q + 1$ be an element of $\mathcal{N}_{q,1}$. By definition of the prolongation, this is equivalent to $\delta(f) \in \mathcal{N}_q \otimes T_x^* \mathcal{X}$ and hence we find for every $1 \leq i \leq n$ that $\frac{\nu_i}{\nu!} f_\nu^\alpha \mathrm{d}x^{\nu-1_i} \otimes \partial_{u^\alpha} \in \mathcal{N}_q$. In other words, the coefficients f_ν^α must satisfy the linear system of equations $\sum_{1 \leq \alpha \leq m, |\nu|=q+1, \nu_i > 0} \frac{\partial \Phi^\tau}{\partial u_{\nu-1_i}^\alpha} f_\nu^\alpha = 0$ with $\tau = 1, \ldots, t$ and $i = 1, \ldots, n$. A comparison with (7.3) shows that this system describes the prolonged symbol $\mathcal{N}_{q+1}$. Hence we have $\mathcal{N}_{q+1} = \mathcal{N}_{q,1}$ as claimed. $\square$

In this proposition we used a sequence of points $\rho_r \in \mathcal{R}_r$ with $\pi_q^r(\rho_r) = \rho_q$ in order to consider the symbols $(\mathcal{N}_r)_{\rho_r}$. Obviously, such a sequence does not necessarily exists, unless we are dealing with a formally integrable equation. However, by the final assertion, the obtained symbolic system is independent of the choice of these points, as we may simply set $\mathcal{N}_{r+1} = \mathcal{N}_{r,1}$ for all $r \geq q$. Hence at each point $\rho \in \mathcal{R}_q$ the symbol $(\mathcal{N}_q)_\rho$ induces a symbolic system which, according to Lemma 2.8, we may alternatively consider as a subcomodule $\mathcal{N}[\rho] \subseteq S(T_x^* \mathcal{X}) \otimes V_\xi \pi$; we then speak of the *symbol comodule* of $\mathcal{R}_q$ at the point ρ. One can now easily verify that the symbolic systems given in Example 2.9 are associated to the their mentioned differential equations.

Remark 7.7 In Proposition 7.6 and in the definition of the symbol comodule $\mathcal{N}$ we simply set the lower-order components $\mathcal{N}_r$ for $0 \leq r < q$ to the full symmetric product $S_r(T_x^* \mathcal{X}) \otimes V_\xi \pi$. In principle, one could use a more precise approach by considering instead the symbols of the projected equations $\mathcal{R}_r^{(q-r)}$. However, for the subsequent involution analysis it only matters what happens in degree q and beyond which is not affected by such changes in lower order. Hence we stick to this simpler approach.

Remark 7.8 The comodules $\mathcal{N}$ arising as symbols are of a special form: their annihilators $\mathcal{N}^0$ possess bases where all generators are homogeneous of the same degree q, namely the order of the underlying differential equation $\mathcal{R}_q$. It follows now immediately from the identification of the degree of involution of $\mathcal{N}$ and the Castelnuovo–Mumford regularity of $\mathcal{N}^0$ that the minimal free resolution of $\mathcal{N}^0$ is linear, i.e. the syzygy modules of any order can be generated by syzygies of degree 1. This was already noted as a "curiosité" by Serre in his letter appended to [21]. As later shown by Eisenbud and Goto [14], this represents in fact a characteristic property of q-regular modules: if $\mathcal{M}$ is q-regular, then the truncation $\mathcal{M}_{\geq q}$ possesses a linear resolution.

Definition 7.9 The symbol $\mathcal{N}_q$ of the differential equation $\mathcal{R}_q \subseteq J_q\pi$ of order q is *involutive* at the point $\rho \in \mathcal{R}_q$, if the symbol comodule $\mathcal{N}[\rho]$ is involutive at degree q.

Choosing local coordinates $(\mathbf{x}, \mathbf{u}^{(q)})$ in a neighbourhood of a given point $\rho \in \mathcal{R}_q$, we can apply Cartan's test (Theorem 3.4) for deciding involution. Recall that it requires only linear algebra computations with the two symbols $\mathcal{N}_q$ and $\mathcal{N}_{q+1}$ and thus is easily performed effectively. In practice, one uses a dual approach exploiting that the annihilator $\mathcal{N}^0 \subseteq S(T_x\mathcal{X}) \otimes V_\xi\pi$ is an $S(T_x\mathcal{X})$-submodule, the *symbol module*. In our chosen coordinates and bases the submodule $\mathcal{N}^0$ is generated by the "polynomials"

$$\sum_{\substack{1 \leq \alpha \leq m \\ |\mu|=q}} \frac{\partial \Phi^\tau}{\partial u_\mu^\alpha} \partial_x^\mu \otimes \partial_{u^\alpha} , \qquad \tau = 1, \ldots, t , \tag{7.5}$$

corresponding to the left hand sides in (7.2). Identifying $S(T_x\mathcal{X})$ with the polynomial ring $\mathcal{P} = \mathbb{R}[\partial_{x^1}, \ldots, \partial_{x^n}]$, one readily recognises in $\mathcal{N}^0$ the polynomial module generated by the rows of the matrix $T[\chi]$ of the principal symbol which already appeared in Remark 7.4. We may now apply the theory of Pommaret bases to the submodule $\mathcal{N}^0$. Then the following result follows immediately from Theorem 5.2.

Proposition 7.10 *The symbol $\mathcal{N}_q$ of the differential equation $\mathcal{R}_q \subseteq J_q\pi$ is involutive at the point $\rho \in \mathcal{R}_q$, if and only if in suitable local coordinates an involutive head autoreduction transforms the generators (7.5) into a Pommaret basis of the symbol module $\mathcal{N}^0$ for a class respecting term order.*

In principle, at some points $\rho \in \mathcal{R}_q$ the symbol could be involutive, whereas at other points on the differential equation this is not the case. For notational simplicity, we will assume throughout this work that all points on $\mathcal{R}_q$ behave uniformly and therefore drop from now on the explicit reference to the point $\rho \in \mathcal{R}_q$.

Proposition 7.10 transforms the Cartan test into an easily applicable effective criterion for an involutive symbol. In order to recover some results in the literature, we express it in a less algebraic language. Recall that the columns of M_q correspond to the unknowns $\dot{u}_\mu^\alpha$; we sort them according to a class respecting term order (it suffices, if we take care that a column corresponding to an unknown $\dot{u}_\mu^\alpha$ is always to the left of a column corresponding to the unknown $\dot{u}_\nu^\beta$, if $\operatorname{cls} \mu > \operatorname{cls} \nu$). Now an involutive head autoreduction is equivalent to determining a row echelon form $M_q^\triangle$ of M_q using only row operations. The unknown $\dot{u}_\mu^\alpha$ corresponding to the column where the first non-vanishing entry of a row sits is called the *leader* of this row. If $\beta_q^{(k)}$ is the number of leaders that are of class k, then we call these numbers the *indices* of the symbol $\mathcal{N}_q$.

The problem of δ-regularity concerns this notion. The class of a derivative is not invariant under coordinate transformations. In different coordinate systems we may thus obtain different values for the indices. δ-regular coordinates are distinguished by the fact that the sum $\sum_{k=1}^{n} k\beta_q^{(k)}$ takes its maximal value. It is not difficult to see that actually we are here only reformulating Remark 4.5. Hence (4.2) immediately implies the following result.

Proposition 7.11 *The symbol $\mathcal{N}_q$ with the indices $\beta_q^{(k)}$ is involutive, if and only if the matrix M_{q+1} of the prolonged symbol $\mathcal{N}_{q+1}$ satisfies* rank $M_{q+1} = \sum_{k=1}^{n} k\beta_q^{(k)}$.

Remark 7.12 A special situation arises, if there is only one dependent variable, as then *any* first-order symbol $\mathcal{N}_1$ is involutive. The symbol module $\mathcal{N}^0$ is now an ideal in $\mathcal{P}$ generated by linear polynomials. Using some linear algebra, we may always assume that all generators have different leading terms (with respect to the degree reverse lexicographic order). Because of the linearity, this implies that all leading terms are relatively prime. It is straightforward to show (in fact, this is nothing but Buchberger's first criterion) that all S-polynomials reduce to zero and hence our generating set is a Gröbner basis. As one easily verifies that the leading terms involutively generate the leading ideal, we have a Pommaret basis of $\mathcal{N}^0$ or equivalently $\mathcal{N}_1$ is involutive.

This observation is the deeper reason for a classification of partial differential equations suggested by Drach (see [59, Chapt. 5]). Using a simple trick due to him, we may transform any differential equation $\mathcal{R}_q$ into one with only one dependent variable. If we first rewrite $\mathcal{R}_q$ as a first-order equation, then the transformed equation will be of second order. Only in special circumstances one can derive a first-order equation in one dependent variables. Thus from a theoretical point of view we may distinguish two basic classes of differential equations: first-order and second-order equations, respectively, in one dependent variable. The first class is much simpler, as its symbol is always involutive (like for ordinary differential equations).

Definition 7.13 The differential equation $\mathcal{R}_q$ is called *involutive*, if it is formally integrable and if its symbol $\mathcal{N}_q$ is involutive.

The term "involution" is often used in a rather imprecise manner. In particular, involution is sometimes taken as a synonym for formal integrability. While Definition 7.13 obviously implies that an involutive equation is also formally integrable, the converse is generally not true: involution is a stronger concept than formal integrability.

Theorem 7.14 *$\mathcal{R}_q$ is an involutive differential equation, if and only if its symbol $\mathcal{N}_q$ is involutive and $\mathcal{R}_q^{(1)} = \mathcal{R}_q$.*

We omit a proof of this theorem, as it follows immediately from Theorem 7.15 below and the finiteness of the Spencer cohomology (Theorem 2.20). Checking whether or not the symbol $\mathcal{N}_q$ is involutive via the Cartan test (Theorem 3.4 or its alternative formulation Proposition 7.11) requires only computations in order q and $q + 1$. Obviously, the same is true for verifying the equality $\mathcal{R}_q^{(1)} = \mathcal{R}_q$. Hence Theorem 7.14

represents indeed a finite criterion for involution. A closer look at the above developed homological theory yields a finite criterion for formal integrability independent of involution.

Theorem 7.15 *The differential equation $\mathcal{R}_q$ is formally integrable, if and only if an integer $r \geq 0$ exists such that the symbolic system $\mathcal{N}$ defined by the symbol $\mathcal{N}_q$ and all its prolongations is 2-acyclic at degree $q + r$ and the equality $\mathcal{R}_{q+r'}^{(1)} = \mathcal{R}_{q+r'}$ holds for all values $0 \leq r' \leq r$.*

Proof. One direction is trivial. For a formally integrable equation $\mathcal{R}_q$ we even have $\mathcal{R}_{q+r'}^{(1)} = \mathcal{R}_{q+r'}$ for all $r' \geq 0$ and by Theorem 2.20 the symbolic system $\mathcal{N}$ must become 2-acyclic at some degree $q + r$. For the converse, we first note that, because of Lemma 2.13, the symbolic system $\mathcal{N}$ is trivially 1-acyclic at degree q. Our assumption says that in addition the Spencer cohomology modules $H^{q+s,2}(\mathcal{N})$ vanish for all $s \geq r$. According to Proposition 2.18, this implies dually that the Koszul homology modules $H_{q+s,1}(\mathcal{N}^0)$ of the symbol module $\mathcal{N}^0$ vanish for all $s > r$.

Recall from Remark 2.15 that the Koszul homology corresponds to a minimal free resolution of $\mathcal{N}^0$ and hence our assumption tells us that the maximal degree of a minimal generator in the first syzygy module $\mathrm{Syz}(\mathcal{N}^0)$ is $q + r$. In Remark 7.4 we have seen that the syzygies of $\mathcal{N}^0$ are related to those integrability conditions arising from generalised cross-derivatives between the highest-order equations. If now the equality $\mathcal{R}_{q+r'}^{(1)} = \mathcal{R}_{q+r'}$ holds for all $0 \leq r' \leq r$, then none of these cross-derivatives can produce an integrability condition. Furthermore, no integrability conditions can arise from lower-order equations. Hence $\mathcal{R}_q$ is formally integrable.   □

An abstract proof of this result was given by Goldschmidt [19]. The proof above is interesting from a computational point of view as it demonstrates that (a generating set of) the Koszul homology module $H_1(\mathcal{N}^0)$ shows us exactly which generalised cross-derivatives may produce integrability conditions (Kruglikov and Lychagin developed recently an alternative approach for the construction of these conditions based on multi-brackets, see [32] and references therein). Of course, we cannot decide solely on the basis of the symbol $\mathcal{N}_q$ whether or not these integrability conditions vanish modulo the equations describing $\mathcal{R}_q$, as this depends on the lower-order terms. Therefore, we must check a finite number of projections $\hat{\pi}_{q+r}^{q+r+1} : \mathcal{R}_{q+r+1} \to \mathcal{R}_{q+r}$ for surjectivity.

Example 7.16 We continue with Janet's partial differential equation $\mathcal{R}_2$ defined by $u_{33} + x^2 u_{11} = u_{22} = 0$. In Example 7.5 above we constructed via the syzygy $\mathbf{S}_1 = \chi_2^2 \mathbf{e}_1 - (\chi_3^2 + x^2 \chi_1^2) \mathbf{e}_2$ the integrability condition $u_{112} = 0$. Geometrically, we have arrived then at the equation $\mathcal{R}_3^{(1)}$ defined by this condition, the original equations and their formal derivatives. The rows of the principal symbol of $\mathcal{R}_3^{(1)}$ generate the ideal $\mathcal{I}_2 = \langle \chi_3^2 + x^2 \chi_1^2, \chi_2^2, \chi_1^2 \chi_2 \rangle$. Its syzygy module is spanned by $\mathbf{S}_1$, $\mathbf{S}_2 = \chi_1^2 \mathbf{e}_2 - \chi_2 \mathbf{e}_3$ and $\mathbf{S}_3 = \chi_1^1 \chi_2 \mathbf{e}_1 - x^2 \chi_2^2 \mathbf{e}_3$. Applying the differential operator corresponding to $\mathbf{S}_2$ yields zero, whereas $\mathbf{S}_3$ leads to a further integrability condition: $u_{1111} = 0$.

Geometrically, we are now dealing with the differential equation $\mathcal{R}_4^{(2)}$ described by the two integrability conditions, the original equations and all prolongations up to order 4. The rows of the principal symbol $T[x]$ define now the ideal $\mathcal{I}_3 = \mathcal{I}_2 + \langle x^4_1 \rangle + \langle x^1_1 \rangle$ and for its syzygy module we need two further generators, namely $\mathbf{S}_4 = x^2_1 \mathbf{e}_3 - x_2 \mathbf{e}_4$ and $\mathbf{S}_5 = (x^3_2 + x^2_1 x^1_1)\mathbf{e}_4 - x^4_1 \mathbf{e}_1$. One easily checks than none of them leads to a new integrability condition so that $\mathcal{R}^{(2)}_4$ is a formally integrable equation.

However, $\mathcal{R}^{(2)}_4$ is not involutive. One way to see this consisting of noting that in the syzygy $\mathbf{S}_5$ the coefficient of $\mathbf{e}_4$ is of degree 2. Since $\mathbf{e}_4$ represents a differential equation of order 4, the corresponding cross-derivative takes place in order 6. According to Remark 7.8, we can always obtain a linear resolution for an involutive symbol. Indeed, we must prolong here once: $\mathcal{R}^{(2)}_5$ is an involutive equation with vanishing symbol.

Another way to prove this goes as follows. Consider a point $\rho \in \mathcal{R}^{(2)}_4$ where $x^2 = a$ for some constant $a \in \mathbb{R}$ and the ideal $\mathcal{I} = \langle (x^3)^2 + a(x^1)^2, (x^2)^2, (x^1)^2 x^2, (x^1)^4 \rangle$. Then one easily verifies that the truncated ideal $\mathcal{I}_{\geq 4}$ is the annihilator of the symbol comodule $\mathcal{N}$ at the point ρ. For a Pommaret basis of $\mathcal{I}$ we must add the generators $(x^2)^2 x^3$, $(x^1)^2 x^2 x^3$ and $(x^1)^4 x^3$. Since the last generator is of degree 5, we find that $\text{reg}\,\mathcal{I} = 5$ and hence according to Remark 4.15 the degree of involution of $\mathcal{N}$ is 5.

8 Conclusions

A central question for any differential equation is the existence of solutions. For formal solutions the existence is equivalent to the formal integrability of the equation. The *Cartan–Kähler theorem* (see [42, 53] and references therein) provides us with an existence and uniqueness theorem for analytic solutions of involutive analytic equations generalising the classical Cauchy–Kovalevskaya theorem. Compared with alternative approaches like Riquier's existence theorem, the proof of the Cartan–Kähler theorem does not require a convergence analysis of power series. This allows us sometimes to extend it to larger function spaces (which is very important for applications), if additional information about the equation is given: a concrete example can be found in [52] where smooth solutions of hyperbolic systems with elliptic constraints are treated.

For deciding the mere existence of solutions, formal integrability is sufficient. If one is interested in the size of the solution space (or equivalently in the number and form of conditions leading to a unique solution), then one needs more information. The simplest approach (implicitly already exploited by Janet) consists of using a complementary decomposition of the symbol module for deriving a (formally) well-posed initial value problem and is trivial for a not only formally integrable but even involutive system. Generally speaking, the main difference between formal integrability and involution is the same as the one between a Gröbner and a Pommaret basis: the former one is concerned only with the first syzygy module (i. e. $H_1(\mathcal{M})$), the latter one with the full syzygy resolution (i. e. the full Koszul homology $H_\bullet(\mathcal{M})$).

In order to apply such results, it is important that one deals with an involutive differential equation. The *Cartan–Kuranishi theorem* (see again [42, 53] and references therein) asserts that any differential equation $\mathcal{R}_q$ satisfying some modest regularity assumptions is either inconsistent or can be completed to an equivalent involutive equation of the form $\mathcal{R}_{q+r}^{(s)}$; a concrete instance of such a completion process was given in Example 7.16 for Janet's equation. The key for proving this result is the observation that any symbol becomes involutive, if it is sufficiently often prolonged, in other words the finiteness of the Spencer cohomology (Theorem 2.20). The power of the homological approach to involution becomes evident in its trivial proof.

For concrete computations, a direct application of the Cartan–Kuranishi procedure becomes quickly cumbersome, as it requires an explicit local representation of every appearing differential equation. In the (small!) Janet example the final involutive system tem $\mathcal{R}_5^{(2)}$ is locally described by 44 equations. However, all relevant information can be extracted from just 7 equations corresponding to the Pommaret basis of the symbol module. In the language of [23] these equations comprise the skeleton of $\mathcal{R}_5^{(2)}$. On the basis of this notion, [23] presents a hybrid completion algorithm that combines the algebraic efficiency of Pommaret bases with the intrinsic geometry of the Cartan–Kuranishi procedure.

For lack of space we could not discuss applications of involutive differential equations in this contribution. Pommaret [43, 44, 45] presents in his books many applications, in particular in mathematical physics and control theory. Some applications in numerical analysis can be found in [53] and references therein. Generally speaking, wherever under- or overdetermined systems of differential equations appear, the theory of involution will make any subsequent analysis significantly easier; in many cases such an analysis will even be impossible without the concept of involution (or at least formal integrability).

A   Multi Indices

As there exist different kinds of multi indices but apparently no standard names for them, we must introduce our own terminology. Let $x^1, \ldots, x^n$ be n variables. For various constructions with them, we distinguish in this article between multi indices and repeated indices. A *multi index* is an element $\mu = [\mu_1, \ldots, \mu_n] \in \mathbb{N}_0^n$; the value $|\mu| = \mu_1 + \cdots + \mu_n$ is its length. A typical use of a multi index is $x^\mu = (x^1)^{\mu_1} \cdots (x^n)^{\mu_n}$ or for a function $u = u(x^1, \ldots, x^n)$

$$\frac{\partial^{|\mu|} u}{\partial x^\mu} = \frac{\partial^{|\mu|} u}{\partial (x^1)^{\mu_1} \cdots \partial (x^n)^{\mu_n}}. \qquad (A.1)$$

We furthermore define $\mu! = \mu_1! \cdots \mu_n!$. If k is the smallest value such that $\mu_k \neq 0$, we call it the *class* of the multi index μ and write cls $\mu = k$.

By convention, we introduce for the special multi index $0 = [0, \ldots, 0]$ that $x^0 = 1$ and $\partial^{|0|} u/\partial x^0 = u$. Obviously, $0! = 1$ and $|0| = 0$. In principle, cls 0 is undefined, but in many situations it is convenient to set cls $0 = n$. Other special multi indices appearing occasionally are

$$\mathbb{1}_i = [0, \ldots, 0, \mathbb{1}, 0, \ldots, 0] \qquad (A.2)$$

where $\mathbb{1} \in \mathbb{N}$ is the ith entry and all other entries vanish. The addition of multi indices is defined componentwise, i. e. $\mu + \nu = [\mu_1 + \nu_1, \ldots, \mu_n + \nu_n]$. If we want to increase the ith entry of a multi index μ by one, we can thus simply write $\mu + 1_i$ using (A.2).

A *repeated index* of length q is an ordered sequence $I = (i_1, \ldots, i_q)$ where each entry i_k is an element of $\{1, \ldots, n\}$. Now x^I is a shorthand for the product $x^{i_1} x^{i_2} \cdots x^{i_q}$ and correspondingly for partial derivatives. Obviously, here the ordering of the entries does not matter. However, our main use of repeated indices is for exterior forms where the ordering determines the sign. In fact, there we only consider indices $I = (i_1, \ldots, i_q)$ with $i_1 < i_2 < \cdots < i_q$, i. e. all entries are different and sorted in ascending order. If I, J are two such repeated indices, then $I \cup J$ denotes the index obtained by first concatenating I and J and then sorting the entries. Obviously, this only yields a valid result, if I and J have no entries in common. We set $\mathrm{sgn}\,(I \cup J) = \pm 1$ depending on whether an even or odd number of transpositions is required for the sorting. If I and J have entries in common, we set $\mathrm{sgn}\,(I \cup J) = 0$; this convention is useful to avoid case distinctions in some sums.

B   Term Orders

Term orders are crucial for the definition of Gröbner bases and thus of involutive bases. As some of our conventions are inverse to those usually used in commutative algebra, we collect them in this short appendix.

Let $\mathcal{P} = \Bbbk[x^1, \ldots, x^n]$ and define the set of *terms* $\mathbb{T} = \{x^\mu \mid \mu \in \mathbb{N}_0^n\}$. Recall that a *term order* is a total order $\prec$ on $\mathbb{T}$ satisfying (i) $1 \preceq t$ and (ii) $r \prec s \Rightarrow rt \prec st$ for all $r, s, t \in \mathbb{T}$. A term order is *degree compatible*, if $\deg s < \deg t$ implies $s \prec t$. Finally, we say that a term order *respects classes*, if $\deg s = \deg t$ and $\mathrm{cls}\,s < \mathrm{cls}\,t$ implies $s \prec t$.

We define now the *lexicographic* order by $x^\mu \prec_{\mathrm{lex}} x^\nu$, if the last non-vanishing entry of $\mu - \nu$ is negative. With respect to the *reverse lexicographic* order, $x^\mu \prec_{\mathrm{revlex}} x^\nu$, if the first non-vanishing entry of $\mu - \nu$ is positive. The latter one is not a term order, as 1 is not the smallest term, but its degree compatible version is a term order: $x^\mu \prec_{\mathrm{degrevlex}} x^\nu$, if $|\mu| < |\nu|$ or if $|\mu| = |\nu|$ and $x^\mu \prec_{\mathrm{revlex}} x^\nu$.

Here we defined the orders inverse to the usual convention in most texts on Gröbner bases: the classical forms arise, if one inverts the order of the variables: $x^1, \ldots, x^n \mapsto x^n, \ldots, x^1$. Our version fits better to the conventions used in differential equations theory, in particular to our definition of the class of a multi index.

Lemma B.1 *Let $\prec$ be degree compatible and the condition* $\mathrm{lt}\,f \in \langle x^1, \ldots, x^k \rangle$ *be equivalent to $f \in \langle x^1, \ldots, x^k \rangle$ for any homogeneous polynomial $f \in \mathcal{P}$, then $\prec$ is the degree reverse lexicographic order $\prec_{\mathrm{degrevlex}}$.*

The proof of this well-known characterisation lemma is left as an easy exercise to the reader. We note the following simple consequence of it: on terms of the same degree any class respecting term order coincides with the degree reverse lexicographic order.

C   Coalgebras and Comodules

Since the notion of a coalgebra and a comodule is still unfamiliar to many mathematicians, we collect here the basic definitions and properties; for an in depth treatment we refer to [7]. Roughly, the idea behind coalgebras is the inversion of certain arrows in diagrams encoding properties of the multiplication in an algebra. Thus, if $\mathcal{A}$ is an algebra over a field $\Bbbk$, then the product is a homomorphism $\mathcal{A} \otimes \mathcal{A} \to \mathcal{A}$ and the unit may be interpreted as a linear map $\Bbbk \to \mathcal{A}$. Correspondingly, a *coalgebra* $\mathcal{C}$ over a field $\Bbbk$ is a vector space equipped with a *coproduct*, a homomorphism $\Delta : \mathcal{C} \to \mathcal{C} \otimes \mathcal{C}$, and a *counit*, a linear map $\epsilon : \mathcal{C} \to \mathcal{R}$. The associativity of the product in an algebra and the defining property of the unit dualise to the requirement that the diagrams

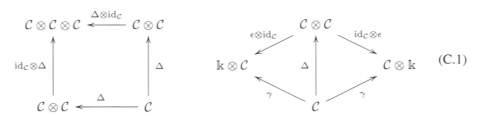

$$(\text{C.1})$$

(where γ maps $c \in \mathcal{C}$ to $1 \otimes c$ or $c \otimes 1$, respectively) commute.

Analogously, $\mathcal{C}$-comodules arise from dualising $\mathcal{A}$-modules: a (right) *comodule* is a vector space $\mathcal{N}$ with a *coaction* $\rho : \mathcal{N} \to \mathcal{N} \otimes \mathcal{C}$ such that the two diagrams

$$
\begin{array}{ccc}
\mathcal{N} & \xrightarrow{\ \rho\ } & \mathcal{N} \otimes \mathcal{C} \\
{\scriptstyle \rho}\downarrow & & \downarrow{\scriptstyle \mathrm{id}_{\mathcal{N}} \otimes \Delta} \\
\mathcal{N} \otimes \mathcal{C} & \xrightarrow[\rho \otimes \mathrm{id}_{\mathcal{C}}]{} & \mathcal{N} \otimes \mathcal{C} \otimes \mathcal{C}
\end{array}
\qquad
\begin{array}{ccc}
\mathcal{N} & \xrightarrow{\ \rho\ } & \mathcal{N} \otimes \mathcal{C} \\
& {\scriptstyle \mathrm{id}_{\mathcal{N}}}\searrow & \downarrow{\scriptstyle \mathrm{id}_{\mathcal{N}} \otimes \epsilon} \\
& & \mathcal{N}
\end{array}
\qquad (\text{C.2})
$$

commute. A special case is a *coideal* $\mathcal{N} \subseteq \mathcal{C}$ where the coaction is the coproduct Δ. The subcomodule $\mathcal{L} \subset \mathcal{N}$ *cogenerated* by a set $\mathcal{G} \subseteq \mathcal{N}$ is by definition the intersection of all subcomodules of $\mathcal{N}$ containing $\mathcal{G}$.

The linear dual $\mathcal{C}^*$ of a coalgebra $\mathcal{C}$ has a natural algebra structure via the *convolution product* $\star$. It is defined for arbitrary elements $\phi, \psi \in \mathcal{C}^*$ by requiring that the relation

$$\langle \phi \star \psi, c \rangle = \langle \phi \otimes \psi, \Delta(c) \rangle \qquad (\text{C.3})$$

holds for all $c \in \mathcal{C}$. The unit element of $\mathcal{C}^*$ is simply the counit ϵ. If $\mathcal{N}$ is a $\mathcal{C}$-comodule with coaction ρ, then its dual space $\mathcal{N}^*$ is naturally a right $\mathcal{C}^*$-module with the action $\rho^* : \mathcal{N}^* \otimes \mathcal{C}^* \to \mathcal{N}^*$ defined in similar manner by requiring that the relation

$$\langle \rho^*(\nu, \psi), n \rangle = \langle \nu \otimes \psi, \rho(n) \rangle \qquad (\text{C.4})$$

holds for all $\nu \in \mathcal{N}^*$, $\psi \in \mathcal{C}^*$ and $n \in \mathcal{N}$. For arbitrary subsets $\mathcal{L} \subseteq \mathcal{N}$ we define in the usual manner the *annihilator* $\mathcal{L}^0 = \{\nu \in \mathcal{N}^* \mid \nu(\ell) = 0 \ \forall \ell \in \mathcal{L}\} \subseteq \mathcal{N}^*$. Similarly, for any subset $\mathcal{L}^* \subseteq \mathcal{N}^*$ the annihilator is $(\mathcal{L}^*)^0 = \{n \in \mathcal{N} \mid \lambda(n) = 0 \,\forall \lambda \in \mathcal{L}^*\} \subseteq \mathcal{N}$. One can show that if $\mathcal{L} \subseteq \mathcal{N}$ is a subcomodule, then $\mathcal{L}^0 \subseteq \mathcal{N}^*$ is a submodule, and conversely if $\mathcal{L}^* \subseteq \mathcal{N}^*$ is a submodule, then $(\mathcal{L}^*)^0 \subseteq \mathcal{N}$ is a subcomodule.

If $\mathcal{V}$ is a finite-dimensional vector space, then the tensor algebra $T\mathcal{V}$ can be given the structure of a coalgebra with the coproduct

$$\Delta(v_1 \otimes \cdots \otimes v_q) = \sum_{i=0}^{q} (v_1 \otimes \cdots \otimes v_i) \otimes (v_{i+1} \otimes \cdots \otimes v_q) \tag{C.5}$$

and the counit $\epsilon : T\mathcal{V} \to \Bbbk$ which is the identity on $T_0\mathcal{V}$ and zero everywhere else. This coalgebra structure is inherited by the symmetric algebra $S\mathcal{V}$ defined in the usual way as a factor algebra of $T\mathcal{V}$. We denote the symmetric coalgebra by $\mathfrak{S}\mathcal{V}$.

If $\{x^1, \ldots, x^n\}$ is a basis of $\mathcal{V}$, then we may use as basis of the symmetric coalgebra $\mathfrak{S}\mathcal{V}$ all monomials x^μ with a multi index $\mu \in \mathbb{N}_0^n$ providing the well known isomorphy of $S\mathcal{V}$ with the polynomial algebra $\Bbbk[x^1, \ldots, x^n]$. In this basis the coproduct of $\mathfrak{S}\mathcal{V}$ is given by "Taylor expansion"

$$\Delta(f) = \sum_{\mu \in \mathbb{N}_0^n} \frac{1}{\mu!} \frac{\partial^{|\mu|} f}{\partial x^\mu} \otimes x^\mu \tag{C.6}$$

for any polynomial $f \in \Bbbk[x^1, \ldots, x^n]$.

By definition, a subset $\mathcal{J} \subset \mathfrak{S}\mathcal{V}$ is a coideal, if and only if $\Delta(\mathcal{J}) \subseteq \mathcal{J} \otimes \mathcal{C}$ which, by (C.6), is equivalent to the condition $\partial^{|\mu|} f / \partial x^\mu \in \mathcal{J}$ for all $f \in \mathcal{J}$. Similarly, a subset $\mathcal{N} \subseteq (\mathfrak{S}\mathcal{V})^m$ is a subcomodule, if and only if this condition holds in each component.

Let $\mathcal{F} \subset \mathfrak{S}_q\mathcal{V}$ be a finite set of homogeneous polynomials of degree q. We are interested in the homogeneous coideal $\mathcal{J}$ cogenerated by $\mathcal{F}$. Obviously, we must take for $\mathcal{J}_q$ the $\Bbbk$-linear span of $\mathcal{F}$. In a given basis $\{x^1, \ldots, x^n\}$ of $\mathcal{V}$, we set for $0 < r \leq q$

$$\mathcal{J}_{q-r} = \left\{ \frac{\partial^{|\mu|} f}{\partial x^\mu} \mid f \in \mathcal{J}_q, \ \mu \in \mathbb{N}_0^n, \ |\mu| = r \right\} . \tag{C.7}$$

It is easy to see that $\mathcal{J} = \bigoplus_{r=0}^{q} \mathcal{J}_r$ satisfies $\Delta(\mathcal{J}) \subseteq \mathcal{J} \otimes \mathcal{C}$ and that it is the smallest subset of $\mathfrak{S}\mathcal{V}$ containing $\mathcal{F}$ with this property. Note that, in contrast to the algebra case, we obtain components of lower degree and $\mathcal{J}$ is finite-dimensional as vector space.

Bibliography

[1] W. W. Adams and P. Loustaunau, *An Introduction to Gröbner Bases*, Graduate Studies in Mathematics 3, American Mathematical Society, Providence, 1994.

[2] D. Bayer and D. Mumford, *What can be Computed in Algebraic Geometry?*, Computational Algebraic Geometry and Commutative Algebra (D. Eisenbud and L. Robbiano, eds.), Symposia Mathematica 34, Cambridge University Press, Cambridge, 1993, pp. 1–48.

[3] D. Bayer and M. Stillman, *A Criterion for Detecting m-Regularity*, Invent. Math. 87 (1987), pp. 1–11.

[4] Th. Becker and V. Weispfenning, *Gröbner Bases*, Graduate Texts in Mathematics 141, Springer-Verlag, New York, 1993.

[5] I. Bermejo and P. Gimenez, *Saturation and Castelnuovo-Mumford Regularity*, J. Alg. 303 (2006), pp. 592–617.

[6] R. L. Bryant, S. S. Chern, R. B. Gardner, H. L. Goldschmidt, and P. A. Griffiths, *Exterior Differential Systems*, Mathematical Sciences Research Institute Publications 18, Springer-Verlag, New York, 1991.

[7] T. Brzezinski and R. Wisbauer, *Corings and Comodules*, London Mathematical Society Lecture Notes Series 309, Cambridge University Press, Cambridge, 2003.

[8] E. Cartan, *Sur la Théorie des Systèmes en Involution et ses Applications à la Relativité*, Bull. Soc. Math. France 59 (1931), pp. 88–118, (also: Œuvres Complètes d'Élie Cartan, Partie II, Vol. 2, pp. 1199–1230).

[9] ———, *Les Systèmes Différentielles Extérieurs et leurs Applications Géométriques*, Hermann, Paris, 1945.

[10] J. Chrastina, *The Formal Theory of Differential Equations*, Folia Mathematica 6, Masaryk University, Brno, 1998.

[11] D. Cox, J. Little, and D. O'Shea, *Ideals, Varieties, and Algorithms*, Undergraduate Texts in Mathematics, Springer-Verlag, New York, 1992.

[12] M. Dubois-Violette, *The Theory of Overdetermined Linear Systems and its Applications to Non-Linear Field Equations*, J. Geom. Phys. 1 (1984), pp. 139–172.

[13] S. Eilenberg and J. C. Moore, *Homology and Fibrations I: Coalgebras, Cotensor Product and its Derived Functors*, Comment. Math. Helv. 40 (1966), pp. 199–236.

[14] D. Eisenbud and S. Goto, *Linear Free Resolutions and Minimal Multiplicity*, J. Alg. 88 (1984), pp. 89–133.

[15] V. P. Gerdt and Yu. A. Blinkov, *Involutive Bases of Polynomial Ideals*, Math. Comp. Simul. 45 (1998), pp. 519–542.

[16] ———, *Minimal Involutive Bases*, Math. Comp. Simul. 45 (1998), pp. 543–560.

[17] V. P. Gerdt, Yu. A. Blinkov, and D. A. Yanovich, *Construction of Janet Bases II: Polynomial Bases*, Computer Algebra in Scientific Computing — CASC 2001 (V.G. Ghanza, E.W. Mayr, and E.V. Vorozhtsov, eds.), Springer-Verlag, Berlin, 2001, pp. 249–263.

[18] H. L. Goldschmidt, *Existence Theorems for Analytic Linear Partial Differential Equations*, Ann. Math. 82 (1965), pp. 246–270.

[19] ———, *Integrability Criteria for Systems of Non-Linear Partial Differential Equations*, J. Diff. Geom. 1 (1969), pp. 269–307.

[20] W. Gröbner, *Moderne Algebraische Geometrie*, Springer-Verlag, Wien, 1949.

[21] V. W. Guillemin and S. Sternberg, *An Algebraic Model of Transitive Differential Geometry*, Bull. Amer. Math. Soc. 70 (1964), pp. 16–47.

[22] M. Hausdorf, M. Sahbi, and W.M. Seiler, *δ- and Quasi-Regularity for Polynomial Ideals*, Global Integrability of Field Theories (J. Calmet, W.M. Seiler, and R.W. Tucker, eds.), Universitätsverlag Karlsruhe, Karlsruhe, 2006, pp. 179–200.

[23] M. Hausdorf and W. M. Seiler, *An Efficient Algebraic Algorithm for the Geometric Completion to Involution*, Appl. Alg. Eng. Comm. Comp. 13 (2002), pp. 163–207.

[24] T. A. Ivey and J. M. Landsberg, *Cartan for Beginners: Differential Geometry via Moving Frames and Exterior Differential Systems*, Graduate Studies in Mathematics 61, American Mathematical Society, Providence, 2003.

[25] M. Janet, *Sur les Systèmes d'Équations aux Dérivées Partielles*, J. Math. Pure Appl. 3 (1920), pp. 65–151.

[26] ———, *Les Modules de Formes Algébriques et la Théorie Générale des Systèmes Différentiels*, Ann. École Norm. Sup. 41 (1924), pp. 27–65.

[27] ———, *Leçons sur les Systèmes d'Équations aux Dérivées Partielles*, Cahiers Scientifiques, Fascicule IV, Gauthier-Villars, Paris, 1929.

[28] J. Johnson, *On Spencer's Cohomology Theory for Linear Partial Differential Equations*, Trans. Amer. Math. Soc. 154 (1971), pp. 137–149.

[29] E. Kähler, *Einführung in die Theorie der Systeme von Differentialgleichungen*, Teubner, Leipzig, 1934.

[30] I. S. Krasilshchik, V. V. Lychagin, and A. M. Vinogradov, *Geometry of Jet Spaces and Nonlinear Partial Differential Equations*, Gordon & Breach, New York, 1986.

[31] H. Kredel and V. Weispfenning, *Computing Dimension and Independent Sets for Polynomial Ideals*, J. Symb. Comp. 6 (1988), pp. 231–247.

[32] B. S. Kruglikov and V. V. Lychagin, *Geometry of Differential Equations*, Preprint IHES/M/07/04, Institut des Hautes Études Scientifiques, Bures-sur-Yvette, 2007.

[33] L. A. Lambe and W. M. Seiler, *Differential Equations, Spencer Cohomology, and Computing Resolutions*, Georg. Math. J. 9 (2002), pp. 723–772.

[34] B. Malgrange, *Cohomologie de Spencer (d'après Quillen)*, Publications du Séminaire Mathématique, Orsay, 1966.

[35] ———, *Cartan Involutiveness = Mumford Regularity*, Contemp. Math. 331 (2003), pp. 193–205.

[36] ———, *Systèmes Différentiels Involutif*, Panoramas et Synthèses 19, Societé Mathématique de France, Paris, 2005.

[37] E. L. Mansfield, *Differential Gröbner Bases*, Ph.D. thesis, Macquarie University, Sydney, 1991.

[38] ———, *A Simple Criterion for Involutivity*, J. London Math. Soc. 54 (1996), pp. 323–345.

[39] Y. Matsushima, *On a Theorem Concerning the Prolongation of a Differential System*, Nagoya Math. J. 6 (1953), pp. 1–16.

[40] ———, *Sur les Algèbres de Lie Linéaires Semi-Involutives*, Colloque de Topologie de Strasbourg, Université de Strasbourg, 1954–55, p. 17.

[41] P. J. Olver, *Applications of Lie Groups to Differential Equations*, Graduate Texts in Mathematics 107, Springer-Verlag, New York, 1986.

[42] J. F. Pommaret, *Systems of Partial Differential Equations and Lie Pseudogroups*, Gordon & Breach, London, 1978.

[43] ———, *Lie Pseudogroups and Mechanics*, Gordon & Breach, London, 1988.

[44] ———, *Partial Differential Control Theory I: Mathematical Tools*, Mathematics and Its Applications 530, Kluwer, Dordrecht, 2001.

[45] ———, *Partial Differential Control Theory II: Control Systems*, Mathematics and Its Applications 530, Kluwer, Dordrecht, 2001.

[46] D. G. Quillen, *Formal Properties of Over-Determined Systems of Linear Partial Differential Equations*, Ph.D. thesis, Harvard University, Cambridge, 1964.

[47] D. Rees, *A Basis Theorem for Polynomial Modules*, Proc. Cambridge Phil. Soc. 52 (1956), pp. 12–16.

[48] C. Riquier, *Les Systèmes d'Équations aux Derivées Partielles*, Gauthier-Villars, Paris, 1910.

[49] C. Ruiz, *Propriétés de Dualité du Prolongement Formel des Systèmes Différentiels Extérieur*, C.R. Acad. Sci. Ser. A 280 (1975), pp. 1625–1627.

[50] W. M. Seiler, *A Combinatorial Approach to Involution and δ-Regularity I: Involutive Bases in Polynomial Algebras of Solvable Type*, Preprint Universität Mannheim, 2002.

[51] ———, *A Combinatorial Approach to Involution and δ-Regularity II: Structure Analysis of Polynomial Modules with Pommaret Bases*, Preprint Universität Mannheim, 2002.

[52] ———, *Completion to Involution and Semi-Discretisations*, Appl. Num. Math. 42 (2002), pp. 437–451.

[53] ———, *Involution — The Formal Theory of Differential Equations and its Applications in Computer Algebra and Numerical Analysis*, Habilitation thesis, Dept. of Mathematics and Computer Science, Universität Mannheim, 2002.

[54] J.-P. Serre, *Local Algebra*, Springer-Verlag, Berlin, 2000.

[55] I. M. Singer and S. Sternberg, *The Infinite Groups of Lie and Cartan I: The Transitive Groups*, J. Anal. Math. 15 (1965), pp. 1–114.

[56] D. C. Spencer, *Deformation of Structures on Manifolds Defined by Transitive, Continuous Pseudogroups: I–II*, Ann. Math. 76 (1962), pp. 306–445.

[57] ———, *Overdetermined Systems of Linear Partial Differential Equations*, Bull. Amer. Math. Soc. 75 (1969), pp. 179–239.

[58] R. P. Stanley, *Hilbert Functions of Graded Algebras*, Adv. Math. 28 (1978), pp. 57–83.

[59] O. Stormark, *Lie's Structural Approach to PDE Systems*, Encyclopedia of Mathematics and its Applications 80, Cambridge University Press, Cambridge, 2000.

[60] B. Sturmfels and N. White, *Computing Combinatorial Decompositions of Rings*, Combinatorica 11 (1991), pp. 275–293.

[61] J. W. Sweeny, *The D-Neumann Problem*, Acta Math. 120 (1968), pp. 223–277.

[62] N. G. Trung, *Evaluations of Initial Ideals and Castelnuovo–Mumford Regularity*, Proc. Amer. Math. Soc. 130 (2002), pp. 1265–1274.

[63] W. T. Wu, *On the Construction of Gröbner Basis of a Polynomial Ideal Based on Riquier-Janet Theory*, Syst. Sci. Math. Sci. 4 (1991), pp. 194–207.

Author information

Werner M. Seiler, AG Computational Mathematics, Universität Kassel, Heinrich-Plett-Straße 40, 34132 Kassel, Germany.
Email: seiler@mathematik.uni-kassel.de

Radon Series Comp. Appl. Math **2**, 217–243

Differential Invariants for Lie Pseudo-groups

Peter Olver and Juha Pohjanpelto

Key words. Pseudo-group, moving frame, differential invariant, Gröbner basis.

AMS classification. 58H05, 53A55, 58J70.

1 Introduction

Lie pseudo-groups, roughly speaking, are the infinite-dimensional counterparts of local Lie groups of transformations. Pseudo-groups were first studied systematically at the end of the 19th century by Sophus Lie, whose great insight in the subject was to postulate the additional condition that pseudo-group transformations form the general solution to a system of partial differential equations, the determining equations for the pseudo-group. In contrast to finite dimensional Lie groups, which since Lie's day have been rigorously formalized and have become a widely used mathematical tool, the foundations of infinite-dimensional pseudo-groups remain to date in a relatively undeveloped stage. Infinite dimensional Lie pseudo-groups can for the most part only be studied through their concrete action on space, which makes the classification problems and analytical foundations of the subject thorny, particularly in the intransitive situation. We refer the reader to the original papers of Lie, Medolaghi, and Vessiot [37, 47, 69, 71] for the classical theory of pseudo-groups, to Cartan [13] for their reformulation in terms of exterior differential systems, and [20, 29, 30, 35, 36, 39, 62, 63, 67, 68] for a variety of modern approaches.

Lie pseudo-groups appear in many fundamental physical and geometrical contexts, including gauge symmetries [6], Hamiltonian mechanics and symplectic and Poisson geometry [51], conformal geometry of surfaces and conformal field theory [19, 21], the geometry of real hypersurfaces [16], symmetry groups of both linear and nonlinear

First author: Supported in part by NSF grant DMS 05-05293
Second author: Supported in part by NSF grants DMS 04-53304 and OCE 06-21134

partial differential equations, such as the Navier-Stokes and Kadomtsev–Petviashvili (KP) equations appearing in fluid and plasma mechanics [5, 18, 51], geometric hydro-dynamics [2], Vessiot's method of group splitting for producing explicit solutions to nonlinear partial differential equations [46, 50, 61, 71], mathematical morphology and computer vision [66, 72], and geometric numerical integration [45]. Pseudo-groups also appear as foliation-preserving groups of transformations, with the associated char-acteristic classes defined by certain invariant forms [24]. Also sufficiently regular local Lie group actions can be regarded as Lie pseudo-groups.

In a series of collaborative papers, starting with [22, 23], the first author has success-fully reformulated the classical theory of moving frames in a general, algorithmic, and equivariant framework that can be readily applied to a wide range of finite-dimensional Lie group actions. Applications have included complete classifications of differential invariants and their syzygies, equivalence and symmetry properties of submanifolds, rigidity theorems, invariant signatures in computer vision [3, 7, 10, 54], joint invariants and joint differential invariants [8, 54], rational and algebraic invariants of algebraic group actions [27, 28], invariant numerical algorithms [31, 55, 72], classical invariant theory [4, 53], Poisson geometry and solitons [42, 43, 44], and the calculus of varia-tions [32]. New applications of these methods to computation of symmetry groups and classification of partial differential equations can be found in [41, 48]. Furthermore, MAPLE software implementing the moving frame algorithms, written by E. Hubert, can be found at [26].

Our main goal in this contribution is to survey the extension of the moving frame theory to general Lie pseudo-groups recently put forth by the authors in [57, 58, 59, 60], and in [14, 15] in collaboration with J. Cheh. Following [32], we develop the theory in the framework of the variational bicomplexes over the bundles of (infinite) jets of mappings $J^\infty(M, M)$ of M into M and of p-dimensional submanifolds $J^\infty(M, p)$ of M, cf. [1, 32, 70]. The interactions between the two bicomplexes provide the key to understanding the moving frame constructions. Importantly, the invariant contact forms on the diffeomorphism jet bundle $\mathcal{D}^{(\infty)} \subset J^\infty(M, M)$ will play the role of Maurer–Cartan forms for the diffeomorphism pseudo-group which enables us to for-mulate explicitly the structure equations for $\mathcal{D}^{(\infty)}$. Restricting the diffeomorphism-invariant forms to the pseudo-group subbundle $\mathcal{G}^{(\infty)} \subset \mathcal{D}^{(\infty)}$ yields a complete system of Maurer–Cartan forms for the pseudo-group. Remarkably, the restricted Maurer–Cartan forms satisfy an "invariantized" version of the linear infinitesimal determining equations for the pseudo-group, which can be used to produce an explicit form of the pseudo-group structure equations. Application of these results to directly determin-ing the structure of symmetry (pseudo-)groups of partial differential equations can be found in [5, 14, 15, 49].

Assuming freeness of the action, the explicit construction of a moving frame for a pseudo-group is based on a choice of local cross-section to the pseudo-group orbits in $J^\infty(M, p)$, [23]. The moving frame induces an invariantization process that projects general differential functions and differential forms on $J^\infty(M, p)$ to their invariant counterparts. In particular, invariantization of the standard jet coordinates results in a complete system of normalized differential invariants, while invariantization of the horizontal and contact one-forms yields an invariant coframe. The corresponding dual invariant total derivative operators will map invariants to invariants of higher order.

The structure of the algebra of differential invariants, including the specification of a finite generating set of differential invariants, and the syzygies or differential relations among the generators, will then follow from the recurrence formulas that relate the differentiated and normalized differential invariants. Remarkably, besides the choice of a cross section, this final step requires only linear algebra and differentiation based on the infinitesimal determining equations for the pseudo-group, and not the explicit formulas for either the differential invariants, the invariant differential operators, or the moving frame. Except possibly for some low order details, the underlying structure of the differential invariant algebra is then entirely governed by two commutative algebraic modules: the symbol module of the infinitesimal determining system of the pseudo-group and a new module, named the prolonged symbol module, that quantifies the symbols of the prolonged action of the pseudo-group on $J^\infty(M, p)$.

2 The Diffeomorphism Pseudo-Group

Let M be a smooth m-dimensional manifold and let $\mathcal{D} = \mathcal{D}(M)$ denote the pseudo-group of all local diffeomorphisms $\varphi \colon M \to M$. For each $0 \leq n \leq \infty$, let $J^n(M, M)$ denote the bundle of nth order jets of smooth mappings $\phi \colon M \to M$ and $\mathcal{D}^{(n)} = \mathcal{D}^{(n)}(M) \subset J^n(M, M)$ the groupoid of nth order jets of local diffeomorphisms [40]. The source map $\sigma^n \colon \mathcal{D}^{(n)} \to M$ and target map $\tau^n \colon \mathcal{D}^{(n)} \to M$ are given by

$$\sigma^n(j_z^n \varphi) = z, \qquad \tau^n(j_z^n \varphi) = \varphi(z), \tag{2.1}$$

respectively, and groupoid multiplication is induced by composition of mappings,

$$j_{\varphi(z)}^n \psi \cdot j_z^n \varphi = j_z^n(\psi \circ \varphi). \tag{2.2}$$

Let $\widehat{\pi}_k^n \colon \mathcal{D}^{(n)} \to \mathcal{D}^{(k)}$, $0 \leq k \leq n$, denote the natural projections.

Given local coordinates $(z, Z) = (z^1, \ldots, z^m, Z^1, \ldots, Z^m)$ on an open subset of $M \times M$, the induced local coordinates of $g^{(n)} = j_z^n \varphi \in \mathcal{D}^{(n)}$ are denoted $(z, Z^{(n)})$, where the components

$$Z_B^a = \frac{\partial^{\#B} \varphi^a}{\partial z^B}(z), \qquad \text{for } 1 \leq a \leq m, \quad 0 \leq \#B \leq n,$$

of $Z^{(n)}$, represent the partial derivatives of the coordinate expression of φ at the source point $z = \sigma^n(g^{(n)})$. We will consistently use lower case letters, $z, x, u, \ldots$ for the source coordinates and the corresponding upper case letters $Z^{(n)}, X^{(n)}, U^{(n)}, \ldots$ for the derivative target coordinates of our diffeomorphisms φ.

The groupoid $\mathcal{D}^{(\infty)} \subset J^\infty(M, M)$ of infinite order jets inherits the structure of a variational bicomplex from $J^\infty(M, M)$, [1, 70]. This provides a natural splitting of the cotangent bundle $T^*\mathcal{D}^{(\infty)}$ into horizontal and vertical (or contact) components [1, 52], and we use $d = d_M + d_G$ to denote the induced splitting of the exterior derivative on $\mathcal{D}^{(\infty)}$. In terms of local coordinates $(z, Z^{(\infty)})$, the horizontal subbundle of $T^*\mathcal{D}^{(\infty)}$ is spanned by the one-forms $dz^a = d_M z^a$, $a = 1, \ldots, m$, while the vertical subbundle is spanned by the basic *contact forms*

$$\Upsilon_B^a = d_G Z_B^a = dZ_B^a - \sum_{c=1}^m Z_{Bc}^a dz^c, \qquad a = 1, \ldots, m, \quad \#B \geq 0. \tag{2.3}$$

Composition by a local diffeomorphism $\psi \in \mathcal{D}$ induces an action by right multiplication on diffeomorphism jets,

$$R_\psi(j_z^n \varphi) = j_{\psi(z)}^n(\varphi \circ \psi^{-1}). \tag{2.4}$$

A differential form μ on $\mathcal{D}^{(n)}$ is *right-invariant* if $R_\psi \mu = \mu$, where defined, for every $\psi \in \mathcal{D}$. Since the splitting of forms on $\mathcal{D}^{(\infty)}$ is invariant under this action, the differentials $d_M \mu$ and $d_G \mu$ of a right-invariant form μ are again invariant. The target coordinate functions Z^a are obviously right-invariant, and hence their horizontal differentials

$$\sigma^a = d_M Z^a = \sum_{b=1}^m Z_b^a dz^b \tag{2.5}$$

form an invariant horizontal coframe, while their vertical differentials

$$\mu^a = d_G Z^a = \Upsilon^a = dZ^a - \sum_{b=1}^m Z_b^a dz^b, \qquad a = 1, \ldots, m, \tag{2.6}$$

are the invariant contact forms of order zero. Let

$$\mathbb{D}_{Z^a} = W_a^b \mathbb{D}_{z^b} \tag{2.7}$$

denote the total total derivative operators on $\mathcal{D}^{(\infty)}$ dual to the horizontal forms (2.5), where

$$\mathbb{D}_{z^b} = \frac{\partial}{\partial z^b} + \sum_{\#B \geq 0} Z_{Bb}^c \frac{\partial}{\partial Z_B^c} = \frac{\partial}{\partial z^b} + Z_b^c \frac{\partial}{\partial Z^c} + Z_{b_1 b}^c \frac{\partial}{\partial Z_{b_1}^c} + Z_{b_1 b_2 b}^c \frac{\partial}{\partial Z_{b_1 b_2}^c} + \cdots, \tag{2.8}$$

$b = 1, \ldots, m$, are the standard total derivative operators on $\mathcal{D}^{(\infty)}$ and where $W_b^a = (Z_b^a)^{-1}$ is the inverse Jacobian matrix. Then the higher-order invariant contact forms are obtained by successively Lie differentiating the invariant contact forms (2.6),

$$\mu_B^a = \mathbb{D}_{Z^B} \mu^a = \mathbb{D}_{Z^B} \Upsilon^a, \tag{2.9}$$

where $\mathbb{D}_{Z^B} = \mathbb{D}_{z^{b_1}} \cdots \mathbb{D}_{z^{b_k}}$, $a = 1, \ldots, m$, $k = \#B \geq 0$.

The next step in our program is to establish the structure equations for the diffeomorphism groupoid $\mathcal{D}^{(\infty)}$, which can be derived efficiently by employing Taylor series. Let

$$Z^a[h] = \sum_{\#B \geq 0} \frac{1}{B!} Z_B^a h^B, \qquad 1 \leq a \leq m, \tag{2.10}$$

be the individual components of the column vector-valued Taylor series $Z^a[h]$, depending on $h = (h^1, \ldots, h^m)$, obtained by expanding a local diffeomorphism $Z = $

$\varphi(z + h)$ at $h = 0$. Further, let $\Upsilon[\![h]\!]$, $\mu[\![H]\!]$ denote the column vectors of contact form-valued and invariant contact form-valued power series with individual components

$$\Upsilon^a[\![h]\!] = \sum_{\#B \geq 0} \frac{1}{B!}\,\Upsilon^a_B\, h^B, \qquad \mu^a[\![H]\!] = \sum_{\#B \geq 0} \frac{1}{B!}\,\mu^a_B\, H^B, \qquad a = 1, \ldots, m,$$

(2.11)

respectively. Equations (2.9) imply that

$$\mu[\![H]\!] = \Upsilon[\![h]\!] \quad \text{when } H = Z[\![H]\!] - Z[\![0]\!],$$

(2.12)

which, after an application of the exterior derivative, can be used to derive the diffeomorphisms pseudo-group structure equations.

Theorem 2.1 *The complete structure equations for the diffeomorphism pseudo-group are obtained by equating coefficients in the power series identity*

$$d\mu[\![H]\!] = \nabla_H \mu[\![H]\!] \wedge (\mu[\![H]\!] - dZ[\![0]\!]), \quad d\sigma = - d\mu[\![0]\!] = \nabla_H \mu[\![0]\!] \wedge \sigma. \ (2.13)$$

Here $\nabla_H \mu[\![H]\!] = \left(\dfrac{\partial Z^a}{\partial H^b}[\![H]\!] \right)$ *denotes the* $m \times m$ *power series Jacobian matrix obtained by differentiating* $\mu[\![H]\!]$ *with respect to* $H = (H^1, \ldots, H^m)$.

Let $\mathcal{X} = \mathcal{X}(M)$ denote the space of locally defined vector fields on M, which we write in local coordinates as

$$\mathbf{v} = \sum_{a=1}^{m} \zeta^a(z) \frac{\partial}{\partial z^a}.$$

(2.14)

We regard $\mathcal{X}$ as the space of infinitesimal generators of the diffeomorphism pseudo-group. Let $\mathcal{X}^{(n)} = J^n TM$, $0 \leq n \leq \infty$, denote the tangent n-jet bundle. The n-jet $j^n_z \mathbf{v} \in \mathcal{X}^{(n)}$ of the vector field (2.14) at a point z is prescribed by all the partial derivatives of its coefficients up to order n, which we denote by

$$\zeta^{(n)} = (\ldots, \zeta^a_B, \ldots), \quad a = 1, \ldots, m, \quad 0 \leq \#B \leq n.$$

(2.15)

3   Lie Pseudo-Groups

Several variants of the precise technical definition of a Lie pseudo-group appear in the literature. Ours is:

Definition 3.1 A sub-pseudo-group $\mathcal{G} \subset \mathcal{D}$ will be called a *Lie pseudo-group* if there exists $n_o \geq 1$ such that for all $n \geq n_o$:

1. the corresponding sub-groupoid $\mathcal{G}^{(n)} \subset \mathcal{D}^{(n)}$ forms a smooth, embedded subbundle,

2. every smooth local solution $Z = \varphi(z)$ to the determining system $\mathcal{G}^{(n)}$ belongs to $\mathcal{G}$,

3. $\mathcal{G}^{(n)} = \mathrm{pr}^{(n-n_o)}\mathcal{G}^{(n_o)}$ is obtained by prolongation.

The minimal value of n_o is called the *order* of the pseudo-group.

Thus on account of conditions (1) and (3), for $n \geq n_o$, the pseudo-group subbundle $\mathcal{G}^{(n)} \subset \mathcal{D}^{(n)}$ is defined in local coordinates by a formally integrable system of nth order partial differential equations

$$F^{(n)}(z, Z^{(n)}) = 0, \tag{3.1}$$

the *determining equations* for the pseudo-group, whose local solutions $Z = \varphi(z)$, by condition (2), are precisely the pseudo-group transformations. Our assumptions, moreover, imply that the isotropy jets $\mathcal{G}_z^{(n)} = \{g^{(n)} \in \mathcal{G}^{(n)} \mid \sigma^n(g^{(n)}) = \tau^n(g^{(n)}) = z\}$ form a finite dimensional Lie group for all $z \in M$.

Given a Lie pseudo-group $\mathcal{G}$, let $\mathfrak{g} \subset \mathcal{X}$ denote the local Lie algebra of infinitesimal generators, i.e., the set of locally defined vector fields whose flows belong to $\mathcal{G}$. Let $\mathfrak{g}^{(n)} \subset \mathcal{X}^{(n)}$ denote their jets. In local coordinates, the subspace $\mathfrak{g}^{(n)} \subset \mathcal{X}^{(n)}$ is defined by a linear system of partial differential equations

$$L^{(n)}(z, \zeta^{(n)}) = 0 \tag{3.2}$$

for the vector field coefficients (2.13), called the *linearized* or *infinitesimal determining equations* for the pseudo-group. Conversely, any vector field $\mathbf{v}$ satisfying infinitesimal determining equations (3.2) is an infinitesimal generator for $\mathcal{G}$, [57]. In practice, the infinitesimal determining equations are constructed by linearizing the nth order determining equations (3.1) at the identity transformation. If $\mathcal{G}$ is the symmetry group of a system of differential equations, then the linearized determining equations (3.2) are (the completion of) the usual determining equations for its infinitesimal generators obtained via Lie's algorithm [51].

Let us explain how the underlying structure of the pseudo-group is explicitly prescribed by its infinitesimal determining equations. As with finite-dimensional Lie groups, the structure of a pseudo-group is described by its Maurer–Cartan forms. A complete system of right-invariant one-forms on $\mathcal{G}^{(\infty)} \subset \mathcal{D}^{(\infty)}$ is obtained by restricting (or pulling back) the Maurer–Cartan forms (2.5), (2.9). For simplicity, we continue to denote these forms by σ^a, μ_B^a. The restricted Maurer–Cartan forms are, of course, no longer linearly independent, but are subject to certain constraints dictated by the pseudo-group. Remarkably, these constraints can be explicitly characterized by an invariant version of the linearized determining equations (3.2), obtained by replacing the source coordinates z^a by the corresponding target coordinates Z^a and the vector field jet coordinates ζ_B^a by the corresponding Maurer–Cartan form μ_B^a.

Theorem 3.2 *The linear system*

$$L^{(n)}(Z, \mu^{(n)}) = 0 \tag{3.3}$$

serves to define the complete set of linear dependencies among the right-invariant Maurer–Cartan forms $\mu^{(n)}$ on $\mathcal{G}^{(n)}$.

In this way, we effectively and efficiently bypass Cartan's more complicated prolongation procedure [9, 13] for accessing the pseudo-group structure equations.

Example 3.3 In this example we derive the structure equations for the pseudo-group $\mathcal{G}$ consisting of transformations $\varphi \colon \mathbb{R}^3 \to \mathbb{R}^3$ with

$$X = f(x), \quad Y = e(x, y) \equiv f'(x)\, y + g(x),$$

$$U = u + \frac{e_x(x, y)}{f'(x)} = u + \frac{f''(x)\, y + g'(x)}{f'(x)}, \tag{3.4}$$

where $f(x) \in \mathcal{D}(\mathbb{R})$ and $g(x)$ is an arbitrary smooth function. The transformations (3.4) form the general solution to the first order system of determining equations

$$X_y = X_u = 0, \quad Y_y = X_x \neq 0, \quad Y_u = 0, \quad Y_x = (U - u)X_x, \quad U_u = 1 \tag{3.5}$$

for $\mathcal{G}$. The infinitesimal generators are given by

$$\mathbf{v} = \xi \frac{\partial}{\partial x} + \eta \frac{\partial}{\partial y} + \varphi \frac{\partial}{\partial u} = a(x) \frac{\partial}{\partial x} + [a'(x)\, y + b(x)] \frac{\partial}{\partial y} + [a''(x)\, y + b'(x)] \frac{\partial}{\partial u}, \tag{3.6}$$

where $a(x), b(x)$ are arbitrary functions, forming the general solution to the first order infinitesimal determining system

$$\xi_x = \eta_y, \quad \xi_y = \xi_u = \eta_u = \varphi_u = 0, \quad \eta_x = \varphi, \tag{3.7}$$

which is obtained by linearizing the determining system (3.5) at the identity jet.

In accordance with Theorem 3.2, the pull-backs of the Maurer–Cartan forms (2.9) satisfy the invariantized version

$$\mu_X^x = \mu_Y^y, \quad \mu_Y^x = \mu_U^x = \mu_U^y = \mu_U^u = 0, \quad \mu_X^y = \mu^u, \tag{3.8}$$

of the linearized determining equations. By a repeated application of the invariant total derivative operators $\mathbb{D}_X, \mathbb{D}_Y, \mathbb{D}_U$ (cf. (2.7)) we find that

$$\mu_{X^n Y}^y = \mu_{X^{n+1}}^x, \quad \mu_{X^n}^u = \mu_{X^{n+1}}^y, \quad \mu_{X^n Y}^u = \mu_{X^{n+2}}^x, \quad n \geq 0, \tag{3.9}$$

while all the other pulled-back basis Maurer–Cartan forms vanish. As a result, the one-forms

$$\sigma^x, \quad \sigma^y, \quad \sigma^u, \quad \mu_n^x = \mu_{X^n}^x, \quad \mu_n^y = \mu_{X^n}^y, \quad n \geq 0, \tag{3.10}$$

form a $\mathcal{G}$–invariant coframe on $\mathcal{G}^{(\infty)}$.

The structure equations are obtained by substituting the expansions

$$\mu^x[\![H]\!] = \sum_{n=0}^{\infty} \frac{1}{n!}\, \mu_n^x\, H^n,$$

$$\mu^y[\![H, K]\!] = \sum_{n=0}^{\infty} \frac{1}{n!}\, \mu_n^y\, H^n + K \sum_{n=0}^{\infty} \frac{1}{n!}\, \mu_{n+1}^x\, H^n, \tag{3.11}$$

$$\mu^u[\![H, K]\!] = \sum_{n=0}^{\infty} \frac{1}{n!}\, \mu_{n+1}^y\, H^n + K \sum_{n=0}^{\infty} \frac{1}{n!}\, \mu_{n+2}^x\, H^n,$$

into (2.13). Those involving $\mu^x_{X^n}$ reduce to the structure equations

$$d\sigma^x = \mu^x_X \wedge \sigma^x, \quad d\mu^x_{X^n} = -\mu^x_{X^{n+1}} \wedge \sigma^x + \sum_{i=0}^{n-1} \binom{n}{i} \mu^x_{X^{i+1}} \wedge \mu^x_{X^{n-i}}$$

$$= \sigma^x \wedge \mu^x_{X^{n+1}} - \sum_{j=0}^{[n+1/2]} \frac{n-2j+1}{n+1} \binom{n+1}{j} \mu^x_{X^j} \wedge \mu^x_{X^{n+1-j}} \qquad (3.12)$$

for $\mathcal{D}(\mathbb{R})$ (cf. Cartan [12], eq. (48)), while

$$d\sigma^y = \mu^y_X \wedge \sigma^x + \mu^x_X \wedge \sigma^y, \qquad d\sigma^u = \mu^y_{X^2} \wedge \sigma^x + \mu^x_{X^2} \wedge \sigma^y,$$

$$d\mu^y_{X^n} = \sigma^x \wedge \mu^y_{X^{n+1}} + \sigma^y \wedge \mu^x_{X^{n+1}} + \sum_{j=0}^{n-1} \left[\binom{n}{j} - \binom{n}{j+1} \right] \mu^y_{X^{j+1}} \wedge \mu^x_{X^{n-j}}.$$

$$(3.13)$$

Additional examples of this procedure can be found in [14, 58]; see also [49] for a comparison with other approaches appearing in the literature.

4   Pseudo-Group Actions on Extended Jet Bundles

In this paper, our primary focus is on the induced action of our pseudo-group on submanifolds of a fixed dimension. For $0 \leq n \leq \infty$, let $J^n = J^n(M,p)$ denote the nth order (extended) jet bundle consisting of equivalence classes of p-dimensional submanifolds $S \subset M$ under the equivalence relation of nth order contact, cf. [52]. We use the standard local coordinates

$$z^{(n)} = (x, u^{(n)}) = (\ldots, x^i, \ldots, u^\alpha_J, \ldots) \qquad (4.1)$$

on J^n induced by a splitting of the local coordinates

$$z = (x, u) = (x^1, \ldots, x^p, u^1, \ldots, u^q)$$

on M into p independent and $q = m - p$ dependent variables [51, 52].

   The choice of independent and dependent variables brings about a decomposition of the differential one-forms on J^∞ into horizontal and vertical components. The basis *horizontal forms* are the differentials $dx^1, \ldots, dx^p$ of the independent variables, while the basis vertical forms are provided by the *contact forms*

$$\theta^\alpha_J = du^\alpha_J - \sum_{i=1}^{p} u^\alpha_{Ji} \, dx^i, \quad \alpha = 1, \ldots, q, \quad \#J \geq 0. \qquad (4.2)$$

This decomposition induces a splitting of the exterior derivative $d = d_H + d_V$ on J^∞ into horizontal and vertical (or contact) components, and locally endows the algebra of differential forms on J^∞ with the structure of a variational bicomplex [1, 32, 70].

Local diffeomorphisms preserve the nth order contact equivalence relation between submanifolds, and thus give rise to an action on the jet bundle J^n, known as the nth prolonged action, which, by the chain rule factors through the diffeomorphism jet groupoid $\mathcal{D}^{(n)}$. It will be useful to combine the two bundles $\mathcal{D}^{(n)}$ and J^n into a new groupoid $\mathcal{E}^{(n)} \to J^n$ by pulling back $\mathcal{D}^{(n)} \to M$ via the standard projection $\widetilde{\pi}_o^n \colon J^n \to M$. Points in $\mathcal{E}^{(n)}$ consists of pairs $(z^{(n)}, g^{(n)})$, where $z^{(n)} \in J^n$ and $g^{(n)} \in \mathcal{G}^{(n)}$ are based at the same point $z = \widetilde{\pi}_o^n(g^{(n)}) = \widetilde{\pi}_o^n(z^{(n)})$.

Local coordinates on $\mathcal{E}^{(n)}$ are written as $\mathbf{Z}^{(n)} = (z^{(n)}, Z^{(n)})$, where

$$z^{(n)} = (x, u^{(n)}) = (\ldots, x^i, \ldots, u_J^\alpha, \ldots)$$

indicate submanifold jet coordinates, while

$$Z^{(n)} = (X^{(n)}, U^{(n)}) = (\ldots, X_A^i, \ldots, U_A^\alpha, \ldots)$$

indicate diffeomorphism jet coordinates. The groupoid structure on $\mathcal{E}^{(n)}$ is induced by the source map, which is merely the projection, $\widetilde{\sigma}^n(z^{(n)}, g^{(n)}) = z^{(n)}$, and the target map $\widetilde{\tau}^n(z^{(n)}, g^{(n)}) = g^{(n)} \cdot z^{(n)}$, which is defined by the prolonged action of $\mathcal{D}^{(n)}$ on J^n. We let $\varphi \in \mathcal{D}$ with domain $\operatorname{dom} \varphi = \mathcal{U} \subset M$ act on the set $\mathcal{E}^{(n)}|_{\mathcal{U}} = \{(z^{(n)}, g^{(n)}) \in \mathcal{E}^{(n)} \mid \widetilde{\pi}_o^n(z^{(n)}) \in \mathcal{U}\}$ by

$$\varphi \cdot (z^{(n)}, g^{(n)}) = (j_z^n \varphi \cdot z^{(n)}, g^{(n)} \cdot j_{\varphi(z)}^n \varphi^{-1}), \tag{4.3}$$

where $\widetilde{\pi}_o^n(z^{(n)}) = z$. The action (4.3) obviously factors into an action of $\mathcal{D}^{(n)}$ on $\mathcal{E}^{(n)}$.

The cotangent bundle $T^*\mathcal{E}^\infty$ naturally splits into jet and group components, spanned, respectively, by the *jet forms*, consisting of the horizontal one-forms dx^i and contact one-forms θ_J^α from the submanifold jet bundle J^∞, and by the contact one-forms Υ_B^α from the diffeomorphism jet bundle $\mathcal{D}^{(\infty)}$. We accordingly decompose the differential on $\mathcal{E}^\infty$ into jet and group components, the former further splitting into horizontal and vertical components:

$$d = d_J + d_G = d_H + d_V + d_G. \tag{4.4}$$

The resulting operators satisfy the tricomplex relations [32],

$$d_J^2 = d_G^2 = d_H^2 = d_V^2 = 0,$$

$$d_J d_G = -d_G d_J, \quad d_H d_V = -d_V d_H, \quad d_H d_G = -d_G d_H, \quad d_V d_G = -d_G d_V.$$

The above splitting determines *lifted total derivative operators*

$$\mathbf{D}_{x^j} = \mathbb{D}_{x^j} + \sum_{\alpha=1}^q u_j^\alpha \, \mathbb{D}_{u^\alpha} + \sum_{\#J \geq 1} u_{Jj}^\alpha \frac{\partial}{\partial u_J^\alpha} \tag{4.5}$$

on $\mathcal{E}^\infty$, where $\mathbb{D}_{x^j}$, $\mathbb{D}_{u^\alpha}$ are the standard total derivative operators (2.8) on $\mathcal{G}^{(\infty)}$. Proceeding in analogy with the construction of the invariant total derivative operators (2.7) on $\mathcal{D}^{(\infty)}$, we define *lifted invariant total derivative operators* on $\mathcal{E}^\infty$ by

$$\mathbf{D}_{X^j} = \sum_{k=1}^p \widehat{W}_j^k \, \mathbf{D}_{x^k}, \qquad \text{where} \qquad \widehat{W}_j^k = (\mathbf{D}_{x^j} X^k)^{-1} \tag{4.6}$$

indicates the entries of the inverse total Jacobian matrix. With this, the chain rule formulas for the higher-order prolonged action of $\mathcal{D}^{(n)}$ on J^n, i.e., coordinates $\widehat{U}_J^\alpha$ of the target map $\widetilde{\tau}^n \colon \mathcal{E}^{(n)} \to J^n$, are obtained by successively differentiating the target dependent variables U^α with respect to the target independent variables X^i, whereby

$$\widehat{U}_J^\alpha = \mathrm{D}_{X^J} U^\alpha = \mathrm{D}_{X^{j_1}} \cdots \mathrm{D}_{X^{j_k}} U^\alpha. \tag{4.7}$$

These are the multi-dimensional versions of the usual implicit differentiation formulas from calculus.

Given a Lie pseudo-group $\mathcal{G}$, let $\mathcal{H}^{(n)} \subset \mathcal{E}^{(n)}$ denote the subgroupoid obtained by pulling back $\mathcal{G}^{(n)} \subset \mathcal{D}^{(n)}$ via the projection $\widetilde{\pi}_o^n \colon J^n \to M$.

Definition 4.1 A *moving frame* $\rho^{(n)}$ of *order* n is a $\mathcal{G}^{(n)}$ equivariant local section of the bundle $\mathcal{H}^{(n)} \to J^n$.

More explicitly, we require $\rho^{(n)} \colon \mathcal{V}^n \to \mathcal{H}^{(n)}$, where $\mathcal{V}^n \subset J^n$ is open, to satisfy

$$\widetilde{\sigma}^n(\rho^{(n)}(z^{(n)})) = z^{(n)}, \qquad \rho^{(n)}(g^{(n)} \cdot z^{(n)}) = g^{(n)} \cdot \rho^{(n)}(z^{(n)}), \tag{4.8}$$

for all $g^{(n)} \in \mathcal{G}^{(n)}|_z$ near the jet $\mathbb{I}_z^{(n)}$ of the identity transformation such that both $z^{(n)}$ and $g^{(n)} \cdot z^{(n)}$ lie in the domain of definition $\mathcal{V}^n$ of $\rho^{(n)}$, where $\mathcal{G}^{(n)}|_z$ denotes the source fibre of $\mathcal{G}^{(n)}$ at $z = \widetilde{\pi}_o^n(z^{(n)})$. Then, with a moving frame at hand, the composition $\widetilde{\tau}^n \circ \rho^{(n)}$, due to equations (4.3), (4.8), is invariant under the action of $\mathcal{G}$ on J^n and, as we will subsequently see, provide differential invariants for the action of $\mathcal{G}$ in J^n .

A moving frame $\rho^{(k)} \colon \mathcal{V}^k \to \mathcal{H}^{(k)}$ of order $k > n$ is *compatible* with a moving frame $\rho^{(n)} \colon \mathcal{V}^n \to \mathcal{H}^{(n)}$ of order n if $\widehat{\pi}_k^n \circ \rho^{(k)} = \rho^{(n)} \circ \widetilde{\pi}_k^n$, where defined. A *complete moving frame* is provided by a mutually compatible collection $\rho^{(k)} \colon \mathcal{V}^k \to \mathcal{H}^{(k)}$ of moving frames of all orders $k \geq n$ with domains $\mathcal{V}^k = (\widetilde{\pi}_n^k)^{-1} \mathcal{V}^n$.

As in the finite-dimensional construction [23], the (local) existence of a moving frame requires that the group action be free and regular.

Definition 4.2 The pseudo-group $\mathcal{G}$ acts *freely* at $z^{(n)} \in J^n$ if its *isotropy subgroup* is trivial, $\mathcal{G}_{z^{(n)}}^{(n)} = \{g^{(n)} \in \mathcal{G}^{(n)}|_z \mid g^{(n)} \cdot z^{(n)} = z^{(n)}\} = \{\mathbb{I}_z^{(n)}\}$, and *locally freely* if $\mathcal{G}_{z^{(n)}}^{(n)}$ is discrete.

According to the standard definition [23] any (locally) free action of a finite-dimensional Lie group satisfies the (local) freeness condition of definition 4.2, but the converse is *not* necessarily valid.

The pseudo-group acts locally freely at $z^{(n)}$ if and only if the dimension of the prolonged pseudo-group orbit through $z^{(n)}$ agrees with the dimension $r_n = \dim \mathcal{G}^{(n)}|_z$ of the source fiber at $z = \widetilde{\pi}_o^n(z^{(n)})$. Thus, freeness of the pseudo-group at order n requires, at the very least, that

$$r_n = \dim \mathcal{G}^{(n)}|_z \leq \dim J^n = p + (m - p)\binom{p + n}{p}. \tag{4.9}$$

Freeness thus provides an alternative and simpler means of quantifying the Spencer cohomological growth conditions imposed in [33, 34]. Pseudo-groups having too large a fiber dimension r_n will, typically, act transitively on (a dense open subset of) J^n, and thus possess no non-constant differential invariants. A key result of [60], generalizing the trivial finite-dimensional case, is the persistence of local freeness.

Theorem 4.3 *Let $\mathcal{G}$ be a Lie pseudo-group acting on an m-dimensional manifold M. If $\mathcal{G}$ acts locally freely at $z^{(n)} \in J^n$ for some $n > 0$, then it acts locally freely at any $z^{(k)} \in J^k$ with $\widetilde{\pi}^k_n(z^{(k)}) = z^{(n)}$, for $k \geq n$.*

As in the finite-dimensional version, [23], moving frames are constructed through a normalization procedure based on a choice of *cross-section* to the pseudo-group orbits, i.e., a transverse submanifold of the complementary dimension.

Theorem 4.4 *Suppose $\mathcal{G}^{(n)}$ acts freely on $\mathcal{V}^n \subset J^n$ with its orbits forming a regular foliation. Let $\mathcal{K}^n \subset \mathcal{V}^n$ be a local cross-section to the pseudo-group orbits. Given $z^{(n)} \in \mathcal{V}^n$, define $\rho^{(n)}(z^{(n)}) \in \mathcal{H}^{(n)}$ to be the unique groupoid jet such that $\widetilde{\sigma}^n(\rho^{(n)}(z^{(n)})) = z^{(n)}$ and $\widetilde{\tau}^n(\rho^{(n)}(z^{(n)})) \in \mathcal{K}^n$ (when such exists). Then $\rho^{(n)} \colon J^n \to \mathcal{H}^{(n)}$ is a moving frame for $\mathcal{G}$ defined on an open subset of $\mathcal{V}^n$ containing $\mathcal{K}^n$.*

In most practical situations, we select a coordinate cross-section of minimal order, defined by fixing the values of r_n of the individual submanifold jet coordinates $(x, u^{(n)})$. We write out the explicit formulas $(X, U^{(n)}) = \mathbf{F}(x, u^{(n)}, g^{(n)})$ using expressions (4.5) for the prolonged pseudo-group action in terms of a convenient system of group parameters $g^{(n)} = (g_1, \ldots, g_{r_n})$. The r_n components corresponding to our choice of cross-section variables serve to define the *normalization equations*

$$F_1(x, u^{(n)}, g^{(n)}) = c_1, \quad \ldots \quad , F_{r_n}(x, u^{(n)}, g^{(n)}) = c_{r_n}. \qquad (4.10)$$

Solving for the group parameters,

$$g_i = \gamma_i^{(n)}(x, u^{(n)}), \qquad i = 1, \ldots, r_n, \qquad (4.11)$$

yields the formula

$$\rho^{(n)}(x, u^{(n)}) = (x, u^{(n)}, \gamma^{(n)}(x, u^{(n)}))$$

for a moving frame section.

The general invariantization procedure introduced in [32] in the finite-dimensional case adapts straightforwardly. To compute the invariantization of a function, differential form, differential operator, etc., one writes out how it explicitly transforms under the pseudo-group, and then replaces the pseudo-group parameters by their moving frame expressions (4.11). Invariantization thus defines a projection, depending upon the choice of cross-section or moving frame, from the spaces of general functions and forms to the spaces of invariant functions and forms. In particular, invariantizing the coordinate functions on J^∞ leads to the *normalized differential invariants*

$$H^i = \iota(x^i), \quad i = 1, \ldots, p, \quad I_J^\alpha = \iota(u_J^\alpha), \quad \alpha = 1, \ldots, q, \quad \#J \geq 0. \qquad (4.12)$$

These naturally split into two species: those appearing in the normalization equations (4.10) will be constant, and are known as the *phantom differential invariants*. The remaining $s_n = \dim J^n - r_n$ components, called the *basic differential invariants*, form a complete system of functionally independent differential invariants of order $\leq n$.

Secondly, invariantization of the basis horizontal one-forms leads to the invariant horizontal one-forms

$$\varpi^i = \iota(dx^i) = \omega^i + \kappa^i, \quad i = 1, \ldots, p, \tag{4.13}$$

where ω^i, κ^i are, respectively, the horizontal and vertical (contact) components. If the pseudo-group acts projectably, then the contact components vanish, $\kappa^i = 0$. Otherwise, the two components are not individually invariant, although the horizontal forms $\omega^1, \ldots, \omega^p$ are, in the language of [52], a contact-invariant coframe on J^∞.

The dual invariant differential operators $\mathcal{D}_1, \ldots, \mathcal{D}_p$ are uniquely defined by the formula

$$dF = \sum_{i=1}^{p} \mathcal{D}_i F \, \varpi^i + \cdots, \tag{4.14}$$

valid for any differential function F, where we omit the contact components (although these do play an important role in the study of invariant variational problems, [32]). The invariant differential operators map differential invariants to differential invariants. In general, they do not commute, but are subject to linear commutation relations of the form

$$[\mathcal{D}_i, \mathcal{D}_j] = \sum_{k=1}^{p} Y_{ij}^k \, \mathcal{D}_k, \quad i, j = 1, \ldots, p, \tag{4.15}$$

where the coefficients Y_{ij}^k are certain differential invariants that must also be determined. Finally, invariantizing the basis contact one-forms

$$\vartheta_K^\alpha = \iota(\theta_K^\alpha), \qquad \alpha = 1, \ldots, q, \quad \#K \geq 0, \tag{4.16}$$

provide a complete system of invariant contact one-forms. The invariant coframe serves to characterize the $\mathcal{G}$ invariant variational complex in the domain of the complete moving frame, [32].

Example 4.5 Consider the action of the pseudo-group (3.4) on surfaces $u = h(x, y)$. The pseudo-group maps the basis horizontal forms dx, dy to the one-forms

$$d_H X = f_x dx, \quad d_H Y = e_x dx + f_x dy. \tag{4.17}$$

By (4.7), the prolonged pseudo-group transformations are found by applying the dual implicit differentiations

$$D_X = \frac{1}{f_x} D_x - \frac{e_x}{f_x^2} D_y, \qquad D_Y = \frac{1}{f_x} D_y$$

successively to $U = u + e_x / f_x$, so that

$$U_X = \frac{u_x}{f_x} + \frac{e_{xx} - e_x u_y}{f_x^2} - 2\frac{f_{xx} e_x}{f_x^3}, \qquad U_Y = \frac{u_y}{f_x} + \frac{f_{xx}}{f_x^2},$$

$$U_{XX} = \frac{u_{xx}}{f_x^2} + \frac{e_{xxx} - e_{xx} u_y - 2 e_x u_{xy} - f_{xx} u_x}{f_x^3}$$

$$+ \frac{e_x^2 u_{yy} + 3 e_x f_{xx} u_y - 4 e_{xx} f_{xx} - 3 e_x f_{xxx}}{f_x^4} + 8\frac{e_x f_{xx}^2}{f_x^5}, \quad (4.18)$$

$$U_{XY} = \frac{u_{xy}}{f_x^2} + \frac{f_{xxx} - f_{xx} u_y - e_x u_{yy}}{f_x^3} - 2\frac{f_{xx}^2}{f_x^4}, \qquad U_{YY} = \frac{u_{yy}}{f_x^2},$$

$$U_{XYY} = \frac{f_x u_{xyy} - e_x u_{yyy} - 2 f_{xx} u_{yy}}{f_x^4}, \qquad U_{YYY} = \frac{u_{yyy}}{f_x^3},$$

and so on. In these formulas, the diffeomorphism jet coordinates f, f_x, f_{xx}, ..., e, e_x, e_{xx}, ... are to be regarded as the independent pseudo-group parameters. The pseudo-group cannot act freely on J^1 since $r_1 = \dim \mathcal{G}^{(1)}|_z = 6 > \dim J^1 = 5$. On the other hand, $r_2 = \dim \mathcal{G}^{(2)}|_z = 8 = \dim J^2$, and the action on J^2 is, in fact, locally free and transitive on the sets $\mathcal{V}_+^2 = J^2 \cap \{u_{yy} > 0\}$ and $\mathcal{V}_-^2 = J^2 \cap \{u_{yy} < 0\}$. Moreover, in accordance with Theorem 4.3, $\mathcal{G}^{(n)}$ acts locally freely on the corresponding open subsets of J^n for any $n \geq 2$.

To construct the moving frame, we adopt the following cross-section normalizations:

$$
\begin{aligned}
X &= 0 &\implies& \quad f = 0, \\
Y &= 0 &\implies& \quad e = 0, \\
U &= 0 &\implies& \quad e_x = -u f_x, \\
U_Y &= 0 &\implies& \quad f_{xx} = -u_y f_x, \\
U_X &= 0 &\implies& \quad e_{xx} = (u u_y - u_x) f_x, \\
U_{YY} &= 1 &\implies& \quad f_x = \sqrt{u_{yy}}, \\
U_{XY} &= 0 &\implies& \quad f_{xxx} = -\sqrt{u_{yy}}(u_{xy} + u u_{yy} - u_y^2), \\
U_{XX} &= 0 &\implies& \quad e_{xxx} = -\sqrt{u_{yy}}(u_{xx} - u u_{xy} - 2 u^2 u_{yy} - 2 u_x u_y + u u_y^2).
\end{aligned}
\qquad (4.19)
$$

At this stage, we have normalized enough pseudo-group parameters to compute the first two basic differential invariants by substituting the normalizations (4.19) into the transformation rules for u_{xyy}, u_{yyy} in (4.18), which yields the expressions

$$I_{12} = \iota(u_{xyy}) = \frac{u_{xyy} + u u_{yyy} + 2 u_y u_{yy}}{u_{yy}^{3/2}}, \qquad I_{03} = \iota(u_{yyy}) = \frac{u_{yyy}}{u_{yy}^{3/2}} \qquad (4.20)$$

for the invariants. Higher order differential invariants are found by continuing this process, or by using the more effective Taylor series method of [59]. Further, substituting the pseudo-group normalizations into (4.17) fixes the invariant horizontal coframe

$$\omega^1 = \iota(dx) = \sqrt{u_{yy}}\, dx, \qquad \omega^2 = \iota(dy) = \sqrt{u_{yy}}\,(dy - u dx).$$

The invariant contact forms are similarly constructed [59]. The dual invariant total derivative operators are

$$\mathcal{D}_1 = \frac{1}{\sqrt{u_{yy}}}\,(\mathrm{D}_x + u\,\mathrm{D}_y), \qquad \mathcal{D}_2 = \frac{1}{\sqrt{u_{yy}}}\,\mathrm{D}_y. \tag{4.21}$$

The higher-order differential invariants can be generated by successively applying these differential operators to the pair of basic differential invariants (4.20). The total derivative operators satisfy the commutation relation

$$[\mathcal{D}_1, \mathcal{D}_2] = -\frac{1}{2}\,I_{03}\mathcal{D}_1 + \frac{1}{2}\,I_{12}\mathcal{D}_2. \tag{4.22}$$

Finally, there is a single basic syzygy

$$\mathcal{D}_1 I_{03} - \mathcal{D}_2 I_{12} = 2 \tag{4.23}$$

among the differentiated invariants from which all others can be deduced by invariant differentiation.

5 Recurrence Formulas

The recurrence formulas [23, 32] connect the differentiated invariants and invariant forms with their normalized counterparts. These formulas are fundamental, since they prescribe the structure of the algebra of (local) differential invariants, underlying a full classification of generating differential invariants, their syzygies (differential identities), as well as the structure of invariant variational problems and, indeed, the local structure of the entire invariant variational bicomplex. As in the finite-dimensional version, the recurrence formulas are established using purely infinitesimal information, requiring only linear algebra and differentiation. In particular, they do *not* require the explicit formulas for either the moving frame, or the Maurer–Cartan forms, or the normalized differential invariants and invariant forms, or even the invariant differential operators! Beyond the formulas for the infinitesimal determining equations, the only additional information required is the specification of the moving frame cross-section.

Under the moving frame map, the pulled-back Maurer–Cartan forms will be denoted $\nu^\infty = (\rho^{(\infty)})^*\mu^\infty$, with individual components

$$\nu_A^b = (\rho^{(\infty)})^*(\mu_A^b), \quad b = 1, \ldots, m, \qquad \#A \geq 0. \tag{5.1}$$

As such, they are invariant one-forms, and so are invariant linear combinations of our invariant coframe elements ω^i, ϑ_K^α defined in (4.13), (4.16), with coefficients that are certain differential invariants.

Fortunately, the precise formulas need not be established a priori, as they will be a direct consequence of the recurrence formulas for the phantom differential invariants. We will extend our invariantization process to vector field coefficient jet coordinates (2.15) by defining

$$\iota(\zeta_A^b) = \nu_A^b, \qquad b = 1, \ldots, m, \qquad \#A \geq 0, \tag{5.2}$$

and, by extension, to their differential function and form-valued linear combinations,

$$\iota\left(\sum_{b=1}^{m}\sum_{\#A\leq n}\zeta_A^b\,\omega_b^A\right)=\sum_{b=1}^{m}\sum_{\#A\leq n}\nu_A^b\wedge\iota\big(\omega_b^A\big),\tag{5.3}$$

where the ω_b^A are k–forms on J^∞, so the result is an invariant differential $(k+1)$-form on J^∞. With this interpretation, the pulled-back Maurer–Cartan forms ν_A^b are subject to the linear relations

$$L^{(n)}(H,I,\nu^{(n)})=\iota\big[L^{(n)}(z,\zeta^{(n)})\big]=0,\qquad n\geq 0,\tag{5.4}$$

obtained by invariantizing the original linear determining equations (3.2). Here

$$(H,I)=\iota(x,u)=\iota(z)$$

are the zero[th] order differential invariants in (4.12).

Given a locally defined vector field

$$\mathbf{v}=\sum_{a=1}^{m}\zeta^a(z)\frac{\partial}{\partial z^a}=\sum_{i=1}^{p}\xi^i(x,u)\frac{\partial}{\partial x^i}+\sum_{\alpha=1}^{q}\varphi^\alpha(x,u)\frac{\partial}{\partial u^\alpha}\ \in\mathcal{X}(M),\tag{5.5}$$

let

$$\mathbf{v}^\infty=\sum_{i=1}^{p}\xi^i(x,u)\frac{\partial}{\partial x^i}+\sum_{\alpha=1}^{q}\sum_{n=\#J\geq 0}\widehat{\varphi}_J^\alpha(x,u^{(n)})\frac{\partial}{\partial u_J^\alpha}\ \in\mathcal{X}(J^\infty)\tag{5.6}$$

denote its infinite prolongation. The coefficients are computed via the usual prolongation formula,

$$\widehat{\varphi}_J^\alpha=\mathrm{D}_J\,Q^\alpha+\sum_{i=1}^{p}\xi^i u_{J,i}^\alpha,\qquad\text{where}\qquad Q^\alpha=\varphi^\alpha-\sum_{i=1}^{p}u_i^\alpha\,\xi^i,\qquad\alpha=1,\ldots,q,\tag{5.7}$$

are the components of the *characteristic* of $\mathbf{v}$, [51, 52].

Consequently, each prolonged vector field coefficient

$$\widehat{\varphi}_J^\alpha=\Phi_J^\alpha(u^{(n)},\zeta^{(n)})\tag{5.8}$$

is a certain universal linear combination of the vector field jet coordinates (2.15), whose coefficients are polynomials in the submanifold jet coordinates u_K^β for $1\leq\#K\leq n$. Therefore, the nth order prolongation of vector fields factors through the nth order vector field jet bundle. Let

$$\eta^i=\iota(\xi^i)=\nu^i,\qquad\widehat{\psi}_J^\alpha=\iota(\widehat{\varphi}_J^\alpha)=\Psi_J^\alpha(I^{(n)},\nu^{(n)}),\tag{5.9}$$

denote the invariantizations of the prolonged infinitesimal generator coefficients (5.8), which are linear combinations of the pulled-back Maurer–Cartan forms (5.2), with polynomial coefficients in the normalized differential invariants I_K^β for $1\leq\#K\leq\#J$.

With all these in hand, the desired *universal recurrence formula* is as follows.

Theorem 5.1 *If ω is any differential form on J^∞, then*

$$d\iota(\omega) = \iota(d\omega + \mathbf{v}^\infty(\omega)), \tag{5.10}$$

where $\mathbf{v}^\infty(\omega)$ denotes the Lie derivative of ω with respect to the prolonged vector field (5.6), and where we use (5.9) to invariantize the result.

Specializing ω in (5.10) to be one of the coordinate functions x^i, u_J^α yields recurrence formulas for the normalized differential invariants (4.12),

$$dH^i = \iota(dx^i + \xi^i) = \varpi^i + \eta^i,$$

$$dI_J^\alpha = \iota(du_J^\alpha + \widehat{\varphi}_J^\alpha) = \iota\left(\sum_{i=1}^p u_{J_i}^\alpha\, dx^i + \theta_J^\alpha + \widehat{\varphi}_J^\alpha\right) = \sum_{i=1}^p I_{J_i}^\alpha\, \varpi^i + \vartheta_J^\alpha + \widehat{\psi}_J^\alpha, \tag{5.11}$$

where $\widehat{\psi}_J^\alpha$ is written in terms of the pulled-back Maurer–Cartan forms ν_A^b as in (5.9), and are subject to the linear constraints (5.4). Each phantom differential invariant is, by definition, normalized to a constant value, and hence has zero differential. Consequently, the phantom recurrence formulas in (5.11) form a system of linear algebraic equations which can, as a result of the transversality of the cross-section, be uniquely solved for the pulled-back Maurer–Cartan forms.

Theorem 5.2 *If the pseudo-group acts locally freely on $V^n \subset J^n$, then the nth order phantom recurrence formulas can be uniquely solved in a neighborhood of the cross section to express the pulled-back Maurer–Cartan forms ν_A^b of order $\#A \leq n$ as invariant linear combinations of the invariant horizontal and contact one-forms $\varpi^i, \vartheta_J^\alpha$.*

Substituting the resulting expressions into the remaining, non-phantom recurrence formulas in (5.11) leads to a complete system of recurrence relations, for both the vertical and horizontal differentials of all the normalized differential invariants.

As the prolonged vector field coefficients $\widehat{\varphi}_J^\alpha$ are polynomials in the jet coordinates u_K^β of order $\#K \geq 1$, their invariantizations are polynomial functions of the differential invariants I_K^β for $\#K \geq 1$. Since the correction terms are constructed by solving a linear system for the invariantized Maurer–Cartan forms, the resulting coefficients are *rational functions* of these differential invariants. Thus, in most cases (including the majority of applications), the algebra of differential invariants for the pseudo-group is endowed with an entirely rational algebraic recurrence structure.

Theorem 5.3 *If $\mathcal{G}$ acts transitively on M, or, more generally, its infinitesimal generators depend polynomially on the coordinates $z = (x, u) \in M$, then the correction terms in the recurrence formulas (5.11) are rational functions of the normalized differential invariants.*

Example 5.4 In this example we let the pseudo-group (3.4) act on surfaces in $\mathbb{R}^3$. We will illustrate how the above ideas can be used to uncover recursion formulas for derivatives of the normalized differential invariants with respect to the invariant total derivative operators (4.21). For this only the horizontal components of equations (5.11) are needed, and we write these as

$$d_H H^1 = \omega^x + \chi^x, \qquad d_H H^2 = \omega^y + \chi^y,$$
$$d_H I_{ij} = I_{i+1,j}\,\omega^x + I_{i,j+1}\,\omega^y + \varkappa_{ij} \qquad i,j \geq 0. \tag{5.12}$$

In (5.12) each $\varkappa_{ij}$ is a linear combination of the horizontal pulled-back Maurer–Cartan forms which we denote

$$\chi^x_{X^a Y^b U^c}, \qquad \chi^y_{X^a Y^b U^c}, \qquad \chi^u_{X^a Y^b U^c}, \qquad a,b,c \geq 0. \tag{5.13}$$

These are again constrained by the invariantized infinitesimal determining equations $L^{(n)}(H, \chi^{(n)}) = 0$, and, as in Example 3.3, we see that a basis for the forms (5.13) is provided by $\chi^x_{X^n}, \chi^y_{X^n}, n \geq 0$, and, moreover, that

$$\chi^y_{X^n Y} = \chi^x_{X^{n+1}}, \qquad \chi^u_{X^n} = \chi^y_{X^{n+1}}, \qquad \chi^u_{X^n Y} = \chi^x_{X^{n+2}}, \qquad n \geq 0, \tag{5.14}$$

while all the other horizontal pulled-back Maurer–Cartan forms vanish.

We choose a cross section by imposing, in addition to (4.19), the normalization equations

$$U_{X^n} = 0, \qquad U_{X^{n-1}Y} = 0, \qquad n \geq 3. \tag{5.15}$$

It now follows from (5.14) that the horizontal correction terms $\varkappa_{ij}$ are precisely the coefficients of the horizontal invariantization of the vector field obtained by first prolonging the vector field

$$\mathbf{v} = \xi(x)\frac{\partial}{\partial x} + \eta(x,y)\frac{\partial}{\partial y} + \eta_x(x,y)\frac{\partial}{\partial u}$$

and then applying the relations

$$\eta_y = \xi_x, \qquad \xi_y = \xi_u = \eta_u = 0$$

and their differential consequences to express the resulting coefficient functions solely in terms of the repeated x-derivatives of ξ and η. With this, the u_{ij}-component $\widehat{\varphi}_{ij}$ of pr $\mathbf{v}$ becomes

$$\widehat{\varphi}_{ij} = \delta_{oj}\,\eta_{x^{i+1}} + \delta_{1j}\,\xi_{x^{i+2}}$$

$$- \sum_{s=1}^{i+1} \frac{i+1+(j-1)s}{i+1}\binom{i+1}{s}\xi_{x^s}u_{i-s+1,j} - \sum_{s=1}^{i}\binom{i}{s}\eta_{x^s}u_{i-s,j+1},$$

yielding the invariantization

$$\varkappa_{ij} = \delta_{oj}\,\chi^y_{X^{i+1}} + \delta_{1j}\,\chi^x_{X^{i+2}}$$

$$- \sum_{s=1}^{i+1} \frac{i+1+(j-1)s}{i+1}\binom{i+1}{s}I_{i-s+1,j}\,\chi^x_{X^s} - \sum_{s=1}^{i}\binom{i}{s}I_{i-s,j+1}\,\chi^y_{X^s}.$$

Taking into account normalizations (4.19), (5.15), the phantom recurrence formulas for the horizontal derivatives reduce to

$$
\begin{aligned}
&0 = d_H H^x = \omega^x + \chi^x, && 0 = d_H H^y = \omega^y + \chi^y, \\
&0 = d_H I = \chi^y_X, && 0 = d_H I_{10} = \chi^y_{X^2}, \\
&0 = d_H I_{01} = \omega^y + \chi^x_{X^2}, && 0 = d_H I_{20} = \chi^y_{X^3}, && (5.16)\\
&0 = d_H I_{11} = I_{12}\omega^y - \chi^y_X + \chi^x_{X^3}, && 0 = d_H I_{02} = I_{12}\omega^x + I_{03}\omega^y - 2\chi^x_X, \\
&0 = d_H I_{30} = \chi^y_{X^4}, && 0 = d_H I_{21} = I_{22}\omega^y + \chi^x_{X^4} - \chi^y_{X^2} - 2I_{12}\chi^y_X,
\end{aligned}
$$

and so on. These can be solved for the Maurer–Cartan forms $\chi^x_{X^n}, \chi^y_{X^n}$ to yield

$$
\begin{aligned}
&\chi^x = -\omega^x, && \chi^y = -\omega^y, && \chi^x_X = \tfrac{1}{2}(I_{12}\omega^x + I_{03}\omega^y), && \chi^y_X = 0, \\
&\chi^x_{X^2} = -\omega^y, && \chi^y_{X^2} = 0, && \chi^x_{X^3} = -I_{12}\omega^y, && \chi^y_{X^3} = 0, && (5.17)\\
&\chi^x_{X^4} = -I_{22}\omega^y, && \chi^y_{X^4} = 0, && \dots
\end{aligned}
$$

Next we substitute the expressions (5.17) in the recurrence formulas for the non-phantom invariants in (5.12). Keeping in mind that

$$
d_H I_{ij} = (\mathcal{D}_1 I_{ij})\omega^x + (\mathcal{D}_2 I_{ij})\omega^y,
$$

the ω^x-, ω^y-components of the resulting equations yield the expressions

$$
\begin{aligned}
&\mathcal{D}_1 I_{12} = I_{22} - \tfrac{3}{2}I_{12}^2, && \mathcal{D}_2 I_{12} = I_{13} - \tfrac{3}{2}I_{12}I_{03} + 2, \\
&\mathcal{D}_1 I_{03} = I_{13} - \tfrac{3}{2}I_{12}I_{03}, && \mathcal{D}_2 I_{03} = I_{04} - \tfrac{3}{2}I_{03}^2, \\
&\mathcal{D}_1 I_{22} = I_{32} - 2I_{12}I_{22}, && \mathcal{D}_2 I_{22} = I_{23} - 2I_{03}I_{22} + 7I_{12}, && (5.18)\\
&\mathcal{D}_1 I_{13} = I_{23} - 2I_{12}I_{13}, && \mathcal{D}_2 I_{13} = I_{14} - 2I_{03}I_{13} + 3I_{03}, \\
&\mathcal{D}_1 I_{04} = I_{14} - 2I_{12}I_{04}, && \mathcal{D}_2 I_{04} = I_{05} - 2I_{03}I_{04},
\end{aligned}
$$

$$
\vdots
$$

It becomes apparent from these formulas that the invariants I_{12} and I_{03} generate the algebra of differential invariants for the pseudo-group (3.4).

A derivation of the recurrence formulas (5.18) using Taylor series methods can be found in [59].

6   Algebra of Differential Invariants

Establishment of the theoretical results that underpin our constructive algorithms relies on the interplay between two algebraic structures associated with the two jet bundles: the first is defined by the symbols of the linearized determining equations for the infinitesimal generator jets; the second from a similar construction for the prolonged generators on the submanifold jet bundle. We begin with the former, which is the more familiar of the two.

We introduce algebraic variables $t = t_1, \ldots, t_m$ and $T = T^1, \ldots, T^m$. Let

$$\mathcal{T} = \left\{ \eta(t, T) = \sum_{a=1}^{m} \eta_a(t) \, T^a \right\} \simeq \mathbb{R}[t] \otimes \mathbb{R}^m \tag{6.1}$$

denote the $\mathbb{R}[t]$ module consisting of real polynomials that are linear in the T's. We grade $\mathcal{T} = \oplus_{n \geq 0} \mathcal{T}^n$, where $\mathcal{T}^n$ consists of the homogeneous polynomials of *degree* n in t. We set $\mathcal{T}^{\leq n} = \oplus_{k=0}^{n} \mathcal{T}^k$ to be the space of polynomials of degree $\leq n$.

Given a subspace $\mathcal{I} \subset \mathcal{T}$, we set $\mathcal{I}^n = \mathcal{I} \cap \mathcal{T}^n$, $\mathcal{I}^{\leq n} = \mathcal{I} \cap \mathcal{T}^{\leq n}$. The subspace is *graded* if $\mathcal{I} = \oplus_{n \geq 0} \mathcal{I}^n$ is the sum of its homogeneous constituents. A subspace $\mathcal{I} \subset \mathcal{T}$ is a *submodule* if the product $\lambda(t) \, \eta(t, T) \in \mathcal{I}$ whenever $\eta(t, T) \in \mathcal{I}$ and $\lambda(t) \in \mathbb{R}[t]$. A subspace $\mathcal{I} \subset \mathcal{T}$ spanned by monomials $t_A T^b = t_{a_1} \cdots t_{a_n} T^b$ is called a *monomial subspace*, and is automatically graded. In particular, a *monomial submodule* is a submodule that is spanned by monomials. A polynomial $\eta \in \mathcal{T}^{\leq n}$ has *degree* $n = \deg \eta$ and *highest order terms* $\mathbf{H}(\eta) \in \mathcal{T}^n$ provided $\eta = \mathbf{H}(\eta) + \widetilde{\eta}$, where $\mathbf{H}(\eta) \neq 0$ and $\widetilde{\eta} \in \mathcal{T}^{\leq n-1}$ is of lower degree. By convention, only the zero polynomial has zero highest order term.

We will assume in the remaining part of the paper that all our functions, vector fields etc. are analytic. We can locally identify the dual bundle to the infinite order jet bundle as $(\mathcal{X}^\infty)^* \simeq M \times \mathcal{T}$ via the pairing $\langle j_z^\infty \mathbf{v} \, ; t_A T^b \rangle = \zeta_A^b$. A parametrized polynomial

$$\eta(z; t, T) = \sum_{b=1}^{m} \sum_{\#A \leq n} h_b^A(z) \, t_A T^b \tag{6.2}$$

corresponds to an nth order linear differential function

$$L(z, \zeta^{(n)}) = \langle j_z^\infty \mathbf{v} \, ; \eta(z; t, T) \rangle = \sum_{b=1}^{m} \sum_{\#A \leq n} h_b^A(z) \, \zeta_A^b. \tag{6.3}$$

Its *symbol* consists, by definition, of the highest order terms in its defining polynomial,

$$\mathbf{\Sigma}(L(z, \zeta^{(n)})) = \mathbf{H}(\eta(z; t, T)) = \sum_{b=1}^{m} \sum_{\#A = n} h_b^A(z) \, t_A T^b. \tag{6.4}$$

The symbol of a linear differential function is well defined by our assumption that the coefficient functions $h_b^A(z)$ be analytic.

Given a pseudo-group $\mathcal{G}$, let $\mathcal{L} = (\mathfrak{g}^\infty)^\perp \subset (\mathcal{X}^\infty)^*$ denote the *annihilator subbundle* of its infinitesimal generator jet bundle. Each equation in the linear determining system (3.2) is represented by a parametrized polynomial (6.2), which cumulatively span $\mathcal{L}$. Let $\mathcal{I} = \mathbf{H}(\mathcal{L}) \subset (\mathcal{X}^\infty)^*$ denote the associated *symbol subbundle* (assuming regularity). On the symbol level, total derivative corresponds to multiplication,

$$\mathbf{\Sigma}(\mathbb{D}_{z^a} L) = t_a \, \mathbf{\Sigma}(L), \qquad a = 1, \ldots, m. \tag{6.5}$$

Thus, formal integrability of (3.2) implies that each fiber $\mathcal{I}_{|z} \subset \mathcal{T}$ forms a graded submodule, known as the *symbol module* of the pseudo-group at the point $z \in M$. On the other hand, the annihilator $\mathcal{L}_{|z} \subset \mathcal{T}$ is typically *not* a submodule.

We now develop an analogous symbol algebra for the prolonged infinitesimal generators. We introduce variables $s = s_1, \ldots, s_p$, $S = S^1, \ldots, S^q$, and let

$$\widehat{\mathcal{S}} = \left\{ \widetilde{\sigma}(s, S) = \sum_{\alpha=1}^{q} \widetilde{\sigma}_\alpha(s) \, S^\alpha \right\} \simeq \mathbb{R}[s] \otimes \mathbb{R}^q \subset \mathbb{R}[s, S] \tag{6.6}$$

be the $\mathbb{R}[s]$ module consisting of polynomials which are linear in S. Let

$$\mathcal{S} = \mathbb{R}^p \oplus \widehat{\mathcal{S}} = \bigoplus_{n=-1}^{\infty} \widehat{\mathcal{S}}^n \tag{6.7}$$

consist of the "extended" polynomials

$$\sigma(s, S, \widetilde{s}) = \sum_{i=1}^{p} c_i \widetilde{s}_i + \widetilde{\sigma}_\alpha(s, S) = \sum_{i=1}^{p} c_i \widetilde{s}_i + \sum_{\alpha=1}^{q} \widetilde{\sigma}_\alpha(s) \, S^\alpha, \tag{6.8}$$

where $\widetilde{s} = \widetilde{s}_1, \ldots, \widetilde{s}_p$ are extra algebraic variables, $c_1, \ldots, c_p \in \mathbb{R}$, and $\widetilde{\sigma}(s, S) \in \widehat{\mathcal{S}}$.

In coordinates, we can identify $T^* J^\infty \simeq J^\infty \times \mathcal{S}$ via the pairing

$$\langle \, \mathbf{V} \, ; \widetilde{s}_i \, \rangle = \xi^i, \qquad \langle \, \mathbf{V} \, ; S^\alpha \, \rangle = Q^\alpha = \varphi^\alpha - \sum_{i=1}^{p} u_i^\alpha \, \xi^i, \tag{6.9}$$

$$\langle \, \mathbf{V} \, ; s_J S^\alpha \, \rangle = \widehat{\varphi}_J^\alpha, \qquad \#J \geq 1,$$

for any tangent vector

$$\mathbf{V} = \sum_{i=1}^{p} \xi^i \frac{\partial}{\partial x^i} + \sum_{\alpha=1}^{q} \sum_{k=\#J \geq 0} \widehat{\varphi}_J^\alpha \frac{\partial}{\partial u_J^\alpha} \in T J^\infty.$$

Every one-form on J^∞ is thereby represented, locally, by a parametrized polynomial

$$\sigma(x, u^{(n)}; \widetilde{s}, s, S) = \sum_{i=1}^{p} h_i(x, u^{(n)}) \, \widetilde{s}_i + \sum_{\alpha=1}^{q} \widetilde{\sigma}_\alpha(x, u^{(n)}; s) \, S^\alpha,$$

depending linearly on the variables $(\widetilde{s}, S) \in \mathbb{R}^m$, polynomially on the variables $s \in \mathbb{R}^p$, and analytically on the jet coordinates $(x, u^{(n)})$ of some finite order $n < \infty$.

Given $z^{(\infty)} \in J^\infty$, let

$$\mathbf{p} = \mathbf{p}^{(\infty)} : J^\infty T M_{|z} \longrightarrow T_{z^{(\infty)}} J^\infty, \qquad \mathbf{p}(j_z^\infty \mathbf{v}) = \mathbf{v}^\infty \, {}_{|z^{(\infty)}}, \tag{6.10}$$

denote the *prolongation map* that takes the jet of a vector field at the base point $z = \widetilde{\pi}_0^\infty(z^{(\infty)}) \in M$ to its prolongation (5.6). Let $\mathbf{p}^* : \mathcal{S} \to \mathcal{T}$ be the corresponding *dual prolongation map*, defined so that

$$\langle \, j^\infty \mathbf{v} \, ; \mathbf{p}^*(\sigma) \, \rangle = \langle \, \mathbf{p}(j^\infty \mathbf{v}) \, ; \sigma \, \rangle \tag{6.11}$$

for all $j_z^\infty \mathbf{v} \in J^\infty TM_{|z}$, $\sigma \in \mathcal{S}$. In general, $\mathbf{p}^*$ is *not* a module morphism. However, on the symbol level it essentially is, as we now explain.

Consider the particular linear polynomials

$$\beta_i(t) = t_i + \sum_{\alpha=1}^{q} u_i^\alpha \, t_{p+\alpha}, \qquad B^\alpha(T) = T^{p+\alpha} - \sum_{i=1}^{p} u_i^\alpha \, T^i, \tag{6.12}$$

for $i = 1, \ldots, p$, $\alpha = 1, \ldots, q$, where $u_i^\alpha = \partial u^\alpha / \partial x^i$ are the first order jet coordinates of our point $z^{(\infty)}$. Note that $B^\alpha(T)$ is the symbol of Q^α, the αth component of the characteristic of $\mathbf{v}$, cf. (6.9), while $\beta_i(t)$ represents the symbol of the total derivative, $\Sigma(\mathrm{D}_i L) = \beta_i(t) \, \Sigma(L)$. The functions $s_i = \beta_i(t), S^\alpha = B^\alpha(T)$ serve to define a linear map $\boldsymbol{\beta} : \mathbb{R}^{2m} \to \mathbb{R}^m$.

Since $\boldsymbol{\beta}$ has maximal rank, the induced pull-back map

$$(\boldsymbol{\beta}^* \sigma)(t_1, \ldots, t_n, T^1, \ldots, T^n) = \sigma(\beta_1(t), \ldots, \beta_p(t), B^1(T) \ldots, B^q(T)) \tag{6.13}$$

defines an injection $\boldsymbol{\beta}^* : \widehat{\mathcal{S}} \to \mathcal{T}$. The algebraic structure of the vector field prolongation map at the symbol level is encapsulated in the following result.

Lemma 6.1 *The symbols of the prolonged vector field coefficients are*

$$\Sigma(\xi^i) = T^i, \quad \Sigma(\varphi^\alpha) = T^{\alpha+p}, \quad \Sigma(Q^\alpha) = \boldsymbol{\beta}^*(S^\alpha), \quad \Sigma(\widehat{\varphi}_J^\alpha) = \boldsymbol{\beta}^*(s_J S^\alpha). \tag{6.14}$$

Now given a Lie pseudo-group $\mathcal{G}$ acting on M, let $\mathfrak{g}^\infty_{|z^{(\infty)}} = \mathbf{p}(J^\infty \mathfrak{g}_{|z}) \subset T_{z^{(\infty)}} J^\infty$ denote the subspace[1] spanned by its prolonged infinitesimal generators. Let

$$\mathcal{Z}_{|z^{(\infty)}} = (\mathfrak{g}^\infty_{|z^{(\infty)}})^\perp = (\mathbf{p}^*)^{-1}(\mathcal{L}_{|z}) \subset \mathcal{S} \tag{6.15}$$

denote the *prolonged annihilator subbundle*, containing those polynomials (6.8) that annihilate all prolonged infinitesimal generators $\mathbf{v}^\infty \in \mathfrak{g}^\infty_{|z^{(\infty)}}$. Further, let $\mathcal{U}_{|z^{(\infty)}} = \mathbf{H}(\mathcal{Z}_{|z^{(\infty)}}) \subset S$ be the subspace spanned by the highest order terms of the prolonged annihilators. In general $\mathcal{U}_{|z^{(\infty)}}$ is *not* a submodule, as, for instance, is the case with the pseudo-group discussed in Example 4.5.

Let

$$\mathcal{J}_{|z^{(1)}} = (\boldsymbol{\beta}^*)^{-1}(\mathcal{I}_{|z}) = \{\sigma(s,S) \mid \boldsymbol{\beta}^*(\sigma)(t,T) = \sigma(\beta(t), B(T)) \in \mathcal{I}_{|z}\} \subset \widehat{\mathcal{S}}, \tag{6.16}$$

where $z = \widetilde{\pi}_o^1(z^{(1)})$, be the *prolonged symbol submodule*, the inverse image of the symbol module under the polynomial pull-back morphism (6.13). In view of (6.15), $\mathcal{U}_{|z^{(\infty)}} \subset \mathcal{J}_{|z^{(1)}}$, where $z^{(1)} = \widetilde{\pi}_1^\infty(z^{(\infty)})$. Moreover, assuming local freeness, these two spaces agree at sufficiently high order, thereby endowing $\mathcal{U}_{|z^{(\infty)}}$ with the structure of an "eventual submodule".

Lemma 6.2 *If $\mathcal{G}^{(n)}$ acts locally freely at $z^{(n)} \in J^n$, then $\mathcal{U}^k_{|z^{(k)}} = \mathcal{J}^k_{|z^{(k)}}$ for all $k > n$ and all $z^{(k)} \in J^k$ with $\widetilde{\pi}_n^k(z^{(k)}) = z^{(n)}$.*

---

[1] In general, $\mathfrak{g}^\infty$ may only be a regular subbundle on a (dense) open subset of jet space. We restrict our attention to this subdomain throughout.

Let us fix a degree compatible term ordering on the polynomial module $\widehat{S}$, which we extend to S by making the extra monomials $\widetilde{s}_i$ be ordered before all the others. At a fixed regular submanifold jet $z^{(\infty)}$, let $\mathcal{N}_{|z^{(\infty)}}$ be the monomial subspace generated by the leading monomials of the polynomials in $\mathcal{Z}_{|z^{(\infty)}}$, or, equivalently, the annihilator symbol polynomials in $\mathcal{U}_{|z^{(\infty)}}$. Let $\mathcal{K}_{|z^{(\infty)}}$ denote the *complementary monomial subspace* spanned by all monomials in S that are *not* in $\mathcal{N}_{|z^{(\infty)}}$, which can be constructed by applying the usual Gröbner basis algorithm [17] to $\mathcal{J}_{|z^{(\infty)}}$ and then possibly supplementing the resulting complementary (or standard) monomials by any additional ones required at orders $\leq n^*$, where J^{n^*} stands for the lowest order jet space on which the pseudo-group $\mathcal{G}$ acts locally freely.

Each monomial in $\mathcal{K}_{|z^{(\infty)}}$ corresponds to a submanifold jet coordinate x^i or u_J^a. For each $0 \leq n \leq \infty$, we let $K^n \subset J^n$ denote the coordinate cross-section passing through $z^{(n)} = \widetilde{\pi}_n^\infty(z^{(\infty)})$ prescribed by the monomials in $\mathcal{K}^{\leq n}_{|z^{(\infty)}}$. For the corresponding "algebraic" moving frame, each normalized differential invariant is indexed by a monomial in S, with H^i corresponding to $\widetilde{s}_i$, and I_J^α to $s_J S^\alpha$. The monomials in $\mathcal{N}_{|z^{(\infty)}}$ index the complete system of functionally independent basic differential invariants, whereas the complementary monomials in $\mathcal{K}_{|z^{(\infty)}}$ index the constant phantom differential invariants.

However, at this stage a serious complication emerges: Because the invariantized Maurer–Cartan forms are found by solving the recurrence formulas for the phantom invariants, their coefficients may depend on $(n+1)^{\text{th}}$ order differential invariants, and hence the correction term in the resulting recurrence formula for dI_J^α can have the same order as the leading term; see, for example, the recurrence formulas for the KP symmetry algebra derived in [15]. As a consequence, the recurrence formulas (5.11) for the non-phantom invariants do not in general directly yield the leading order normalized differential invariants in terms of lower order invariants and their invariant derivatives. However, this complication, which does not arise in the finite dimensional situation [23], can be circumvented by effectively invariantizing all the constructs in this section and by introducing an alternative collection of generating invariants that is better adapted to the underlying algebraic structure of the prolonged symbol module.

To each parametrized symbol polynomial

$$\widetilde{\sigma}(\mathbf{I}^{(k)}; s, S) = \sum_{\alpha=1}^q \sum_{\#J \leq n} h_\alpha^J(\mathbf{I}^{(k)}) \, s_J S^\alpha \in \widehat{S}, \tag{6.17}$$

whose coefficients depend on differential invariants $\mathbf{I}^{(k)} = \iota(x, u^{(k)})$ of order $\leq k$, we associate a differential invariant

$$I_{\widetilde{\sigma}} = \sum_{\alpha=1}^q \sum_{\#J \leq n} h_\alpha^J(\mathbf{I}^{(k)}) \, I_J^\alpha. \tag{6.18}$$

Moreover, let $\widetilde{\mathcal{J}} = \iota(\mathcal{J}_{|z^{(1)}})$ denote the *invariantized prolonged symbol submodule* obtained by invariantizing all the polynomials in $\mathcal{J}_{|z^{(1)}}$. If $\mathcal{G}$ acts transitively on an open subset of J^1, then $\widetilde{\mathcal{J}}$ is a fixed module, since the map β only depends on first order jet coordinates.

As a consequence of Lemma 6.2, every *homogeneous* polynomial $\widetilde{\sigma}(\mathbf{I}^{(1)}; s, S) \in \widetilde{\mathcal{J}}$, for $n > n^*$, is the leading term of an annihilating polynomial

$$\widetilde{\tau}(\mathbf{I}^{(1)}; s, S) = \widetilde{\sigma}(\mathbf{I}^{(1)}; s, S) + \widetilde{\nu}(\mathbf{I}^{(1)}; s, S) \tag{6.19}$$

contained in the invariantized version $\widetilde{\mathcal{Z}}$ of the prolonged annihilator bundle $\mathcal{Z}$. For such polynomials, the recurrence formula (5.11) reduces to

$$d_H I_{\widetilde{\sigma}} = \sum_{i=1}^{p} (I_{s_i \widetilde{\sigma}} + I_{\mathcal{D}_i \widetilde{\sigma}}) \omega^i - \langle \psi^\infty ; \widetilde{\nu} \rangle. \tag{6.20}$$

In contrast to the original recurrence formulas, for $n > n^*$, the *correction term*

$$\sum_{i=1}^{p} I_{\mathcal{D}_i \widetilde{\sigma}} \omega^i - \langle \psi^\infty ; \widetilde{\nu} \rangle \tag{6.21}$$

in the algebraically adapted recurrence formula (6.20) *is* of lower order than the leading term $I_{s_i \widetilde{\sigma}}$. Equating the coefficients of the forms ω^i in (6.20) leads to individual recurrence formulas

$$\mathcal{D}_i I_{\widetilde{\sigma}} = I_{s_i \widetilde{\sigma}} + R_{\widetilde{\sigma}, i}, \tag{6.22}$$

in which, as long as $n = \deg \sigma_i > n^*$, the *leading term* $I_{s_i \widetilde{\sigma}}$ is a differential invariant of order $= n + 1$, while the *correction term* $R_{\widetilde{\sigma}, i}$ is of strictly lower order.

With this in hand, we arrive at the Constructive Basis and Syzygy Theorems governing the differential invariant algebra of an eventually locally freely acting pseudo-group [60].

Theorem 6.3 *Let $\mathcal{G}$ be a Lie pseudo-group that acts locally freely on the submanifold jet bundle at order n^*. Then the following constitute a finite generating system for its differential invariant algebra:*

1. *the differential invariants $I_\nu = I_{\sigma_\nu}$, where $\sigma_1, \ldots, \sigma_l$ form a Gröbner basis for the submodule $\widetilde{\mathcal{J}}$ relative to our chosen term ordering, and, possibly,*

2. *a finite number of additional differential invariants of order $\leq n^*$.*

Theorem 6.4 *Every differential syzygy among the generating differential invariants is either a syzygy among those of order $\leq n^*$, or arises from an algebraic syzygy among the Gröbner basis polynomials in $\widetilde{\mathcal{J}}$. Therefore, the differential syzygies are all generated by a finite number of rational syzygies, corresponding to the generators of the syzygy module of $\widetilde{\mathcal{J}}$ plus possibly a finite number of additional syzygies of order $\leq n^*$.*

Applications of these results can be found in our papers listed in the references.

Bibliography

[1] I. M. Anderson, *The Variational Bicomplex*, Utah State Technical Report, 1989, http://math.usu.edu/~fg_mp.

[2] V. I. Arnold, B. A. Khesin, *Topological hydrodynamics*, Springer, New York, 1998.

[3] P.–L. Bazin, M. Boutin, *Structure from motion: theoretical foundations of a novel approach using custom built invariants* SIAM J. Appl. Math. 64 (2004), pp. 1156–1174.

[4] I. A. Berchenko, P. J. Olver, *Symmetries of polynomials*, J. Symb. Comp., 29 (2000), pp. 485–514.

[5] N. Bílă, E. L. Mansfield, P. A. Clarkson, *Symmetry group analysis of the shallow water and semi-geostrophic equations*, Quart. J. Mech. Appl. Math., to appear.

[6] D. Bleecker, *Gauge Theory and Variational Principles*, Addison-Wesley, Reading, Mass., 1981.

[7] M. Boutin, *Numerically invariant signature curves*, Int. J. Computer Vision 40 (2000), pp. 235–248.

[8] ———, *On orbit dimensions under a simultaneous Lie group action on n copies of a manifold*, J. Lie Theory 12 (2002), pp. 191–203.

[9] R. L. Bryant, S.-S. Chern, R. B. Gardner, H. L. Goldschmidt, P. A. Griffiths, *Exterior Differential Systems*, Math. Sci. Res. Inst. Publ., Vol. 18, Springer-Verlag, New York, 1991.

[10] E. Calabi, P. J. Olver, C. Shakiban, A. Tannenbaum, S. Haker, *Differential and numerically invariant signature curves applied to object recognition*, Int. J. Computer Vision 26 (1998), pp. 107–135.

[11] É. Cartan, *La Méthode du Repère Mobile, la Théorie des Groupes Continus, et les Espaces Généralisés*, Exposés de Géométrie, no. 5, Hermann, Paris, 1935.

[12] ———, *Sur la structure des groupes infinis de transformations*, Oeuvres Complètes; Part. II, Vol. 2, Gauthier-Villars, Paris, 1953, pp. 571–714.

[13] ———, *La structure des groupes infinis*, Oeuvres Complètes; part. II, vol. 2, Gauthier-Villars, Paris, 1953, pp. 1335–1384.

[14] J. Cheh, P. J. Olver, J. Pohjanpelto, *Maurer-Cartan equations for Lie symmetry pseudo-groups of differential equations*, J. Math. Phys. 46 (2005), 023504.

[15] ———, *Algorithms for differential invariants of symmetry groups of differential equations*, Found. Comput. Math., to appear.

[16] S.-S. Chern, J. K. Moser, *Real hypersurfaces in complex manifolds*, Acta Math. 133 (1974), pp. 219–271; also *Selected Papers*, vol. 3, Springer-Verlag, New York, 1989, pp. 209–262.

[17] D. Cox, J. Little, D. O'Shea, *Ideals, Varieties, and Algorithms*, 2nd ed., Springer-Verlag, New York, 1996.

[18] D. David, N. Kamran D. Levi, P. Winternitz, *Subalgebras of loop algebras and symmetries of the Kadomtsev-Petviashivili equation*, Phys. Rev. Lett. 55 (1985), pp. 2111–2113.

[19] P. Di Francesco, P. Mathieu, D. Sénéchal, *Conformal Field Theory*, Springer-Verlag, New York, 1997.

[20] C. Ehresmann, *Introduction à la théorie des structures infinitésimales et des pseudo-groupes de Lie*, Géometrie Différentielle; Colloq. Inter. du Centre Nat. de la Rech. Sci., Strasbourg, 1953, pp. 97–110.

[21] C. Fefferman, C. R. Graham, *Conformal invariants*, Élie Cartan et les Mathématiques d'aujourd'hui; Astérisque, hors série, Soc. Math. France, Paris, 1985, pp. 95–116.

[22] M. Fels, P. J. Olver, *Moving coframes. I. A practical algorithm*, Acta Appl. Math. 51 (1998), pp. 161–213.

[23] ———, *Moving coframes. II. Regularization and theoretical foundations*, Acta Appl. Math. 55 (1999), pp. 127–208.

[24] D. B. Fuchs, A. M. Gabrielov, I. M. Gel'fand, *The Gauss-Bonnet theorem and Atiyah-Patodi-Singer functionals for the characteristic classes of foliations*, Topology 15 (1976), pp. 165–188.

[25] H. W. Guggenheimer, *Differential Geometry*, McGraw-Hill, New York, 1963

[26] E. Hubert, *The AIDA Maple package*, http://www.inria.fr/cafe/Evelyne.Hubert/aida, 2006.

[27] E. Hubert, I. A. Kogan, *Rational invariants of an algebraic group action. Construction and rewriting*, J. Symb. Comp., to appear.

[28] E. Hubert, I. A. Kogan, *Smooth and algebraic invariants of a group action. Local and global constructions*, Found. Comput. Math., to appear.

[29] N. Kamran, *Contributions to the study of the equivalence problem of Elie Cartan and its applications to partial and ordinary differential equations*, Mém. Cl. Sci. Acad. Roy. Belg. 45 (1989) Fac. 7.

[30] N. Kamran, T. Robart, *A manifold structure for analytic isotropy Lie pseudogroups of infinite type*, J. Lie Theory 11 (2001), pp. 57–80.

[31] P. Kim, *Invariantization of Numerical Schemes for Differential Equations Using Moving Frames*, Ph.D. Thesis, University of Minnesota, Minneapolis, (2006).

[32] I. A. Kogan, P. J. Olver, *Invariant Euler-Lagrange equations and the invariant variational bicomplex*, Acta Appl. Math. 76 (2003), pp. 137–193.

[33] B. Kruglikov, V. Lychagin, *Invariants of pseudogroup actions: homological methods and finiteness theorem*, arXiv: math.DG/0511711, 2005.

[34] A. Kumpera, *Invariants différentiels d'un pseudogroupe de Lie*, J. Diff. Geom. 10 (1975), pp. 289–416.

[35] M. Kuranishi, *On the local theory of continuous infinite pseudo groups I*, Nagoya Math. J. 15 (1959), pp. 225–260.

[36] ———, *On the local theory of continuous infinite pseudo groups II*, Nagoya Math. J. 19 (1961), pp. 55–91.

[37] S. Lie, *Die Grundlagen für die Theorie der unendlichen kontinuierlichen Transformationsgruppen*, Leipzig. Ber. 43 (1891), pp. 316–393; also *Gesammelte Abhandlungen*, Vol. 6, B. G. Teubner, Leipzig, 1927, pp. 300–364.

[38] I. G. Lisle, G. J. Reid, *Geometry and structure of Lie pseudogroups from infinitesimal defining systems*, J. Symb. Comp. 26 (1998), pp. 355–379.

[39] ———, *Cartan structure of infinite Lie pseudogroups*, Geometric Approaches to Differential Equations (P. J. Vassiliou and I. G. Lisle, eds.), Austral. Math. Soc. Lect. Ser., 15, Cambridge Univ. Press, Cambridge, 2000, pp. 116–145.

[40] K. Mackenzie, *Lie Groupoids and Lie Algebroids in Differential Geometry*, London Math. Soc. Lecture Notes, vol. 124, Cambridge Univ. Press, Cambridge, 1987.

[41] E. L. Mansfield, *Algorithms for symmetric differential systems*, Found. Comput. Math. 1 (2001), pp. 335–383.

[42] G. Marí Beffa, *Relative and absolute differential invariants for conformal curves*, J. Lie Theory, 13 (2003), pp. 213–245.

[43] ———, *Poisson geometry of differential invariants of curves in some nonsemisimple homogeneous spaces*, Proc. Amer. Math. Soc., 134 (2006), pp. 779–791.

[44] G. Marí Beffa, P. J. Olver, *Differential invariants for parametrized projective surfaces*, Commun. Anal. Geom. 7 (1999), pp. 807–839.

[45] R. I. McLachlan, G. R. W. Quispel, *What kinds of dynamics are there? Lie pseudogroups, dynamical systems and geometric integration*, Nonlinearity 14 (2001), pp. 1689–1705.

[46] L. Martina, M. B. Sheftel, P. Winternitz, *Group foliation and non-invariant solutions of the heavenly equation*, J. Phys. A, 34 (2001), pp. 9243–9263.

[47] P. Medolaghi, *Classificazione delle equazioni alle derivate parziali del secondo ordine, che ammettono un gruppo infinito di trasformazioni puntuali*, Ann. Mat. Pura Appl. 1 (3) (1898), pp. 229–263.

[48] O. Morozov, *Moving coframes and symmetries of differential equations*, J. Phys. A 35 (2002), pp. 2965–2977.

[49] ———, *Structure of symmetry groups via Cartan's method: survey of four approaches*, SIGMA: Symmetry Integrability Geom. Methods Appl. 1 (2005), paper 006.

[50] Y. Nutku, M. B. Sheftel, *Differential invariants and group foliation for the complex Monge-Ampère equation*, J. Phys. A; 34 (2001), pp. 137–156.

[51] P. J. Olver, *Applications of Lie Groups to Differential Equations*, Second Edition, Graduate Texts in Mathematics, vol. 107, Springer-Verlag, New York, 1993.

[52] ———, *Equivalence, Invariants, and Symmetry*, Cambridge University Press, Cambridge, 1995.

[53] ———, *Classical Invariant Theory*, London Math. Soc. Student Texts, vol. 44, Cambridge University Press, Cambridge, 1999.

[54] ———, *Joint invariant signatures*, Found. Comput. Math. 1 (2001), pp. 3–67.

[55] ———, *Geometric foundations of numerical algorithms and symmetry*, Appl. Alg. Engin. Commun. Comput. 11 (2001), pp. 417–436.

[56] ———, *Generating differential invariants*, J. Math. Anal. Appl., to appear.

[57] P. J. Olver, J. Pohjanpelto, *Regularity of pseudogroup orbits*, Symmetry and Perturbation Theory (G. Gaeta, B. Prinari, S. Rauch-Wojciechowski, S. Terracini, eds.), World Scientific, Singapore, 2005, pp. 244–254.

[58] ———, *Maurer-Cartan forms and the structure of Lie pseudo-groups*, Selecta Math. 11 (2005), pp. 99–126.

[59] ———, *Moving frames for Lie pseudo-groups*, Canadian J. Math., to appear.

[60] ———, *Differential invariant algebras of Lie pseudo-groups*, University of Minnesota, 2007.

[61] L. V. Ovsiannikov, *Group Analysis of Differential Equations*, Academic Press, New York, 1982.

[62] J. F. Pommaret, *Systems of Partial Differential Equations and Lie Pseudogroups*, Gordon and Breach, New York, 1978.

[63] T. Robart, N. Kamran, *Sur la théorie locale des pseudogroupes de transformations continus infinis I*, Math. Ann. 308 (1997), pp. 593–613.

[64] W. Seiler, *On the arbitrariness of the solution to a general partial differential equation*, J. Math. Phys. 35 (1994), pp. 486–498.

[65] ———, *Involution*, in preparation.

[66] J. Serra, *Image Analysis and Mathematical Morphology*. Academic Press, London 1982.

[67] I. M. Singer, S. Sternberg, *The infinite groups of Lie and Cartan. Part I (the transitive groups)*, J. Analyse Math. 15 (1965), pp. 1–114.

[68] O. Stormark, *Lie's Structural Approach to PDE Systems*, Cambridge University Press, Cambridge, 2000.

[69] A. Tresse, *Sur les invariants différentiels des groupes continus de transformations*, Acta Math. 18 (1894), pp. 1–88.

[70] T. Tsujishita, *On variational bicomplexes associated to differential equations*, Osaka J. Math. 19 (1982), pp. 311–363.

[71] E. Vessiot, *Sur l'intégration des systèmes différentiels qui admettent des groupes continues de transformations*, Acta. Math. 28 (1904), pp. 307–349.

[72] Welk, M., Kim, P., Olver, P.J., Numerical invariantization for morphological PDE schemes, in: Scale Space and Variational Methods in Computer Vision; F. Sgallari, A. Murli and N. Paragios, eds., Lecture Notes in Computer Science, Springer-Verlag, New York, 2007.

Author information

Peter Olver, School of Mathematics, University of Minnesota, Minneapolis, MN 55455, USA.
Email: olver@math.umn.edu

Juha Pohjanpelto, Department of Mathematics, Oregon State University, Corvallis, OR 97331, USA.
Email: juha@math.oregonstate.edu

Radon Series Comp. Appl. Math **2**, 245–265 © de Gruyter 2007

Invariant Theory and Differential Operators

William N. Traves

Key words. Invariant theory, Reynolds operator, Derksen's algorithm, Weyl algebra, differential operator, Grassmann variety.

AMS classification. 13A50, 16S32.

Constructive invariant theory was a preoccupation of many nineteenth century mathematicians, but the topic fell out of fashion in the early twentieth century. In the latter twentieth century the topic enjoyed a resurgence, partly due to its connections with the construction of moduli spaces in algebraic geometry and partly due to the development of computational algorithms suitable for implementation in modern symbolic computation packages. In this survey paper we briefly discuss some of the history and applications of invariant theory and apply one particular algorithm that uses Gröbner bases to find invariants of linearly reductive algebraic groups acting on the Weyl algebra. After showing how we can present the ring of invariant differential operators in terms of generators and relations, we turn to the operators on the invariant ring itself. The theory is particularly nice for finite groups acting on polynomial rings, but we also compute an example involving an $SL_2\mathbb{C}$-action. In this example, we give a description of the generators and relations of $D(G(2,4))$, the ring of differential operators on the Grassmannian of 2-planes in 4-space (or on the affine cone over the Grassmannian of lines in projective 3-space).

This paper is based on my talk in the workshop on Gröbner Bases in Symbolic Analysis held at RISC and RICAM in May 2006. Many of the technical details are omitted. The interested reader can find them in my paper [34] or in Derksen and Kemper's monograph [8], as indicated in the text.

This work was conducted during the Special Semester on Gröbner Bases, February 1 - July 31, 2006, organized by RICAM, Austrian Academy of Sciences, and RISC, Johannes Kepler University, Linz, Austria. The author would like to thank all the organizing parties for their hospitality and support for the conference. Finally, the author would also like to thank Gregor Kemper and the staff at the Technical University of München for their warm hospitality in May, 2006.

1 Invariant Theory

When a group G acts on an affine algebraic variety X, then it makes sense to ask whether the orbits of G form an algebraic variety in their own right. This is the basic question at the heart of geometric invariant theory and the answer is subtle [24]. To make matters much easier, we restrict ourselves to the **non-modular case**: throughout this paper we work with complex varieties but all the results hold over any field of characteristic zero or in any situation where the characteristic of the field does not divide the order of a finite group G. Two simple examples suffice to introduce the theory.

Example 1.1 If $G = \mathbb{Z}_2 = \{-1, 1\}$ acts on the affine plane $X = \mathbb{A}_{\mathbb{C}}^2$ by scalar multiplication, $g \bullet (x, y) = (gx, gy)$, then all the orbits consist of two points except for the orbit of the origin, which is a fixed point of the group action. If the orbits do form an algebraic variety X/G then the natural projection map $X \to X/G$ that sends each point to its orbit is surjective and corresponds to an injective map of the coordinate rings $\mathbb{C}[X/G] \hookrightarrow \mathbb{C}[X] = \mathbb{C}[x, y]$. So $\mathbb{C}[X/G]$ can be identified with the subring of $\mathbb{C}[x, y]$ consisting of functions that are constant on each orbit. In our example, this just consists of those polynomials $f(x, y)$ such that $f(x, y) = f(-x, -y)$ and so $\mathbb{C}[X/G] = \mathbb{C}[x^2, xy, y^2] \cong \mathbb{C}[a, b, c]/(b^2 - ac)$. Though the space X and the G-action were about as nice as possible, the quotient variety X/G is a singular surface, a cone with vertex at the origin.

Generalizing this example, when G acts on a variety X there is a natural left action on $f \in R = \mathbb{C}[X]$ given by $(g \bullet f)(x) = f(g^{-1} \bullet x)$ and

$$R^G = \{f \in R : g \bullet f = f \text{ for all } g \in G\}$$

is the ring of G-invariant functions on X. The variety $X//G = \text{Spec}(R^G)$ is called the categorical quotient of X by G. However, the categorical quotient may not be the quotient X/G as the next example demonstrates.

Example 1.2 If $G = \mathbb{C}^* = \mathbb{C} \setminus \{0\}$ acts on $X = \mathbb{A}_{\mathbb{C}}^2$ by scalar multiplication then most of the orbits have the form $L \setminus \{(0, 0)\}$, where L is a line in X passing through the origin. The sole exception is the orbit of the fixed point, $(0, 0)$. However, since any continuous function that is constant on an orbit must also take the same value on its closure, the fact that $(0, 0)$ is in the closure of all orbits forces $\mathbb{C}[X//G] = \mathbb{C}$. That is, $X//G = \text{Spec}(\mathbb{C})$ is a point. Since this doesn't seem to reflect the structure of the orbit space, the common approach is to restrict our attention to an open subset of points $Y \subset X$ on which G acts; these form a ringed space and on each affine chart $U \subset Y$ we can consider the ring of invariants $\mathbb{C}[U]^G$. Patching together the $\text{Spec}(\mathbb{C}[U]^G)$ gives a variety Y/G. For instance, in the case of the torus acting on the plane, the algebraic variety $Y = X \setminus \{(0, 0)\}$ is covered by two affine charts $Y_1 = \{(x, y) : x \neq 0\}$ and $Y_2 = \{(x, y) : y \neq 0\}$. Now it is not hard to see that on each chart the slope parameterizes the orbits – $Y_1/G = \text{Spec}(\mathbb{C}[y/x])$ and $Y_2/G = \text{Spec}(\mathbb{C}[x/y])$. Since these are also the charts for the projective line, $Y/G \cong \mathbb{P}_{\mathbb{C}}^1$.

Generalizing the method of the last example, we call a point x in a projective variety X semi-stable (and write $x \in X^{ss}$) if there is an affine neighborhood U of x on which there is an invariant $f \in \mathbb{C}[U]^G$ such that $f(x) \neq 0$. The quotient $X^{ss}//G$ is a projective variety, called the geometric invariant theory (G.I.T.) quotient of X under the G-action. It may still occur that the points in $X^{ss}//G$ do not correspond to the orbits of G on X^{ss} (roughly speaking, the invariants may fail to separate orbits in X^{ss}), but even in this case, the variety $X^{ss}//G$ enjoys many functorial properties that we would expect of a quotient. A trivial example of this construction occurs when G is finite; then every point in X is semi-stable and $X/G = X//G$. Here we ought to be clear that we are omitting many details of the G.I.T. construction. The interested reader is encouraged to consult [24] for the full story (or [9, chapters 6 and 8] for a cogent précis).

Let's look at some more complicated examples to further illustrate the power and applicability of the invariant theory viewpoint.

Example 1.3 One of the great tools in algebraic geometry is the construction of moduli spaces whose points parameterize varieties of interest. For instance, consider the variety $\mathcal{M}_{0,d}(\mathbb{P}^2)$ that parameterizes degree-d rational curves in the plane. Geometric invariant theory appears in the description of this space: we'd like to describe each curve using an explicit parametrization $\mathbb{P}^1 \to \mathbb{P}^2$ but then we need to identify those curves that differ only by a linear change of coordinates on the domain $\mathbb{P}^1$. To do this we take the quotient of the space of parameterizations by an $\text{Aut}(\mathbb{P}^1) = PGL_2$-group action.

Example 1.4 Another important example of the G.I.T. method involves the construction of the Hilbert scheme parameterizing subvarieties of projective space with given Hilbert polynomial. A simple example is the Hilbert scheme parameterizing two points in $\mathbb{P}^1$, corresponding to the constant Hilbert polynomial with value 2. It is easy to parameterize pairs of points, just take $(a, b) \in \mathbb{P}^1 \times \mathbb{P}^1$. However, since the order of the points doesn't matter we should identify (a, b) with (b, a). Taking the quotient by the $\mathbb{Z}_2$-action that swaps the points, we obtain the Hilbert scheme for pairs of points in $\mathbb{P}^1$: $(\mathbb{P}^1 \times \mathbb{P}^1)/\mathbb{Z}_2$. Though it is a standard exercise in a first course in algebraic geometry to show that $\mathbb{P}^1 \times \mathbb{P}^1 \ncong \mathbb{P}^2$, it is less common to explain that once we quotient by the $\mathbb{Z}_2$ action we do get $\mathbb{P}^2$. Indeed, if we think of the points on $\mathbb{P}^1 \times \mathbb{P}^1$ as pairs of polynomials $(a_1 x + a_2 y, b_1 x + b_2 y)$ the multiplication map sends this pair to a degree two homogeneous polynomial, which is identified with an element of $\mathbb{P}^2$. The multiplication map is generically 2-to-1 but since we identify the pre-images of $(a_1 x + a_2 y)(b_1 x + b_2 y)$ in $(\mathbb{P}^1 \times \mathbb{P}^1)/\mathbb{Z}_2$ the induced map to $\mathbb{P}^2$ is an isomorphism. The book [25] contains a detailed exposition on Hilbert schemes.

Example 1.5 Another interesting example involves the Grassmannian $G(k, n)$, a variety whose points parameterize k-dimensional subspaces of n-space. Equivalently one could consider $\mathbb{G}(k - 1, n - 1) = \mathbb{P}(G(k, n))$, a projective variety parameterizing the projective $(k - 1)$-dimensional linear spaces in $\mathbb{P}^{n-1}$. To describe each $(k - 1)$-dimensional linear space $\mathbb{P}(V)$ in $\mathbb{P}^{n-1}$, we choose a basis $\{b_1, \dots, b_k\}$ of V and associate to their span the k by n matrix whose rows consist of the b_i's. Among all k by n matrices, only the full-rank matrices correspond to $(k-1)$-dimensional spaces $\mathbb{P}(V)$, so we only consider the open set of $\mathbb{A}^{nk}$ consisting of full-rank matrices. Moreover, there are several parameterizations for each $\mathbb{P}(V)$, one for each choice of basis for $\mathbb{P}(V)$. To identify these copies we quotient by an SL_k-action, where SL_k acts on the k by n matrices by left multiplication. The quotient is precisely the Grassmannian $\mathbb{G}(k-1, n-1)$. The common way to describe this space is to compute $\mathbb{C}[X^n//SL_k\mathbb{C}] = \mathbb{C}[X^n]^{SL_k\mathbb{C}}$ where $X = \mathbb{C}^k$ and set $\mathbb{G}(k-1, n-1) = \mathbb{P}(X^n//SL_k\mathbb{C})$ (see [8, section 4.4] or [32, chapter 3] for details). In section 6 we compute the ring $\mathbb{C}[G(2, 4)] = \mathbb{C}[(\mathbb{C}^2)^4//\mathrm{SL}_2\mathbb{C}]$ of functions on the Grassmannian $G(2, 4)$ and describe the ring of differential operators on $G(2, 4)$.

2 Structural Properties of Rings of Invariants

In general it is difficult to compute the ring of invariants $R^G = \mathbb{C}[X//G]$. Indeed, this was a major field of research for mathematicians in the nineteenth century. In 1868 the acknowledged "king of invariant theory" Paul Gordan proved that when $G = \mathrm{SL}_2\mathbb{C}$ acts on a finite dimensional $\mathbb{C}$-vector space X, the ring of invariants R^G is a finitely generated $\mathbb{C}$-algebra. Moreover, his proof was constructive so that – at least in principle – it was possible to compute a set of generators. In 1890 David Hilbert stunned the mathematical community by giving a nonconstructive proof that whenever a linearly reductive group G acts on a finite dimensional $\mathbb{C}$-vector space, the ring of invariants R^G is a finitely generated $\mathbb{C}$-algebra. Hilbert's nonconstructive proof met with serious opposition. Gordan even described it as "Theologie und nicht Mathematik!". Hilbert continued to consider invariant theory a major area of mathematics: his 14th problem [12] is related to the question of whether R^G is finitely generated for any group acting on a finite dimensional vector space. Masayoshi Nagata answered this question – and Hilbert's 14th problem – in the negative [26], providing an example where G is not linearly reductive and R^G fails to be finitely generated. For details, see the expository article [23].

In today's mathematical culture it may seem hard to believe that nonconstructive methods like those used by Hilbert met with such fierce resistance. Perhaps in order to counter his critics, Hilbert provided a constructive method to compute the generators for R^G just three years after the publication of his controversial proof [11]. However, nearly a hundred years went by before Harm Derksen turned Hilbert's ideas into something that could actually be used for symbolic computation. We'll describe Derksen's algorithm in the next section. For now, let's examine Hilbert's proof that R^G is finitely generated. The proof depends on a certain map $\mathcal{R} : R \to R^G$ called the Reynolds operator.

Recall that an algebraic group G is called linearly reductive if every G-invariant subspace W of a G-vector space V has a G-invariant complement: $V = W \oplus W^C$. Examples of linearly reductive groups in characteristic zero are GL_n, all semi-simple groups including SL_n, O_n and Sp_n, finite groups and tori. Finite groups are also linearly reductive in prime characteristic when the characteristic does not divide the order of the group. Now let a linearly reductive group G act on a finite dimensional vector space X. Since the induced G-action on $R = k[X]$ preserves degree we see that the inclusion $R^G \hookrightarrow R$ is a graded map of R^G-algebras. Restricting to the degree d piece, the G-invariant subspace R_d^G of R_d has a G-invariant complement and for each d we can project R_d onto R_d^G. The Reynolds operator $\mathcal{R}$ is the R^G-linear map $R \to R^G$ that agrees with this projection in each degree. Note that the Reynolds operator is a splitting of the inclusion $R^G \to R$ as a map of R^G-algebras.

In general it can be quite difficult to compute the Reynolds operator for a given group action. However, when G is a finite group the Reynolds operator just averages the group action:

$$\mathcal{R}(f) = \frac{1}{|G|} \sum_{g \in G} g \bullet f.$$

When G is infinite then we can compute the Reynolds operator by integrating over a compact subgroup. In particular when G is a connected semi-simple group there are explicit algebraic algorithms [8, Algorithm 4.5.19] to compute the value of the Reynolds operator on any element of R, though no simple closed form algebraic expression for $\mathcal{R}$ is known in these cases. In the special case of $G = SL_n$ or $G = GL_n$, Cayley's Omega process does give a closed form expression for the Reynolds operator (see [8, section 4.5.3]).

Theorem 2.1 (Hilbert (1890)) *If G is a linearly reductive group acting on a Noetherian k-algebra R, then R^G is a finitely generated k-algebra.*

Proof. Let I be the Hilbert ideal of R, the ideal generated by all the G-invariant functions of positive degree: $I = (f \in R_{>0}^G)R$. Since R is Noetherian, I is a finitely generated ideal in R. Moreover, I is a homogeneous ideal, so we can find homogeneous elements $f_1, \ldots, f_t$ in $R_{>0}^G$ generating the R-ideal I. Now $k[f_1, \ldots, f_t] \subseteq R^G$, but we claim that we actually have equality. We prove this for each graded piece of R^G by induction. The base case is trivial since $k[f_1, \ldots, f_t]_0 = R_0^G = k$. Now assume that the rings agree in degree less than d and let $g \in R_d^G$. Then $g \in I$ so there exist homogeneous elements $h_1, \ldots, h_t$ of R such that $\deg(h_i) = d - \deg(f_i) < d$ and

$$g = h_1 f_1 + \cdots + h_t f_t.$$

Applying the R^G-linear Reynolds operator $\mathcal{R}$ gives

$$g = \mathcal{R}(g) = \mathcal{R}(h_1)f_1 + \cdots + \mathcal{R}(h_t)f_t. \tag{2.1}$$

Now since $\mathcal{R}(h_i) \in R^G$ has degree less than d, $\mathcal{R}(h_i) \in k[f_1, \ldots, f_t]$. Now (2.1) shows that $g \in k[f_1, \ldots, f_t]$. This completes the inductive step so $R^G = k[f_1, \ldots, f_t]$. $\square$

It is possible to use the Reynolds operator, together with the theory of tight closure, to give an elegant proof [14, Theorem 3.6] of a theorem due to Hochster and Roberts [13].

Theorem 2.2 *If a linearly reductive group G acts on a Noetherian $\mathbb{C}$-algebra R, then R^G is Cohen-Macaulay. That is, there is a homogeneous system of parameters $f_1, \ldots, f_d$ in R^G such that $\mathbb{C}[f_1, \ldots, f_d]$ is a polynomial ring, and R^G is a finite*

$\mathbb{C}[f_1, \ldots, f_d]$-*module. The parameters* f_i *are said to be primary invariants and the module generators are called secondary invariants.*

Finding primary and secondary invariants tends to require significant computation, but the amount of computation is reduced if we know the number and degree in which these invariants occur. This is precisely the information contained in the classical statement of Molien's theorem, which deals with finite group actions.

If G is a group acting on $R = \mathbb{C}[x_1, \ldots, x_n]$, then the Molien series is the Hilbert series for the ring R^G, a series that encodes the dimensions of the graded pieces of R^G:

$$H(R^G, t) = \sum_{d=0}^{\infty} \left(\dim_{\mathbb{C}} R_d^G \right) t^d.$$

In 1897 Molien proved that it is possible to compute $H(R^G, t)$ without first computing R^G.

Theorem 2.3 (Molien's theorem) *If G is a finite group of order $|G|$ acting on $R = \mathbb{C}[V] = \mathbb{C}[x_1, \ldots, x_n]$ via the representation $\rho : G \to GL(V)$ then the Molien series can be expressed as*

$$H(R^G, t) = \frac{1}{|G|} \sum_{g \in G} \frac{1}{\det(1 - t\rho(g))}.$$

We refer the reader to Sturmfels's account [32, Theorem 2.2.1] for a very readable proof that only relies on elementary linear algebra. Replacing the sum by an integral, Molien's theorem can be extended to algebraic groups (see [8] for details).

The Molien series can be expressed in the form

$$H(R^G, t) = \frac{P(t)}{\prod_{i=1}^{p}(1 - t^{d_i})}.$$

The degrees d_i of the primary invariants can be read off this expression, as can the degrees k_i and number in each degree m_i of the secondary invariants: these are encoded by the polynomial $P(t) = \sum m_i t^{k_i}$. There are algorithms to compute the primary invariants (see [5]). Once these are found, we can apply the Reynolds operator to a basis for R_d until the results (together with the polynomials of degree d in the polynomial algebra generated by the primary invariants) span a vector space of dimension $\dim_{\mathbb{C}}(R_d^G)$, as predicted by the Molien series.

We end this section with a short example to illustrate Molien's theorem.

Example 2.4 Let $G = \mathbb{Z}/2\mathbb{Z} \times \mathbb{Z}/4\mathbb{Z} = \langle \gamma, \delta : \gamma^2 = \delta^4 = \mathrm{id}_G \rangle$ act on $X = \mathbb{C}^3$ so that γ is a reflection in the $x_2 x_3$-plane and δ is a 90-degree rotation about the x_1-axis: the representation $\rho : G \to \mathrm{End}_{\mathbb{C}}(\mathbb{C}^3)$ is given by $\rho(\gamma) = \begin{bmatrix} -1 & 0 & 0 \\ 0 & 1 & 0 \\ 0 & 0 & 1 \end{bmatrix}$ and

$$\rho(\delta) = \begin{bmatrix} 1 & 0 & 0 \\ 0 & 0 & -1 \\ 0 & 1 & 0 \end{bmatrix}.$$ The Molien series is

$$H(R^G, t) = \frac{1 + t^4}{(1 - t^2)^2(1 - t^4)}.$$

It is not hard to see that $x_1^2, x_2^2 + x_3^2, x_2^4 + x_3^4$ is a system of parameters of the degrees required by the Molien series. These form the primary invariants. There is a single secondary invariant in degree 4. Using the Reynolds operator, we find the secondary invariant to be $x_2 x_3^3 - x_2^3 x_3$. We will return to this example throughout the paper.

3 Computing Rings of Invariants

There are a variety of algorithms to compute rings of invariants. One of the oldest is Gordan's symbolic calculus [27], which deals with the important case where $G = SL_n(\mathbb{C})$ acts on n-ary d-forms. Cayley's Omega process [32] uses differential operators to compute invariants ([8, section 4.5.3], [32, section 4.3]) and when G is a Lie group, we also have access to infinitesimal methods[1] based on the induced Lie algebra action [32, section 4.3]. Additionally, in many circumstances we can use Molien's theorem to help search for generators, as described above. If we can find a homogeneous system of parameters for R^G to serve as the primary invariants then we can reduce the problem of finding the secondary invariants to a large linear algebra problem. This is a very appealing approach but it is not always easy to find a set of primary invariants. Kemper [16] gives a good exposition describing many methods to compute rings of invariants (also see [6]).

Instead of describing these approaches, we return to Hilbert's original construction of the finite set of generators. This algorithm was generally dismissed as being far too computationally expensive, but in 1999 Harm Derksen surprised many mathematicians by finding an elegant way to recast Hilbert's ideas into a simple algorithm [7]. Though other algorithms may be faster than Derksen's algorithm, it is appealing because it can be applied in a wide variety of contexts. We choose to describe it in detail since it uses Gröbner bases and fits in well with the theme of these conference proceedings.

Let G be a linearly reductive group acting on a vector space $X = \text{Spec}(R)$. Derksen's algorithm is based on the observation that the zero set of the Hilbert ideal I of $R = \mathbb{C}[x_1, \ldots, x_n]$, the ideal generated by all positive degree invariants, is precisely the non-semi-stable points of X (see [8, Lemma 2.4.2]). The collection of these points $\mathbb{V}(I) = X \setminus X^{ss}$ is called the nullcone of X and denoted $\mathcal{N}_X$. To describe the algorithm we first parameterize G so that we can think of G as an algebraic variety. If G is a finite group then we can identify the elements of G with a finite set of points and

---

[1] Recently Bedratyuk [1, 2] produced invariants and co-variants for binary forms in previously inaccessible cases by solving the differential equations coming from the infinitesimal action of $SL_2(\mathbb{C})$. These very interesting papers are only peripherally related to the material in this paper but they are highly recommended.

if G is an algebraic group then this parametrization is implicit in the definition of G. Now let $\psi : G \times X \to X \times X$ be the map of varieties given by $\psi(g, x) = (x, g \bullet x)$. Let Z be the image of ψ and let $\overline{Z}$ be its Zariski-closure. Identify $\mathbb{C}[X \times X]$ with $\mathbb{C}[\mathbf{x}, \mathbf{y}] = \mathbb{C}[x_1, \ldots, x_n, y_1, \ldots, y_n]$. Now we claim that

$$\overline{Z} \cap (X \times \{0\}) = \mathcal{N}_X \cap \{0\}. \tag{3.1}$$

If $(w, 0) \in \mathcal{N}_X \cap \{0\}$ then w is not a semi-stable point and so $0 \in \overline{Gw}$. Thus $(w, 0)$ is in the closure of the image Z. For the other inclusion, we prove that if $(w, 0) \in \overline{Z}$ then $w \in \mathbb{V}(I) = \mathcal{N}_X$. Suppose that $f \in I$ has positive degree. Then $f(\mathbf{x}) - f(\mathbf{y})$ vanishes on all of Z because $f(x) - f(g \cdot x) = 0$. But then $f(\mathbf{x}) - f(\mathbf{y})$ must also vanish on the closure of Z. In particular, $f(w) - f(0) = f(w) = 0$. Thus $w \in \mathbb{V}(I)$, as desired.

Derksen [8, Theorem 4.1.3] used the Reynolds operator to show that the equality (3.1) of sets actually descends to an equality of ideals. If $B = \mathbb{I}(\overline{Z})$ then

$$B + (y_1, \ldots, y_n) = I + (y_1, \ldots, y_n).$$

Now we can compute the ideal B by elimination using Gröbner basis methods and then setting each of $y_1, \ldots, y_n$ to zero we get the generators for the ideal I.

These observations lead to the following algorithm to compute R^G:

Algorithm 3.1 (Derksen's algorithm) INPUT: A linearly reductive algebraic group G acting on a finite dimensional complex vector space X by the representation ρ. OUTPUT: A generating set for $\mathbb{C}[X]^G$. STEP 1: Parameterize the group G by the zero set of an ideal $J \subset \mathbb{C}[\mathbf{t}] = \mathbb{C}[t_1, \ldots, t_k]$. As well, express the representation ρ in as a matrix A whose entries are polynomials in $\mathbb{C}[\mathbf{t}]$. STEP 2: Construct the ideal $\mathbb{I}(\Gamma)$ describing the graph Γ of $\psi : G \times X \to X \times X$ as follows. Identify the first copy of X in the range with the copy of X in the domain and, writing $\mathbf{x}$ for the column vector containing the variables $x_1, \ldots, x_n$, construct the ideal

$$\mathbb{I}(\Gamma) = (y_1 - (A\mathbf{x})_1, \ldots, y_n - (A\mathbf{x})_n) + J\mathbb{C}[\mathbf{t}, \mathbf{x}, \mathbf{y}]$$

in the ring $\mathbb{C}[\mathbf{t}, \mathbf{x}, \mathbf{y}]$. STEP 3: Compute a Gröbner basis for $I(\Gamma)$ in an elimination order on $\mathbb{C}[\mathbf{t}, \mathbf{x}, \mathbf{y}]$ that gives the parameters $\mathbf{t}$ higher weight that the $\mathbf{x}$'s and $\mathbf{y}$'s (see [4] or [17] for details on elimination). Intersecting this basis with $\mathbb{C}[\mathbf{x}, \mathbf{y}]$ gives generators for the ideal B. STEP 4: Set $y_1 = \cdots = y_n = 0$ to get generators for the Hilbert ideal I of R. STEP 5: The generators from step 4 may fail themselves to be invariants. So apply the Reynolds operator to each of them to get invariants that generate the Hilbert ideal I. These invariants also generate the ring R^G, as described in Theorem 2.

Example 3.2 Let $G = \mathbb{Z}/2\mathbb{Z} \times \mathbb{Z}/4\mathbb{Z} = \langle < \gamma, \delta : \gamma^2 = \delta^4 = \mathrm{id}_G \rangle$ act on $X = \mathbb{C}^3$ as in Example 2.4 so that γ is a reflection in the $x_2 x_3$-plane and δ is a 90-degree rotation about the x_1-axis. We parameterize G by the pairs (s, t) where s is a square root of 1

($s = -1$ corresponds to γ) and t is a fourth root of 1 ($t = i$ corresponds to δ). Then interpolating the representation matrices gives a parametrization of the representation,

$$\rho(s,t) = \begin{bmatrix} s & 0 & 0 \\ 0 & 1 & 0 \\ 0 & 0 & 1 \end{bmatrix} \begin{bmatrix} 1 & 0 & 0 \\ 0 & \frac{t^3+t}{2} & \frac{(t-t^3)i}{2} \\ 0 & \frac{(t^3-t)i}{2} & \frac{t^3+t}{2} \end{bmatrix} = \begin{bmatrix} s & 0 & 0 \\ 0 & \frac{t^3+t}{2} & \frac{(t-t^3)i}{2} \\ 0 & \frac{(t^3-t)i}{2} & \frac{t^3+t}{2} \end{bmatrix}.$$

We compute the ring of invariants using Derksen's algorithm. We write $\mathbb{I}(\Gamma) = (s^2 - 1, t^4 - 1, y_1 - (sx_1), y_2 - (\frac{t^3+t}{2}x_2 + \frac{(t^3-t)i}{2}x_3), y_3 - (\frac{(t-t^3)i}{2}x_2 + \frac{t^3+t}{2}x_3))$ and compute a Gröbner basis in an elimination order designed to eliminate s and t. For example, we can use a product order, refined by degree lex order $\prec$, in which the first block of variables is $s \prec t$ and the second block of variables is $x_1 \prec x_2 \prec x_3 \prec y_1 \prec y_2 \prec y_3$. The Gröbner basis $\mathcal{G}$ contains 22 polynomials. Considering only $\mathcal{G} \cap \mathbb{C}[x_1, x_2, x_3, y_1, y_2, y_3]$ gives seven polynomials and setting $y_1 = y_2 = y_3 = 0$ kills 3 of these, leaving $\{x_2^2 + x_3^2, x_1^2, x_3^4, x_2x_3^3 + ix_3^4\}$. Applying the Reynolds operator to these four polynomials produces a Gröbner basis for the Hilbert ideal I: $I = (x_2^2 + x_3^2, x_1^2, x_2^4 + x_3^4, ix_2^4 - x_2^3x_3 + x_2x_3^3 + ix_3^4)$. Cleaning this up shows that $I = (x_2^2 + x_3^2, x_1^2, x_2^4 + x_3^4, x_2^3x_3 - x_2x_3^3)$. So $\mathbb{C}[x_1, x_2, x_3]^G = \mathbb{C}[x_2^2 + x_3^2, x_1^2, x_2^4 + x_3^4, x_2^3x_3 - x_2x_3^3]$, as in Example 2.4. Now another elimination computation shows that the quotient variety is a singular hypersurface: setting $a = x_2^2 + x_3^2$, $b = x_1^2$, $c = x_2^4 + x_3^4$ and $d = x_2^3x_3 - x_2x_3^3$ gives

$$\mathbb{C}[x_1, x_2, x_3]^G \cong \mathbb{C}[a, b, c, d]/(a^4 - 3b^2c + 2c^2 + 2d^2).$$

Note that the singularities lie along the line $a = c = d = 0$, which corresponds to the quotient of the x_1-axis by the group action.

4 Group Actions on the Weyl Algebra

The Weyl algebra is the algebra of differential operators on affine n-space. It can be used to formulate quantum mechanics (see [3]) and to study systems of differential equations in an algebraic manner (see, for example, [30]). To be precise, if $R = \mathbb{C}[x_1, \ldots, x_n]$ is the coordinate ring of $X = \mathbb{A}^n_{\mathbb{C}}$ then the Weyl algebra $D(R)$ is the ring $\mathbb{C}\langle x_1, \ldots, x_n, \partial_1, \ldots, \partial_n \rangle$ in which the variables $x_1, \ldots, x_n$ commute among themselves, the variables $\partial_1, \ldots, \partial_n$ commutate among themselves, and the ∂_i's and the x_j's interact via the commutator relation $[\partial_i, x_j] := \partial_i x_j - x_j \partial_i = \delta_{ij}$, where $\delta_{ij} = 1$ if $i = j$ and 0 otherwise. The variables ∂_i should be thought of as the operators $\partial/\partial x_i$ on the ring $R = \mathbb{C}[x_1, \ldots, x_n]$ and the variables $x_j \in D(R)$ should be thought of as the operators that multiply functions in R by x_j. Under this interpretation the rule for commuting ∂_i and x_i corresponds to the product rule in multi-variable calculus:

$$\begin{aligned} (\partial_i x_i) \bullet f(x_1, \ldots, x_n) &= \frac{\partial}{\partial x_i}(x_i f(x_1, \ldots, x_n)) \\ &= x_i \frac{\partial f}{\partial x_i}(x_1, \ldots, x_n) + f(x_1, \ldots, x_n) \\ &= (x_i \partial_i + \delta_{ii}) \bullet f(x_1, \ldots, x_n). \end{aligned}$$

When G acts on affine space $X = \mathbb{A}^n_{\mathbb{C}}$, it not only induces an action on the coordinate ring $R = \mathbb{C}[X]$ but also on the Weyl algebra $D(R)$. For $g \in G$, $\theta \in D(R)$ and $f \in R$,

$$(g \bullet \theta)(f) = g \bullet (\theta(g^{-1} \bullet f)).$$

Those readers familiar with differential geometry will not find it surprising that G acts on the operators $\partial_1, \ldots, \partial_n$ via the contragredient representation: if $g \in G$ acts on $R = \mathbb{C}[x_1, \ldots, x_n]$ via the matrix A, $g \bullet \mathbf{x} = A\mathbf{x}$ (all vectors are represented by column matrices), then g acts on $\mathbb{C}[\partial] = \mathbb{C}[\partial_1, \ldots, \partial_n]$ via $\left(A^T\right)^{-1}$, where T stands for the Hermitian transpose. However, many readers might enjoy an explicit proof of this fact communicated to the author by Harrison Tsai. To establish this claim, it is enough to show that the defining identities $[\partial_i, x_j] := \partial_i x_j - x_j \partial_i = \delta_{ij}$ for the Weyl algebra $D(R)$ are preserved under the proposed group action.

We first observe that the identities can be written in the matrix formulation

$$[\partial, \mathbf{x}] = \partial\mathbf{x}^T - (\mathbf{x}\partial^T)^T = \mathbf{1}.$$

At first sight this may seem odd because we are familiar with the formula $(AB)^T = B^T A^T$ in $GL_n(\mathbb{C})$ but such a formula depends on the commutativity of multiplication in $\mathbb{C}$, while here the x's and the ∂'s do not commute.

Now we show that the identity is preserved under the group action. For ease of notation, let B stand for $\left(A^T\right)^{-1}$, then

$$
\begin{aligned}
[g \bullet \partial, g \bullet \mathbf{x}] &= (g \bullet \partial)(g \bullet \mathbf{x}) - (g \bullet \mathbf{x})(g \bullet \partial) \\
&= B\partial \left(A\mathbf{x}\right)^T - \left(A\mathbf{x}\left(B\partial\right)^T\right)^T \\
&= B\partial\mathbf{x}^T A^T - \left(A\mathbf{x}\partial^T B^T\right)^T \\
&= B\partial\mathbf{x}^T A^T + B\left(-\mathbf{x}\partial^T\right)^T A^T \\
&= B\partial\mathbf{x}^T A^T + B\left(\mathbf{1} - \partial\mathbf{x}^T\right) A^T \\
&= B\left(\partial\mathbf{x}^T + \mathbf{1} - \partial\mathbf{x}^T\right) A^T \\
&= B\mathbf{1}A^T \\
&= (A^T)^{-1}A^T = \mathbf{1}.
\end{aligned}
$$

In an earlier paper [34] it was shown how to extend Derksen's algorithm to the Weyl algebra in order to compute the ring of invariant differential operators $D(R)^G$. For this we exploit the close connection between $D(R)$ and the commutative ring $GrD(R)$. To introduce $GrD(R)$, note that $D(R)$ is a filtered ring: if we assign degree 1 to each ∂_i and degree 0 to each x_j we say that an operator in $D(R)$ has order $\leq n$ if some representation of the operator has degree no greater than n. Note that it we need to be cautious when determining the order of an operator: for example, the operator $1 = \partial_i x_i - x_i \partial_i$ seems to have order 0 or 1, depending on its representation. Of course, 1 is an operator of order ≤ 0. If F_n consists of those operators of order $\leq n$, it is immediate that (1) $F_n \subset F_{n+1}$, (2) F_n is closed under addition and (3) $F_n \cdot F_m \subset F_{n+m}$. These properties ensure that the F_n define a filtration on the algebra $D(R) = \cup_{n \geq 0} F_n$.

Whenever we have a filtered ring such as $D(R)$, we can form its graded ring,

$$GrD(R) = \bigoplus_{n \geq 0} \frac{F_n}{F_{n+1}}.$$

The graded ring comes equipped with a symbol map, $\sigma : D(R) \to GrD(R)$, assigning $\sigma(\theta) = \theta \mod F_{n+1}$ to each $\theta \in F_n$. If we write ξ_i for $\sigma(\partial_i)$ (and abuse notation by writing x_j for $\sigma(x_j) \in GrD(R)$ too) it is easy to see that $GrD(R)$ is generated by $x_1, \ldots, x_n, \xi_1, \ldots, \xi_n$. Moreover, $GrD(R)$ is a commutative ring since the commutation relation $\partial_i x_j - x_j \partial_i = \delta_{ij}$ in $D(R)$ becomes $\xi_i x_j - x_j \xi_i = 0 \mod F_0$ in $GrD(R)$. Indeed, this shows that $GrD(R) = \mathbb{C}[x_1, \ldots, x_n, \xi_1, \ldots, \xi_n]$ is a polynomial ring in $2n$ variables.

The group G preserves the order filtration when it acts on $D(R)$, so there is an induced action on the graded ring $GrD(R)$. Indeed, the action of G on $D(R)$ is compatible with the symbol map, so if $g \in G$ acts on $x_1, \ldots, x_n$ via the matrix A then as in $D(R)$, g acts on $\xi_1, \ldots, \xi_n$ via the matrix $(A^T)^{-1}$. Moreover, the filtration $\{F_n\}$ restricts to a filtration on R^G, giving rise to the graded ring $Gr(D(R)^G)$. Since the action is compatible with the filtration, it should come as no surprise that $Gr(D(R)^G) = [GrD(R)]^G$; see [34, Theorem 1] for a proof.

Now we can apply Derksen's algorithm to the polynomial ring $GrD(R)$ to compute $[GrD(R)]^G = Gr(D(R)^G)$. Then we can lift the generators of $Gr(D(R)^G)$ to elements of $D(R)^G$. It is not hard to prove that if S is a filtered $\mathbb{C}$-algebra then any lifting of a set of generators for GrS is a set of generators for S, so the lifts of the generators of $Gr(D(R)^G)$ generate the ring of invariant differential operators $D(R)^G$.[1]

Example 4.1 We compute generators for $D(R)^G$ where G and R are as in Example 3.2. Listing the generators for $GrD(R)$ in the order $x_1, x_2, x_3, \xi_1, \xi_2, \xi_3$, the action of G on $GrD(R)_1$ is given by $\tilde{\rho} : G \to \text{Aut}_{\mathbb{C}}(GrD(R)_1)$, where

$$
\tilde{\rho}(s,t) = \rho(s,t) \oplus \rho(s,t) =
\begin{bmatrix}
s & 0 & 0 & 0 & 0 & 0 \\
0 & \frac{t^3+t}{2} & \frac{(t-t^3)i}{2} & 0 & 0 & 0 \\
0 & \frac{(t^3-t)i}{2} & \frac{t^3+t}{2} & 0 & 0 & 0 \\
0 & 0 & 0 & s & 0 & 0 \\
0 & 0 & 0 & 0 & \frac{t^3+t}{2} & \frac{(t-t^3)i}{2} \\
0 & 0 & 0 & 0 & \frac{(t^3-t)i}{2} & \frac{t^3+t}{2}
\end{bmatrix}.
$$

Following Derksen's algorithm, we write $\mathbb{I}(\Gamma) = (s^2 - 1, t^4 - 1, y_1 - (sx_1),$ $y_2 - (\frac{t^3+t}{2}x_2 + \frac{(t^3-t)i}{2}x_3), y_3 - (\frac{(t-t^3)i}{2}x_2 + \frac{t^3+t}{2}x_3), \eta_1 - (s\xi_1), \eta_2 - (\frac{t^3+t}{2}\xi_2 + \frac{(t^3-t)i}{2}\xi_3),$ $\eta_3 - (\frac{(t-t^3)i}{2}\xi_2 + \frac{t^3+t}{2}\xi_3))$ and compute a Gröbner basis in an elimination order designed to eliminate s and t. The Gröbner basis $\mathcal{G}$ consists of 92 polynomials. But $\mathcal{G} \cap \mathbb{C}[x_1, x_2, x_3, y_1, y_2, y_3, \xi_1, \xi_2, \xi_3, \eta_1, \eta_2, \eta_3]$ consists of only 48 polynomials. After setting $y_1 = y_2 = y_3 = \eta_1 = \eta_2 = \eta_3 = 0$, we recover only 17 polynomials and applying the Reynolds operator to these gives seventeen generators for $Gr(D(R)^G)$. Replacing ξ_i with ∂_i and clearing fractions, et cetera, we get the following seventeen

---

[1] It is possible to simplify the previous discussion using the Poincare-Birkhoff-Witt theorem on normal orderings in $D(R)$; however, it is not clear how to apply Derksen's algorithm directly to $D(R)$, so we've taken a more elementary approach in this paper.

generators for the ring of invariant differential operators $D(R)^G$:

$$
\left\{
\begin{array}{lll}
\partial_2^2 + \partial_3^2, & x_3\partial_2 - x_2\partial_3, & x_2\partial_2 + x_3\partial_3, \\
\partial_1^2, & x_1\partial_1, & -x_2x_3\partial_2^2 + x_2x_3\partial_3^2, \\
x_1^2, & \partial_2^4 + \partial_3^4, & -\partial_2^3\partial_3 + \partial_2\partial_3^3, \\
x_2\partial_2^3 + x_3\partial_3^3, & -x_3\partial_2^3 + x_2\partial_3^3, & x_2^2\partial_2^2 + x_3^2\partial_3^2, \\
x_2^2 + x_3^2, & x_2^3\partial_2 + x_3^3\partial_3, & -x_2^2x_3\partial_2 + x_2x_3^2\partial_3, \\
x_2^4 + x_3^4, & -x_2^3x_3 + x_2x_3^3 &
\end{array}
\right\}.
$$

It is worth noting that the Molien series for $(GrD(R))^G$ is

$$
\frac{1 + 2t^2 + 10t^4 + 2t^6 + t^8}{(1-t^2)^5(1-t^4)}.
$$

Thus, it requires 16 secondary generators to generate $(GrD(R))^G$ as a module over a polynomial ring generated by 6 primary invariants. In this example, Derksen's algorithm finds fewer generators of $(GrD(R))^G$ and $D(R)^G$ than Molien's method, but they are *algebra generators* rather than *module generators*. Perhaps this trade-off is inevitable: we seem to need a larger number of generators if we require them to enjoy better structural properties.

Not only can we compute the generators for $D(R)^G$, but we can also compute the relations among these generators. Using elimination we can compute the relations among the generators of $[GrD(R)]^G = Gr(D(R)^G)$. Each of these can be lifted to a relation in $D(R)^G$ (see below for an example). The complete set of relations among the generators in $D(R)^G$ is the two-sided ideal of $D(R)^G$ generated by these lifted relations and the commutator relations among the generators. For details, see [34, Algorithm 10]).

Example 4.2 We continue Example 4.1 and find the relations among the 17 generators of $D(R)^G$. To start we perform an elimination computation to compute the relations among the 17 generators of $Gr(D(R)^G)$. This computation is surprisingly fast (under 6 seconds on a Pentium III 933 MHz computer with 376 MB of RAM), but yields 221 relations among the generators in the graded ring. For instance, one of these relations indicates that

$$
(x_1\xi_1)^2 - (x_1^2)(\xi_1^2) = 0 \text{ in } Gr(D(R)^G).
$$

This only means that $(x_1\partial_1)^2 - (x_1^2)(\partial_1^2)$ is an operator of order less than 2 in $D(R)^G$. Performing the computation in the Weyl algebra[2] the operator equals $x_1\partial_1$ so the graded relation lifts to the relation

$$
(x_1\partial_1)^2 - (x_1^2)(\partial_1^2) - (x_1\partial_1) = 0 \text{ in } D(R)^G.
$$

The 221 lifted relations, together with the commutator relations among the 17 generators (there are 82 nontrivial commutator relations) generate the two-sided ideal in $D(R)^G$ of relations among the given generators.

---

[2] Many computer algebra systems can compute in the Weyl algebra: in SINGULAR we can use the PLURAL package; in Macaulay2 we can use the Dmodules package; in MAPLE we can use the Ore algebra package and both RISA/ASIR and CoCoA also support such computations.

At this stage the reader might well wonder whether there is a smaller set of generators for the ring $D(R)^G$. In fact, Levasseur and Stafford [19] prove that for finite groups G, the ring $D(R)^G$ is generated *as a noncommutative algebra* by the operators generating $\mathbb{C}[x_1, \ldots, x_n]^G$ and the operators generating $\mathbb{C}[\partial_1, \ldots, \partial_n]^G$. This greatly reduces the number of generators, but at the moment there is no good way to determine the relations among these generators. As well, the symbols of the generators that Levasseur and Stafford provide are not sufficient to generate $Gr(D(R)^G)$. There seems to be a need for a noncommutative version of Derksen's algorithm – one that works directly in $D(R)$ and not through $GrD(R)$ – though it remains an open problem to generalize Derksen's work in this direction.

5   Rings of Differential Operators

Alexander Grothendieck [10] introduced rings of differential operators associated to algebraic varieties. Suppose that $X \subset \mathbb{A}_\mathbb{C}^n$ is an algebraic variety and that X is the vanishing set of the ideal $I \subset R = \mathbb{C}[x_1, \ldots, x_n]$. Then the ring of differential operators can be described in terms of the Weyl algebra $D(R)$ (see [21, Chapter 15] for details):

$$D(X) := D(R/I) := \frac{\{\theta \in D(R) : \theta \bullet I \subseteq I\}}{ID(R)}.$$

The ring $D(X)$ inherits a filtration from the ring $D(R)$ and, just as for the Weyl algebra, $GrD(X)$ is a commutative ring (see [21] or [22] for a nice explanation of these facts). The rings $D(R)$ have been the subject of intense study for many years. Levasseur and Stafford's monograph [18] is a good description of rings of differential operators and their connection to invariant theory.

We apply these definitions to the case where $X = \mathbb{A}_\mathbb{C}^n//G$ for some linearly reductive group G. We first realize X as an embedded variety in an affine space $\mathbb{A}_\mathbb{C}^d$ by presenting the ring $R^G = \mathbb{C}[X]$ as a finitely generated algebra: $R^G \cong \mathbb{C}[t_1, \ldots, t_n]/J$. Then $D(R^G) = \{\theta \in D(\mathbb{C}[t_1, \ldots, t_d]) : \theta \bullet J \subseteq J\}/JD(\mathbb{C}[t_1, \ldots, t_d])$. Now we need to be cautious: the ring of differential operators on the quotient variety is not the same thing as the ring of invariant differential operators! However, the natural map $\pi : X \to X//G$ induces the inclusion $R^G \hookrightarrow R = \mathbb{C}[\mathbb{A}_\mathbb{C}^n] = \mathbb{C}[x_1, \ldots, x_n]$ and in turn this induces a map $\pi_* : D(R)^G \to D(R^G)$ given by restriction. To be precise, if $\theta \in D(R)^G$, then $\pi_*(\theta)$ is the map that makes the diagram commute.

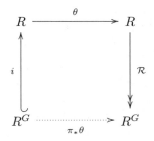

If $\theta \in D(R)^G$, and $r \in R^G$ then $(\pi_* \theta)(r) = \theta(r)$. We check that $\theta(r) \in R^G$: for any $g \in G$,

$$g \bullet (\theta(r)) = g \bullet (\theta(g^{-1} \bullet r)) = (g \bullet \theta)(r) = \theta(r).$$

In general $\pi_* \theta = \mathcal{R}(\theta \circ i)$ is a differential operator on R^G of no higher order than θ.

5.1   Finite Group Actions

We turn to the case of a finite group action on a polynomial algebra over a characteristic zero field.

Theorem 5.1 (Kantor [15], Levasseur [20]) *When G is a finite group acting on a polynomial ring R, the map $\pi_* : D(R)^G \to D(R^G)$ is injective.*

Proof. See [34, Theorem 2].   □

Example 5.2 The map π_* can fail to be surjective. Consider the group $G = \mathbb{Z}_2$ acting on $R = \mathbb{C}[x]$ so that the generator of G sends x to $-x$. Then $R^G = \mathbb{C}[x^2]$ is a polynomial ring and $D(R^G)$ is a Weyl algebra. However, $D(R)^G = \mathbb{C}\langle x^2, x\partial, \partial^2 \rangle$ so $D(R)^G$ is not isomorphic to $D(R^G)$. Thus π_* is not a surjection. Schwarz [31, Example 5.7] gives a more detailed argument.

The group action in Example 5.2 was generated by a reflection. We say that an element $g \in G$ acts as a pseudoreflection if it acts on X such that the eigenvalues of the action of g are all 1 except for a single value (which must be a root of unity since G is assumed to be finite). Equivalently, $g \in G$ is a pseudoreflection when the action of G fixes (point-wise) a codimension 1 hypersurface; in our case, the fixed set is a hyperplane since the action of G is linear. We call a group a reflection group if G is generated by pseudoreflections. The celebrated Sheppard-Todd-Chevalley theorem shows that $D(R^G)$ is a Weyl algebra precisely when G is a reflection group, as illustrated in Example 5.2.

Theorem 5.3 (Sheppard-Todd-Chevalley) *Let G be a finite group acting on a polynomial ring $R = \mathbb{C}[x_1, \ldots, x_n]$. Then R^G is a polynomial ring (and $D(R^G)$ is a Weyl algebra) if and only if G is a reflection group.*

Kantor [15, Theorem 4 in section 3.3.1] showed that the other extreme case – when G contains no pseudoreflections – characterizes the case where π_* is surjective.

Theorem 5.4 (Kantor) *When G is a finite group acting on a polynomial ring $R = \mathbb{C}[x_1, \ldots, x_n]$, the map π_* is a surjection precisely when G contains no pseudoreflections. In this case, $D(R^G) = D(R)^G$.*

When G acting on a polynomial ring R contains some pseudoreflections, but G is not a reflection group, we factor the action of G on $X = \mathrm{Spec}(R)$ as follows. First note that the subgroup P generated by the pseudoreflections is a normal subgroup of G. To see this, it is enough to check that if p is a pseudoreflection and $g \in G$, then $gpg^{-1} \in P$;

this follows since $(g \cdot X^p) \subset X^{gpg^{-1}}$ so $\mathrm{codim}(X^{gpg^{-1}}) \leq \mathrm{codim}(g \cdot X^p) = 1$. Now G/P acts[1] on the polynomial ring R^P and $R^G = (R^P)^{G/P}$.

Since G/P contains no pseudoreflections, the map $\pi_* : D(R^P)^{G/P} \to D(R^G)$ is an isomorphism, so $D(R^G)$ can be described as the ring of invariant differential operators of the group G/P acting on the Weyl algebra $D(R^P)$. It is in this sense that we will be able to describe $D(R^G)$ for finite groups G.

Example 5.5 We return to Example 3.2 and compute a presentation for $D(R^G)$ in terms of generators and relations. First note that the subgroup $P \lhd G$ of pseudoreflections is generated by γ, the reflection in the x_2x_3-plane. Direct observation shows that $R^P = \mathbb{C}[x_1^2, x_2, x_3]$. Write $z = x_1^2$. The quotient G/P is generated by the image of δ and this element acts on R^P by sending z to itself, x_2 to x_3, and x_3 to $-x_2$. Applying Derksen's algorithm to G/P acting on $D(R^P) = \mathbb{C}\langle z, x_2, x_3, \partial_z, \partial_2, \partial_3 \rangle$ gives six generators for $D(R^P)^{G/P} = D(R^G)$:

$$\left\{ \partial_z, z, \partial_2^2 + \partial_3^2, x_3\partial_2 - x_2\partial_3, x_2\partial_2 + x_3\partial_3, x_2^2 + x_3^2 \right\}.$$

Let a, b, c, d, e, f denote these six operators and let $[a], [b], [c], [d], [e], [f]$ denote their symbols. The symbols of these six operators generate $Gr(D(R^P))^{G/P}$ and an elimination computation shows that there is only one syzygy on these generators, $[d]^2 + [e]^2 - [c][f] = 0$. This lifts to a single syzygy on the generators of $D(R^P)^{G/P} = D(R^G)$,

$$d^2 + e^2 - cf + 4e + 4 = 0. \tag{5.1}$$

There are also four nontrivial commutator relations among the generators:

$$[a, b] = 1, \quad [c, e] = 2c, \quad [c, f] = 4e + 4, \quad [e, f] = 2f.$$

The third commutator relation shows that the syzygy (5.1) has a nicer form,

$$d^2 + e^2 + fc = 0. \tag{5.2}$$

The commutator relations, together with the syzygy (5.2) generate the two-sided ideal of relations in among the generators of $D(R^G)$.

6   Differential Operators on G(2,4)

We now give an example involving $G = \mathrm{SL}_2\mathbb{C}$. If V is a 2-dimensional complex vector space, then $\mathbb{C}[V^4]^{\mathrm{SL}_2\mathbb{C}}$ is the coordinate ring of the affine cone over the Grassmannian $G(2, 4)$ of 2-planes in $\mathbb{C}^4$. Let $\{x_{1i}, x_{2i}\}$ be coordinate functions on the ith copy of V in V^4, then the Fundamental Theorems of Invariant Theory for SL_n (see

---

[1] I'm grateful to Gregor Kemper who provided the following short proof that the action of G/P on R^P is linear in the non-modular case. The vector space $(R_{>0}^P)^2$ has a G/P-complement U with basis B. Then G/P acts by linear transformations on the vectors of B. But by the homogeneous version of Nakayama's lemma, B generates R^P minimally.

DK, Theorems 4.4.4 and 4.4.5) imply that $\mathbb{C}[V^4]^{\mathrm{SL}_2\mathbb{C}}$ is generated by six polynomials $[12], [13], [14], [23], [24], [34]$, where $[ij] = x_{1i}x_{2j} - x_{1j}x_{2i}$ is the 2×2 minor of the matrix

$$\begin{bmatrix} x_{11} & x_{12} & x_{13} & x_{14} \\ x_{21} & x_{22} & x_{23} & x_{24} \end{bmatrix}.$$

The ideal of relations on these generators is generated by the Plücker relation

$$[12][34] - [13][24] + [14][23] = 0.$$

We apply Derksen's algorithm to compute the ring of differential operators on the affine cone over the Grassmannian $G(2,4)$.

Example 6.1 We represent the group $\mathrm{SL}_2\mathbb{C}$ as the vanishing set of $a_1a_3 - a_2a_4 - 1$, where the point (a_1, a_2, a_3, a_4) corresponds to the matrix $\begin{bmatrix} a_1 & a_2 \\ a_3 & a_4 \end{bmatrix} \in \mathrm{SL}_2\mathbb{C}$. The group G acts on the x_{ij} by matrix multiplication on the left. This induces an action on $GrD(\mathbb{C}[V^4])$; writing ξ_{ij} for the symbol of $\partial/\partial x_{ij}$, the matrix corresponding to (a_1, a_2, a_3, a_4) acts on

$$\begin{bmatrix} x_{11} & x_{12} & x_{13} & x_{14} \\ x_{21} & x_{22} & x_{23} & x_{24} \\ \xi_{11} & \xi_{12} & \xi_{13} & \xi_{14} \\ \xi_{21} & \xi_{22} & \xi_{23} & \xi_{24} \end{bmatrix} \quad \text{to give} \quad \begin{bmatrix} a_1 & a_2 & 0 & 0 \\ a_3 & a_4 & 0 & 0 \\ 0 & 0 & a_4 & -a_3 \\ 0 & 0 & -a_2 & a_1 \end{bmatrix} \begin{bmatrix} x_{11} & x_{12} & x_{13} & x_{14} \\ x_{21} & x_{22} & x_{23} & x_{24} \\ \xi_{11} & \xi_{12} & \xi_{13} & \xi_{14} \\ \xi_{21} & \xi_{22} & \xi_{23} & \xi_{24} \end{bmatrix}.$$

Applying Derksen's algorithm we obtain 28 generators for $(GrD(\mathbb{C}[V^4]))^{\mathrm{SL}_2\mathbb{C}}$. Each of these is already invariant under $\mathrm{SL}_2\mathbb{C}$, so there is no need to apply the Reynolds operator. Lifting these operators gives generators for $D(\mathbb{C}[V])^{\mathrm{SL}_2\mathbb{C}}$:

$$\left\{ \begin{array}{llll} \partial_{14}\partial_{23} - \partial_{13}\partial_{24}, & \partial_{14}\partial_{22} - \partial_{12}\partial_{24}, & \partial_{13}\partial_{22} - \partial_{12}\partial_{23}, & \partial_{14}\partial_{21} - \partial_{11}\partial_{24}, \\ \partial_{13}\partial_{21} - \partial_{11}\partial_{23}, & \partial_{12}\partial_{21} - \partial_{11}\partial_{22}, & x_{14}\partial_{14} + x_{24}\partial_{24}, & x_{13}\partial_{14} + x_{23}\partial_{24}, \\ x_{12}\partial_{14} + x_{22}\partial_{24}, & x_{11}\partial_{14} + x_{21}\partial_{24}, & x_{14}\partial_{13} + x_{24}\partial_{23}, & x_{13}\partial_{13} + x_{23}\partial_{23}, \\ x_{12}\partial_{13} + x_{22}\partial_{23}, & x_{11}\partial_{13} + x_{21}\partial_{23}, & x_{14}\partial_{12} + x_{24}\partial_{22}, & x_{13}\partial_{12} + x_{23}\partial_{22}, \\ x_{12}\partial_{12} + x_{22}\partial_{22}, & x_{11}\partial_{12} + x_{21}\partial_{22}, & x_{14}\partial_{11} + x_{24}\partial_{21}, & x_{13}\partial_{11} + x_{23}\partial_{21}, \\ x_{12}\partial_{11} + x_{22}\partial_{21}, & x_{11}\partial_{11} + x_{21}\partial_{21}, & x_{14}x_{23} - x_{13}x_{24}, & x_{14}x_{22} - x_{12}x_{24}, \\ x_{13}x_{22} - x_{12}x_{23}, & x_{14}x_{21} - x_{11}x_{24}, & x_{13}x_{21} - x_{11}x_{23}, & x_{12}x_{21} - x_{11}x_{22} \end{array} \right\}.$$

Furthermore, in an important paper about the behavior of π_* [31] Gerald Schwarz showed that the LS-alternative holds for $\mathrm{SL}_2\mathbb{C}$: either $\mathbb{C}[V^4]^{\mathrm{SL}_2\mathbb{C}}$ is regular or the map $\pi_* : D(\mathbb{C}[V^4])^{\mathrm{SL}_2\mathbb{C}} \to D(\mathbb{C}[V^4]^{\mathrm{SL}_2\mathbb{C}})$ is surjective. Since $\mathbb{C}[V]^{\mathrm{SL}_2\mathbb{C}}$ represents a cone it is not a regular ring so π_* is surjective. It follows that the generators for $D(\mathbb{C}[V^4])^{\mathrm{SL}_2\mathbb{C}}$ generate $D(\mathbb{C}[V^4]^{\mathrm{SL}_2\mathbb{C}})$, when restricted to $\mathbb{C}[V^4]^{\mathrm{SL}_2\mathbb{C}}$.

This example illustrates the power and the generality of the Gröbner basis techniques, but the result also follows from the Fundamental Theorems of Invariant Theory for $SL_n\mathbb{C}$ (for details see see [29, sections 9.3 and 9.4]). We now explain this connection.

Let V be an n-dimensional complex vector space and let V^* be the dual space of V. Then $\mathbb{C}[V^r \oplus (V^*)^s]$ is generated by the coordinates x_{ij} and ξ_{ij} ($1 \leq i \leq r, 1 \leq j \leq r$; here $\xi_{ij} = \mathbf{x}_{ij}^*$). If $\langle \cdot, \cdot \rangle : V \times V^* \to \mathbb{C}$ is the canonical pairing, for each $i \leq r$ and $j \leq s$ we have an invariant $\langle ij \rangle : V^r \oplus (V^*)^s \to \mathbb{C}$ that sends $(v_1, \ldots, v_r, w_1, \ldots, w_s)$ to $\langle v_i, w_j \rangle$. In coordinates $\langle ij \rangle = \sum_{k=1}^n x_{ki}\xi_{kj}$.

There are other invariants too. If $1 \leq i_1 < i_2 < \cdots < i_n \leq r$, we have a bracket invariant $[i_1 i_2 \cdots i_n] : V^r \oplus (V^*)^s \to \mathbb{C}$ given by

$$(v_1, \ldots, v_r, w_1, \ldots, w_s) \to \det(v_{i_1} v_{i_2} \cdots v_{i_n}).$$

This is an operator of degree n that only involves the x_{ij}. As well, if $1 \leq j_1 < j_2 < \cdots < j_n \leq s$, we have an invariant $|j_1 j_2 \cdots j_n| : V^r \oplus (V^*)^s \to \mathbb{C}$ given by

$$(v_1, \ldots, v_r, w_1, \ldots, w_s) \to \det(w_{j_1} w_{j_2} \cdots w_{j_n}).$$

This is an operator of total degree n that only involves the ξ_{ij}.

Theorem 6.2 (Fundamental Theorem of Invariant Theory for $SL_n\mathbb{C}$) *Let V be an n-dimensional complex vector space. The invariant ring*

$$\mathbb{C}[V^r \oplus (V^*)^s]^{SL_n\mathbb{C}}$$

is generated by all $\langle ij \rangle$ ($1 \leq i \leq r, 1 \leq j \leq s$), all $[i_1 i_2 \cdots i_n]$ ($1 \leq i_1 < i_2 < \cdots < i_n \leq r$) and all $|j_1 j_2 \cdots j_n|$ ($1 \leq j_1 < j_2 < \cdots < j_n \leq s$). The relations among these generators are of five types:

(a) *For $1 \leq i_1 < i_2 < \cdots < i_n \leq r$ and $1 \leq j_1 < j_2 < \cdots < j_n \leq s$:*
$$\det((\langle i_k j_\ell \rangle))_{k,\ell=1}^n = [i_1 i_2 \cdots i_n]|j_1 j_2 \cdots j_n|$$

(b) *For $1 \leq i_1 < i_2 < \cdots < i_{n+1} \leq r$ and $1 \leq j \leq s$:*
$$\sum_{k=1}^{n+1} (-1)^{k-1} [i_1 i_2 \cdots \hat{i_k} \cdots i_{n+1}] \langle i_k j \rangle = 0$$

(c) *For $1 \leq j_1 < j_2 < \cdots < j_{n+1} \leq s$ and $1 \leq i \leq r$:*
$$\sum_{k=1}^{n+1} (-1)^{k-1} \langle ij_k \rangle |j_1 j_2 \cdots \hat{j_k} \cdots j_{n+1}| = 0$$

(d) *For $1 \leq i_1 < i_2 < \cdots < i_{n-1} \leq r$ and $1 \leq j_1 < j_2 < \cdots < j_{n+1} \leq r$:*
$$\sum_{k=1}^{n+1} (-1)^{k-1} [i_1 i_2 \cdots i_{n-1} j_k][j_1 j_2 \cdots \hat{j_k} \cdots j_{n+1}]$$

(e) *For $1 \leq i_1 < i_2 < \cdots < i_{n-1} \leq s$ and $1 \leq j_1 < j_2 < \cdots < j_{n+1} \leq s$:*
$$\sum_{k=1}^{n+1} (-1)^{k-1} |i_1 i_2 \cdots i_{n-1} j_k| |j_1 j_2 \cdots \hat{j_k} \cdots j_{n+1}|.$$

Now $(GrD(\mathbb{C}[V^4]))^{SL_2\mathbb{C}} = \mathbb{C}[V^4 \oplus (V^*)^4]^{SL_2\mathbb{C}}$ so we can apply Theorem 6.2 in the case $r = s = 4$. We see that $(GrD(\mathbb{C}[V^4]))^{SL_2\mathbb{C}}$ is generated by twenty eight operators: the six $[ij]$, the six $|ij|$ and the sixteen $\langle ij \rangle$. These are precisely the operators found in Example 6.1.

There are 156 relations among the generators of $(GrD(\mathbb{C}[V^4]))^{SL_2\mathbb{C}}$, 36 each of types (a), (d) and (e) and 24 each of types (b) and (c). Each of these extends to an ordered relation on $D(\mathbb{C}[V^4]^{SL_2\mathbb{C}}) = (GrD(\mathbb{C}[V^4]))^{SL_2\mathbb{C}}$. In most cases no modification of the formula is needed, if we take care to write the relations in the order given by Theorem 6.2. However, the relations in part (a) need to be properly interpreted. We explain how to do this for the case $\mathrm{SL}_2\mathbb{C}$. Each term in the determinant $\det(\langle i_k j_\ell \rangle)_{k,\ell=1}^n$ involves the product of two terms $\langle i_k j_\ell \rangle$. When possible we write these products in an order where the last entry of the first term does not coincide with the first entry of the second term. If this can be achieved, then no modification to the formula in part (a) is necessary. If not, then we have a term $\langle ab \rangle \langle ba \rangle$ in the expansion of the determinant and to compensate we must add $\langle aa \rangle$ to the right-hand side of the relation:

$$\det \begin{pmatrix} \langle aa \rangle & \langle ab \rangle \\ \langle ba \rangle & \langle bb \rangle \end{pmatrix} + \langle aa \rangle = \langle aa \rangle \langle bb \rangle - \langle ab \rangle \langle ba \rangle + \langle aa \rangle = [ab]|ab|.$$

The commutator relations among the 28 generators also give rise to relations. Unfortunately, many of these are non-trivial, $[a_i, a_j] \neq 0$ in 156 of 406 cases. However, we do have a compact description of the commutator relations:

$$[|ij|, [ij]] = \langle ii \rangle + \langle jj \rangle + 2,$$
$$[|ij|, [ik]] = \langle kj \rangle,$$
$$[[ij], \langle ii \rangle] = [ji],$$
$$[|ij|, \langle ii \rangle] = |ij|,$$
$$[[ij], \langle kj \rangle] = [ki],$$
$$[|ij|, \langle ik \rangle] = |kj|,$$
$$[\langle ij \rangle, \langle ji \rangle] = \langle ii \rangle - \langle jj \rangle,$$
$$[\langle ij \rangle, \langle ii \rangle] = -\langle ij \rangle,$$
$$[\langle ij \rangle, \langle jj \rangle] = \langle ij \rangle,$$
$$[\langle ij \rangle, \langle ki \rangle] = -\langle kj \rangle,$$
$$[\langle ij \rangle, \langle jk \rangle] = \langle ik \rangle.$$

The relations described so far are enough to determine $D(R)^G$. However, the map $\pi_* : D(R)^G \to D(R^G)$ is not injective. It is known [31] that the kernel of π_* consists of the G-stable part of the left ideal of $D(R)$ generated by the Lie algebra $\mathfrak{g} = \mathfrak{sl}_2\mathbb{C}$. The Lie algebra $\mathfrak{sl}_2$ is generated by three elements g_{12}, g_{21} and $g_{11} - g_{22}$, where g_{ij} corresponds to the adjoint action of the matrix E_{ij} with a 1 in the $(i,j)^{\text{th}}$ position and zero elsewhere. Explicitly,

$$\begin{aligned}
g_{12} &= x_{11}\partial_{21} + x_{12}\partial_{22} + x_{13}\partial_{23} + x_{14}\partial_{24}, \\
g_{21} &= x_{21}\partial_{11} + x_{22}\partial_{12} + x_{23}\partial_{13} + x_{24}\partial_{14}, \\
g_{11} - g_{22} &= x_{11}\partial_{11} + x_{12}\partial_{12} + x_{13}\partial_{13} + x_{14}\partial_{14} \\
&\quad - x_{21}\partial_{21} - x_{22}\partial_{22} - x_{32}\partial_{32} - x_{42}\partial_{42}.
\end{aligned}$$

We can compute the part of the left ideal generated by g_{11}, g_{22} and $g_{11} - g_{22}$ that is G-invariant by intersecting with the subalgebra generated by the invariants. This is a gigantic computation that was performed in SINGULAR using the nctools package. In an extension of $GrD(R)$, $\mathbb{C}[x_{ij}, \xi_{ij}, [ij], |ij|, \langle ij \rangle]$, we form an ideal containing $g_{12}, g_{21}, g_{11} - g_{22}$, and the relations that describe $[ij], |ij|$, and $\langle ij \rangle$ in terms of the x_{ij} and ξ_{ij}. Imposing the block order that places the x_{ij} and ξ_{ij} in the first block and the $[ij], |ij|$ and $\langle ij \rangle$ in the second block, we compute a Gröbner basis of the ideal. After intersecting with $\mathbb{C}[[ij], |ij|, \langle ij \rangle]$ we have 191 polynomials in the Gröbner basis. These polynomials generate the graded kernel K of the map $\pi_* : GrD(R)^G \to GrD(R^G)$. Each of these graded generators extend to an element in $D(R)^G$.

As a result, we've shown that the ring of differential operators on the Grassmannian $D(G(2,4)) = D(R^G)$ is generated by 28 operators satisfying a two-sided ideal of relations generated by the commutator relations and the extensions of the relations from K.

Among the extensions of the generators of K is the interesting element

$$\theta(\theta + 2) - 4 \sum_{i<j} [ij]|ij|, \tag{6.1}$$

where θ is the operator

$$\langle 11 \rangle + \langle 22 \rangle + \langle 33 \rangle + \langle 44 \rangle.$$

The generator (6.1) is a multiple of the Casimir operator of $\mathfrak{sl}_2\mathbb{C}$. This is easily verified by an explicit computation as follows. The Lie algebra $\mathfrak{sl}_2\mathbb{C}$ has inner product given by the Killing form

$$\kappa(\delta, \gamma) = \mathrm{Tr}(\mathrm{ad}(\delta), \mathrm{ad}(\gamma)),$$

where Tr is the trace and $\mathrm{ad}(\delta)(\gamma) = \delta\gamma - \gamma\delta$ is the adjoint action of $\delta \in \mathfrak{sl}_2\mathbb{C}$ on itself. A dual basis for $\mathfrak{sl}_2\mathbb{C}$ with respect to the Killing form is given by $g_{12}/4, g_{21}/4$, and $(g_{11} - g_{22})/8$. Then the Casimir operator [8, Definition 4.5.10] is just given by

$$\frac{g_{12}g_{21}}{4} + \frac{g_{21}g_{12}}{4} + \frac{(g_{11} - g_{22})^2}{8}.$$

Explicit computation in $D(R)$ then shows that the Casimir operator is equal to the operator (6.1) divided by 8.

At one time I conjectured that the kernel of π_* was a two-sided ideal of $D(R)^G$ generated by the Casimir operator. Many people suggested that this should be the case since the Casimir operator generates the center of $\mathfrak{sl}_2\mathbb{C}$; however, it turns out that the Casimir operator does not generate the kernel of π_* (see Traves [35] for details).

7 Conclusion

This paper dealt with constructive techniques in invariant theory for rings of differential operators. Derksen's algorithm was applied to $GrD(R)$ in order to compute $Gr(D(R)^G)$ and then the relationship between $D(R)^G$ and $D(R^G)$ was used to find generators and relations for the ring of differential operators on the quotient variety, $D(R^G)$. In particular, the generators and relations for $D(G(2,4))$ were described.

Levasseur and Stafford [18] work out many other cases of invariant rings of differential operators for the classical groups. As well, Schwarz's work on lifting differential operators [31] is crucial in understanding the relation between $D(R)^G$ and $D(R^G)$.

The ring of invariants R^G is a module over the invariant differential operators $D(R)^G$. Of course, in many cases R^G is a simple $D(R)^G$ module, but if we restrict ourselves to looking at R^G as a module over a subalgebra of $D(R)^G$, then it may well be possible to find many fewer module generators for R^G. This topic is central to invariant theory in prime characteristic, where the subalgebra of choice is the Steenrod algebra (see Smith [33] for details). Pleskin and Robertz [28] investigate the characteristic zero case, but one gets the feeling that much more can be said about the theory of invariant rings R^G as modules over appropriately chosen submodules of $D(R)^G$.

Bibliography

[1] L. Bedratyuk, *On differential equation of invariants of binary forms*, preprint, arXiv:math.AG/0602373.

[2] L. Bedratyuk, *On complete system of invariants for the binary form of degree 7*, preprint, arXiv:math.AG/0611122.

[3] S. C. Coutinho, *The many avatars of a simple algebra*, Amer. Math. Monthly, 104 (1997), pp. 593–604.

[4] D. Cox, J. Little and D. O'Shea, *Ideals, varieties and algorithms*, 2nd ed., Springer-Verlag, New York, 1996.

[5] W. Decker, A. E. Heydtmann, and F.-O. Schreyer, *Generating a noetherian normalization of the invariant ring of a finite group*, J. Symbolic Comput., 25 (1998), pp. 727–731.

[6] W. Decker and T. de Jong, *Gröbner bases and invariant theory*, Gröbner bases and applications (B. Buchberger and F. Winkler, eds.), London Math Society LNS, 251, Cambridge University Press, 1998, pp. 61–89.

[7] H. Derksen, *Computation of invariants for reductive groups,* Adv. Math., 141 (1999), pp. 366–384.

[8] H. Derksen and G. Kemper, *Computational invariant theory*, Invariant Theory and Algebraic Transformation Groups, 1, Springer-Verlag, Berlin, 2002.

[9] I. Dolgachev, *Lectures on invariant theory*, London Math. Soc. LNS, 296, Cambridge University Press, Cambridge, 2003.

[10] A. Grothendieck, *Éléments de géométrie algébrique IV, Étude locale des schémas et des morphismes de schémas IV*, Inst. Hautes Études Sci. Publ. Math., 32 (1967).

[11] D. Hilbert, *Über die Theorie der algebraischen Formen*, Math. Ann., 42 (1893), pp. 313–370.

[12] D. Hilbert, *Mathematische Probleme*, Archiv für Math. und Physik 1 (1901), pp. 44–63.

[13] M. Hochster and J. L. Roberts, *Rings of invariants of reductive groups acting on regular rings are Cohen-Macaulay*, Advances in Math., 13 (1984), pp. 115–175.

[14] C. Huneke, *Tight closure and its applications*, C.B.M.S. Regional Conf. Ser. in Math. 88 (1996), Amer. Math. Soc., Providence, R.I., 1986.

[15] J.-M. Kantor, *Formes et opérateurs différentiels sur les espaces analytiques complexes*, Bull. Soc. Math. France Mém., 53 (1977), pp. 5–80.

[16] G. Kemper, *Computational invariant theory*, The Curves Seminar at Queen's XII, Queen's Papers in Pure and Applied Mathematics 114, Queen's University Press, Kingston, 1998, pp. 5–26.

[17] M. Kreuzer and L. Robbiano, *Computational commutative algebra 1*, Springer-Verlag, Berlin, 2000.

[18] T. Levasseur and J. T. Stafford, *Rings of differential operators on classical rings of invariants* Mem. Amer. Math. Soc., 412, Am. Math. Soc., Providence, R.I., 1989.

[19] T. Levasseur and J. T. Stafford, *Invariant differential operators and an homomorphism of Harish-Chandra*, Journal of the Amer. Math. Soc., 8 (1995), pp. 365–372.

[20] T. Levasseur, *Anneaux d'opérateurs différentiels*, Lecture Notes in Math., 867, Springer, Berlin, 1981, pp. 157–173.

[21] J. C. McConnell and J. C. Robson, *Noncommutative noetherian rings*, John Wiley and Sons, New York, 1987.

[22] D. Miličić, *Lectures on algebraic theory of D-modules*. University of Utah, 1986. Available online at: http://www.math.utah.edu/ milicic/.

[23] D. Mumford, *Hilbert's fourteenth problem – the finite generation of subrings such as rings of invariants*, Proc. Symp. in Pure Math., 28 (1974), pp. 431–444.

[24] D. Mumford, J. Fogarty and F. Kirwan, *Geometric invariant theory*, third ed., Springer-Verlag, Berlin, 2002.

[25] H. Nakajima, *Lectures on Hilbert schemes of points on surfaces*, American Mathematical Society, Providence, R.I., 1999.

[26] M. Nagata, *On the 14th problem of Hilbert*, Am. J. Math., 81 (1959), pp. 766–772.

[27] P. Olver, *Classical invariant theory*, Cambridge University Press, Cambridge, 1999.

[28] W. Plesken and D. Robertz, *Constructing invariants for finite groups*, Experiment. Math., 14 (2005), pp. 175–188.

[29] V. L. Popov and E. B. Vinberg, *Invariant theory*, Algebraic Geometry IV, (N. N. Parshin, I. R. Shafarevich, eds.), Encyclopedia of Mathematical Sciences, 55, Springer-Verlag, Berlin, 1994.

[30] M. Saito, B. Sturmfels, and N. Takayama, *Gröbner deformations of hypergeometric differential equations*, Algorithms and Computation in Mathematics, 6, Springer-Verlag, New York, 2000.

[31] G. W. Schwarz, *Lifting differential operators from orbit spaces*, Ann. Sci. École Norm. Sup. (4), 28 (1995), pp. 253–305.

[32] B. Sturmfels, *Algorithms in invariant theory*. Texts and Monographs in Symbolic Computation. Springer-Verlag, Vienna, 1993.

[33] L. Smith, *Polynomial invariants of finite groups*, A.K. Peters, Wellesley, Mass., 1995.

[34] W. N. Traves, *Differential operators on orbifolds*, Journal of Symbolic Computation, 41 (2006), pp. 1295 – 1308.

[35] W. N. Traves, *Differential operators on Grassmann varieties*, preprint (2007).

Author information

William N. Traves, Mathematics Department, U.S. Naval Academy, Mail stop 9E, Annapolis, MD, 21402, USA.
Email: traves@usna.edu

Radon Series Comp. Appl. Math **2**, 267–281 © de Gruyter 2007

Compatibility Complexes for Overdetermined Boundary Problems

Katsiaryna Krupchyk and Jukka Tuomela

Key words. Overdetermined system of partial differential equations, involution, boundary problem, compatibility complex, normalised operator.

AMS classification. 35J55, 35N10, 13P10.

1 Introduction

When analysing general systems of partial differential equations (PDEs) it is important to check if the system is involutive, and if not then transform it to the involutive form. For example, in [6] we proved that some systems may not be elliptic initially (even in the general sense), but their involutive forms are elliptic. The technical definition of the involutive form is quite complicated (see [10], [13] and [12] and for the actual definition). However, essentially the involutivity means that one has to find all integrability conditions (or differential consequences) of the given system up to some order.

Now the involutive form is in general overdetermined. To study overdetermined systems one needs to find all solvability conditions, or more generally, to construct a compatibility complex for the corresponding overdetermined operator. We show that for linear partial differential operators with constant coefficients one can compute the compatibility complex by simply computing the free resolution of the module generated by the rows of operators. Incidentally this shows that the length of the compatibility complex is at most the number of independent variables.

However, to study boundary value problems one needs to compute the compatibility operators involving the boundary operators. To perform this task it is convenient to further transform the involutive system to a normalised system. Roughly speaking, an

---

First author: Supported by the Academy of Finland, grant 108394

operator is normalised if it is a first order involutive operator and there are no (explicit or implicit) algebraic (i.e., non-differential) relations between dependent variables. A boundary value problem operator is normalised if the system is normalised and the boundary conditions contain only differentiation in directions tangent to the boundary. Computing the compatibility complex for a normalised boundary problem operator is not as straightforward as the simple free resolution, but anyway we show that the problem can be formulated again with modules, and choosing suitable module orderings we can compute the necessary information by Gröbner basis techniques. We explain how to construct the compatibility complex for a general boundary problem operator using the compatibility complex for a corresponding normalised boundary problem operator.

The construction of compatibility complexes is useful and even necessary when investigating the well-posedness of overdetermined boundary value problems. In [8] and [7] we have used compatibility complexes to study well-posedness of elliptic problems and moreover, in [9] compatibility complexes are even used in the numerical solution of PDEs. Note that constructions given in this paper are also essential in the theory of overdetermined parabolic and hyperbolic systems of PDEs.

2   Preliminaries

2.1   Formal Theory of PDEs

Let us consider a smooth[1] manifold $\mathcal{X}$. Let $\pi : V \to \mathcal{X}$ be a vector bundle over $\mathcal{X}$ and let $\pi^q : J_q(V) \to \mathcal{X}$ be the bundle of q-jets of the bundle V. Let us also introduce the canonical projections

$$\pi^q_r : J_q(V) \to J_r(V),$$

for $r < q$. Let y be a section of the bundle V. Then its qth prolongation, a section of $J_q(V)$, is denoted by $j^q y$. We write $C^\infty(V)$ for the space of smooth sections of the bundle V.

Definition 2.1 A (partial) differential equation of order q on V is a subbundle $\mathcal{R}_q$ of $J_q(V)$. Solutions of $\mathcal{R}_q$ are its (local) sections.

We will only consider linear problems, so $\mathcal{R}_q$ will be a vector bundle. Suppose V^0 and V^1 are two (vector) bundles. A linear qth order differential operator A can be thought of as a linear map $C^\infty(V^0) \to C^\infty(V^1)$. Then we can associate to A a bundle map $\varphi_A : J_q(V^0) \to V^1$ by the formula $A = \varphi_A j^q$. Now with φ_A one can represent a differential equation as a zero set of a bundle map, $\mathcal{R}_q = \ker \varphi_A$, or $\varphi_A(x, j^q y(x)) = 0$.

Definition 2.2 The differential operator $j^r A : C^\infty(V^0) \to C^\infty(J_r(V^1))$ is said to be the rth prolongation of A. The associated bundle map is denoted by $p_r(\varphi_A)$.

---

[1] In this paper smooth means infinitely differentiable.

Then we can define the rth prolongation of $\mathcal{R}_q$ by $\mathcal{R}_{q+r} = \ker p_r(\varphi_A)$. We also define $\mathcal{R}_{q+r}^{(s)} = \pi_{q+r}^{q+r+s}(\mathcal{R}_{q+r+s})$ for all $s \geq 0$. Note that $\mathcal{R}_{q+r}^{(s)} \subset \mathcal{R}_{q+r}$, but in general these sets are not equal.

Definition 2.3 A differential operator A is sufficiently regular if $\mathcal{R}_{q+r}^{(s)}$ is a vector bundle for all $r \geq 0$ and $s \geq 0$.

If $\mathcal{X} \subset \mathbb{R}^n$ and the operator A has constant coefficients, then A is sufficiently regular.

Definition 2.4 A differential operator A of order q is formally integrable if A is sufficiently regular and $\mathcal{R}_{q+r}^{(1)} = \mathcal{R}_{q+r}$ for all $r \geq 0$.

The formal integrability of an operator A of order q means that for any $r \geq 1$, all the differential consequences of order $q + r$ of the relations $Ay = 0$ may be obtained by means of differentiations of order no greater than r, and application of linear algebra.

The formal integrability cannot in general be checked in practice because there is an infinite number of conditions. Hence we need a stronger property, the *involutivity* of the system, which implies formal integrability, and can be checked in a finite number of steps. For the actual definition of involutivity we refer to [12], [10], [11]. There is the following important result.

Theorem 2.5 *For a given sufficiently regular system $\mathcal{R}_q$ there are numbers r and s such that $\mathcal{R}_{q+r}^{(s)}$ is involutive,*

In practice to complete a system to the involutive form one may use DETOOLS package [2] in the computer algebra system MuPAD [4].

In the context of the formal theory the principal symbol of the system is defined as follows. Let us first define the embedding ε_q by requiring that the following complex be exact

$$0 \longrightarrow S^q(T^*\mathcal{X}) \otimes V \xrightarrow{\varepsilon_q} J_q(V) \xrightarrow{\pi_{q-1}^q} J_{q-1}(V) \longrightarrow 0.$$

Here S^q is the bundle of symmetric tensors of order q. Recall that in a complex a composition of two consecutive maps is zero, and the exactness means that image of each map is the kernel of the following map.

Definition 2.6 Let $\mathcal{R}_q \subset J_q(V^0)$ be a sufficiently regular differential equation given by $\mathcal{R}_q = \ker \varphi_A$. The principal symbol σA of A is the map $S^q(T^*\mathcal{X}) \otimes V^0 \to V^1$ defined by $\sigma A = \varphi_A \varepsilon_q$.

To see that this actually coincides with the classical definition we need to introduce a coordinate system on $\mathcal{X}$. Then a linear qth order partial differential equation $\mathcal{R}_q$ is given by

$$Ay := \sum_{|\alpha| \leq q} a_\alpha(x) D^\alpha y = f,$$

where $x \in U \subset \mathbb{R}^n$ and $a_\alpha(x)$ are of size $k \times m$. Fixing any one form ξ we get a bundle map $\sigma A(x, \xi) : V^0|_x \to V^1|_x$ which in coordinates is given by

$$\sigma A(x, \xi) = \sum_{|\alpha|=q} a_\alpha(x) \xi^\alpha.$$

A differential operator A is said to be *elliptic* if $\sigma A(x, \xi)$ is injective for all $x \in \mathcal{X}$ and $\xi \in \mathbb{R}^n \setminus \{0\}$.

2.2 Compatibility Complexes

A linear partial differential operator $A : C^\infty(V^0) \to C^\infty(V^1)$ is *overdetermined* if there is a non-zero differential operator A^1 such that $A^1 A = 0$. Hence $A^1 f = 0$ is a necessary condition for the solvability of the system $Ay = f$.

A classical example of an overdetermined operator in $\mathcal{X} \subset \mathbb{R}^3$ is the gradient which maps a scalar function y to $\nabla y = (\partial y / \partial x_1, \partial y / \partial x_2, \partial y / \partial x_3)$. A necessary solvability condition for the system $\nabla y = f$ is

$$\nabla \times f = \left(\frac{\partial f^3}{\partial x_2} - \frac{\partial f^2}{\partial x_3}, \frac{\partial f^1}{\partial x_3} - \frac{\partial f^3}{\partial x_1}, \frac{\partial f^2}{\partial x_1} - \frac{\partial f^1}{\partial x_2} \right) = 0.$$

The operator $\nabla \times$ is itself overdetermined because $\nabla \cdot \nabla \times = 0$, where

$$\nabla \cdot h = \frac{\partial h^1}{\partial x_1} + \frac{\partial h^2}{\partial x_2} + \frac{\partial h^3}{\partial x_3}.$$

Hence setting $V^0 = \mathcal{X} \times \mathbb{R}$ and $V^1 = \mathcal{X} \times \mathbb{R}^3$ we get a complex

$$0 \longrightarrow C^\infty(V^0) \xrightarrow{\nabla} C^\infty(V^1) \xrightarrow{\nabla \times} C^\infty(V^1) \xrightarrow{\nabla \cdot} C^\infty(V^0) \longrightarrow 0.$$

The main problem in studying overdetermined systems consists in finding all solvability conditions for a given system $Ay = f$. The following definition explains the meaning of the words "all solvability conditions".

Definition 2.7 Let $A^0 : C^\infty(V^0) \to C^\infty(V^1)$ be a differential operator. A differential operator $A^1 : C^\infty(V^1) \to C^\infty(V^2)$ is a *compatibility operator* for A^0 if

(i) $A^1 A^0 = 0$ and

(ii) for any differential operator $\tilde{A}^1 : C^\infty(V^1) \to C^\infty(\tilde{V}^2)$ such that $\tilde{A}^1 A^0 = 0$, there is a differential operator $T : C^\infty(V^2) \to C^\infty(\tilde{V}^2)$ such that $\tilde{A}^1 = TA^1$.

This idea leads naturally to

Definition 2.8 A complex

$$\mathcal{C} : \quad 0 \longrightarrow C^\infty(V^0) \xrightarrow{A^0} C^\infty(V^1) \xrightarrow{A^1} C^\infty(V^2) \xrightarrow{A^2} \cdots$$

is a *compatibility complex* for A^0 if every differential operator A^i for $i \geq 1$ is a compatibility operator for A^{i-1}.

The following theorem gives the main result about the existence of compatibility complexes (see [3], [10] and [13] for more details).

Theorem 2.9 *Every sufficiently regular differential operator has a compatibility complex.*

2.3 Cochain Equivalence

We want to construct the compatibility complex for a given operator. However, it turns out that to do this we must first complete the system into involutive form, and then reduce it to a certain first order system. These other systems should be equivalent to the original one in order that this construction makes sense. The following definition gives the appropriate meaning of equivalence.

Definition 2.10 Two complexes

$$\mathcal{C} \quad : \quad \cdots \longrightarrow C^\infty(V^i) \xrightarrow{\Phi^i} C^\infty(V^{i+1}) \longrightarrow \cdots$$

$$\tilde{\mathcal{C}} \quad : \quad \cdots \longrightarrow C^\infty(\tilde{V}^i) \xrightarrow{\tilde{\Phi}^i} C^\infty(\tilde{V}^{i+1}) \longrightarrow \cdots$$

are *cochain equivalent* if the following conditions are satisfied:

1. there are differential operators M^i and N^i such that the following diagram commutes for all i

$$
\begin{array}{ccc}
C^\infty(V^i) & \xrightarrow{\Phi^i} & C^\infty(V^{i+1}) \\
{\scriptstyle M^i}\big\downarrow\big\uparrow{\scriptstyle N^i} & & {\scriptstyle M^{i+1}}\big\downarrow\big\uparrow{\scriptstyle N^{i+1}} \\
C^\infty(\tilde{V}^i) & \xrightarrow{\tilde{\Phi}^i} & C^\infty(\tilde{V}^{i+1})
\end{array}
$$

2. there are differential operators Ψ^i and $\tilde{\Psi}^i$ such that for all i

$$\Psi^i \Phi^i + \Phi^{i-1}\Psi^{i-1} = \mathrm{id} - N^i M^i$$

$$\tilde{\Psi}^i \tilde{\Phi}^i + \tilde{\Phi}^{i-1}\tilde{\Psi}^{i-1} = \mathrm{id} - M^i N^i$$

Definition 2.11 Operators $\Phi : C^\infty(V^0) \to C^\infty(V^1)$ and $\tilde{\Phi} : C^\infty(\tilde{V}^0) \to C^\infty(\tilde{V}^1)$ are *cochain equivalent* if the complexes

$$0 \longrightarrow C^\infty(V^0) \xrightarrow{\Phi} C^\infty(V^1)$$

$$0 \longrightarrow C^\infty(\tilde{V}^0) \xrightarrow{\tilde{\Phi}} C^\infty(\tilde{V}^1)$$

are cochain equivalent.

If we know a compatibility complex for some operator, we can construct a compatibility complex for a cochain equivalent operator as follows.

Theorem 2.12 *Let Φ^0 and $\tilde{\Phi}^0$ be cochain equivalent differential operators. If there is a compatibility complex for $\tilde{\Phi}^0$, then there is also a compatibility complex for Φ^0. Moreover their compatibility complexes are cochain equivalent.*

Proof. Suppose that we know a compatibility complex for $\tilde{\Phi}^0$ and that a compatibility operator Φ^i, operators M^{i+1}, N^{i+1}, and Ψ^i have already been constructed as in Definition 2.10. Then we can construct Φ^{i+1} by the following formula.

$$\Phi^{i+1} \; : \; C^\infty(V^{i+1}) \to C^\infty(V^{i+2}) = C^\infty(\tilde{V}^{i+2} \oplus V^{i+1})$$
$$\Phi^{i+1} = \left(\tilde{\Phi}^{i+1} M^{i+1}\right) \oplus \left(\mathsf{id} - N^{i+1} M^{i+1} - \Phi^i \Psi^i\right). \tag{2.1}$$

For the details of the proof we refer to [13, p. 28]. ☐

3 Compatibility Complexes for Differential Operators

Consider a differential operator $A^0 : C^\infty(V^0) \to C^\infty(V^1)$ with constant coefficients on an open set $\mathcal{X} \subset \mathbb{R}^n$ where $V^0 = \mathcal{X} \times \mathbb{R}^{k_0}$ and $V^1 = \mathcal{X} \times \mathbb{R}^{k_1}$.

Let us introduce the *full symbol* of A^0:

$$\sigma_{\mathsf{F}}(A^0) = \sum_{|\alpha| \leq q} a_\alpha \xi^\alpha.$$

Let $\mathbb{A} = \mathbb{K}[\xi_1, \ldots, \xi_n]$ be a polynomial ring in n variables where $\mathbb{K}$ is some field of characteristic zero that contains the coefficients of the differential operator A^0. Denoting by $a^1, \ldots, a^{k_1}$ the rows of $\sigma_{\mathsf{F}}(A_0)$ we may construct a free resolution of the module $M = \mathbb{A}^{k_0}/M_0$ where $M_0 = \langle a^1, \ldots, a^{k_1} \rangle$:

$$0 \longrightarrow \mathbb{A}^{k_r} \xrightarrow{\mathsf{A}_r^T} \cdots \xrightarrow{\mathsf{A}_1^T} \mathbb{A}^{k_1} \xrightarrow{\sigma_{\mathsf{F}}(A^0)^T} \mathbb{A}^{k_0} \longrightarrow M \longrightarrow 0 \; .$$

Let A^i be the differential operator with the full symbol matrix $\sigma_{\mathsf{F}}(A^i) = \mathsf{A}_i$. In this case we say that the differential operator A^i is associated to the syzygy matrix A_i^T. Now a complex $\mathcal{C}$ consisting of trivial bundles $V^i = \mathcal{X} \times \mathbb{R}^{k_i}$ and operators A^i is said to be a *Hilbert complex*, if the operators A^i are associated to the syzygy matrices of the free resolution of $\mathbb{A}$-module M.

Theorem 3.1 *[13, p. 31] Let $\mathcal{C}$ be a complex of differential operators with constant coefficients. $\mathcal{C}$ is a compatibility complex for A^0 if and only if $\mathcal{C}$ is a Hilbert complex associated with the $\mathbb{A}$-module M.*

Hence the compatibility complex for a differential operator with constant coefficients on an open set in $\mathbb{R}^n$ can be constructively computed using Gröbner basis techniques.

Example 3.2 Consider the following system $Ay = (\nabla \times y, \nabla \cdot y)$. Let $b^1, \ldots, b^4$ be the rows of the full symbol[1] of A

$$\sigma_F(A) = \begin{pmatrix} 0 & -\xi_3 & \xi_2 \\ \xi_3 & 0 & -\xi_1 \\ -\xi_2 & \xi_1 & 0 \\ \xi_1 & \xi_2 & \xi_3 \end{pmatrix}, \quad M_0 = \langle b^1, \ldots, b^4 \rangle.$$

Computing the syzygy of module M_0, we get

$$S = (\xi_1, \xi_2, \xi_3, 0), \quad M_1 = \langle S \rangle.$$

Computing the syzygy of M_1, we get $M_2 = 0$. Hence we have the following free resolution for $\mathbb{A}^3/M_0$:

$$0 \longrightarrow \mathbb{A}^1 \xrightarrow{S^T} \mathbb{A}^4 \xrightarrow{A^T} \mathbb{A}^3 \longrightarrow \mathbb{A}^3/M_0 \longrightarrow 0 .$$

Thus, the compatibility complex for A is

$$0 \longrightarrow C^\infty(\mathcal{X} \times \mathbb{R}^3) \xrightarrow{A} C^\infty(\mathcal{X} \times \mathbb{R}^4) \xrightarrow{(\nabla \cdot, 0)} C^\infty(\mathcal{X} \times \mathbb{R}) \longrightarrow 0 .$$

4 Compatibility Complexes for Boundary Problem Operators

From now on we suppose that $\mathcal{X}$ is a manifold with boundary $\mathcal{Y}$. To consider boundary value problems let us introduce two bundles W^0 and W^1 whose base manifold is the boundary $\mathcal{Y}$. The bundle $V^i|_y \to \mathcal{Y}$ is the restriction of $V^i \to \mathcal{X}$ to the boundary. If y is a section of $V^i \to \mathcal{X}$, then γy is the corresponding section of $V^i|_y \to \mathcal{Y}$. The map γ is called the trace map.

Definition 4.1 An operator $\mathcal{A} : C^\infty(V^0) \times C^\infty(W^0) \to C^\infty(V^1) \times C^\infty(W^1)$ of the form

$$\mathcal{A}(y, w) = \begin{pmatrix} \mathcal{A}_{1,1} & 0 \\ \gamma \mathcal{A}_{2,1} & \mathcal{A}_{2,2} \end{pmatrix} \begin{pmatrix} y \\ w \end{pmatrix},$$

where $\mathcal{A}_{i,j}$ are differential operators, is called a *boundary problem operator*.

---

[1] In fact in this case $\sigma_F(A) = \sigma A$.

If $W^0 = 0$, we obtain in this way an operator $\mathcal{A}(y) = (Ay, \gamma By)$ which defines a classical boundary problem on $\mathcal{X}$.

It turns out that one can construct the compatibility complex for a boundary problem operator $\mathcal{A} = (A, \gamma B)$ using a certain equivalent first order system.

Definition 4.2 A differential operator $A : C^\infty(V^0) \to C^\infty(V^1)$ is *normalised* if

  (i) A is a first order operator;

 (ii) A is involutive;

(iii) the principal symbol $\sigma A : T^*\mathcal{X} \otimes V^0 \to V^1$ is surjective.

Condition (iii) means that there are no (explicit or implicit) algebraic (i.e., non-differential) relations between dependent variables in the system. If such relations exist, then we may use them to reduce the number of dependent variables.

Theorem 4.3 *Every sufficiently regular operator A can be transformed in a finite number of steps into an equivalent normalised operator.*

Definition 4.4 A boundary problem operator $\mathcal{A}$ is normalised if $\mathcal{A}_{1,1}$ is normalised and $\gamma \mathcal{A}_{2,1}$ contains only differentiation in directions tangent to the boundary.

Theorem 4.5 *Every boundary problem operator $\mathcal{A}$ whose component $\mathcal{A}_{1,1}$ is sufficiently regular is cochain equivalent to a normalised boundary problem operator.*

For the proofs of the above theorems we refer to [13].

To construct compatibility operators, we introduce the tangent part of a first order differential operator $A : C^\infty(V^0) \to C^\infty(V^1)$. Let us choose a coordinate system $x = (x_1, \ldots, x_n)$ such that the boundary is given by the equation $x_n = 0$. Then in these coordinates there is a part of A which contains differentiations only with respect to $x_1, \ldots, x_{n-1}$. This part is denoted by $A^\tau : C^\infty(V^0|_\mathcal{Y}) \to C^\infty(V^{1\tau})$ where $V^{1\tau}$ is a certain bundle over $\mathcal{Y}$, and it is called the *tangent part* of A. It can be shown that if A is normalised, then so is A^τ [3].

If we have a system $Ay = f$, then for the tangent part we get a system $A^\tau y = f^\tau$. This defines a projection

$$\mathrm{pr}^\tau : V^1 \to V^{1\tau} \quad , \quad \mathrm{pr}^\tau(f) = f^\tau .$$

Now we can rewrite any normalised boundary problem operator in the form

$$\mathcal{A} : C^\infty(V^0) \times C^\infty(W^0) \to C^\infty(V^1) \times C^\infty(W^1)$$

$$\mathcal{A}(y, w) = \begin{pmatrix} \mathcal{A}_{1,1} & 0 \\ \mathcal{A}_{2,1}\gamma & \mathcal{A}_{2,2} \end{pmatrix} \begin{pmatrix} y \\ w \end{pmatrix} \tag{4.1}$$

where $\mathcal{A}_{2,1}$ is a differential operator on the boundary $\mathcal{Y}$. Then on $\mathcal{Y}$ we define a differential operator $\mathcal{A}^\tau : C^\infty(V^0|_\mathcal{Y}) \times C^\infty(W^0) \to C^\infty(V^{1\tau}) \times C^\infty(W^1)$,

$$\mathcal{A}^\tau(y^\tau, w) = \begin{pmatrix} \mathcal{A}_{1,1}^\tau & 0 \\ \mathcal{A}_{2,1} & \mathcal{A}_{2,2} \end{pmatrix} \begin{pmatrix} y^\tau \\ w \end{pmatrix} . \tag{4.2}$$

Definition 4.6 A normalised boundary problem operator $\mathcal{A}$ is *regular* if the differential operator $\mathcal{A}^\tau$ is sufficiently regular.

Definition 4.7 A boundary problem operator $\mathcal{A}$ is *regular* if the differential operators $\mathcal{A}_{1,1}$ and $\mathcal{A}^\tau_{\text{norm}}$ are sufficiently regular where $\mathcal{A}^\tau_{\text{norm}}$ corresponds to an equivalent normalised operator.

We start the construction of a compatibility operator. It suffices to consider regular normalised boundary problem operators. Then, we can use Theorems 4.5 and 2.12, which enable us to construct a compatibility operator for an arbitrary regular boundary problem operator.

Let $\mathcal{A}$ be a regular normalised boundary problem operator given by (4.1), $\mathcal{A}^\tau_{1,1}$ be the tangent part of $\mathcal{A}_{1,1}$ and $\mathcal{A}^\tau$ the operator defined by (4.2). As $\mathcal{A}^\tau$ is sufficiently regular, by Theorem 2.9 it has a compatibility operator $\mathcal{A}^{\tau 1}$ which can always be written in the form

$$\mathcal{A}^{\tau 1}(f^\tau, g) = \begin{pmatrix} \mathcal{A}^{\tau 1}_{1,1} f^\tau \\ \Upsilon^\tau(f^\tau, g) \end{pmatrix}. \tag{4.3}$$

Here $\mathcal{A}^{\tau 1}_{1,1}$ is a compatibility operator for $\mathcal{A}^\tau_{1,1}$ and Υ^τ does not contain any relations involving only the components of f^τ. Let us then finally define

$$\mathcal{A}^1 : C^\infty(V^1) \times C^\infty(W^1) \to C^\infty(V^2) \times C^\infty(W^2),$$

$$\mathcal{A}^1(f, g) = \begin{pmatrix} \mathcal{A}^1_{1,1} f \\ \Upsilon^\tau(\mathsf{pr}^\tau f, g) \end{pmatrix}, \tag{4.4}$$

where $\mathcal{A}^1_{1,1}$ is a compatibility operator for $\mathcal{A}_{1,1}$. We will need the following important result [3, p. 40].

Theorem 4.8 *Let $\mathcal{A}$ be a regular normalised boundary operator whose component $\mathcal{A}_{1,1}$ is elliptic. Then $\mathcal{A}^1$ defined by (4.4) is a compatibility operator for $\mathcal{A}$.*

The operator $\mathcal{A}^1$ is itself regular and normalised. This together with ellipticity of $\mathcal{A}_{1,1}$ enables us to construct the whole compatibility complex for $\mathcal{A}$

$$0 \longrightarrow C^\infty(V^0) \times C^\infty(W^0) \xrightarrow{\mathcal{A}} C^\infty(V^1) \times C^\infty(W^1) \xrightarrow{\mathcal{A}^1} \cdots$$

If $\mathcal{A}$ is regular but not normalised, then it needs to be replaced by an equivalent normalised operator for which the compatibility complex is constructed. But then by Theorem 2.12 we can construct the compatibility complex for $\mathcal{A}$ using the compatibility complex of the corresponding normalised operator.

5 Computations

Here we show that on each step of the construction of a compatibility operator for a normalised boundary problem operator one may effectively use Gröbner basis computations.

- **Computation of the tangent part A^τ of a differential operator A.**

  Let $M \subset \mathbb{A}^m$ be the module generated by the rows of the full symbol of A. We choose a product ordering such that ξ_n is bigger than all other ξ_i. Then we define a TOP module ordering using this ordering and compute the Gröbner basis of M. Now A^τ is defined by the elements of the Gröbner basis that do not contain ξ_n. This follows from the fact [1, p. 156] that if G is a Gröbner basis for M, then $G \cap \tilde{\mathbb{A}}^m$ is a Gröbner basis for $M \cap \tilde{\mathbb{A}}^m$ where $\tilde{\mathbb{A}} = \mathbb{K}[\xi_1, \ldots, \xi_{n-1}]$.

- **Computation of a compatibility operator $A^{\tau 1}$ for A^τ defined in** (4.2).

  Since A^τ is a differential operator on the manifold $\mathcal{Y}$ without boundary, one can compute its compatibility operator by computing the syzygy module of the module generated by the rows of the full symbol of A^τ, see Theorem 3.1.

- **Computation of operator Υ^τ in** (4.3)

  In the previous step we computed $A^{\tau 1}$; now we would like to eliminate rows which contain only f^τ. To do this we choose a POT module ordering (and any convenient monomial ordering) and compute the Gröbner basis G of the module generated by the rows of $\sigma_\mathsf{F}(A^{\tau 1})$. The elements of G correspond to differential operators which operate on (f^τ, g). The full symbol of Υ^τ is now obtained by discarding those elements whose corresponding differential operators operate on f^τ only.

  The correctness of this construction follows from the following fact. Let M be a submodule of $\mathbb{A}^m = \mathbb{A}^i \oplus \mathbb{A}^{m-i}$. We choose a POT module ordering for $\mathbb{A}^m$, and any monomial ordering in $\mathbb{A}$. Then if G is a Gröbner basis for M, then $G \cap \mathbb{A}^{m-i}$ is a Gröbner basis for $M \cap \mathbb{A}^{m-i}$ [5, p. 177].

6 Example

Let us construct a compatibility complex for the stationary Stokes problem in two dimensions. Consider the boundary problem in $\mathbb{R}^2_+ = \{x \in \mathbb{R}^2 : x_2 > 0\}$

$$
A : \begin{cases} -\Delta u + \nabla p = 0, \\ \nabla \cdot u = 0 \end{cases} \quad \text{in } \mathcal{X} = \mathbb{R}^2_+, \quad B : \begin{cases} -u^2_{20} + p_{01} = 0, \\ u^1_{11} = 0 \end{cases} \quad \text{on } \mathcal{Y} = \partial\mathbb{R}^2_+,
$$

where $u = (u^1, u^2)$ is the velocity field and p is the pressure. Here we have used for the derivatives the notation

$$
u^i_{k\ell} = \frac{\partial^{k+\ell} u^i}{\partial x_1^k \partial x_2^\ell}.
$$

By completing the above system to the involutive form we arrive at an overdetermined system

$$
A_0 : \quad
\begin{cases}
-\Delta u + \nabla p &= 0, \\
\Delta p &= 0, \\
\nabla \cdot u &= 0, \\
u_{20}^1 + u_{11}^2 &= 0, \\
u_{11}^1 + u_{02}^2 &= 0
\end{cases}
\tag{6.1}
$$

Introducing nine new variables (unknown functions)

$$
\begin{aligned}
v^{1,00} &= u^1, & v^{1,10} &= u_{10}^1, & v^{1,01} &= u_{01}^1, \\
v^{2,00} &= u^2, & v^{2,10} &= u_{10}^2, & v^{2,01} &= u_{01}^2, \\
v^{3,00} &= p, & v^{3,10} &= p_{10}, & v^{3,01} &= p_{01}
\end{aligned}
$$

and substituting them into (6.1), and also adding the compatibility equations we get the first order system

$$
A_0' : \quad
\begin{cases}
-v_{10}^{1,10} - v_{01}^{1,01} + v^{3,10} &= 0, \\
-v_{10}^{2,10} - v_{01}^{2,01} + v^{3,01} &= 0, \\
v_{10}^{3,10} + v_{01}^{3,01} &= 0, \\
v^{1,10} + v^{2,01} &= 0, \\
v_{10}^{1,10} + v_{10}^{2,01} &= 0, \\
v_{01}^{1,10} + v_{01}^{2,01} &= 0, \\
v_{10}^{j,00} - v^{j,10} &= 0, \\
v_{01}^{j,00} - v^{j,01} &= 0, \\
v_{01}^{j,10} - v_{10}^{j,01} &= 0
\end{cases}
$$

for $j = 1, 2, 3$. This system is not normalised since there is an algebraic relation $v^{1,10} + v^{2,01} = 0$ between the dependent variables. Using this relation we can now eliminate the unknown function $v^{2,01}$ from the system and obtain the following normalised system

$$
A_0'' : \quad
\begin{cases}
-v_{10}^{1,10} - v_{01}^{1,01} + v^{3,10} &= 0, \\
v_{01}^{1,10} - v_{10}^{2,10} + v^{3,01} &= 0, \\
v_{10}^{3,10} + v_{01}^{3,01} &= 0, \\
v_{10}^{j,00} - v^{j,10} &= 0, \\
v_{10}^{2,00} - v^{2,10} &= 0, \\
v_{01}^{j,00} - v^{j,01} &= 0, \\
v_{01}^{2,00} + v^{1,10} &= 0, \\
v_{01}^{j,10} - v_{10}^{j,01} &= 0, \\
v_{10}^{1,10} + v_{01}^{2,10} &= 0
\end{cases}
$$

for $j = 1, 3$. Finally, substituting the new unknown functions in the boundary conditions, we obtain

$$B'' : \quad \begin{cases} -v_{10}^{2,10} + v^{3,01} &= 0, \\ v_{10}^{1,01} &= 0. \end{cases}$$

Hence it follows that the classical boundary problem operator $\mathcal{A}'' = (A_0'', B'')$ is normalised. Using Gröbner basis computations, we find the tangent part of A_0''

$$A_0''^{\tau} : \quad \begin{cases} v_{10}^{2,10} - v_{10}^{1,01} - v^{3,01} &= 0, \\ v_{10}^{3,00} - v^{3,10} &= 0, \\ v_{10}^{2,00} - v^{2,10} &= 0, \\ v_{10}^{1,00} - v^{1,10} &= 0. \end{cases}$$

Let us compute a compatibility operator for the operator $\mathcal{A}''^{\tau} = (A_0''^{\tau}, B'')$ defined on the boundary $\mathcal{Y}$. Computing the syzygy module for the module generated by the rows of the full symbol matrix of the operator $\mathcal{A}''^{\tau}$, we get

$$\mathcal{A}''^{\tau 1}(f^{\tau}, g) = f^{1^{\tau}} + g^1 + g^2$$

where $f^{\tau} = (f^{1^{\tau}}, \ldots, f^{4^{\tau}})$ and $g = (g^1, g^2)$. Now let us compute a compatibility operator for A_0''. Computing the syzygy module of the module generated by the rows of the full symbol of A_0'', we have

$$A_0''^{1}(f'') = \begin{pmatrix} f_{01}''^4 - f_{10}''^7 + f''^{10} \\ f_{01}''^5 - f_{10}''^8 + f''^{11} \\ f_{01}''^6 - f_{10}''^9 + f''^{12} \\ -f_{10}''^1 - f_{01}''^2 + f''^3 + f_{01}''^{10} - f_{10}''^{12} \end{pmatrix}$$

where $f'' = (f''^1, \ldots, f''^{12})$. Note that the projection $\mathrm{pr}^{\tau} : V''^1 \to V''^{1^{\tau}}$ is given by

$$\mathrm{pr}^{\tau}(f'') = (f''^{10}|_{\mathcal{Y}} - f''^2|_{\mathcal{Y}}, f''^5|_{\mathcal{Y}}, f''^6|_{\mathcal{Y}}, f''^4|_{\mathcal{Y}}).$$

Using (4.4) we find a compatibility operator for the normalised boundary problem operator $\mathcal{A}''$,

$$\mathcal{A}''^{1}(f'', g) = \begin{pmatrix} f_{01}''^4 - f_{10}''^7 + f''^{10} \\ f_{01}''^5 - f_{10}''^8 + f''^{11} \\ f_{01}''^6 - f_{10}''^9 + f''^{12} \\ -f_{10}''^1 - f_{01}''^2 + f''^3 + f_{01}''^{10} - f_{10}''^{12} \\ f''^{10}|_{\mathcal{Y}} - f''^2|_{\mathcal{Y}} + g^1 + g^2 \end{pmatrix}.$$

Now we compute that the tangent part of $\mathcal{A}''^1_{1,1}$ is the zero operator. Then (4.2) implies that

$$\mathcal{A}''^{1^{\tau}}(f''^{\tau}, g) = \begin{pmatrix} 0 \\ f''^{10^{\tau}} - f''^{2^{\tau}} + g^1 + g^2 \end{pmatrix}.$$

It is clear that the compatibility operator for $\mathcal{A}''^{1^\top}$ is the zero operator. Since the compatibility operator for $\mathcal{A}''^1_{1,1}$ is equal to zero, the compatibility operator for $\mathcal{A}''^1$ is the zero operator as well. Hence, we arrive at the following compatibility complex for the normalised boundary problem operator $\mathcal{A}''$,

$$
\begin{array}{ccc}
0 \longrightarrow C^\infty(V''^0) \xrightarrow{\ \mathcal{A}''\ } & C^\infty(V''^1) \times C^\infty(W^1) \\
& \Big\downarrow {\scriptstyle \mathcal{A}''^1} \\
& C^\infty(V''^2) \times C^\infty(W''^2) \longrightarrow 0
\end{array}
$$

Now we will construct a compatibility complex for the original boundary problem operator $\mathcal{A} = (A, B)$. First simple computations show that boundary problem operators $\mathcal{A}''$ and $\mathcal{A}$ are cochain equivalent with the following operators in Definition 2.11,

$$
\begin{array}{ccc}
0 \longrightarrow C^\infty(V^0) & \xrightarrow{\ \mathcal{A}\ } & C^\infty(V^1) \times C^\infty(W^1) \\
{\scriptstyle M^0}\Big\downarrow\Big\uparrow{\scriptstyle N^0} & & {\scriptstyle M^1}\Big\downarrow\Big\uparrow{\scriptstyle N^1} \\
0 \longrightarrow C^\infty(V''^0) & \xrightarrow{\ \mathcal{A}''\ } & C^\infty(V''^1) \times C^\infty(W^1)
\end{array}
$$

$M^0(u^1, u^2, p) = (u^1, u^2, p, u^1_{10}, u^2_{10}, p_{10}, u^1_{01}, p_{01}),$

$N^0(v^{1,00}, v^{2,00}, v^{3,00}, v^{1,10}, v^{2,10}, v^{3,10}, v^{1,01}, v^{3,01}) = (v^{1,00}, v^{2,00}, v^{3,00}),$

$M^1(f^1, f^2, f^3, g^1, g^2) =$

$\qquad (f^1, f^2 + f^3_{01}, f^1_{10} + f^2_{01} + \Delta f^3, 0, 0, 0, 0, 0, f^3, 0, 0, f^3_{10}, g^1, g^2),$

$N^1(f''^1, \ldots, f''^{12}, g^1, g^2) =$

$\qquad (f''^1 - f''^4_{10} - f''^7_{01} + f''^5, f''^2 - f''^6_{10} - f''^9_{01} + f''^8, f''^4 + f''^9, g^1, g^2),$

$\Psi^0(f^1, f^2, f^3, g^1, g^2) = 0,$

$\tilde\Psi^0(f''^1, \ldots, f''^{12}, g^1, g^2) = (0, 0, 0, -f''^4, -f''^6, -f''^5, -f''^7, -f''^8).$

Then using formula (2.1), we define compatibility operator for $\mathcal{A}$ by

$$
\mathcal{A}^1 : C^\infty(V^1) \times C^\infty(W^1) \to C^\infty(V''^2 \oplus V^1) \times C^\infty(W''^2 \oplus W^1),
$$

$$
\mathcal{A}^1 = \mathcal{A}''^1 M^1 \oplus (\mathrm{id} - N^1 M^1).
$$

Computing $\mathcal{A}^1$, we get

$$
\mathcal{A}^1(f^1, f^2, f^3, g^1, g^2) = \begin{pmatrix} 0 & 0 & 0 & 0 & -f^2|_y - f^3_{01}|_y + g^1 + g^2 & 0 & 0 & 0 & 0 & 0 \end{pmatrix}.
$$

Hence, we arrive at the following compatibility complex for the original boundary problem operator $\mathcal{A}$,

$$0 \longrightarrow C^\infty(V^0) \xrightarrow{\ \mathcal{A}\ } C^\infty(V^1) \times C^\infty(W^1)$$

$$\downarrow \mathcal{A}^1$$

$$C^\infty(V''^2 \oplus V^1) \times C^\infty(W''^2 \oplus W^1) \longrightarrow 0\ .$$

Bibliography

[1] W. Adams and P. Loustaunau, *An introduction to Gröbner bases*, Graduate Studies in Mathematics, vol. 3, American Mathematical Society, 1994.

[2] J. Belanger, M. Hausdorf, and W. Seiler, *A MuPAD Library for Differential Equations*, Computer Algebra in Scientific Computing — CASC 2001 (V.G. Ghanza, E.W. Mayr, and E.V. Vorozhtsov, eds.), Springer-Verlag, Berlin/Heidelberg, 2001, pp. 25–42.

[3] P. I. Dudnikov and S. N. Samborski, *Linear Overdetermined Systems of Partial Differential Equations. Initial and Initial-Boundary Value Problems*, Partial Differential Equations VIII (M.A. Shubin, ed.), Encyclopaedia of Mathematical Sciences 65, Springer-Verlag, Berlin/Heidelberg, 1996, pp. 1–86.

[4] J. Gerhard, W. Oevel, F. Postel, and S. Wehmeier, MuPAD TUTORIAL, Springer, 2000, http://www.mupad.de/.

[5] G.-M. Greuel and G. Pfister, *A **Singular** introduction to commutative algebra*, Springer, 2002.

[6] K. Krupchyk, W. Seiler, and J. Tuomela, *Overdetermined elliptic PDEs*, Found. Comp. Math. 6 (2006), pp. 309–351.

[7] K. Krupchyk, N. Tarkhanov, and J. Tuomela, *Elliptic Quasicomplexes in Boutet de Monvel Algebra*, J. Funct. Anal. 247 (2007), no. 1, pp. 202–230.

[8] K. Krupchyk and J. Tuomela, *The Shapiro-Lopatinskij condition for elliptic boundary value problems*, LMS J. Comput. Math. 9 (2006), pp. 287–329.

[9] B. Mohammadi and J. Tuomela, *Simplifying numerical solution of constrained PDE systems through involutive completion*, M2AN 39 (2005), pp. 909–929.

[10] J. F. Pommaret, *Systems of Partial Differential Equations and Lie Pseudogroups*, Mathematics and its applications, vol. 14, Gordon and Breach Science Publishers, 1978.

[11] W. M. Seiler, *Involution — The Formal Theory of Differential Equations and its Applications in Computer Algebra and Numerical Analysis*, Habilitation thesis, Dept. of Mathematics, Universität Mannheim, 2001, (manuscript accepted for publication by Springer-Verlag).

[12] D. Spencer, *Overdetermined systems of linear partial differential equations*, Bull. Am. Math. Soc 75 (1969), pp. 179–239.

[13] N. N. Tarkhanov, *Complexes of differential operators*, Mathematics and its Applications, vol. 340, Kluwer Academic Publishers Group, 1995, Translated from the 1990 Russian original by P. M. Gauthier and revised by the author.

Author information

Katsiaryna Krupchyk, Department of Physics and Mathematics, University of Joensuu, P.O. Box 111, FI-80101 Joensuu, Finland.
Email: katya.krupchyk@joensuu.fi

Jukka Tuomela, Department of Physics and Mathematics, University of Joensuu, P.O. Box 111, FI-80101 Joensuu, Finland.
Email: jukka.tuomela@joensuu.fi

Radon Series Comp. Appl. Math **2**, 283–340

On Initial Value Problems for Ordinary Differential-Algebraic Equations

F. Leon Pritchard and William Y. Sit

Key words. Differential algebraic equation, initial value problem, quasi-linear, Gröbner basis, differential algebra, prolongation, differential index, algebraic index, existence and uniqueness, essential degree.

AMS classification. 34A12, 34A09, 34A34, 65L05, 65L80, 53-04, 12H05.

1   Introduction

Differential algebraic equations arise in many applications such as electrical networks, dynamics of incompressible fluids, and constrained robotic systems. Many authors have worked on both theoretical and practical aspects of solving initial value problems of differential algebraic equations. There are several concerns.

1.1   Concern 1: Singularities

An implicit differential algebraic system is often studied initially as an algebraic system and the first concern is that of singularities. Singularity analysis is a difficult problem and computer algebra systems now exist to specifically study this topic (see for example, the system SINGULAR in Greuel [27]). In a series of papers, Campbell [11, 12, 13] introduced the notion of an index as a measure of singularity for systems of differential equations. Gear and Petzold [24, 25], Gear [22], and Reich [52, 53] studied linear index and differentiation index. Thomas [62] discussed the problem of defining singular points for differential equations that are *quasi-linear* namely, first order equations which are linear in the first order derivatives. Impasse points were studied in Rabier and Rheinboldt [49]. Tuomela [63] used the notion of an integral manifold of a distribution to resolve certain singularities and he [64] proposed a method to regularize singular systems by using jet spaces. In the case that the equations are first order and quasi-linear, a property that is commonly assumed to avoid singular points is that the matrices involved have constant rank. This, however, is only a sufficient condition, and Kunkel and Mehrmann [39] gave an example which has a unique solution for consistent initial values but which does not satisfy the constant rank condition. Rabier and Rheinboldt [50, 51] also studied discontinuous solutions that extend past impasse points for semi-linear systems.

The computation of the differentiation index for a system of differential algebraic equations has been approached both by a combination of symbolic and numeric methods (Campbell and Griepentrog [15]). Recently, probabilistic methods of the Monte Carlo type are used by Matera and Sedoglavic [43], where the non-linear equations of the system are assumed to be differentially algebraically independent. Thomas [61] proposed a symbolic method for computing the differentiation index for a quasi-linear system

$$A(t, \mathbf{x})\mathbf{p} = b(t, \mathbf{x}) \tag{1.1}$$

where A and b have entries polynomial in t and $\mathbf{x}$ (here $\mathbf{x}$ is a vector function of t, $\mathbf{p} = \dot{\mathbf{x}} = d\mathbf{x}/dt$, A is a matrix and b is a vector). His method is based on algebraic geometry and he viewed (1.1) as an algebraic system in $(t, \mathbf{x}, \mathbf{p})$. In a lemma, he asserted that the irreducible components of the associated algebraic set are defined by what he called simple systems. These are then prolonged by differentiating the algebraic equations that involve only the variables $t, \mathbf{x}$. Repeating the decompose-prolong process produces a tree of simple systems which must have finite depth. It seems to us what he defined as *the* geometric index may depend on the sequence of irreducible varieties occurring in the reduction processs (that is, on a choice of path in the tree).

There are many other notions of index for differential algebraic systems. For a discussion on how some of these relate, see Campbell and Gear [14].

1.2   Concern 2: Numerical Solutions

The second concern is obtaining numerical solutions. Standard numerical methods (mostly based on a backward difference formula for the first order derivatives) for implicit systems often failed when the constant rank condition is not satisfied, or when the differentiation index is higher than 1. On the one hand, implicit systems of the form $F(t, \mathbf{x}, \mathbf{p}) = 0$ are generally not solvable by these standard numerical methods. On the other hand, while reduction processes may rewrite the system explicitly for the first order derivatives, numerical methods may not preserve the algebraic constraints that are present because of round-off errors (the "drift"), and for high index systems, the numerical methods may not be stable when the step sizes are small. Campbell [13] applied repeated differentiations to the system to reduce the index. Nielsen [45] described an implementation of the singly implicit Runge-Kutta method for certain implicit systems and differential algebraic equations. For an extensive survey of modern advances with numerical methods that address these and other difficulties, we refer readers to Petzold [46] and to the references therein. An introductory graduate textbook on the subject is Ascher and Petzold [2].

A new breed of numerical methods called geometric integration that aims at preserving certain algebraic constraints in an explicit system by controlling the errors has been studied by Iserles and Zanna [33]. In special cases, these methods can be combined with the method of the present paper to provide accurate numerical solutions for implicit systems.

1.3   Concern 3: Existence and Uniqueness

The third concern is existence and uniqueness, which usually are obtained locally by reducing the given implicit system to an explicit system. We describe two developments that illustrate the theoretical difficulties.

Kunkel and Mehrmann [37, 38, 39] (see also [40]) studied linear differential algebraic equations of the form

$$E(t)\dot{\mathbf{x}}(t) = A(t)\mathbf{x}(t) + f(t) \tag{1.2}$$

in terms of three numbers: the rank $r(t)$ of $E(t)$, and the ranks $a(t)$ and $s(t)$ of two matrices derived *pointwisely* from $E(t)$ and $A(t)$. They showed that these ranks are invariant under a certain local equivalence which allows linear but perhaps discontinuous transformation of the variables $\mathbf{x}(t)$ and $\dot{\mathbf{x}}(t)$ in an almost independent way. Under this equivalence, the system may be transformed into one in which the n components of the vector function $\mathbf{x}$ are partitioned into 5 subvectors $\mathbf{x}_1, \mathbf{x}_2, \mathbf{x}_3, \mathbf{x}_4, \mathbf{x}_5$, with dimensions (which may possibly be zero) respectively $s(t), r(t) - s(t), a(t), s(t)$, and $n - r(t) - a(t) - s(t)$. With corresponding parts of f given by f_1, f_2, f_3, f_4, f_5, the transformed system becomes

$$\begin{aligned}
\dot{\mathbf{x}}_1(t) &= 0, \\
\dot{\mathbf{x}}_2(t) &= 0, \\
0 &= \mathbf{x}_3(t) + f_3(t), \\
0 &= \mathbf{x}_1(t) + f_4(t), \\
0 &= f_5(t).
\end{aligned} \tag{1.3}$$

The number $s(t)$ is called the *strangeness* of the system. When the ranks $r(t), a(t)$, and $s(t)$ are independent of t, and $E(t)$ and $A(t)$ are sufficiently smooth, they showed that the system (1.2) is globally equivalent (that is, $\mathbf{x}(t)$ is transformed continuously and

$\dot{\mathbf{x}}(t)$ is transformed by differentiating $\mathbf{x}(t)$) to one similar to (1.3):

$$
\begin{aligned}
\dot{\mathbf{x}}_1(t) &= A_{12}(t)\mathbf{x}_2(t) + A_{14}(t)\mathbf{x}_4(t) + A_{15}(t)\mathbf{x}_5(t) + f_1(t), \\
\dot{\mathbf{x}}_2(t) &= A_{24}(t)\mathbf{x}_4(t) + A_{25}(t) + f_2(t), \\
0 &= \mathbf{x}_3(t) + f_3(t), \\
0 &= \mathbf{x}_1(t) + f_4(t), \\
0 &= f_5(t).
\end{aligned}
\tag{1.4}
$$

By eliminating $\dot{\mathbf{x}}_1(t)$ through differentiation of the fourth equation above, a new differential algebraic system in which the rank of $E(t)$ is reduced by the strangeness is obtained. *If this reduction step can be carried out repeatedly* (that is, if at each step, the ranks $r(t), a(t), s(t)$ are independent of t, for example), the minimum number of steps required to obtain a system with zero strangeness is called the *strangeness index* of the original system.

In our opinion, the results of Kunkel and Mehrmann are limited since they deal with a semi-linear system, as compared to a quasi-linear system (which is much more general, see Chapter 2), and the constant rank conditions are extremely restrictive (in particular, the strangeness index need not be defined). The consistency of initial conditions, and hence also the existence and uniqueness results depend either on an *a priori* knowledge of the differentiation index or on the assumption that the strangeness index is well defined.

A very general existence and uniqueness theory for implicit differential algebraic systems was developed by Rabier and Rheinboldt [47, 48]. In [48], a given implicit autonomous system $F(\mathbf{x}, \mathbf{p}) = 0$ of n first order ordinary differential equations in n dependent variables $\mathbf{x}$ was studied through the zero set $\mathcal{M} = F^{-1}(0)$, which is assumed to be a smooth submanifold of the tangent bundle $T\mathbb{R}^n$ of $\mathbb{R}^n$. Reich [52] defined an index as follows. Let $\mathcal{W} = \pi(\mathcal{M})$ be the image of $\mathcal{M}$ under the canonical projection π from $\mathbb{R}^{2n}$ to the first n-coordinates. Under suitable assumptions, $\mathcal{W}$ may be a submanifold of $\mathbb{R}^n$ enjoying the property that for any solution $\varphi : I \longrightarrow \mathbb{R}^n$ of $F(\mathbf{x}, \mathbf{p}) = 0$ defined on an open interval $I \subseteq \mathbb{R}$, the point $(\varphi(t), \dot{\varphi}(t))$ belongs to $\mathcal{M}_1 = T\mathcal{W} \cap \mathcal{M}$ for all $t \in I$ (here $T\mathcal{W}$ denotes the tangent bundle of $\mathcal{W}$). The reduction from $\mathcal{M}$ to $\mathcal{M}_1$ may be repeated if the first reduction $\mathcal{M}_1$ and its projection $\pi(\mathcal{M}_1)$ are both submanifolds and $\mathcal{M}_1$ satisfies the same assumptions that $\mathcal{M}$ does. Under suitable conditions, then, a descending chain of manifolds $\mathcal{M}_1 \supseteq \mathcal{M}_2 \supseteq \cdots$ may be constructed, and if this chain becomes stationary, the first natural number p such that $\mathcal{M}_p = \mathcal{M}_{p+1}$ is called the *Reich index* of the system.

Rabier and Rheinboldt [48] pointed out that Reich did not prove any existence theorem on $\mathcal{M}_p$, and to do so, they passed from a global to a local theory. For this development, Rabier and Rheinboldt need to restrict $\mathcal{M}$ to be a π-submanifold (roughly speaking, the restriction map $\pi_{|\mathcal{M}}$ is a local subimmersion). The reduction $\mathcal{M}_1$ is replaced by the set of points $(\mathbf{x}, \mathbf{p})$ which has a local neighborhood $\mathcal{V}$ with $\mathbf{p}$ belonging to the tangent space $T_{\mathbf{x}}\mathcal{W}$ of $\mathcal{W} = \pi(\mathcal{V})$. The π-submanifold $\mathcal{M}$ is said to be *reducible* if certain additional conditions on rank hold at each point of $\mathcal{M}_1$, and is said to be *completely reducible* if there is a descending chain of reducible π-submanifolds

$$
\mathcal{M}_0 = \mathcal{M} \supseteq \mathcal{M}_1 = (\mathcal{M}_0)_1 \supseteq \cdots \supseteq \mathcal{M}_{j+1} = (\mathcal{M}_j)_1 \supseteq \cdots.
$$

The system $F(\mathbf{x}, \mathbf{p}) = 0$ is *non-singular* if $\mathcal{M} = F^{-1}(0)$ is a completely reducible π-submanifold of $\mathcal{T}\mathbb{R}^n$. Rabier and Rheinboldt proved that if F is non-singular, the descending chain is stationary. When this is the case, the intersection $C(\mathcal{M}) = \bigcap_{j \geqslant 0} \mathcal{M}_j$ is called the *core* of $\mathcal{M}$. An important property of $C(\mathcal{M})$ is that for every point $(\mathbf{x}, \mathbf{p}) \in C(\mathcal{M})$, there exist a local projection $\mathcal{W} = \pi(\mathcal{V})$ of $C(\mathcal{M})$ at $(\mathbf{x}, \mathbf{p})$ and a section $\eta : \mathcal{W} \longrightarrow \mathcal{T}\mathcal{W}$ such that $\mathcal{V} = \eta(\mathcal{W})$ (see [48, Theorem 6.1]). The section η defines a smooth vector field and hence the implicit system is locally equivalent to an explicit system, which is locally solvable.

Recently, for special systems with index $\leqslant 3$ and some structural properties, Schwarz [55], Schwarz and Lamour [56], Hanke and Lamour [28] described methods to compute a consistent initial value point.

1.4 Our Approach

This article is much influenced by Rabier and Rheinboldt [48]. We consider systems of ordinary differential equations which are polynomial in the unknown functions and their derivatives. For a given system, we are interested in computing algebraic constraints on the initial values such that on the algebraic set determined by the constraint equations, the initial value problem of the original system of differential equations has a unique solution. By restricting our attention to systems which are polynomial in the derivatives, we avoid the main difficulty encountered by their topological approach, which requires proving that $\mathcal{W} = \pi(\mathcal{M})$ and $\mathcal{T}\mathcal{W} \cap \mathcal{M}$ are submanifolds. We are able also to avoid any direct discussion of singular points in the set of initial values. Here we present a purely algebraic theory using an algebraic geometry approach similar to, but different from Thomas [61], and develop computation algorithms based on ideal-theoretical results. While our results are not as general as Rabier and Rheinboldt, contrary to previous methods, our reduction process is both algebraic and general. The process can always be carried out to its completion. It requires no assumptions such as constant rank, irreducibility, non-singularity, differential algebraic independence of the equations, and even the number of given equations may be arbitrary.

This is accomplished through a new concept of *essential degree* and an algorithmic process called *prolongation*. The prolongation process may be repeated at most a finite number of times, at the end of which the original system is replaced by an equivalent system (which we define as *complete*). The length of the prolongation process (defined as the number of repetitions to reach a complete system) is an invariant. We are thus able to define the concept of an index as an algebraic invariant called the *algebraic index* for such non-linear systems. Over-determined and under-determined systems, whether constrained or not, are studied from one unified ideal theoretic approach. This algebraic viewpoint leads to a more elementary treatment, and naturally yields algorithms for computing prolongations and the index. For the case when the system is quasi-linear, and even for many non-linear systems, we are able to compute algebraic constraints on the initial values that are necessary for the initial value problem to have any solution. This set of initial values satisfying these constraints forms a Zariski closed subset M of $\mathbb{C}^n$. Furthermore, we can compute the constraints on the initial values that are sufficient for the system to be transformable into an explicit system, and on this constrained set M^0, which is the intersection of an open dense submanifold $\mathcal{O}$ of $\mathbb{C}^n$ with M, compute an explicit form to define a rational vector field on $\mathcal{O}$, thus leading to the existence and uniqueness of solutions on $\mathcal{O}$. These solutions automatically satisfy the constraints whenever the initial values lie in M^0. We have no need to worry about the singular points of the set M. Readers may compare our results in Chapter 6 with Theorem 6.1 of Rabier and Rheinboldt [48].

The effective computation of this constrained set M^0 for initial values supersedes a method in the index-zero case for semi-explicit systems studied by Brown, Hindmarsh, and Petzold [9]. They described an initialization method that completes a partially given set of initial conditions to a consistent set which can then be used as a starting point to numerically solve the system. Whereas their paper is concerned with finding consistent initial conditions for index-1 systems, we start from any implicitly given system of arbitrary and unknown algebraic index and compute the algebraic constraints and an explicit form of algebraic index 0 that guarantees existence and uniqueness of a solution. Our method treats all dependent variables equally, without the need to distinguish some as algebraic and others as differential as in Kunkel and Mehrmann [37]. Moreover, the dependent variables are not transformed. For the initialization problem, our final form avoids all the special treatments necessary for the application of the method by Brown, Hindmarsh, and Petzold [9] in case a system is more general than semi-explicit. Since our resulting vector field is always given by rational expressions, and we shall show that the contraints are automatically satisfied by any local solutions, we believe that our preliminary treatment of a differential algebraic system should simplify subsequent numerical methods and may avoid the "drift" problem.

We would like to point out that this paper attempts neither to solve the three concerns raised earlier nor to study complexity issues. Our purpose is to provide a solid algebraic theory to support the use of symbolic computation for differential algebraic equations. We have implemented these algorithms in the computer algebra system *Axiom* developed by Jenks and Sutor [34][1] and our tests show that at least for problems of reasonably small size, the algorithms worked as predicted by our theory. In writing up this paper, we have deliberately removed our earlier use of projective geometry and analysis to prove some key results. This theory is thus accessible to any one familiar with Gröbner basis theory (see Becker and Weispfenning [3] and Cox, Little, and O'Shea [16]) and no background in differential manifold theory is needed. Except for following comparisons with differential algebraic methods (mainly Chapter 8 and the next paragraph), the reader does not need to know differential algebra. Throughout we shall give many examples to illustrate our theory and implementation. Our main contributions are in the conceptual simplicity and general applicability of the algorithms to any arbitrarily given polynomial differential first order system. Research in impasse points, in numerical methods in our specialized setting, and in complexity of our algorithms are all worthwhile but clearly beyond the scope of this paper.

One referee pointed out that, in the review version, we did not compare our method with the characteristic set method used in the Ritt-Kolchin algorithm [54, 35, 59] and

---

[1] *Axiom* became an open source project in 2002 under the leadership of Tim Daly [18] and is now available freely on many operation systems and computer platforms. However, not all components of the former *Axiom*, as marketed and available before October, 2001 by the Numerical Algorithm Group, have been ported successfully yet.

improvements in the closely related Gröbner-Rosenfeld algorithm by Boulier *et al.* [7], Hubert [29], and Bouziane *et al.* [8]. Another referee asked whether our method may be applied to problems in control theory. We shall discuss in Chapter 8 the relation of our method with characteristic set methods after our full exposition. The applicability of our method to control theory, however, requires a lengthy discussion of the subject and also dimension concepts in differential algebra that would be beyond the scope of this paper. We do want to point out that researchers (notably Fliess [20]) in non-linear control theory have indeed applied differential algebraic methods to clarify fundamental concepts and have studied differential algebraic equations. While the goal remains to find an explicit system equivalent to the given implicit one, the method used in Fliess *et al.* [21] involves infinite dimensional differential geometry (infinite jets and prolongations), Lie-Bächlund morphisms, and exterior algebra. They proved an existence and uniqueness result assuming that the given equations and their derivatives are functionally independent. Under similar assumptions (which are common for control theory applications), D'Alfonso, Jeronimo, and Solerno [17], following Matera and Sedoglavic [43], applied probabilistic methods to compute the differential index (and other quantities related to dimensions) for general systems arising from control theory. Our method by contrast is very elementary and does not depend on any *a priori* assumptions of independence.

Our paper is organized as follows. After setting up notations and a brief discussion on quasi-linear systems in the next chapter, we begin our study of a differential polynomial (and not necessarily quasi-linear) autonomous system in Chapter 3 through the polynomial ideal it generates. We introduce a new concept called the *essential* **P**-*degree basis* of an ideal and provide an algorithm (based on Gröbner basis) to compute it. The algorithm allows us to decide whether a given differential algebraic system is equivalent to a quasi-linear system (in the generalized sense as will be defined later in this paper). In Chapter 4, we define the notions of prolongation, completion, and index for a polynomial ideal and prove basic results. In particular, the prolongation and completion algorithms will allow us to compute the algebraic constraints on initial values of the differential algebraic system. We introduce the notion of the quasi-linear ideal associated with an ideal in Chapter 5, and describe several generalized notions of quasi-linearity. This is followed by a study of initial values and their zero-dimensional fibers. We begin Chapter 6 with a result on quasi-linear ideals where we show that Cramer's rule may be used to compute the explicit form, thus obtaining a well-defined vector field. We then prove existence and uniqueness theorems by showing that the local solutions satisfy the constraints on initial values whenever the initial values do. For "unconstrained" or "under-determined" non-linear differential algebraic systems (these are precisely defined in our paper), we propose in Chapter 7 the method of quasi-linearization, which enables us to apply our results. In Chapter 8 we discuss our method from the view point of differential algebra. Examples are given in Chapter 9, which also illustrates our implementation in *Axiom*. We end the paper with some comments in the last chapter.

2   Notations and Basic Transformations

Throughout this paper, we adopt a consistent set of notations that distinguish algebraic indeterminates from solutions to algebraic equations, as well as differential indeterminates from solutions to differential equations.

2.1   Implicit Autonomous System

Let $\mathbf{z} = (z_1, \ldots, z_n)$ be a vector of n indeterminate complex-valued functions of an independent real variable t. Let $\dot{\mathbf{z}} = (\dot{z}_1, \ldots, \dot{z}_n)$ be its derivative with respect to t, and let $\mathbf{x}_0$ be a point in $\mathbb{C}^n$, where $\mathbb{C}$ is the field of complex numbers (or some computable algebraically closed field of characteristic zero). Let $f_1, \ldots, f_m$ be polynomials in $2n$ indeterminates $(\mathbf{X}, \mathbf{P}) = (X_1, \ldots X_n, P_1, \ldots, P_n)$ over $\mathbb{C}$. We consider the initial value problem $\mathbf{z}(0) = \mathbf{x}_0$ for the following implicit autonomous system of differential equations:

$$f_1(z_1, \ldots, z_n, \dot{z}_1, \ldots, \dot{z}_n) = 0,$$

$$\vdots \qquad\qquad (2.1)$$

$$f_m(z_1, \ldots, z_n, \dot{z}_1, \ldots, \dot{z}_n) = 0.$$

Definition 2.1.1 *A system* (2.1) *is said to be* quasi-linear (in the traditional sense) *if the total degree of each f_i in the indeterminates $\mathbf{P}$ is at most one.*

Definition 2.1.2 *For the purpose of this paper, a system* (2.1) *is said to be an* explicit system *or an* explicitly given system *if $m \geqslant n$ and $f_1, \ldots, f_m$ have the form*

$$
\begin{aligned}
f_i &= P_i - r_i(\mathbf{X}), & &\text{where } r_i(\mathbf{X}) \in \mathbb{C}(\mathbf{X}) \text{ for } 1 \leqslant i \leqslant n, \\
f_{n+k} &\in \mathbb{C}[\mathbf{X}], & &\text{for } 1 \leqslant k \leqslant m - n.
\end{aligned}
\qquad (2.2)
$$

For autonomous differential equations, it is customary to say a system is explicitly given if the derivatives $\dot{z}_i$ are given as functions $r_i(z_1, \ldots, z_n)$. Since we are dealing with differential polynomial equations, in Definition 2.1.2, we restrict these functions to be rational functions. Technically, we have $f_i \in \mathbb{C}(\mathbf{X})[\mathbf{P}]$ instead of $f_i \in \mathbb{C}[\mathbf{X}, \mathbf{P}]$, but by means of a trick of Rabinowitsch, we can put the system back into polynomial form as follows. We introduce new indeterminates X_0, P_0. Writing the rational functions $r_i(\mathbf{X}) = R_i(\mathbf{X})/S(\mathbf{X})$ with a common denominator $S(\mathbf{X})$, where $R_i(\mathbf{X}), S(\mathbf{X}) \in \mathbb{C}[\mathbf{X}]$, the system (2.1) with f_i as given by (2.2) is equivalent to:

$$g_1(z_0, z_1, \ldots, z_n, \dot{z}_0, \dot{z}_1, \ldots, \dot{z}_n) = 0,$$

$$\vdots \qquad\qquad (2.3)$$

$$g_{m+2}(z_0, z_1, \ldots, z_n, \dot{z}_0, \dot{z}_1, \ldots, \dot{z}_n) = 0,$$

where $g_i \in \mathbb{C}[X_0, \mathbf{X}, P_0, \mathbf{P}]$ for $1 \leqslant i \leqslant m + 2$, and

$$
\begin{aligned}
g_i &= P_i - X_0 R_i(\mathbf{X}) && \text{for } 1 \leqslant i \leqslant n, \\
g_{n+k} &= f_{n+k} && \text{for } 1 \leqslant k \leqslant m - n, \\
g_{m+1} &= X_0 S(\mathbf{X}) - 1, && \\
g_{m+2} &= P_0 + X_0^3 \sum_{i=1}^{n} \frac{\partial S(\mathbf{X})}{\partial X_i} R_i(\mathbf{X}). &&
\end{aligned} \tag{2.4}
$$

After a suitable renumbering, the system (2.3) given by (2.4) represents an equivalent explicit system in polynomial form. It is clearly preferable to adopt the convention of allowing rational functions in $\mathbf{X}$ in an explicit system.

2.2   Transformations

The trick of Rabinowitsch can also be applied to transform a (not necessarily explicit) system of first order ordinary differential equations (2.1) where $f_i(\mathbf{X}, \mathbf{P}) \in \mathbb{C}(\mathbf{X})[\mathbf{P}]$ to one which is polynomial in all the indeterminates. To show that differential polynomial systems actually represent a very general class of systems, we will discuss briefly how more general systems can be transformed into polynomial systems. Some transformations are well-known.

A general system can often be transformed in stages: (1) to an autonomous system, (2) to a first order autonomous system, (3) to a first order polynomial autonomous system, and (4) to a quasi-linear system. First, a *non-autonomous system* (in which t may explicitly appear) can be transformed into an autonomous system: the variable t is replaced by a new variable z_{n+1} and the linear differential equation $\dot{z}_{n+1} = 1$ with the initial condition $z_{n+1}(0) = 0$ is added to the system. Second, for any system of order $k > 1$, we may reduce the order by 1 by introducing, for each z_i that appears to order k, a new indeterminate function w_i, replacing, for $1 \leqslant j \leqslant k$, $z_i^{(j)}$ in the given system by $w_i^{(j-1)}$, and adding the linear differential equations $\dot{z}_i = w_i$ with initial conditions $w_i^{(j-1)}(0) = z_i^{(j)}(0)$. Thus it suffices to consider only general first order systems. Third, let

$$
\begin{aligned}
g_1(z_1, \ldots, z_n, \dot{z}_1, \ldots, \dot{z}_n) &= 0, \\
&\;\;\vdots \\
g_m(z_1, \ldots, z_n, \dot{z}_1, \ldots, \dot{z}_n) &= 0
\end{aligned} \tag{2.5}
$$

be a general (first-order) autonomous system, not necessarily quasi-linear or polynomial. Suppose an expression E appearing in some g_j is a polynomial in some elementary functions like $\sin(u), \cos(u), \exp(u), \log u, 1/u$, where $u = U(z_1, \ldots, z_n)$ for some polynomial $U = U(\mathbf{X})$. Because any such function either is the unique solution to a specific initial value problem for a linear homogeneous differential equation with constant coefficients, or satisfies a (polynomial) quasi-linear equation which can be easily integrated, it is not difficult to see that by replacing such a function with a new indeterminate function, and introducing an additional *quasi-linear* equation with a corresponding initial condition, we can transform the expression E to a polynomial. For example, terms like $\sin(u), \cos(u)$ may be replaced by w_1, w_2, and appending to the system the equations

$$
\begin{aligned}
0 &= \dot{w}_1 - w_2 \dot{u}, \\
0 &= \dot{w}_2 + w_1 \dot{u}, \\
\dot{u} &= \sum_{i=1}^{n} \frac{\partial U}{\partial X_i}(z_1, \ldots, z_n) \dot{z}_i
\end{aligned}
$$

and the initial conditions $u(0) = U(\mathbf{x}_0)$, $w_1(0) = \sin(U(\mathbf{x}_0))$, $w_2(0) = \cos(U(\mathbf{x}_0))$.

Finally, to see how a polynomial autonomous system may be transformed into a quasi-linear system, suppose for now that (2.5) represents a general polynomial autonomous system. Let $\mathbf{w} = (w_1, \ldots, w_{2n})$ be new indeterminate functions of t. Intuitively, $w_k = z_k$, $w_{n+k} = \dot{z}_k$ for $1 \leqslant k \leqslant n$. Formally, the quasi-linear system is defined by polynomials in $\mathbb{C}[\mathbf{Y}, \mathbf{Q}]$ where $\mathbf{Y} = (Y_1, \ldots, Y_{2n})$ and $\mathbf{Q} = (Q_1, \ldots, Q_{2n})$ are indeterminates. The defining polynomials $f_1, \ldots, f_{n+m}$ of the transformed system are given by:

$$
f_k = \begin{cases} Q_k - Y_{n+k} & \text{if } 1 \leqslant k \leqslant n; \\ g_{k-n}(Y_1, \ldots, Y_{2n}) & \text{if } n+1 \leqslant k \leqslant n+m. \end{cases} \tag{2.6}
$$

It is obvious that the resulting system $f_j(\mathbf{w}, \dot{\mathbf{w}}) = 0$ for $1 \leqslant j \leqslant n+m$ is linear in the variables $\dot{w}_k$ ($1 \leqslant k \leqslant 2n$). The reader should note that (2.6) does *not* represent an explicit system since the new variables $Q_{n+1}, \ldots, Q_{2n}$ do not appear at all.

Definition 2.2.1 *The system* $\{ f_1, \ldots, f_{n+m} \}$ *of* (2.6) *is called the* quasi-linearization *of the system* $\{ g_1, \ldots, g_m \}$ *of* (2.5).

The two systems are equivalent in the following sense.

Proposition 2.2.2 *With notations as above, let B be some open interval on the real line. There is a bijection between the set of twice differentiable curves $\varphi : B \longrightarrow \mathbb{C}^n$ such that $\mathbf{z} = \varphi(t)$ satisfies the system of differential equations (2.5) for all $t \in B$, and the set of differentiable curves $\sigma : B \longrightarrow \mathbb{C}^{2n}$ such that $\mathbf{w} = \sigma(t)$ satisfies the system of differential equations*

$$
f_k(\mathbf{w}, \dot{\mathbf{w}}) = 0, \qquad 1 \leqslant k \leqslant n+m \tag{2.7}
$$

with f_k defined in (2.6).

Proof. The correspondence is given by $\sigma_k = \varphi_k$, $\sigma_{n+k} = \dot{\varphi}_k$ for $1 \leqslant k \leqslant n$. Then $\mathbf{z} = \varphi(t)$ satisfies (2.5) if and only if $\mathbf{w} = \sigma(t)$ satisfies the system (2.7). □

The original initial conditions on φ can now be imposed on the new functions σ by requiring $\sigma_k(0) = \varphi_k(0)$ and $\sigma_{n+k}(0) = \dot{\varphi}_k(0)$ for $1 \leqslant k \leqslant n$. Note that in general, the values of $\dot{\varphi}_i(0)$ are only implicitly defined in terms of $\varphi_i(0)$ by the algebraic constraints

$$
g_1(\varphi(0), \dot{\varphi}(0)) = 0, \quad \ldots, \quad g_m(\varphi(0), \dot{\varphi}(0)) = 0. \tag{2.8}
$$

These are *necessary* algebraic constraints on initial values when the given polynomials g_j are non-linear in $\mathbf{P}$. By Proposition 2.2.2, an existence and uniqueness theorem for quasi-linear systems will lead to one for non-linear systems when the algebraic constraints (2.8) are satisfied by the initial values. In Chapter 7, we shall investigate the application of quasi-linearization further.

2.3   Constructible Sets

We recall some notations. The empty set is $\emptyset$. For any ring $\mathcal{R}$, $J \lhd \mathcal{R}$ means J is an ideal in $\mathcal{R}$. The radical of J is written as $\sqrt{J}$. If $F \subseteq \mathcal{R}$, (F) is the ideal in $\mathcal{R}$ generated by F. If $\mathbf{X} = (X_1, \ldots, X_n)$ is a family of indeterminates over $\mathbb{C}$, then $\mathbb{C}[\mathbf{X}]$ is the polynomial ring $\mathbb{C}[X_1, \ldots, X_n]$. For $f \in \mathbb{C}[\mathbf{X}]$, let

$$D(f) = \{\mathbf{x} \in \mathbb{C}^n \mid f(\mathbf{x}) \neq 0\}.$$

The collection $\{\, D(f) \mid f \in \mathbb{C}[\mathbf{X}] \,\}$ is a basis for open sets of the *Zariski topology* on $\mathbb{C}^n$. For $F \subseteq \mathbb{C}[\mathbf{X}]$, $Z(F)$ is the (Zariski closed) *algebraic set*

$$\{\, \mathbf{x} \in \mathbb{C}^n \mid f(\mathbf{x}) = 0 \text{ for all } f \in F \,\}.$$

The set $Z(F)$ depends only on the radical of the ideal generated by F. Let $W \subseteq \mathbb{C}^n$ be an algebraic set. Recall that the *dimension* $\dim W$ of W is the supremum of all integers k such that there exists a chain

$$W_0 \subset W_1 \subset \cdots \subset W_k$$

of distinct irreducible closed subsets of W. Equivalently, $\dim W$ is the Krull dimension of the coordinate ring $\mathbb{C}[\mathbf{X}]/I(W)$, where $I(W)$ is the *defining ideal* of W.

In this paper, all finite subsets of a polynomial ring are assumed to be non-empty unless specified otherwise. We say a Zariski closed set W (resp. an ideal $J \lhd \mathbb{C}[\mathbf{X}]$) is *effectively computable* (from some suitable input set S) if there is an algorithm (using S) to compute a finite set of polynomials F such that $Z(F) = W$ (resp. $(F) = J$). Typically, S is a set of polynomials generating an ideal I on which W (or J) depends. We frequently omit mentioning S if S is clear from the context. A Zariski open set is considered effectively computable when its complement is. More generally, a subset of the Zariski topological space $\mathbb{C}^n$ is (effectively computable and) *locally closed* if it is the intersection of an (effectively computable) open subset with an (effectively computable) closed subset. A locally closed subset of $\mathbb{C}^n$ can often be viewed as a Zariski closed subset in $\mathbb{C}^{n+1}$. A subset is (effectively computable and) *constructible* if it can be written as a finite union of (effectively computable) locally closed subsets. If $\pi : V \longrightarrow W$ is a continuous map (relative to the induced Zariski topologies) of affine sets, the inverse image of any constructible subset of W is a constructible subset of V, and by the well-known theorem of Chevalley, if π is a morphism of affine subsets, the image of any constructible subset of V is a constructible subset of W (see for example, Matsumura [44, Ch. 2, Sec. 6]). Since we shall be addressing the effective computability of these sets directly, readers need not be familiar with these results. For algorithms on computations involving constructible sets, see Sit [58].

3   Essential P-Degree Basis

It is possible for a polynomial autonomous system as given by (2.5) to be equivalent to
a quasi-linear system without introducing the type of transformations discussed in the
previous chapter.

3.1   Essential P-Degree and P-Strong Sets

Let $\mathbf{P} = (P_1, \ldots, P_n)$ be indeterminates and $\mathbb{C}[\mathbf{X}, \mathbf{P}] = \mathbb{C}[X_1, \ldots, X_n, P_1, \ldots, P_n]$.
We present an algorithm to test whether the ideal generated by a set $G = \{\, g_1, \ldots, g_\mu \,\}$
of polynomials in $\mathbb{C}[\mathbf{X}, \mathbf{P}]$ can be generated by another set $F = \{\, f_1, \ldots, f_\nu \,\}$ of polynomials that are of degree at most one in the indeterminates $\mathbf{P}$. As it turns out, the
algorithm is not limited to the degree at most one case, nor do we require the two sets
of indeterminates $\mathbf{X}$ and $\mathbf{P}$ have the same cardinality. This relaxed assumption will be
made for *this* chapter only.

Definition 3.1.1 *Let $m \geqslant 0$ and $n > 0$ be natural numbers, and let $\mathbb{C}[\mathbf{X}, \mathbf{P}]$ be the
polynomial ring $\mathbb{C}[X_1, \ldots, X_m, P_1, \ldots, P_n]$. If $f \in \mathbb{C}[\mathbf{X}, \mathbf{P}]$, $f \neq 0$, let $\deg_{\mathbf{P}} f$ be the
degree of f in the indeterminates $P_1, \ldots, P_n$. The zero polynomial has $\mathbf{P}$-degree $-\infty$
by convention. If F is a finite subset of $\mathbb{C}[\mathbf{X}, \mathbf{P}]$, we define the $\mathbf{P}$-degree of F by*

$$\deg_{\mathbf{P}} F = \max\{\deg_{\mathbf{P}} f \mid f \in F\}.$$

If $J \neq (0)$ is an ideal in $\mathbb{C}[\mathbf{X}, \mathbf{P}]$, we define the essential $\mathbf{P}$-degree of J to be

$$\mathrm{edeg}_{\mathbf{P}}(J) = \min\{\deg_{\mathbf{P}} F \mid (F) = J, F \text{ is finite}\}.$$

*For the zero ideal, we define its essential $\mathbf{P}$-degree to be $-\infty$. Any finite subset F of an
ideal J that generates J and has $\deg_{\mathbf{P}} F$ equal to the essential $\mathbf{P}$-degree of J is called
an essential $\mathbf{P}$-degree basis of J.*

The notion of essential $\mathbf{P}$-degree basis is our main tool to develop an ideal-theoretical
model for differential algebraic systems. Moreover, such a basis is very useful in *presenting* a system in a "minimal" way: for in general, a Gröbner basis contains far too
many polynomials that are not essential to defining the system. The notion can be applied even for $m = 0$, in which case it is more convenient to rename the indeterminates
$\mathbf{P}$ to $\mathbf{X}$ and an essential $\mathbf{X}$-degree basis may be referred to simply as an essential basis.

Definition 3.1.2 *Let J be an ideal with essential $\mathbf{P}$-degree d and let F be a finite subset
of J. We say F is $\mathbf{P}$-strong if it has the following property:*

$(\ast)$ *Every $f \in J$ of $\mathbf{P}$-degree $\leqslant d$ has a representation $f = \sum_{j=1}^{N} h_j f_j$ for some N,
    where for each $j = 1, \ldots, N$, $h_j \in \mathbb{C}[\mathbf{X}, \mathbf{P}]$, $h_j \neq 0$, $f_j \in F$ and the $\mathbf{P}$-degree of
    each $h_j f_j$ is at most the $\mathbf{P}$-degree of f.*

Remark 3.1.3 Clearly, when F is $\mathbf{P}$-strong, the subset F_d of F of $\mathbf{P}$-degree $\leqslant d$ generates the subset of J of $\mathbf{P}$-degree $\leqslant d$. In particular, $(F) = (F_d) = J$, F_d is $\mathbf{P}$-strong,

and $F_d \subseteq F$ is an essential **P**-degree basis of J. Thus, given any **P**-strong subset F, we may always replace it by F_d and assume that it is an essential **P**-degree basis as well.

Example 3.1.4 This example shows that an essential **P**-degree basis, even a minimal one, need not be **P**-strong. Let $f_1, f_2 \in \mathcal{R} = \mathbb{C}[X_1, X_2, P_1, P_2]$ be given by

$$f_1 = X_1 P_1 - X_2, \quad f_2 = X_1 P_2 - X_1,$$

let $F = \{ f_1, f_2 \}$, and let $J = (F)$. Clearly F is an essential **P**-degree basis of J. Let

$$f = P_2 f_1 - P_1 f_2 = X_1 P_1 - X_2 P_2.$$

Then $f \in J$, has **P**-degree 1, but cannot be represented as $h_1 f_1 + h_2 f_2$ for any $h_1, h_2 \in \mathcal{R}$ such that the **P**-degrees of $h_1 f_1$ and $h_2 f_2$ are at most 1. Assuming the contrary, we would have $h_1, h_2 \in \mathbb{C}[X_1, X_2]$, and $f = h_1 X_1 P_1 + h_2 X_1 P_2 - (h_1 X_2 + h_2 X_1)$. Since P_1, P_2 are algebraically independent over $\mathbb{C}[X_1, X_2]$, we would have $h_1 X_1 = X_1$ and $h_2 X_1 = -X_2$, which is clearly impossible. Thus F is not **P**-strong.

3.2   Algorithm for Essential P-Degree

The notions of essential **P**-degree basis and **P**-strong are *independent of any term-ordering*. Nonetheless, a **P**-strong essential **P**-degree basis can always be found as part of a reduced Gröbner basis relative to an elimination term-ordering compatible with **P**-degree. Specifically, we give an algorithm whose input is some finite subset F of $\mathbb{C}[\mathbf{X}, \mathbf{P}]$, and whose output is the essential **P**-degree of (F) together with a **P**-strong essential **P**-degree basis of the ideal (F). For this algorithm, let $<$ be any term order on the terms of $\mathbb{C}[\mathbf{X}, \mathbf{P}]$ such that if u is a term in the indeterminates $X_1, \ldots, X_m$ and $v \neq 1$ is a term in the indeterminates $P_1, \ldots, P_n$, then $u < v$. This requirement is often indicated notationally by $\mathbf{X} \ll \mathbf{P}$ (Becker and Weispfenning [3, p. 256]). In addition we will require that if the **P**-degree of a term v_1 is less than the **P**-degree of a term v_2, then $v_1 < v_2$. We can always find such term orders. For instance, let $<_\mathbf{X}$ be an arbitrary term order on the terms of $\mathbb{C}[\mathbf{X}]$ and let $<_\mathbf{P}$ be an arbitrary term order on the terms of $\mathbb{C}[\mathbf{P}]$ with the property that $\deg v_1 < \deg v_2$ implies $v_1 <_\mathbf{P} v_2$ (for example degree lexicographically). Now combine $<_\mathbf{P}$ and $<_\mathbf{X}$ lexicographically to obtain $<$ as required. With such a term order, consider the Algorithm in Fig. 3.1.

Proposition 3.2.1 *Given any finite, nonempty, subset F of $\mathbb{C}[\mathbf{X}, \mathbf{P}]$, the algorithm* ESSENTIAL_**P**-DEGREE_BASIS(F) *in Fig. 3.1 computes a pair (d, E) where d is the essential **P**-degree of (F) and E is a **P**-strong essential **P**-degree basis of (F).*

Proof. The algorithm clearly terminates. Let $J = (F)$. If (d, E) is returned, the essential **P**-degree of J is at most d since E is a generating set for J with $\deg_\mathbf{P} E = d$. We must show that the algorithm terminates when d is the essential **P**-degree of J. Assume that J is generated by some set $F' = \{ f_1', \ldots, f_\nu' \} \subseteq \mathbb{C}[\mathbf{X}, \mathbf{P}]$, where $\deg_\mathbf{P} F'$ is the

```
begin
    if F = {0} then
        return (-∞, {0})
    else
        E ⟵ ∅
        G ⟵ a Gröbner basis for (F) with respect to <
        d ⟵ -1
        repeat
            d ⟵ d + 1
            E ⟵ E ∪ { g ∈ G | deg_P(g) = d }
        until  G ⊆ (E)
        return (d, E)
    endif
end
```

Figure 3.1 Algorithm ESSENTIAL_P-DEGREE_BASIS(F) of Proposition 3.2.1

essential **P**-degree d' of J. Let $\{ g_1, \ldots, g_\mu \}$ be the elements of a Gröbner basis G for J. Fix i and let

$$f_i' = \sum_{j=1}^{\mu} h_j g_j, \qquad h_j \in \mathbb{C}[\mathbf{X}, \mathbf{P}]$$

be a standard representation of f_i' with respect to G. By the definition of standard representation (see Becker and Weispfenning [3, Sect. 5.4]) and by our choice of term order, we must have $h_j = 0$ for every j such that g_j has **P**-degree greater than d'. It follows that if E is the set of elements of G that are of **P**-degree at most d' then E is a generating set for J, and $\deg_\mathbf{P} E = d'$. Thus the algorithm terminates when d is the essential **P**-degree of J. It is clear from the above proof that E is **P**-strong, too.   □

Remark 3.2.2 We note that the algorithm as given is quite inefficient in that the test $G \subseteq (E)$ requires repeated Gröbner basis computations of an ascending chain of ideals. This presentation is simply to facilitate the proof. It is easy to improve the efficiency through implementation. For example, let G be written as the union of disjoint subsets G_i consisting of members of G that have **P**-degree i, and let $E_i = \cup_{k=0}^{i} G_k$. We only need to test membership in (E_i) of the elements of G_{i+1} through G_λ, where λ is the minimum of (a) the **P**-degree of G and (b) an *a priori* bound of the essential **P**-degree such as the **P**-degree of F. Note also that once G is computed, any term-ordering may be used for the test $G_k \subseteq (E_i)$, $i + 1 \leqslant k \leqslant \lambda$. An interesting open problem is: Find a more direct algorithm to compute an essential **P**-degree basis from a set of generators.

Example 3.2.3 In this algorithm, even if G is chosen to be a *reduced* Gröbner basis, the elements of G may involve polynomials of **P**-degree strictly more than the essential **P**-degree of J. Let $f_1, f_2 \in \mathbb{C}[X_1, P_1, P_2]$ be given by $f_1 = X_1 P_2 + P_2$, $f_2 = X_1 P_1$, and let $J = (f_1, f_2)$. Let terms be ordered first by degree lex in **P** (with $P_1 < P_2$) and then by degree in X_1. Clearly J has a generating set of degree at most one in **P**, but it is easy to see that J has a reduced Gröbner basis $G = \{f_1, f_2, f_3\}$ where $f_3 = P_1 P_2$.

4 Prolongation

Rabier and Rheinboldt [48] described a reduction process that produces, for a given implicit system of differential equations, a finite sequence $M_0 \supset M_1 \supset \cdots \supset M_p = M_{p+1}$ of manifolds. The intersection M_p of these manifolds is called the "core" and p is called the *differentiation index*. It is on this "core" that the implicit system has a solution. See Section 1.3 for a brief review. Our principal tools in transforming these ideas to the current setting of algebraic geometry will be the ideal-theoretic notions of prolongation, completion, and algebraic index.

4.1 Prolongation Ideal

We now resume the assumption each of $\mathbf{X}$ and $\mathbf{P}$ is a set of n indeterminates. For arbitrary $h \in \mathbb{C}[\mathbf{X}]$, let

$$\nabla h = \nabla_{\mathbf{X}}(h) = \sum_{j=1}^{n} \frac{\partial h}{\partial X_j} P_j.$$

For any set $R \subseteq \mathbb{C}[\mathbf{X}]$, we define ∇R to be the set consisting of ∇h for all $h \in R$.

Definition 4.1.1 *Let J be an ideal in $\mathbb{C}[\mathbf{X}, \mathbf{P}]$. Let*

$$R = R(J) = \sqrt{J} \cap \mathbb{C}[\mathbf{X}] = \sqrt{J \cap \mathbb{C}[\mathbf{X}]}.$$

The ideal J^* generated by $J \cup R \cup \nabla R$ will be called the prolongation ideal *(or simply,* prolongation*) of J. By construction, J^* depends only on the ideal J.*

Remark 4.1.2 Let $J_1 \subseteq J_2$ be ideals. Then $J_1^* \subseteq J_2^*$.

It is easy to show that the solutions to the system defined by J are the same as those to the prolonged system J^* in the following sense.

Proposition 4.1.3 *Let J be an ideal in $\mathbb{C}[\mathbf{X}, \mathbf{P}]$, let J^* be its prolongation, let $\mathcal{B}$ be some open interval on the real line, and let $\varphi : \mathcal{B} \longrightarrow \mathbb{C}^n$ be a differentiable map. Then $\mathbf{z} = \varphi(t)$ satisfies the system of differential equations for all $t \in \mathcal{B}$:*

$$f(\mathbf{z}, \dot{\mathbf{z}}) = 0, \quad f \in J \tag{4.1}$$

if and only if $\mathbf{z} = \varphi(t)$ satisfies the system of differential equations for all $t \in \mathcal{B}$:

$$f(\mathbf{z}, \dot{\mathbf{z}}) = 0, \quad f \in J^*. \tag{4.2}$$

Proof. It is clear that if $\mathbf{z} = \varphi(t)$ satisfies the system (4.2), then it also satisfies the system (4.1) since $J \subseteq J^*$. Now suppose that $f(\varphi(t), \dot{\varphi}(t)) = 0$ for every $f \in J$ and $t \in \mathcal{B}$. For any $h \in R(J)$ we have a positive integer s such that $h^s \in J \cap \mathbb{C}[\mathbf{X}] \subseteq J$ and hence $h(\varphi(t)) = 0$ for every $t \in \mathcal{B}$. Finally $\nabla h(\varphi(t), \dot{\varphi}(t)) = 0$ follows by differentiating $h(\varphi(t)) = 0$ with respect to t. Since J^* is generated by $J \cup R(J) \cup \nabla R(J)$, the system (4.2) is satisfied by $\mathbf{z} = \varphi(t)$ for every $t \in \mathcal{B}$. $\square$

A generating set for $J \cap \mathbb{C}[\mathbf{X}]$ may be computed as a by-product of the essential **P**-basis of J or more simply using Algorithm ELIMINATION in Becker and Weispfenning [3, p. 258] (see also the Elimination Theorem in Cox, Little and O'Shea [16, p. 114] and Gianni, Trager and Zacharias [26, p. 18]). A Gröbner basis for the radical ideal $\sqrt{J} \cap \mathbb{C}[\mathbf{X}]$ can then be computed using Algorithm RADICAL in [3, Sect. 8.7]. Using these algorithms, the next lemma shows that we can effectively compute the prolongation J^* of an ideal J when a set of generators for J is given.

Lemma 4.1.4 Let $f_1, \ldots, f_m \in \mathbb{C}[\mathbf{X}, \mathbf{P}]$ be any set of generators of an ideal J, and let $q_1, \ldots, q_N$ be any set of generators of $R(J)$. Then

$$J^* = (f_1, \ldots, f_m, q_1, \ldots, q_N, \nabla q_1, \ldots, \nabla q_N). \tag{4.3}$$

Proof. It suffices to show that for any $h \in R(J)$, ∇h belongs to the ideal on the right hand side of (4.3). Let $h_i \in \mathbb{C}[\mathbf{X}]$ be such that $h = \sum_{i=1}^N h_i q_i$. We have

$$\nabla h = \sum_{i=1}^N \left[\sum_{k=1}^n \frac{\partial(h_i q_i)}{\partial X_k} P_k \right] = \sum_{i=1}^N (h_i \nabla q_i + q_i \nabla h_i),$$

which belongs to that ideal. □

4.2   Complete Ideal

Definition 4.2.1 An ideal J of $\mathbb{C}[\mathbf{X}, \mathbf{P}]$ is said to be complete if $J = J^*$.

Example 4.2.2 The two ideals (0) and $\mathbb{C}[\mathbf{X}, \mathbf{P}]$ are clearly complete. If $J \cap \mathbb{C}[\mathbf{X}]$ is the zero ideal or $\mathbb{C}[\mathbf{X}]$, then J is complete.

Lemma 4.2.3 If $R(J) \subseteq J$ (for example, if J is complete), then $J \cap \mathbb{C}[\mathbf{X}] = R(J)$ and hence is a radical ideal.

Proof. Clearly, if $\sqrt{J} \cap \mathbb{C}[\mathbf{X}] = R(J) \subseteq J$, then $R(J) \subseteq J \cap \mathbb{C}[\mathbf{X}]$. □

Note that $J \cap \mathbb{C}[\mathbf{X}]$ is a radical ideal if J is complete even though J itself need not be radical (for example, $J = (X_1^2 P_1)$).

Lemma 4.2.4 The intersection of an arbitrary family of complete ideals of $\mathbb{C}[\mathbf{X}, \mathbf{P}]$ is complete.

Proof. Let $\{ J_k \}_{k \in K}$ be a family of complete ideals and let J be their intersection. Then

$$
\begin{aligned}
R(J) &= \sqrt{J \cap \mathbb{C}[\mathbf{X}]} \\
&= \sqrt{\cap_{k \in K}(J_k \cap \mathbb{C}[\mathbf{X}])} \\
&\subseteq \cap_{k \in K} \sqrt{J_k \cap \mathbb{C}[\mathbf{X}]} \\
&= \cap_{k \in K} R(J_k) \\
&\subseteq \cap_{k \in K} J_k = J.
\end{aligned}
$$

Moreover, for any $h \in R(J)$, $h \in R(J_k)$ for every $k \in K$, and hence $\nabla h \in J_k$ for every $k \in K$ and therefore $\nabla h \in J$. This shows that J is complete. $\quad\square$

Remark 4.2.5 The ideal generated by the union of two complete ideals need not be complete. For example, in $\mathbb{C}[X, P]$, let $J_1 = (X + P^2)$ and $J_2 = (X - P^2)$. Then each J_i is complete since $J_i \cap \mathbb{C}[X] = (0)$. The ideal J generated by J_1 and J_2 is clearly (X, P^2) which is not complete. Indeed, $J^* = (X, P)$.

Definition 4.2.6 *Let J be an ideal of $\mathbb{C}[\mathbf{X}, \mathbf{P}]$. The* completion ideal *(or simply,* completion *of J is the smallest complete ideal containing J. By Lemma 4.2.4, the completion ideal of J exists and is unique. We denote the completion of J by $\tilde{J}$.*

The following proposition shows that the completion ideal $\tilde{J}$ is effectively computable from J.

Proposition 4.2.7 *Let J be an ideal in $\mathbb{C}[\mathbf{X}, \mathbf{P}]$. Then the sequence of prolongation ideals defined by $J_0 = J$ and $J_k = (J_{k-1})^*$ is stationary. If $p = p(J)$ is the smallest index k such that $J_k = J_{k+1}$, then $J_p = \tilde{J}$.*

Proof. The sequence is an ascending chain of ideals: $J_0 \subseteq J_1 \subseteq \cdots$ and hence is stationary since the ring $\mathbb{C}[\mathbf{X}, \mathbf{P}]$ is Noetherian. Clearly, J_p is complete and contains $\tilde{J}$. Since $J \subseteq \tilde{J}$, we have $J_1 \subseteq \tilde{J}$ by Remark 4.1.2, and hence by induction, $J_p \subseteq \tilde{J}$. Thus they are equal. $\quad\square$

Definition 4.2.8 *The integer $p = p(J)$ is called the* algebraic index *(or simply* index*) of J and will be denoted by $\mathrm{ind}\,(J)$. If J is generated by $f_1, \ldots, f_m$, we say that the system (2.1) has algebraic index p if J has, and that the system is* complete *if J is.*

Corollary 4.2.9 *If J is an ideal in $\mathbb{C}[\mathbf{X}, \mathbf{P}]$, then $\tilde{J} = \widetilde{J^*}$.*

The algorithm in Fig. 4.1 summarizes the computation for the completion of an ideal and its algebraic index. It should be emphasized that in the algorithm, H (resp. Q) may be *any* ideal basis, not necessarily a Gröbner basis. In particular, we can use an essential **P**-degree basis. Because the notions are ideal-theoretic, considerable flexibility is allowed in implementations.

Proposition 4.2.10 *For any finite, nonempty subset F of $\mathbb{C}[\mathbf{X}, \mathbf{P}]$ generating a given ideal J, the algorithm* COMPLETE(J) *in Fig. 4.1 computes the completion $\tilde{J}$ and index of the ideal J by returning a pair (p, F'), where p is the index, and F' is a finite, nonempty subset of $\mathbb{C}[\mathbf{X}, \mathbf{P}]$ generating the completion $\tilde{J}$.*

Proof. The result returned is clearly correct if $J = (0)$, which is complete and has index 0. Suppose $J \neq (0)$ so that the **while**-loop will be executed at least once. Suppose by induction hypothesis, after the **while**-loop has completed k times, $k \geqslant 1$, (F') is the $(k - 1)^{\mathrm{st}}$ prolongation J_{k-1} of $J = J_0$, $p = k - 1$, and $(G) = J_{k-1}^*$. If the condition to enter the **while**-loop is next tested, and if the test succeeds, then $J_{k-1} \neq J_{k-1}^*$ and (F') at the end of next iteration is $J_{k-1}^* = J_k$, $p = k$, and $(G) = J_k^*$ (by Lemma 4.1.4),

```
begin
    if F = {0} then return (0, {0})
    p ⟵ -1
    G ⟵ F
    F' ⟵ {0}
    while G ⊄ (F') do
        F' ⟵ G
        H ⟵ a basis for (G) ∩ ℂ[X]
        Q ⟵ a basis for the radical of (H)
        G ⟵ G ∪ Q ∪ ∇Q
        p ⟵ p + 1
    end
    return (p, F')
end
```

Figure 4.1 Algorithm COMPLETE(J) of Proposition 4.2.10

maintaining the induction hypothesis. If the test fails, $(F') = J_{k-1} = J^*_{k-1}, p = k - 1$, and by Proposition 4.2.7, the index of J is p and $J_p = \tilde{J}$. $\quad\square$

4.3   Properties and Examples

We collect now some results on prolongations and completions relating to operations on ideals.

Proposition 4.3.1 *Let* I, J *be ideals in* $\mathbb{C}[\mathbf{X}, \mathbf{P}]$. *Then we have*

1. $I^* + J^* \subseteq (I + J)^*$,
2. $\widetilde{(I^* + J^*)} = \widetilde{(I + J)}$, *and* $\operatorname{ind}(I + J) - 1 \leqslant \operatorname{ind}(I^* + J^*) \leqslant \operatorname{ind}(I + J)$,
3. $(I \cap J)^* \subseteq I^* \cap J^*$,
4. $\widetilde{(I \cap J)} \subseteq \tilde{I} \cap \tilde{J}$.

Proof. Items 1 and 3 follow from Remark 4.1.2, and hence also

$$I + J \subseteq I^* + J^* \subseteq (I + J)^* \subseteq (I^* + J^*)^*. \tag{4.4}$$

Let $p = \operatorname{ind}(I + J)$ and $q = \operatorname{ind}(I^* + J^*)$. By Proposition 4.2.7, prolonging the first three ideals of (4.4) p times gives $\widetilde{(I + J)} = (I^* + J^*)_p$ and hence $q \leqslant p$. Prolonging the last three ideals q times gives $\widetilde{(I^* + J^*)} = (I + J)_{q+1}$ and hence $p \leqslant q + 1$. This proves 2. To prove 4, note that $(I \cap J) \subseteq \tilde{I} \cap \tilde{J}$ and by Proposition 4.2.4, the latter is complete. Hence $\widetilde{(I \cap J)} \subseteq \tilde{I} \cap \tilde{J}$. $\quad\square$

Remark 4.3.2 Proposition 4.3.1 may be used to provide flexibility (though not necessarily more efficiency) in implementation of algorithms for the completion of an ideal

J generated by a finite set $F = \{f_1, \ldots, f_m\}$. Let $I_k = (f_1, \ldots, f_k)$ for $1 \leqslant k \leqslant m$. Then $I_{k+1} = (I_k + (f_{k+1}))$. Thus we can inductively compute the completion of I_k. We can also break F into two arbitrary non-empty disjoint subsets F_1, F_2 and prolong (F_1) and (F_2) in parallel until one of them is complete before combining the resulting ideals and continuing to prolong to completion. Such a strategy may be useful in particular when the system F is in some triangular form or block triangular form.

Example 4.3.3 We now give examples that show that all the inequalities in Proposition 4.3.1 may be strict. The last two shows that there is no general inclusion relation among $(I \cdot J)^*$ and $I^* \cdot J^*$.

1. Let $n = 1$, $I = (P)$, $J = (X^2 + P)$. Then $\widetilde{I} = I^* = I$ and $\widetilde{J} = J^* = J$. However $I^* + J^* = (X^2, P)$ and $(I + J)^* = (X, P)$. Thus $I^* + J^* \subset (I + J)^*$. We also have $\operatorname{ind}(I + J) - 1 < \operatorname{ind}(I^* + J^*)$.

2. Let $n = 1$ and $I = J = (X)$. Then $\operatorname{ind}(I^* + J^*) < \operatorname{ind}(I + J)$.

3. Let $n = 2$, $I = (X_1^2, P_1)$, $J = (X_2^3)$. Then $\widetilde{I} = I^* = (X_1, P_1)$ and $\widetilde{J} = J^* = (X_2, P_2)$. We have $I \cap J = (X_1^2 X_2^3, X_2^3 P_1)$. Then

$$(I \cap J)^* = (X_1 X_2, X_2 P_1, X_1 P_2) = \widetilde{I \cap J}$$

    but $I^* \cap J^* = (X_1 X_2, X_2 P_1, X_1 P_2, P_1 P_2) = \widetilde{I} \cap \widetilde{J}$ is strictly larger. This shows both inequalities in 3 and 4 in Proposition 4.3.1 may be strict.

4. Let $n = 1$, $I = J = (X)$. Then $I^* = J^* = (X, P)$. We have $I \cdot J = (X^2)$ and $(I \cdot J)^* = (X, P) \supset I^* \cdot J^*$.

5. Let $n = 2$, $I = (X_1)$, $J = (X_2)$. Then $I^* = (X_1, P_1)$ and $J^* = (X_2, P_2)$. We have $I \cdot J = (X_1 X_2)$ and

$$(I \cdot J)^* = (X_1 X_2, X_1 P_2 + X_2 P_1) \subset (X_1 X_2, X_1 P_2, X_2 P_1, P_1 P_2) = I^* \cdot J^*.$$

4.4   Initial and Jet Domains

We end this chapter by examining the geometric meaning of prolongation and completion. First we recall the Closure Theorem (see Cox, Little and O'Shea [16, p. 123]) in algebraic geometry, which we restate as Theorem 4.4.1 in the present context. Note that for any ideal J in $\mathbb{C}[\mathbf{X}, \mathbf{P}]$, the ideal $J \cap \mathbb{C}[\mathbf{X}]$ is the n-th elimination ideal of J and that in [16], an affine variety is the same as an algebraic set, not necessarily irreducible. In what follows, if $\mathbf{x} = (x_1, \ldots, x_n) \in \mathbb{C}^n$ and $\mathbf{p} = (p_1, \ldots, p_n) \in \mathbb{C}^n$, then $(\mathbf{x}, \mathbf{p})$ denotes the point $(x_1, \ldots, x_n, p_1, \ldots, p_n)$ in $\mathbb{C}^{2n}$.

Theorem 4.4.1 *Let $J \subseteq \mathbb{C}[\mathbf{X}, \mathbf{P}]$ be an ideal, let $V = Z(J) \subseteq \mathbb{C}^{2n}$, and let $W = Z(R) \subseteq \mathbb{C}^n$, where $R = \sqrt{J \cap \mathbb{C}[\mathbf{X}]}$. Let $\pi : \mathbb{C}^{2n} \longrightarrow \mathbb{C}^n$ be the canonical projection onto the first n coordinates. Then*

1. *W is the Zariski closure of $\pi(V)$, and*

2. *When $V \neq \emptyset$, there is a non-empty Zariski open subset O of W such that $O \subseteq \pi(V)$.*

Here V plays the role of the differential manifold $F^{-1}(0)$ in Rabier and Rheinboldt [48]. The pullback mapping $\pi^* : \mathbb{C}[\mathbf{X}] \longrightarrow \mathbb{C}[\mathbf{X}, \mathbf{P}]$ is just the inclusion mapping. However, the image $\pi(V)$ is not necessarily all of W. The algebraic analog of the tangent bundle of the differential manifold is the tangent variety $T(W)$ of the algebraic set W, defined as a subset of $\mathbb{C}^{2n}$ by

$$T(W) = Z(R \cup \nabla R).$$

Note that $T(W)$ has the property that for each $\mathbf{x} \in W$, the tangent space to W at $\mathbf{x}$ is given by

$$T_{\mathbf{x}}(W) = \{\mathbf{p} \in \mathbb{C}^n \mid (\mathbf{x}, \mathbf{p}) \in T(W)\}.$$

In differential geometry, some authors identify the *tangent space* $T_{\mathbf{x}}(W)$ with the fiber

$$\pi^{-1}(\mathbf{x}) = \{(\mathbf{x}, \mathbf{p}) \in \mathbb{C}^{2n} \mid \mathbf{p} \in \mathbb{C}^n, (\mathbf{x}, \mathbf{p}) \in T(W)\}$$

over $\mathbf{x}$ in the *tangent bundle*.

Remark 4.4.2 When J is complete, we have $V = Z(J) = Z(J^*) \subseteq T(W)$.

The above discussion motivates the definitions below related to the geometric aspects of the algebraic sets defined by an ideal J.

Definition 4.4.3 *Let J be an ideal of $\mathbb{C}[\mathbf{X}, \mathbf{P}]$. We will call $V = Z(J)$ the (first) jet domain and $W = Z(J \cap \mathbb{C}[\mathbf{X}]) = Z(R(J))$ the initial domain of J. The projection map $V \longrightarrow W$ will be denoted by π. For any point $\mathbf{x} \in W$, we say $\mathbf{x}$ is a k-point of the ideal J (more correctly, of π) if the fiber $\pi^{-1}(\mathbf{x}) \subseteq V$ has dimension k (as an algebraic set; the empty set has dimension -1 by convention). Note that the fiber defined differs slightly from the fiber in differential geometry since π is now restricted to V rather than to $T(W)$. The set of k-points of J will be denoted by W^k or $W^k(J)$.*

Remark 4.4.4 By a result of Cartan and Chevalley, $W^k(J)$ is a constructible set for every k. A 0-point $\mathbf{x}$ has a finite, not-empty fiber. The above definitions are very general, but in this paper (see the next two chapters), we shall study mainly the set W^0 of 0-points for a complete quasi-linear ideal. We shall have occasions to decorate J with notational ornaments and we will apply the same decorations for corresponding initial and jet domains and its sets of k-points. For example, let $\widetilde{J}$ be the completion of J. Then $\widetilde{V}$ (resp. $\widetilde{W}$) will denote the jet (resp. initial) domain of $\widetilde{J}$, $\widetilde{\pi} : \widetilde{V} \longrightarrow \widetilde{W}$ will denote the natural projection to the $\mathbf{X}$-coordinates, and $\widetilde{W}^k$ will denote the set of k-points of $\widetilde{J}$. These notations will be used without further comments. We have clearly the following inclusions:

1. $V \supseteq V^* \supseteq \cdots \supseteq \widetilde{V}$.

2. $W \supseteq W^* \supseteq \cdots \supseteq \widetilde{W}$.

3. For any $\mathbf{x} \in \widetilde{W}$, $\pi^{-1}(\mathbf{x}) \supseteq (\pi^*)^{-1}(\mathbf{x}) \supseteq \cdots \supseteq \widetilde{\pi}^{-1}(\mathbf{x})$.

5 Quasi-Linearities

In this chapter, we study a differential algebraic system through its implied quasi-linear subsystem. If the system is to be constrained algebraically, then surely after prolongation, the system will contain one or more quasi-linear equations. If furthermore some existence and uniqueness theorem is to be valid (as would be desirable for systems arising from applications), there will be a well-defined vector field, which at least locally, will be given by some formulae (hopefully rational in the dependent variables) for the first order derivatives. If for some domain of definition, an implicitly given system (2.1) has an explicit form (2.2), then using the notations immediately after Definition 2.1.2, the polynomials $S(\mathbf{X})P_i - R_i(\mathbf{X})$, where $r_i(\mathbf{X}) = R_i(\mathbf{X})/S(\mathbf{X})$ defines the explicit form (2.2), are quasi-linear. It is then probable that these polynomials are *algebraic*[1] consequences of the original system, and thus lie in the ideal generated by the original system (or its completion). The methods to be developed in this and later chapters will recognize systems with these properties and compute the vector field in such cases. We begin by introducing several generalized notions of quasi-linearity.

5.1 Notions of Quasi-Linearity

Definition 5.1.1 *Given an ideal J of $\mathbb{C}[\mathbf{X},\mathbf{P}]$, we define the* quasi-linear ideal *associated with J to be the ideal J^ℓ generated by the set $L = L(J)$ of all polynomials of $\mathbf{P}$-degree at most 1 in J. An ideal J is called* essentially quasi-linear *or simply* quasi-linear *(resp. strictly quasi-linear) if it has essential $\mathbf{P}$-degree at most (resp. exactly) 1. We say J is* eventually quasi-linear *if its completion $\tilde{J}$ is quasi-linear.*

A differential system given by (2.1) may not be quasi-linear in the traditional sense of Definition 2.1.1 if some f_i has $\mathbf{P}$-degree more than 1, while the *ideal J* generated by the set $F = \{\, f_1, \ldots, f_m \,\}$ may still be essentially quasi-linear. By applying Algorithm ESSENTIAL_$\mathbf{P}$-DEGREE_BASIS(F) of Proposition 3.2.1, essential quasi-linearity is effectively decidable. The algorithm returns a pair (d, E), where d is the essential $\mathbf{P}$-degree and E is a $\mathbf{P}$-strong essential $\mathbf{P}$-degree basis of J. Thus J is essentially quasi-linear if and only if $d \leqslant 1$.

Similarly, it may be the case that J is not essentially quasi-linear, but its completion is (see variation at end of Example 9.2.3). The class of differential systems given by eventually quasi-linear ideals represents a much wider class of systems than the traditionally quasi-linear systems. Using the algorithms in Propositions 3.2.1 and 4.2.10, eventual quasi-linearity is also decidable. By Propositions 4.2.7 and 4.1.3, we can reduce the problem of solving an essentially quasi-linear or eventually quasi-linear system to the problem of solving a quasi-linear system.

For an ideal J that is not eventually quasi-linear, we shall work with its associated quasi-linear ideal. It is obvious that $J^\ell \subseteq J$ and that equality holds if and only if J is quasi-linear.

The following results hold for any ideal J, and show that J^ℓ is quasi-linear (obviously) and is effectively computable from J.

---

[1] As will be explained in Chapter 8, a more correct adjective is *differential algebraic*.

Proposition 5.1.2 *Let J^* be the prolongation of an ideal $J \subset \mathbb{C}[\mathbf{X}, \mathbf{P}]$. If the essential $\mathbf{P}$-degree of J is $\geqslant 1$, the essential $\mathbf{P}$-degree of J^* is at most the essential $\mathbf{P}$-degree of J. If J is quasi-linear, so are J^* and $\widetilde{J}$, and in particular, J is eventually quasi-linear.*

Proof. The construction of the prolongation ideal J^* only introduces polynomials of essential $\mathbf{P}$-degree 0 or 1. The completion procedure uses only prolongation.   □

Proposition 5.1.3 *Let J be an ideal with essential $\mathbf{P}$-degree d in $\mathbb{C}[\mathbf{X}, \mathbf{P}]$ and let E be a $\mathbf{P}$-strong subset of J. Then J^ℓ is generated by the set E^ℓ of polynomials in E of $\mathbf{P}$-degree at most 1, J^ℓ is quasi-linear, and E^ℓ is a $\mathbf{P}$-strong essential $\mathbf{P}$-degree basis of J^ℓ.*

Proof. The proposition is clear if $d \leqslant 1$ since in that case, $J = J^\ell$. Suppose then $d > 1$. Let f be a polynomial in J with $\mathbf{P}$-degree at most 1. Since E is $\mathbf{P}$-strong, we can write f in its representation with respect to E as $f = \sum h_i g_i$, where $h_i \in \mathbb{C}[\mathbf{X}, \mathbf{P}]$ is non-zero, $g_i \in E$ and the $\mathbf{P}$-degree of each $h_i g_i$ is at most 1. Hence the $\mathbf{P}$-degree of g_i must be at most 1 and $g_i \in E^\ell$. This shows that J^ℓ is generated by E^ℓ and hence is quasi-linear. This also shows that E^ℓ is $\mathbf{P}$-strong.

Let L be the set of all polynomials of $\mathbf{P}$-degree at most 1 in J. We observe that

$$L \cap \mathbb{C}[\mathbf{X}] \subseteq J^\ell \cap \mathbb{C}[\mathbf{X}] \subseteq J \cap \mathbb{C}[\mathbf{X}] \subseteq L \cap \mathbb{C}[\mathbf{X}] \tag{5.1}$$

and hence these are all equal. If J^ℓ has essential $\mathbf{P}$-degree exactly 1, then clearly, E^ℓ is an essential $\mathbf{P}$-degree basis. If the essential $\mathbf{P}$-degree of J^ℓ is 0, then the equalities (5.1) show that E^ℓ is an essential $\mathbf{P}$-degree basis.   □

Corollary 5.1.4 *For any ideal J of $\mathbb{C}[\mathbf{X}, \mathbf{P}]$, we have $V \subseteq V^\ell$ and $W = W^\ell$.*

Proof. Clearly, $V = Z(J) \subseteq Z(J^\ell) = V^\ell$. By (5.1), $J \cap \mathbb{C}[\mathbf{X}] = J^\ell \cap \mathbb{C}[\mathbf{X}]$, showing that $W = W^\ell$, that is, the initial domains of J and J^ℓ are the same.   □

Proposition 5.1.5 *For any ideal J of $\mathbb{C}[\mathbf{X}, \mathbf{P}]$, we have*

1. $(J^\ell)^* \subseteq (J^*)^\ell$,

2. $\widetilde{J^\ell} \subseteq \widetilde{J}^\ell$,

3. J is complete if and only if J^ℓ is complete, and

4. $\operatorname{ind} J^\ell \leqslant \operatorname{ind} J$.

Proof. We have $J^\ell \subseteq J$ and hence $(J^\ell)^* \subseteq J^* \subseteq \widetilde{J}$ and $\widetilde{J^\ell} \subseteq \widetilde{J}$. Since $(J^\ell)^*$ is also quasi-linear, $(J^\ell)^* \subseteq (J^*)^\ell$ and $\widetilde{J^\ell} \subseteq \widetilde{J}^\ell$. Let $R = R(J)$. Then $R = \sqrt{J \cap \mathbb{C}[\mathbf{X}]} = \sqrt{J^\ell \cap \mathbb{C}[\mathbf{X}]}$ by the equalities (5.1). For any $f \in R$, we have $f \in J$ (resp. $\nabla f \in J$) if and only if $f \in J^\ell$ (resp. $\nabla f \in J^\ell$) since both f and ∇f have $\mathbf{P}$-degree at most 1. This shows that J is complete if and only if J^ℓ is complete. Finally, let p be the index of J. Since $J^\ell \subseteq J$, we have $(J^\ell)_p \subseteq J_p$ and since $(J^\ell)_p$ is quasi-linear, we have $(J^\ell)_p \subseteq (J_p)^\ell$. The ideal $(J_p)^\ell$ is complete since J_p is. By minimality, we have $(J^\ell)_p \subseteq \widetilde{J^\ell} \subseteq (J_p)^\ell$ and thus $\operatorname{ind} J^\ell \leqslant p$.   □

Example 5.1.6 In general, the inequalities $(J^\ell)^* \subseteq (J^*)^\ell$ and $\operatorname{ind} J^\ell \leqslant \operatorname{ind} J$ are strict. Let $n = 4$ and let $J = (1 + P_2 P_4, X_2 + P_1 P_3, X_1)$. Then $(J^\ell)^* = (X_1, P_1) = \widetilde{J^\ell}$, J^ℓ has index 1, but $J^* = (1 + P_2 P_4, X_2, X_1, P_1)$, and $(J^*)^\ell = (X_1, X_2, P_1)$. Moreover, $J^{**} = \mathbb{C}[\mathbf{X}, \mathbf{P}] = \widetilde{J}$ and J has index 2. This example also shows that while the system represented by $\widetilde{J}$ has no solution, the one represented by $\widetilde{J^\ell}$ has solutions.

5.2 Linear Rank

In preparation for the next chapter, which deals with existence and uniqueness properties, we will introduce the notion of the linear rank of an ideal J at a point $\mathbf{x} \in \mathbb{C}^n$ and develop some results that relates this rank to the set W^0 of 0-points. In the literature on differential algebraic quasi-linear equations, it is customary to consider the rank at $\mathbf{x}$ based on the generators (the "given" system) instead of the ideal. Typically, systems are assumed to be of index 1 and given by k quasi-linear equations and $n - k$ algebraic equations, and there is often an unstated assumption that the equations are "independent" in some sense (so that for example, perhaps no new algebraic equation may be derived). Points in W^0 (or rather, in a set analogous to W^0 but defined using only the given system) are often referred to as *regular points*, and points in W which are not in W^0 are referred to as *impasse points*, which are related to the singularities of W (see for example, Tuomela [63]). Our notion of rank presented below is more general and intrinsic because it does not depend on quasi-linearity of the given generators and yet can be effectively computed from any set of generators. It is applicable to an arbitrary ideal (and therefore to any system) and accurately reflects the singular or non-singular behavior of the system at $\mathbf{x}$.

Let J be an ideal in $\mathbb{C}[\mathbf{X}, \mathbf{P}]$ and let $L = L(J)$ be the set of all polynomials in J of $\mathbf{P}$-degree at most 1. For each $f \in L$, let

$$f^1 = \sum_{i=1}^{n} P_i \frac{\partial f}{\partial P_i}$$

be the $\mathbf{P}$-homogeneous form of $\mathbf{P}$-degree 1 of f. Clearly, for any $f, g \in L$, $(f + g)^1 = f^1 + g^1$ and if further $fg \in L$, we also have $(fg)^1 = fg^1 + f^1 g$. For each $\mathbf{x} \in \mathbb{C}^n$, $f^1(\mathbf{x}, \mathbf{P})$ is a linear homogeneous polynomial in $\mathbb{C}[\mathbf{P}]$. The set of all $f^1(\mathbf{x}, \mathbf{P})$ with $f \in L$ is a $\mathbb{C}$-vector subspace $H(\mathbf{x})$ of $\mathbb{C}[\mathbf{P}]$ of dimension at most n.

Definition 5.2.1 *The $\mathbb{C}$-dimension of $H(\mathbf{x})$ is called the* (linear) rank *of J at $\mathbf{x}$ and will be denoted by* $\operatorname{rank} J(\mathbf{x})$. *If F is any subset of $\mathbb{C}[\mathbf{X}, \mathbf{P}]$, we define the* (linear) rank *of F at $\mathbf{x}$ by* $\operatorname{rank} F(\mathbf{x}) = \operatorname{rank} J(\mathbf{x})$ *where J is the ideal generated by F.*

Remark 5.2.2 Clearly, we have $\operatorname{rank} J(\mathbf{x}) = \operatorname{rank} J^\ell(\mathbf{x})$. If $J_1 \subseteq J_2$, then $\operatorname{rank} J_1(\mathbf{x}) \leqslant \operatorname{rank} J_2(\mathbf{x})$. In particular, $\operatorname{rank} J(\mathbf{x}) \leqslant \operatorname{rank} J^*(\mathbf{x}) \leqslant \operatorname{rank} \widetilde{J}(\mathbf{x})$.

In general, the computation of $\operatorname{rank} J(\mathbf{x})$ for arbitrary $\mathbf{x} \in \mathbb{C}^n$ cannot be effected using generators of J^ℓ (see Example 5.2.5). The next proposition shows that our rank coincides with the usual rank when $\mathbf{x} \in W$ and the generators form a $\mathbf{P}$-strong essential

P-degree basis of J^ℓ. Thus it is computable for any ideal J. First, we set up some notations.

Let $F = \{ f_1, \ldots, f_m \} \subset L(J)$. Each f_i may be written in the form

$$f_i = \sum_{j=1}^n \alpha_{i,j}(\mathbf{X})P_j - \gamma_i(\mathbf{X}), \qquad \alpha_{i,j}(\mathbf{X}), \gamma_i(\mathbf{X}) \in \mathbb{C}[\mathbf{X}]. \tag{5.2}$$

We often write the equations $f_1 = 0, \ldots, f_m = 0$ in matrix notation:

$$\begin{bmatrix} \alpha_{1,1}(\mathbf{X}) & \alpha_{1,2}(\mathbf{X}) & \cdots & \alpha_{1,n}(\mathbf{X}) \\ \alpha_{2,1}(\mathbf{X}) & \alpha_{2,2}(\mathbf{X}) & \cdots & \alpha_{2,n}(\mathbf{X}) \\ \vdots & \vdots & \cdots & \vdots \\ \alpha_{m,1}(\mathbf{X}) & \alpha_{m,2}(\mathbf{X}) & \cdots & \alpha_{m,n}(\mathbf{X}) \end{bmatrix} \begin{bmatrix} P_1 \\ P_2 \\ \vdots \\ P_n \end{bmatrix} = \begin{bmatrix} \gamma_1(\mathbf{X}) \\ \gamma_2(\mathbf{X}) \\ \vdots \\ \gamma_m(\mathbf{X}) \end{bmatrix} \tag{5.3}$$

or simply

$$A(\mathbf{X})\mathbf{P}^T = c(\mathbf{X}), \tag{5.4}$$

where $\mathbf{P}^T$ denotes the transpose of $\mathbf{P}$. Let $\rho_F(\mathbf{x})$ be the usual rank of the matrix $A(\mathbf{x})$.

Remark 5.2.3 It should be noted that when $f \in J \cap \mathbb{C}[\mathbf{X}]$, $f^1(\mathbf{X}, \mathbf{P}) = 0$ and hence f cannot contribute to $\operatorname{rank} J(\mathbf{x})$ or $\rho_F(\mathbf{x})$ for any $\mathbf{x}$, and any F containing f. Thus in the computation of $\rho_F(\mathbf{x})$, we can safely remove any $f \in F \cap \mathbb{C}[\mathbf{X}]$.

Proposition 5.2.4 *Let J be an ideal, let $\pi : V \longrightarrow W$ be the projection of the jet domain to the initial domain of J, and let $\mathbf{x} \in \mathbb{C}^n$. Let $F = \{ f_1, \ldots, f_m \}$ be a finite subset of $L(J)$. Then $\operatorname{rank} J(\mathbf{x}) \geqslant \rho_F(\mathbf{x})$ and equality holds if either $\mathbf{x} \in W$ and F is **P**-strong for J^ℓ, or $\mathbf{x} \in \pi(V)$ and F is a essential **P**-degree basis of J^ℓ.*

Proof. The rank $\rho_F(\mathbf{x})$ of the matrix $A(\mathbf{x})$ is equal to the $\mathbb{C}$-dimension of the $\mathbb{C}$-subspace $N(\mathbf{x})$ of $H(\mathbf{x})$ generated by $\{ f_j^1(\mathbf{x}, \mathbf{P}) \mid 1 \leqslant j \leqslant m \}$. Thus $\operatorname{rank} J(\mathbf{x}) \geqslant \rho_F(\mathbf{x})$. Suppose first that $\mathbf{x} \in W$ and F is P-strong for J^ℓ. For any $f \in L(J)$, let $f = \sum_{j=1}^m h_j f_j$ be a representation of f, where $h_j \in \mathbb{C}[\mathbf{X}, \mathbf{P}]$ and the **P**-degree of $h_j f_j$ is at most the **P**-degree of f. If f has **P**-degree 0, then $f^1 = 0$ and $f^1(\mathbf{x}, \mathbf{P}) = 0$ (because $\mathbf{x} \in W$). Otherwise, f has **P**-degree 1, and

$$f^1(\mathbf{x}, \mathbf{P}) = \sum_{j=1}^m (h_j(\mathbf{x}, \mathbf{P})f_j^1(\mathbf{x}, \mathbf{P}) + h_j^1(\mathbf{x}, \mathbf{P})f_j(\mathbf{x}, \mathbf{P})).$$

Since at least one of h_j, f_j has **P**-degree 0, either $h_j^1 = 0$ or $f_j^1 = 0$. In the second case, $f_j \in R(J)$ and $f_j(\mathbf{x}, \mathbf{P}) = 0$ (because $\mathbf{x} \in W$). In any case, $f^1(\mathbf{x}, \mathbf{P})$ belongs to the $\mathbb{C}$-subspace $N(\mathbf{x})$. This shows that $H(\mathbf{x}) = N(\mathbf{x})$ and hence $\operatorname{rank} J(\mathbf{x}) = \rho_F(\mathbf{x})$.

Suppose next that $\mathbf{x} \in \pi(V)$ and F is an essential **P**-degree basis of J^ℓ. Now $\rho_F(\mathbf{x})$ is the codimension of the solution space $S_\mathbf{x}$ of the homogeneous linear system $A(\mathbf{x})\mathbf{P}^T = 0$ and there is a $\mathbf{p} \in \mathbb{C}^n$ such that $(\mathbf{x}, \mathbf{p}) \in V \subseteq V^\ell$. For any $\mathbf{p}' \in \mathbb{C}^n$, since F generates J^ℓ, we have $(\mathbf{x}, \mathbf{p}') \in V^\ell$ if and only if $A(\mathbf{x})\mathbf{p}'^T = c(\mathbf{x})$. Thus the fiber

$$(\pi^\ell)^{-1}(\mathbf{x}) = \{ (\mathbf{x}, \mathbf{p}') \mid \mathbf{p}' \in \mathbf{p} + S_\mathbf{x} \}.$$

Since the fiber depends only on $\mathbf{x}$ and J^ℓ, the $\mathbb{C}$-dimension of $S_\mathbf{x}$, which is $n - \rho_F(\mathbf{x})$, is independent of the choice of F. But $\mathbf{x} \in W$ and by Algorithm 3.1, we may compute a $\mathbf{P}$-strong subset E of $L(J)$ which is also an essential $\mathbf{P}$-degree basis of J^ℓ. By the first case, $\rho_E(\mathbf{x}) = \operatorname{rank} J(\mathbf{x})$ and the $\mathbb{C}$-dimension of $S_\mathbf{x}$ is therefore also $n - \operatorname{rank} J(\mathbf{x})$. Hence $\rho_F(\mathbf{x}) = \operatorname{rank} J(\mathbf{x})$. $\qquad\square$

Example 5.2.5 Note that when $\mathbf{x} \notin \pi(V)$, the inequality in Proposition 5.2.4 may be strict. In Example 3.1.4, F is an essential $\mathbf{P}$-degree basis but is not $\mathbf{P}$-strong. With set up as in that example, let $E = \{ f_1, f_2, X_2 P_2 - X_2 \}$, which is easily verified to be a $\mathbf{P}$-strong subset of J (using Algorithm 3.1). Since $J \cap \mathbb{C}[\mathbf{X}] = (0)$, J is complete. For the point $\mathbf{x} = (0, 1) \in W = \mathbb{C}^n$, we have $\rho_E(\mathbf{x}) = 1$ while $\rho_F(\mathbf{x}) = 0$. Thus the conditions stated in Proposition 5.2.4 are best possible.

Lemma 5.2.6 *Let J be any ideal in $\mathbb{C}[\mathbf{X}, \mathbf{P}]$. Let W and V be its initial and jet domain, respectively. Let $\mathbf{x} \in W$ be such that $\operatorname{rank} J(\mathbf{x}) = n$. Then there exists a unique $\mathbf{p} \in \mathbb{C}^n$ such that $(\mathbf{x}, \mathbf{p}) \in V$. If this is the case, $\mathbf{x} \in W^0$.*

Proof. Since the $\mathbb{C}$-dimension of $H(\mathbf{x})$ is n, there exist $f_1, \ldots, f_n \in L(J)$ such that the set $\{ f_j^1(\mathbf{x}, \mathbf{P}) \mid 1 \leqslant j \leqslant n \}$ is linearly independent over $\mathbb{C}$. Let $f_j^1 = \sum_{k=1}^n \alpha_{j,k}(\mathbf{X})P_k$ and consider the linear system $(\alpha_{j,k}) \cdot \overline{\mathbf{P}}^T = \mathbf{0}$, where $\overline{P_j} = P_j + (f_1^1, \ldots, f_n^1)$ belongs to the $\mathbb{C}[\mathbf{X}]$-module $(\sum_{j=1}^n \mathbb{C}[\mathbf{X}]P_j)/(f_1^1, \ldots, f_n^1)$. Let Δ be the determinant of the $n \times n$ matrix $(\alpha_{j,k})$. By applying Lang [41, Corol., p. 335] (which is basically Cramer's Rule), we have for each j, $\Delta \cdot P_j$ is a linear combination of $f_1^1, \ldots, f_n^1$ with coefficients in $\mathbb{C}[\mathbf{X}]$. Hence for any $g \in J$ of $\mathbf{P}$-degree d, we may write (not necessarily uniquely) $\Delta^d \cdot g = h + \sum_{j=1}^n h_j f_j$ where $h \in \mathbb{C}[\mathbf{X}]$ and $h_j \in \mathbb{C}[\mathbf{X}, \mathbf{P}]$ for $1 \leqslant j \leqslant n$. Clearly, $h \in J$ and since $\mathbf{x} \in W$, $h(\mathbf{x}) = 0$. Since $\Delta(\mathbf{x}) \neq 0$, let $\mathbf{p}$ be the unique solution to the linear system $f_j(\mathbf{x}, \mathbf{P}) = 0, (1 \leqslant j \leqslant n)$. It follows that $g(\mathbf{x}, \mathbf{p}) = 0$. Since g is any element in J, we have $(\mathbf{x}, \mathbf{p}) \in V$, which is also the only point in $\pi^{-1}(\mathbf{x})$. $\qquad\square$

Note that the converse of Lemma 5.2.6 is not true (that is, if $\mathbf{x} \in W^0$, then $\operatorname{rank} J(\mathbf{x})$ need not be n, see Example 5.2.9 below) unless J is quasi-linear. In the next proposition we prove this restricted converse. We will further show that the set W^0 of 0-points for a quasi-linear ideal J is effectively computable in the next chapter.

Proposition 5.2.7 *Let J be a quasi-linear ideal. Then for any $\mathbf{x} \in W$, $\mathbf{x} \in W^0$ if and only if $\operatorname{rank} J(\mathbf{x}) = n$.*

Proof. The sufficiency follows from Lemma 5.2.6. Suppose then $\mathbf{x} \in W^0$ and hence the fiber $\pi^{-1}(\mathbf{x})$ is finite and non-empty. So $\mathbf{x} \in \pi(V)$ and $\pi = \pi^\ell$ (by assumption, J is quasi-linear). As in the proof of Proposition 5.2.4, $\pi^{-1}(\mathbf{x}) = \{ (\mathbf{x}, \mathbf{p}') \mid \mathbf{p}' \in \mathbf{p} + S_\mathbf{x} \}$, where $(\mathbf{x}, \mathbf{p})$ is any point in the fiber, and $S_\mathbf{x}$ is a vector subspace of $\mathbb{C}^n$. Thus $S_\mathbf{x}$ must have dimension 0 and hence $\operatorname{rank} J(\mathbf{x}) = n$. $\qquad\square$

In the corollary below, the reader is reminded that while $W = W^\ell$, the notation W^0 depends on J or π, not just W.

Corollary 5.2.8 *For any ideal J in $\mathbb{C}[\mathbf{X}, \mathbf{P}]$, we have $W^0 \supseteq (W^\ell)^0$.*

Proof. Let $\mathbf{x} \in W = W^\ell$. The fiber $\pi^{-1}(\mathbf{x}) \subseteq V \subseteq V^\ell$ and hence $\pi^{-1}(\mathbf{x}) \subseteq (\pi^\ell)^{-1}(\mathbf{x})$. Suppose $\mathbf{x} \in (W^\ell)^0$. By Proposition 5.2.7, $\operatorname{rank} J(\mathbf{x}) = \operatorname{rank} J^\ell(\mathbf{x}) = n$. By Lemma 5.2.6, $\mathbf{x} \in W^0$. □

Example 5.2.9 The inequalities in Proposition 5.2.8 may be strict, even if J is complete. For example, let $n = 2$ and let $J = (P_1 - X_2, P_2^2 - 1)$. Clearly, J is complete and $J^\ell = (P_1 - X_2)$. For any $\mathbf{x} = (x_1, x_2) \in \mathbb{C}^2$, $\pi^{-1}(\mathbf{x})$ consists of two points $(x_1, x_2, x_2, \pm 1)$ while $(\pi^\ell)^{-1}(\mathbf{x})$ consists of all points (x_1, x_2, x_2, p_2) with p_2 arbitrary. So $W = W^0 = W^\ell = \mathbb{C}^2$ and $(W^\ell)^0 = \emptyset$. Since $\operatorname{rank} J(\mathbf{x}) = 1$ for any $\mathbf{x} \in W^0$, this example also shows that Proposition 5.2.7 is false if J is not quasi-linear.

Corollary 5.2.10 *For any quasi-linear ideal J in $\mathbb{C}[\mathbf{X}, \mathbf{P}]$, we have*

$$W^0 \cap \widetilde{W} \subseteq (W^*)^0 \cap \widetilde{W} \subseteq \widetilde{W}^0.$$

Proof. Let $\mathbf{x} \in \widetilde{W} \subseteq W^* \subseteq W$. Then $\mathbf{x} \in W^0$ if and only if $\operatorname{rank} J(\mathbf{x}) = n$. If this is the case, then by Remark 5.2.2, $\operatorname{rank} J^*(\mathbf{x}) = n$ too and hence $\mathbf{x} \in (W^*)^0$. The second inequality follows by induction using Proposition 4.2.7. □

6 Existence and Uniqueness Theorem

In this chapter, we shall prove results regarding the existence and uniqueness of the solution to an initial value problem for differential algebraic systems. We shall use an algebraic approach, except, of course, we shall rely on the classical existence and uniqueness theorem.

6.1 Main Theorem on Initial Domain

We begin with a result which is basically just Cramer's rule. The way we present this enables us to rigorously show that there is a unique well-defined vector field and an explicit representation may be obtained locally using any submatrix that works. By applying the theory and algorithms of parametric linear systems as developed in Sit [57, 58], we can compute the least number of local explicit formulae.

Theorem 6.1.1 *Let J be a quasi-linear ideal in $\mathbb{C}[\mathbf{X}, \mathbf{P}]$. There exist a computable non-negative integer ν, and ν non-empty, affine, effectively computable, Zariski basic open subsets $U_1, \ldots, U_\nu$ of the initial domain W of J with these properties:*

1. *$W^0 = \bigcup_{k=1}^\nu U_k$; in particular, W^0 is an effectively computable constructible subset of $\mathbb{C}^n$.*

2. *For each k, $1 \leqslant k \leqslant \nu$, the set $Y_k = \pi^{-1}(U_k)$ is an affine, non-empty, Zariski basic open subset of the jet domain V of J.*

3. *For each k, $1 \leqslant k \leqslant \nu$, the restriction π_k of π to Y_k is an isomorphism from Y_k to U_k as affine sets.*

4. *For each k, $1 \leqslant k \leqslant \nu$, the inverse isomorphism $\eta_k : U_k \longrightarrow Y_k$ is an everywhere defined rational map.*

5. *There is a unique isomorphism $\eta : W^0 \longrightarrow \pi^{-1}(W^0)$ of affine schemes such that $\pi(\eta(\mathbf{x})) = \mathbf{x}$ for $\mathbf{x} \in W^0$. Morever, $\eta|_{U_k} = \eta_k$ for $1 \leqslant k \leqslant \nu$.*

Proof. Let $F = \{ f_1, \ldots, f_m \}$ be any essential **P**-degree basis of J as given by (5.2) and let $A(\mathbf{X})\mathbf{P}^T = c(\mathbf{X})$ be the associated matrix system as in (5.4). Let $\mu = \binom{m}{n}$ and let $\Delta_1(\mathbf{X}), \ldots, \Delta_\mu(\mathbf{X})$ be the determinants corresponding to an enumeration of all non-singular $n \times n$ submatrices of $A(\mathbf{X})$. For $1 \leqslant k \leqslant \mu$, let $U_k = \{\mathbf{x} \in W \mid \Delta_k(\mathbf{x}) \neq 0\}$. If $\mathbf{x} \in U_k$, then $\rho_F(\mathbf{x}) = n = \text{rank } J(\mathbf{x})$ by Proposition 5.2.4 and $\mathbf{x} \in W^0$ by Proposition 5.2.7. Conversely, if $\mathbf{x} \in W^0$, then $\text{rank } J(\mathbf{x}) = n$ by Proposition 5.2.7, $\mathbf{x} \in \pi(V)$ by Proposition 5.2.6, $\rho_F(\mathbf{x}) = \text{rank } J(\mathbf{x}) = n$ by Proposition 5.2.4, and hence $\mathbf{x} \in U_k$ for some k. We have proved that $W^0 = \cup_{k=1}^\mu U_k$ and hence W^0 is a Zariski open subset of W. Without loss of generality, we may assume that $U_1, \ldots, U_\nu$, $0 \leqslant \nu \leqslant \mu$, form a minimal irredundant cover for W^0 (the case $\nu = 0$ happens when $W^0 = \emptyset$). Fix $k, 1 \leqslant k \leqslant \nu$, let F_k be an n-subset of F such that the $n \times n$ submatrix $A_k(\mathbf{X})$ of $A(\mathbf{X})$ corresponding to the polynomials in F_k defines $\Delta_k(\mathbf{X}) = \det(A_k(\mathbf{X}))$ and similarly let $c_k(\mathbf{X})$ be the corresponding subvector of $c(\mathbf{X})$. Clearly U_k is a non-empty basic Zariski open subset of W and is affine. By Lemma 5.2.6, the inverse image

$$Y_k = \pi^{-1}(U_k) = \{ (\mathbf{x}, \mathbf{p}) \in V \mid \Delta_k(\mathbf{x}) \neq 0 \}$$

is non-empty. It is a basic Zariski open subset of V and is also affine. The restriction $\pi_k : Y_k \longrightarrow U_k$ of π is a morphism of affine sets and we shall show it is an isomorphism by showing that it has an inverse η_k which is given by an everywhere defined rational map on U_k. Let $C_k(\mathbf{X})$ be the transpose of the cofactor matrix of $A_k(\mathbf{X})$, and for each $\mathbf{x} \in U_k$, by Cramer's rule, let $\eta_k(\mathbf{x}) = (\mathbf{x}, \mathbf{r}_k(\mathbf{x}))$ where

$$\mathbf{r}_k(\mathbf{x}) = A_k^{-1}(\mathbf{x})c_k(\mathbf{x}) = \frac{1}{\Delta_k(\mathbf{x})} \cdot C_k(\mathbf{x})c_k(\mathbf{x}). \tag{6.1}$$

We have $A_k(\mathbf{x})\mathbf{r}_k(\mathbf{x}) = c_k(\mathbf{x})$, that is, $f(\mathbf{x}, \mathbf{r}_k(\mathbf{x})) = 0$ for $f \in F_k$. Since $\text{rank } A_k(\mathbf{x}) = n$, $\mathbf{r}_k(\mathbf{x})$ is the only solution to the linear system $f(\mathbf{x}, \mathbf{P}) = 0$ ($f \in F_k$). By Lemma 5.2.6, the system $f(\mathbf{x}, \mathbf{P}) = 0$ ($f \in J$) has a unique solution, which therefore must be $\mathbf{r}_k(\mathbf{x})$. Thus $\eta_k(\mathbf{x}) = (\mathbf{x}, \mathbf{r}_k(\mathbf{x})) \in V$ and since $\mathbf{x} \in U_k$, we have $\eta_k(\mathbf{x}) \in Y_k$. It is clear that $\pi_k \circ \eta_k$ (resp. $\eta_k \circ \pi_k$) is the identity map on U_k (resp. Y_k). Since the solution to $f(\mathbf{x}, \mathbf{P}) = 0$ ($f \in J$) is unique, $\mathbf{r}_k(\mathbf{x})$ is independent of the choice of F and of U_k which contains $\mathbf{x}$ and this proves 5.

Finally, the effective computability of $U_1, \ldots, U_\nu$ follows from a more general result on parametric linear systems of Sit [57, Prop. 4.5, Theorem 6.4, and Algorithm 2]. $\square$

Remark 6.1.2 If m is the number of polynomials in any essential **P**-degree basis of J, then $0 \leqslant \nu \leqslant \binom{m}{n}$. Using results in Sit [58], we can further compute $U_1, \ldots, U_\nu$ so that their union is irredundant without using any radical ideal computation. The isomorphisms can also be explicitly computed on each U_k.

Remark 6.1.3 It should be noted that it is possible for W^0 to be the empty set even though $V \neq \emptyset$. For example, the ideal $J = (X_1^2, P_1, X_1 P_2)$ is quasi-linear, but W consists of the X_2-axis, and every point $\mathbf{x} = (x_1, x_2) = (0, x_2) \in W$ has rank $J(\mathbf{x}) = 1$ and so $W^0 = \emptyset$.

6.2   Existence and Uniqueness

We are now ready to prove that the solutions guaranteed by the classical existence and uniqueness theorem automatically satisfy all algebraic constraints. For the convenience of our readers, we restate a version of Cauchy's existence theorem below. In what follows, for any real $\epsilon > 0$, we denote the real interval $(-\epsilon, \epsilon)$ by $\mathcal{B}_\epsilon$. The symbol $\mathbf{z} = \mathbf{z}(t) = (z_1(t), \ldots, z_n(t))$ denotes a vector of n indeterminate complex-valued functions $z_i(t)$ defined for t in some real interval. Similarly, the symbol $\mathbf{w} = \mathbf{w}(t) = (w_1(t), \ldots, w_\ell(t))$ denotes ℓ indeterminate complex-valued functions.

Theorem 6.2.1 *Let $\mathcal{D}$ be an open subset of $\mathbb{C}^n$, and let the system $\mathbf{v}$ on $\mathcal{D}$ be given by $\dot{\mathbf{z}}(t) = \mathbf{r}(\mathbf{z}(t))$ for $t \in \mathbb{R}$, where $\mathbf{r} : \mathcal{D} \longrightarrow \mathbb{C}^n$ is some analytic map. Then for any $\mathbf{x}_0 \in \mathcal{D}$, there exist some $\epsilon > 0$, some open neighborhood $\mathcal{O}$ of $\mathbf{x}_0$, and an analytic map $\psi : \mathcal{B}_\epsilon \times \mathcal{O} \longrightarrow \mathcal{D}$ such that $\mathcal{O} \subseteq \mathcal{D}$, and for every $\mathbf{x} \in \mathcal{O}$, we have*

   1. $\psi(0, \mathbf{x}) = \mathbf{x}$,

   2. the map $\psi_\mathbf{x} : \mathcal{B}_\epsilon \longrightarrow \mathcal{D}$ defined by $t \mapsto \psi(t, \mathbf{x})$ is the unique solution defined on $\mathcal{B}_\epsilon$ satisfying the system $\mathbf{v}$ and the initial condition $\mathbf{z}(0) = \mathbf{x}$.

Proof. This is basically a specialized version to autonomous systems of Dieudonné [19, Theorems (10.8.1) and (10.8.2)]. To apply these results, we let I be the complex time domain. Referring to notations in (10.8.1) of [19], for the base field K, we use $\mathbb{C}$; for the Banach space E, we use $\mathbb{C}^n$; for the open subset H in E, we use $\mathcal{D}$; for the continuously differentiable map $f : \mathrm{I} \times \mathrm{H} \longrightarrow \mathrm{E}$, we define $f(t, x) = \mathbf{r}(x)$ for all $t \in \mathrm{I}$ and $x \in \mathrm{H}$; for the point $(a, b) \in \mathrm{I} \times \mathrm{H}$, we use $(0, \mathbf{x}_0)$; and for (t_0, x_0), we use $(0, \mathbf{x})$. Our $\mathcal{B}_\epsilon$ is the intersection of $\mathbb{R}$ with the open ball $\mathrm{J} \subseteq \mathrm{I}$ centered at a; our $\mathcal{O}$ is the open ball $\mathrm{V} \subseteq \mathrm{H} = \mathcal{D}$ centered at b; and our $\psi(t, \mathbf{x})$ is $u(t, 0, x_0)$, that is, ψ is the restriction of u to $\mathcal{B}_\epsilon \times \{0\} \times \mathcal{O}$ (identified with $\mathcal{B}_\epsilon \times \mathcal{O}$). By (10.8.2) of [19], we may choose J and V such that the map $(t, t_0, x_0) \mapsto u(t, t_0, x_0)$ is analytic in $\mathrm{J} \times \mathrm{J} \times \mathrm{V}$. Thus our map $(t, \mathbf{x}) \mapsto \psi(t, \mathbf{x}) = u(t, 0, \mathbf{x})$ is analytic on $\mathcal{B}_\epsilon \times \mathcal{O}$ by (9.9.4) of [19]. □

Similar versions of Cauchy's theorem for differential manifolds may be found in many books, for example, Arnol'd [1], Lang [42], Spivak [60], and Warner [65]. The map $\mathbf{r}$ in Theorem 6.2.1 then defines an analytic vector field $\mathbf{v} : \dot{\mathbf{z}}(t) = \mathbf{r}(\mathbf{z}(t))$, the map ψ is a local analytic flow generated by $\mathbf{v}$ and for each $\mathbf{x}$, the image of the map $\psi_\mathbf{x}$ is an integral curve for $\mathbf{v}$ through $\mathbf{x}$. Since we are dealing with constructible sets in algebraic geometry rather than differential manifolds, we need analogs of these concepts in the algebraic setting.

Definition 6.2.2 *Let J be an ideal in $\mathbb{C}[\mathbf{X}, \mathbf{P}]$, let $\mathcal{B}_\epsilon$ be an open interval in $\mathbb{R}$, let M be a constructible subset of $\mathbb{C}^n$, and let $\mathbf{x} \in \mathbb{C}^n$. By a differentiable map $\varphi : \mathcal{B}_\epsilon \longrightarrow M$,* *we mean a differentiable map $\varphi : \mathcal{B}_\epsilon \longrightarrow \mathbb{C}^n$ whose image is contained in M. By a* solution to the initial value problem $(J, \mathbf{x})$ on $\mathcal{B}_\epsilon$ in M, *we mean a differentiable map* $\varphi : \mathcal{B}_\epsilon \longrightarrow M$ *such that* $\varphi(0) = \mathbf{x}$ *and* $f(\varphi(t), \dot{\varphi}(t)) = 0$ *for all* $t \in \mathcal{B}_\epsilon$ *and for all* $f \in J$. *Alternatively, we say* φ *satisfies the initial value problem* $(J, \mathbf{x})$, *or that* $(J, \mathbf{x})$ *admits a solution in* M, *or that the image of* φ *is an* integral curve of J through $\mathbf{x}$.

Clearly if J is generated by $f_1, \ldots, f_m$ as an ideal in $\mathbb{C}[\mathbf{X}, \mathbf{P}]$, then a differentiable map $\varphi : \mathcal{B}_\epsilon \longrightarrow M$ satisfies the initial value problem $(J, \mathbf{x})$ if and only if $\varphi(0) = \mathbf{x}$ and for $1 \leqslant i \leqslant m$, and $t \in \mathcal{B}_\epsilon$, we have $f_i(\varphi(t), \dot{\varphi}(t)) = 0$.

We now state and prove our result on existence and uniqueness.

Theorem 6.2.3 *Let J be a complete quasi-linear ideal in $\mathbb{C}[\mathbf{X}, \mathbf{P}]$, let V and W be respectively the jet and initial domain of J, and let $\mathbf{x}_0 \in W^0$. Then there exist some Euclidean open subset $\mathcal{U}$ of W^0 containing $\mathbf{x}_0$, some $\epsilon > 0$ and a mapping*

$$\varphi : \mathcal{B}_\epsilon \times \mathcal{U} \longrightarrow W^0,$$

such that for every $\mathbf{x} \in \mathcal{U}$, we have

1. *$\varphi(0, \mathbf{x}) = \mathbf{x}$;*

2. *the map $\varphi_{\mathbf{x}}(t)$ defined by $t \mapsto \varphi(t, \mathbf{x})$ is the unique solution on $\mathcal{B}_\epsilon$ in W^0 to the initial value problem $(J, \mathbf{x})$; and*

3. *the map φ is the restriction of an analytic map $\psi : \mathcal{B}_\epsilon \times \mathcal{O} \longrightarrow \mathbb{C}^n$ where $\mathcal{O}$ is a Euclidean open subset of $\mathbb{C}^n$ containing $\mathbf{x}_0$ such that $\mathcal{U} = W^0 \cap \mathcal{O}$.*

Proof. Let $F = \{ f_1, \ldots, f_m \}$ be an essential $\mathbf{P}$-degree basis of J. Consider the following two systems of differential equations. The first system is defined for $\mathbf{z}(t) = (z_1(t), \ldots, z_n(t))$. Using notations as in Theorem 6.1.1, fix some k such that $\Delta_k(\mathbf{x}_0) \neq 0$, let $U_k = \{\mathbf{x} \in W \mid \Delta_k(\mathbf{x}) \neq 0\}$ and let $\mathbf{r}_k(\mathbf{X}) = A_k^{-1}(\mathbf{X})c_k(\mathbf{X})$. For convenience of notation we will drop the subscript k from Δ_k, U_k and $\mathbf{r}_k$. Let $\mathcal{D} = D(\Delta) = \{\mathbf{x} \in \mathbb{C}^n \mid \Delta(\mathbf{x}) \neq 0\}$. Note that $\mathcal{D}$ is an open subset of $\mathbb{C}^n$, $\mathbf{x}_0 \in U = W \cap \mathcal{D}$ and $\mathbf{r}(\mathbf{X})$ is defined and analytic on all of $\mathcal{D}$. The first system $\mathbf{v}_1$ is given by $\dot{\mathbf{z}}(t) = \mathbf{r}(\mathbf{z}(t))$ on $\mathcal{D}$ for $t \in \mathbb{R}$.

For the second system, let $\{ q_1, \ldots, q_\ell \}$ be a set of generators for $J \cap \mathbb{C}[\mathbf{X}]$. Without loss of generality, we may assume that $f_1, \ldots, f_n$ are the polynomials such that $f_1^1, \ldots, f_n^1$ correspond to the rows of the submatrix A_k whose determinant is Δ and is non-zero. By Cramer's rule as in the proof of Lemma 5.2.6, we have for each j, $\Delta \cdot P_j$ is a linear combination of $f_1^1, \ldots, f_n^1$ with coefficients in $\mathbb{C}[\mathbf{X}]$. Hence for each $i, 1 \leqslant i \leqslant \ell$, $\Delta \cdot \nabla q_i$ is also such a linear combination. Suppose $\Delta \cdot \nabla q_i = \sum_{j=1}^n q_{ij} f_j^1$, where $q_{ij} \in \mathbb{C}[\mathbf{X}]$. Let $g_i = \sum_{j=1}^n q_{ij}(f_j^1 - f_j) \in \mathbb{C}[\mathbf{X}]$. Then $\Delta \cdot \nabla q_i = g_i + \sum_{j=1}^n q_{ij} f_j$. By the completeness property of J, $g_i \in J \cap \mathbb{C}[\mathbf{X}]$. Thus we may write $g_i = \sum_{i'=1}^\ell h_{ii'} q_{i'}$, where $h_{ii'} \in \mathbb{C}[\mathbf{X}]$. Let the indeterminate functions for the second system be $(\mathbf{z}(t), \mathbf{w}(t))$. The second system $\mathbf{v}_2$ is given by:

$$\dot{\mathbf{z}}(t) = \mathbf{r}(\mathbf{z}(t)), \quad \dot{w}_i(t) = \sum_{i'=1}^m \frac{h_{ii'}(\mathbf{z}(t))}{\Delta(\mathbf{z}(t))} w_{i'}(t) \quad 1 \leqslant i \leqslant \ell, \quad t \in \mathbb{R}.$$

Clearly, $\mathbf{v}_2$ is defined and analytic on all of $\mathcal{D} \times \mathbb{C}^\ell$.

We observe that on $\mathcal{D}$, the system of equations $f_1(\mathbf{X}, \mathbf{P}) = 0, \ldots, f_n(\mathbf{X}, \mathbf{P}) = 0$ is equivalent to system $\mathbf{P} = \mathbf{r}(\mathbf{X})$.

By Theorem 6.2.1 applied to these systems, there exist some Euclidean open neighborhood $\mathcal{O} \subseteq \mathcal{D}$ (hence also an open subset of $\mathbb{C}^n$) of $\mathbf{x}_0$, some Euclidean open neighborhood $\mathcal{O}' \subseteq \mathbb{C}^\ell$ of $\mathbf{0}$, some $\epsilon > 0$, some analytic map $\psi_1 : \mathcal{B}_\epsilon \times \mathcal{O} \longrightarrow \mathcal{D}$ satisfying 1 and 2 in Theorem 6.2.1 with respect to $\mathbf{v} = \mathbf{v}_1$, and finally some analytic map $\psi_2 : \mathcal{B}_\epsilon \times \mathcal{O} \times \mathcal{O}' \longrightarrow \mathcal{D} \times \mathbb{C}^\ell$ satisfying 1 and 2 in Theorem 6.2.1 with respect to $\mathbf{v} = \mathbf{v}_2$ (when $\mathcal{D}$ and $\mathbf{r}$ there are suitably replaced). If necessary, shrink $\mathcal{O}$ so that $W \cap \mathcal{O} \subseteq W^0$. Let $\mathcal{U} = W \cap \mathcal{O} = W^0 \cap \mathcal{O} \subseteq U$ and let $\varphi : \mathcal{B}_\epsilon \times \mathcal{U} \longrightarrow \mathcal{D}$ be the restriction of ψ_1 to $\mathcal{B}_\epsilon \times \mathcal{U}$. Since $W \cap \mathcal{D} = U \subseteq W^0$, it suffices to show that $\varphi(t, \mathbf{x}) \in W$ for all $t \in \mathcal{B}_\epsilon$ and $\mathbf{x} \in \mathcal{U}$. Fix $\mathbf{x} \in \mathcal{U}$.

Clearly the image of the map $\xi_\mathbf{x} : \mathcal{B}_\epsilon \longrightarrow \mathcal{D} \times \mathbb{C}^\ell$ given by $t \mapsto (\varphi_\mathbf{x}(t), \mathbf{0})$ is an integral curve through $(\mathbf{x}, \mathbf{0})$ for the second vector field $\mathbf{v}_2$. We claim that another integral curve through $(\mathbf{x}, \mathbf{0})$ for $\mathbf{v}_2$ is given by the image of the map

$$\chi_\mathbf{x} : t \mapsto (\varphi_\mathbf{x}(t), q_1(\varphi_\mathbf{x}(t)), \ldots, q_\ell(\varphi_\mathbf{x}(t))).$$

To verify this, we clearly have $\chi_\mathbf{x}(0) = (\mathbf{x}, \mathbf{0})$ since $\mathbf{x} \in W$ and for $1 \leqslant i \leqslant \ell$,

$$
\begin{aligned}
\frac{dq_i(\varphi_\mathbf{x}(t))}{dt} &= \sum_{j=1}^n \frac{\partial q_i}{\partial X_j}(\varphi_\mathbf{x}(t)) \frac{d\varphi_\mathbf{x}(t)}{dt} \\
&= \nabla q_i(\varphi_\mathbf{x}(t), \dot\varphi_\mathbf{x}(t)) \\
&= \frac{1}{\Delta(\varphi_\mathbf{x}(t))} \left[g_i(\varphi_\mathbf{x}(t)) + \sum_{j=1}^n q_{ij}(\varphi_\mathbf{x}(t)) f_j(\varphi_\mathbf{x}(t), \dot\varphi_\mathbf{x}(t)) \right] \\
&= \frac{1}{\Delta(\varphi_\mathbf{x}(t))} \sum_{i'}^\ell h_{ii'}(\varphi_\mathbf{x}(t)) q_{i'}(\varphi_\mathbf{x}(t)).
\end{aligned}
$$

Thus, by uniqueness, we have $\xi_\mathbf{x} = \chi_\mathbf{x} = (\psi_2)_\mathbf{x}$ and $q_i(\varphi_\mathbf{x}(t)) = q_i(\varphi(t, \mathbf{x})) = 0$ for $1 \leqslant i \leqslant \ell$ and for all $t \in \mathcal{B}_\epsilon$. Hence $\varphi(t, \mathbf{x}) \in W$. This establishes that the image of φ is a subset of W^0. Part 1 clearly holds. The map $\varphi_\mathbf{x}(t)$ defined in Part 2 is a solution to the initial value problem $(J, \mathbf{x})$ because for every $t \in \mathcal{B}_\epsilon$, $\varphi_\mathbf{x}(t) \in W \cap \mathcal{D}$, $f(\varphi_\mathbf{x}(t), \dot\varphi_\mathbf{x}(t)) = 0$ for $f = f_1, \ldots, f_n$, and hence by Lemma 5.2.6, also for all $f \in J$. For the uniqueness part of 2, we note that any other solution to the initial value problem $(J, \mathbf{x})$ must satisfy the system $f_i(\mathbf{z}(t), \dot{\mathbf{z}}(t)) = 0, 1 \leqslant i \leqslant n$, and since $\mathbf{x} \in \mathcal{D}$, it must satisfy the first system $\mathbf{v}_1$. Part 3 is clear. $\square$

Example 6.2.4 This trivial example illustrates, as guaranteed by the theorem, that we have uniqueness even at points of W^0 that are singular points of W. Let $n = 2$ and $J = (P_1, P_2, X_1 X_2)$. Then J is complete, $W = W^0$ consists of the two axes, the origin is a singular point of W, and every point in W defines a unique equilibrium solution.

As is customary for an existence theorem, the previous result is local in nature. However, because of Theorem 6.1.1, the vector field $\mathbf{v}_1$ defined in the proof of Theorem

6.2.3 can be analytically and uniquely continued to all of W^0 (the vector field $\mathbf{v}_1$, restricted to W^0, is determined by the isomorphism η of Theorem 6.1.1) and hence by the usual extension theorems, any integral curve to the vector field $\mathbf{v}_1$ through $\mathbf{x} \in W^0$ can be extended analytically and uniquely either to the boundary of any compact subset of W^0 containing $\mathbf{x}$ or indefinitely (see for example, Arnol'd [1, Chap. 2, Section 6]). It should be noted, however, that when the domain of definition of the vector field $\mathbf{v}_1$ extends beyond W^0, the integral curve may extend beyond W^0.

At this point we summarize our constructive and existence results for a complete quasi-linear ideal. In Theorem 6.2.7, we extend these to an arbitrary ideal.

Theorem 6.2.5 *Let J be a complete quasi-linear ideal in $\mathbb{C}[\mathbf{X}, \mathbf{P}]$ with a given set of generators (not necessarily quasi-linear). Let V (resp. W) be the first jet (resp. initial) domain of J and let $\pi : V \longrightarrow W$ be the projection onto the $\mathbf{X}$-coordinates. Then we can effectively compute:*

1. *V, W and some non-negative integer ν;*

2. *for each k, $1 \leqslant k \leqslant \nu$, a non-empty Zariski open subset U_k of the initial set W;*

3. *for each k, $1 \leqslant k \leqslant \nu$, an n-dimensional vector $\mathbf{r}_k = (r_{k,1}, \ldots, r_{k,n})$ of rational functions in $\mathbb{C}(\mathbf{X})^n$, everywhere defined on U_k*

   such that

4. *the sets $U_1, \ldots, U_\nu$ form an irredundant Zariski-open cover of W^0;*

5. *for every $\epsilon > 0$, for every $\mathbf{x} \in W^0$ and differentiable map $\psi_{\mathbf{x}} : \mathcal{B}_\epsilon \longrightarrow W^0$, the image of $\psi_{\mathbf{x}}$ is an integral curve of J through $\mathbf{x}$ if and only if $\psi_{\mathbf{x}}(0) = \mathbf{x}$ and for every k, $1 \leqslant k \leqslant \nu$ such that $\mathbf{x} \in U_k$, we have $\dot{\psi}_{\mathbf{x}}(t) = \mathbf{r}_k(\psi_{\mathbf{x}}(t))$;*

6. *for every $\mathbf{x}_0 \in W^0$, there exist some $\epsilon > 0$, some open neighborhood $\mathcal{U}$ of $\mathbf{x}_0$ in W^0 and a map*

$$\varphi : \mathcal{B}_\epsilon \times \mathcal{U} \longrightarrow W^0$$

   such that for every $\mathbf{x} \in \mathcal{U}$, the image of the map $\varphi_{\mathbf{x}} : \mathcal{B}_\epsilon \longrightarrow W^0$ defined by $t \mapsto \varphi(t, \mathbf{x})$ is an integral curve of J through $\mathbf{x}$; and

7. *for any $\mathbf{x} \notin \pi(V)$, the initial value problem $(J, \mathbf{x})$ does not admit a solution on $\mathcal{B}_\epsilon$ in $\mathbb{C}^n$ for any $\epsilon > 0$.*

Proof. We can replace the set of generators by an essential $\mathbf{P}$-degree basis. Parts 1 through 6 follow at once from Theorems 6.1.1 and 6.2.3. To prove 7, consider any solution $\psi_{\mathbf{x}} : \mathcal{B}_\epsilon \longrightarrow \mathbb{C}^n$ to $(J, \mathbf{x})$ for $\mathbf{x} \in \mathbb{C}^n$. Then $(\psi_{\mathbf{x}}(0), \dot{\psi}_{\mathbf{x}}(0)) \in V$ and $\mathbf{x} = \psi_{\mathbf{x}}(0) \in \pi(V)$. $\qquad \square$

Remark 6.2.6 In Theorems 6.2.3 and 6.2.5, we assume that J is complete and quasi-linear. However, in practice, we do not have to know this *a priori*. Given any set of generators for the ideal, we can decide algorithmically whether the ideal is quasi-linear and if it is, we can complete it. The assumption of completeness is crucial to the proof, and indeed shows that under the completeness assumption, we can solve for $\mathbf{p}$ in local coordinates for any given $\mathbf{x} \in W^0$ using Cramer's rule (or any other means). This provides an equivalent explicit system of first order ordinary differential equations to

(5.4). Thus we have provided the theoretical foundation for a rather simple algorithm. It should be emphasized that while the formula (6.1) is a local one, it is independent of the choice of k such that $\mathbf{x} \in U_k$. A vector field on all of W^0 may be defined piecewise on each U_k and in a numerical integration scheme, we may examine, at every $\mathbf{x}$, the condition numbers of these predetermined submatrices $A_k(\mathbf{x})$ (for which $\Delta_k(\mathbf{x}) \neq 0$) and use the one that is least ill-conditioned. The parametric linear equation algorithm of Sit [57] can be used to obtain a least number of pieces.

By passing to the completion and replacing the ideal with the associated quasi-linear ideal, we can generalize the above theorem. A consequence of this result is: if after we complete the ideal, there are n linearly independent quasi-linear equations, we may simply solve these for $\mathbf{p}$ given any $\mathbf{x} \in \widetilde{W}$ whenever this $\mathbf{p}$ is unique (and ignore any equation of higher $\mathbf{P}$-degree). The initial conditions $\mathbf{z}(0) = \mathbf{x}$ for such $\mathbf{x}$ (for all possible choices of the n equations) are the only ones that guarantee a unique solution. Example 6.2.8 illustrates this situation.

Theorem 6.2.7 *Let* $J = (g_1, \ldots, g_m)$ *be an ideal in* $\mathbb{C}[\mathbf{X}, \mathbf{P}]$, *and consider the system of differential algebraic equations*

$$g_1(z_1, \ldots, z_n, \dot{z}_1, \ldots, \dot{z}_n) = 0,$$

$$\vdots$$

$$g_m(z_1, \ldots, z_n, \dot{z}_1, \ldots, \dot{z}_n) = 0.$$

Then we can effectively compute

1. *a Zariski-closed subset M of $\mathbb{C}^n$ and some integer $\nu \geqslant 0$;*

2. *for each k, $1 \leqslant k \leqslant \nu$, a non-empty Zariski open subset U_k of M;*

3. *for each k, $1 \leqslant k \leqslant \nu$, an n-dimensional vector $\mathbf{r}_k = (r_{k,1}, \ldots, r_{k,n})$ of rational functions in $\mathbb{C}(\mathbf{X})^n$, everywhere defined on U_k*

   such that

4. *the union $M^0 = \bigcup_{k=1}^{\nu} U_k$ is irredundant;*

5. *for every $\epsilon > 0$ and for every $\mathbf{x} \in M^0$ and differentiable map $\psi_{\mathbf{x}} : \mathcal{B}_\epsilon \longrightarrow M^0$, the image of $\psi_{\mathbf{x}}$ is an integral curve of J through $\mathbf{x}$ if and only if $\psi_{\mathbf{x}}(0) = \mathbf{x}$ and for every k, $1 \leqslant k \leqslant \nu$, such that $\mathbf{x} \in U_k$, we have $\dot{\psi}_{\mathbf{x}}(t) = \mathbf{r}_k(\psi_{\mathbf{x}}(t))$;*

6. *for every $\mathbf{x}_0 \in M^0$, there exist some $\epsilon > 0$, some open neighborhood $\mathcal{U}$ of $\mathbf{x}_0$ in M^0 and a map $\varphi : \mathcal{B}_\epsilon \times \mathcal{U} \longrightarrow M^0$ such that for every $\mathbf{x} \in \mathcal{U}$, the image of the map $\varphi_{\mathbf{x}} : \mathcal{B}_\epsilon \longrightarrow M^0$ defined by $t \mapsto \varphi(t, \mathbf{x})$ is an integral curve of J through $\mathbf{x}$; and*

7. *for any $\mathbf{x} \notin M$, the initial value problem $(J, \mathbf{x})$ does not admit a solution on $\mathcal{B}_\epsilon$ in $\mathbb{C}^n$ for any $\epsilon > 0$.*

Proof. Let $\widetilde{J}$ be the completion of J and let $\widetilde{J}^\ell$ be the associated quasi-linear ideal of $\widetilde{J}$, which is complete by Proposition 5.1.5. Let $M = \widetilde{W}^\ell = \widetilde{W}$. The set M^0 of 0-points of $\widetilde{J}^\ell$ is an open subset of M. Let $f_1, \ldots, f_m$ be an essential $\mathbf{P}$-degree basis of $\widetilde{J}^\ell$, and let ν, $U_1, \ldots, U_\nu$, $\mathbf{r}_1, \ldots, \mathbf{r}_\nu$, ϵ and φ be as given by Theorem 6.2.5 for the complete quasi-linear ideal $\widetilde{J}^\ell$. Then clearly properties 1–4 hold. By Propositions 4.1.3 and 4.2.7, a differentiable map $\psi_{\mathbf{x}} : \mathcal{B}_\epsilon \longrightarrow M^0$ is a solution to $(J, \mathbf{x})$ if and only if it is a solution to $(\widetilde{J}, \mathbf{x})$; in which case, $\mathbf{z}(t) = \psi_{\mathbf{x}}(t)$ satisfies

$$f(z(t), \dot{z}(t)) = 0 \text{ for all } f \in \widetilde{J}^\ell. \tag{6.2}$$

Conversely, suppose $\mathbf{z}(t) = \psi_{\mathbf{x}}(t)$ satisfies (6.2). Since $\psi_{\mathbf{x}}(t) \in M^0$, rank $\widetilde{J}^\ell(\psi_{\mathbf{x}}(t)) =$ rank $\widetilde{J}(\psi_{\mathbf{x}}(t)) = n$ and hence by Lemma 5.2.6, $\psi_{\mathbf{x}}(t)$ is a solution of $(\widetilde{J}, \mathbf{x})$. Properties 5 and 6 now follow from corresponding properties of Theorem 6.2.5 for $\widetilde{J}^\ell$. Since a solution to $(J, \mathbf{x})$ is a solution to $(\widetilde{J}, \mathbf{x})$, $(\psi_{\mathbf{x}}(0), \dot{\psi}_{\mathbf{x}}(0)) \in \widetilde{V}$ and hence $\mathbf{x} = \psi_{\mathbf{x}}(0) \in \widetilde{\pi}(\widetilde{V}) \subseteq \widetilde{W} = \widetilde{W}^\ell = M$. This proves 7. $\qquad\square$

Example 6.2.8 With t as the independent variable and using a more traditional notation, let $p = \dot{x}$, $q = \dot{y}$. Consider the following non-quasi-linear system:

$$
\begin{aligned}
pq &= xy, \\
-yp + 3xq &= 3x^2 + 6, \\
4q^2 &= 9x^2, \\
p^2 &= x^2 - 4.
\end{aligned}
$$

A solution is $(2\cosh t, 3\sinh t)$. The ideal J corresponding to this system is complete and has essential **P**-degree 2. From Gröbner basis computations, we obtain the equivalent system:

$$
\begin{aligned}
q^2 &= y^2 + 9, \\
27p + 6xyq &= 4y^3 + 54y, \\
(4y^2 + 54)q &= 6xy^2 + 81x, \\
0 &= 9x^2 - 4y^2 - 36.
\end{aligned}
$$

By Theorem 6.2.7, we may drop the non-quasi-linear equation $q^2 = y^2 + 9$ and solve the two quasi-linear equations to yield an explicit form:

$$
\begin{aligned}
9x^2 - 4y^2 - 36 &= 0, \\
2y^2 + 27 &\neq 0,
\end{aligned}
$$

$$\mathbf{v}: \quad p = \frac{2y}{3}, \qquad q = \frac{3x}{2}.$$

It is easy to verify that the integral curve to the vector field $\mathbf{v}$ satisfying $x(0) = x_0, y(0) = y_0$ is given by the graph of

$$x = x_0 \cosh(t) + \frac{2}{3} y_0 \sinh(t), \quad y = y_0 \cosh(t) + \frac{3}{2} x_0 \sinh(t),$$

and this solution exists and lies on the hyperbola $9x^2 - 4y^2 - 36 = 0$ whenever (x_0, y_0) does. Moreover, the solution satisfies the non-quasi-linear equation $q^2 = y^2 + 9$ for all t.

Note however, that when $2y_0^2 + 27 = 0$, we have $x_0^2 + 2 = 0$ and while any one of the four points $(\pm\sqrt{-2}, \pm 3\sqrt{-3/2})$ defined by these two constraints provide an equilibrium solution to the corresponding associated quasi-linear ideal J^ℓ, the equilibrium solutions are *not* solutions to the non-quasi-linear ideal J and are *not* equilibrium points of the vector field $\mathbf{v}$. This example thus shows that it is possible to have non-uniqueness of solutions for the associated quasi-linear ideal while there is uniqueness for the ideal itself and that solutions for the associated quasi-linear ideal need not be solutions for the ideal unless certain additional constraints are satisfied.

It should be noted also that using the *Axiom* package IDECOMP for primary decomposition of polynomial ideals, we found that the primary decomposition of J, when regarded as an ideal in $\mathbb{Q}[\mathbf{X}, \mathbf{P}]$, is $J_1 \cap J_2$, where

$$J_1 = \left(q + \frac{3}{2}x, p + \frac{2}{3}y, y^2 + \frac{27}{2}, x^2 + 2 \right), \quad J_2 = \left(q - \frac{3}{2}x, p - \frac{2}{3}y, y^2 - \frac{9}{4}x^2 + 9 \right)$$

and the primary decomposition of J^ℓ over $\mathbb{Q}$ is $J_3 \cap J_2$, where

$$J_3 = \left(q + \frac{1}{6}pxy, y^2 + \frac{27}{2}, x^2 + 2 \right).$$

The ideal J_2 is complete, while $\widetilde{J_1} = (1)$ and $\widetilde{J_3} = \left(p, q, y^2 + \frac{27}{2}, x^2 + 2 \right)$.

Example 6.2.9 Consider the ideal I corresponding to the system

$$\begin{aligned}
pq &= xy, \\
4q^2 &= 9x^2, \\
p^2 &= x^2 - 4
\end{aligned}$$

which is obtained by removing one quasi-linear equation from Example 6.2.8. Then I has essential $\mathbf{P}$-degree 2 and index 1. The completion $\widetilde{I}$ is given by the system

$$\begin{aligned}
4q^2 &= 9x^2, \\
pq &= xy, \\
p^2 &= x^2 - 4, \\
9xp - 4yq &= 0, \\
xyp - (x^2 - 4)q &= 0, \\
0 &= 9x^2 - 4y^2 - 36.
\end{aligned}$$

The determinant of the associated quasi-linear system is $x(4y^2 - 9x^2 + 36)$, which is always zero on $M = W = W^\ell$, the common initial domain of $\widetilde{I}$ and $\widetilde{I}^\ell$. We have rank $\widetilde{I}((x, y)) = 1$ and by Proposition 5.2.7 and Corollary 5.2.8, we have $M^0 = \emptyset$. It is not hard to find that at each point (x_0, y_0) on the hyperbola $9x^2 - 4y^2 = 36$, there are two solutions:

$$x = x_0 \cosh(t) + \frac{2}{3}y_0 \sinh(t), \quad y = y_0 \cosh(t) + \frac{3}{2}x_0 \sinh(t),$$

and

$$x = x_0 \cosh(t) - \frac{2}{3} y_0 \sinh(t), \quad y = y_0 \cosh(t) - \frac{3}{2} x_0 \sinh(t).$$

The primary decomposition of I over $\mathbb{Q}$ is given by $I_1 \cap I_2 \cap J_2$, where

$$I_1 = \left(p^2 + 4, pq - xy, xyp + 4q, q^2, xq, x^2 \right), \quad I_2 = \left(p + \frac{2}{3} y, q + \frac{3}{2} x, y^2 - \frac{9}{4} x^2 + 9 \right).$$

The component I_2 and J_2 are complete while $\widetilde{I}_1 = (1)$. The primary decomposition of $\widetilde{I}^\ell$ over $\mathbb{Q}$ is

$$\widetilde{I}^\ell = \left(xp - \frac{4}{9} yq, y^2 - \frac{9}{4} x^2 + 9 \right)$$

itself, which again shows that $(\widetilde{W}^\ell)^0 = \emptyset$.

7   Unconstrained or Underdetermined Systems

In Chapters 4 and 5, we study ideals J defined by differential algebraic systems through their completion $\widetilde{J}$ and their associated quasi-linear ideals J^ℓ; both processes can be computed by symbolic manipulations. However, when J has the property that $J \cap \mathbb{C}[\mathbf{X}] = (0)$, the ideal J itself is complete already. In this case, we say the ideal J is *unconstrained*. Whether J is constrained or not, after completion, $\widetilde{J}$ may contain at least n quasi-linear polynomials, and if furthermore the determinant of some $n \times n$ submatrix does not vanish identically on the constrained algebraic set, then $M^0 = (\widetilde{W}^\ell)^0 \neq \emptyset$, and the results of the previous chapter apply. Formally, we say J is *underdetermined* if $(\widetilde{W}^\ell)^0 = \emptyset$, and J is *overdetermined* if otherwise. Either property is clearly algorithmically decidable. Overdetermined systems should be the "generic" situation and our method should succeed for well-posed systems modelling scientific processes.

7.1   Quasi-Linearization Revisited

In this section we briefly discuss how we can adapt earlier results to unconstrained systems or underdetermined systems by the method of quasi-linearization introduced in Chapter 2. We generalize this notion by using an ideal-theoretic definition.

Definition 7.1.1 *Let* $\mathbf{Y} = (Y_1, \ldots, Y_{2n})$ *and* $\mathbf{Q} = (Q_1, \ldots, Q_{2n})$ *be two families of indeterminates over* $\mathbb{C}$. *Let* J *be an ideal in* $\mathbb{C}[\mathbf{X}, \mathbf{P}]$. *Let* $\lambda : \mathbb{C}[\mathbf{X}, \mathbf{P}] \longrightarrow \mathbb{C}[\mathbf{Y}]$ *be the isomorphism defined by* $\lambda(g) = g(Y_1, \ldots, Y_{2n})$ *for any* $g \in \mathbb{C}[\mathbf{X}, \mathbf{P}]$. *Let* $q\ell\,(J)$ *be the ideal in* $\mathbb{C}[\mathbf{Y}, \mathbf{Q}]$ *generated by the polynomials*

$$\begin{aligned} f_k &= Q_k - Y_{n+k} \quad (1 \leqslant k \leqslant n), & (7.1) \\ \lambda(g) &= g(Y_1, \ldots, Y_{2n}) \quad (g \in J). & (7.2) \end{aligned}$$

We call $q\ell\,(J)$ the quasi-linearization (ideal) *of J. We use the notation $\lambda(J)$ to denote the image under λ of an ideal J. Note that $\lambda(J)$ is an ideal in $\mathbb{C}[\mathbf{Y}]$.*

Clearly, if $J_1 \subseteq J_2$ are ideals in $\mathbb{C}[\mathbf{X}, \mathbf{P}]$, then $q\ell\,(J_1) \subseteq q\ell\,(J_2)$ and equality holds if and only if $J_1 = J_2$. The next proposition shows that the quasi-linearization ideal is effectively computable and that if J is unconstrained, then one cannot expect to obtain new algebraic constraints in $\mathbf{X}$ by quasi-linearization alone, even though $q\ell\,(J)$ is constrained if $J \neq (0)$ and its prolongations may provide new algebraic constraints in $\mathbf{X}$ (see Example 7.1.3).

Proposition 7.1.2 *Let J be an ideal in $\mathbb{C}[\mathbf{X}, \mathbf{P}]$ generated by $g_1, \ldots, g_m$ as in (2.5). Then $q\ell\,(J)$ is generated by $f_1, \ldots, f_{n+m}$ as in (2.6), that is, by the set E consisting of $f_1, \ldots, f_n$ of (7.1) and $\lambda(g_1), \ldots, \lambda(g_m)$. Moreover,*

$$q\ell\,(J) \cap \mathbb{C}[\mathbf{Y}] = q\ell\,(J) \cap \mathbb{C}[Y_1, \ldots, Y_{2n}] = \lambda(J),$$

$$q\ell\,(J) \cap \mathbb{C}[Y_1, \ldots, Y_n] = \lambda(J \cap \mathbb{C}[\mathbf{X}]),$$

and E is a $\mathbf{Q}$-strong essential $\mathbf{Q}$-degree basis for $q\ell\,(J)$.

Proof. The first claim follows since λ is an isomorphism of polynomial rings. Clearly, $\lambda(J) \subseteq q\ell\,(J) \cap \mathbb{C}[\mathbf{Y}]$. To show that E is a $\mathbf{Q}$-strong essential $\mathbf{Q}$-degree basis for $q\ell\,(J)$, let $h \in q\ell\,(J)$ be of $\mathbf{Q}$-degree at most 1 and write

$$h = \sum_{k=1}^{2n} h_k(\mathbf{Y})Q_k + h_0(\mathbf{Y})$$

for some $h_0, h_1, \ldots, h_{2n} \in \mathbb{C}[\mathbf{Y}]$. Then

$$h = \sum_{k=1}^{n} h_k(\mathbf{Y})(Q_k - Y_{n+k}) + \sum_{k=1}^{n} h_{n+k}(\mathbf{Y})Q_{n+k} + f(\mathbf{Y})$$

for some $f \in \mathbb{C}[\mathbf{Y}]$. Now $\sum_{k=1}^{n} h_{n+k}(\mathbf{Y})Q_{n+k} + f(\mathbf{Y}) \in q\ell\,(J)$ and hence it may be written as

$$\sum_{k=1}^{n} h_{n+k}(\mathbf{Y})Q_{n+k} + f(\mathbf{Y}) = \sum_{k=1}^{n} s_k(\mathbf{Y}, \mathbf{Q})(Q_k - Y_{n+k}) + \sum_{j=1}^{m} t_j(\mathbf{Y}, \mathbf{Q})\lambda(g_j) \quad (7.3)$$

where $s_k, t_j \in \mathbb{C}[\mathbf{Y}, \mathbf{Q}]$. Substituting Y_{n+k} for Q_k and 0 for Q_{n+k}, $1 \leqslant k \leqslant n$ in (7.3), we obtain

$$f(\mathbf{Y}) = \sum_{j=1}^{m} \lambda(\lambda^{-1}(\bar{t}_j(\mathbf{Y})))\lambda(g_j)$$

where $\bar{t}_j \in \mathbb{C}[\mathbf{Y}]$, showing that $f(\mathbf{Y}) \in \lambda(J)$ and hence $f(\mathbf{Y}) \in q\ell\,(J)$. It follows that $\sum_{k=1}^{n} h_{n+k}(\mathbf{Y})Q_{n+k} \in q\ell\,(J)$ and we can write

$$\sum_{k=1}^{n} h_{n+k}(\mathbf{Y})Q_{n+k} = \sum_{k=1}^{n} s'_k(\mathbf{Y}, \mathbf{Q})(Q_k - Y_{n+k}) + \sum_{j=1}^{m} t'_j(\mathbf{Y}, \mathbf{Q})\lambda(g_j) \quad (7.4)$$

where $s'_k, t'_j \in \mathbb{C}[\mathbf{Y}, \mathbf{Q}]$. Again, substituting Y_{n+k} for Q_k, $1 \leqslant k \leqslant n$ in (7.4), we obtain

$$\sum_{k=1}^{n} h_{n+k}(\mathbf{Y}) Q_{n+k} = \sum_{j=1}^{m} t'_j(\mathbf{Y}, Y_{n+1}, \ldots, Y_{2n}, Q_{n+1}, \ldots, Q_{2n}) \lambda(g_j)$$

from which we may clearly choose the coefficient of each $\lambda(g_j)$ to be $\mathbf{Q}$-homogeneous of $\mathbf{Q}$-degree 1. This shows that E is $\mathbf{Q}$-strong. Finally, for the special case when $h \in \mathbb{C}[\mathbf{Y}]$, we have $h = h_0 = f \in \lambda(J)$ and we have thus also shown that $q\ell(J) \cap \mathbb{C}[\mathbf{Y}] \subseteq \lambda(J)$. Since $\lambda(J \cap \mathbb{C}[\mathbf{X}]) = \lambda(J) \cap \mathbb{C}[Y_1, \ldots, Y_n]$ (in fact, $\lambda(J_1 \cap J_2) = \lambda(J_1) \cap \lambda(J_2)$) for any two ideals J_1, J_2 of $\mathbb{C}[\mathbf{X}, \mathbf{P}]$ because λ is an isomorphism), this completes the proof. $\qquad\square$

Example 7.1.3 Let $n = 1$ and $J = (P_1^2 + X_1^2, X_1 P_1)$. Then J is complete and unconstrained. The quasi-linearization is $q\ell(J) = (Y_2^2 + Y_1^2, Y_1 Y_2, Q_1 - Y_2)$ and we have $q\ell(J)^* = (Y_1, Y_2, Q_1, Q_2)$ since $\sqrt{\lambda(J)} = \sqrt{((Y_2 + Y_1)^2, (Y_2 - Y_1)^2)} = (Y_1, Y_2)$. Thus we not only obtained a new algebraic constraint $Y_1 = 0$ ($X_1 = 0$), but also $Y_2 = 0$ ($P_1 = 0$). The unique solution to $(J, \mathbf{x})$ is the zero function.

Example 7.1.4 Let $n = 2$ and $J = (X_1 X_2 - 1, P_1^2 - X_1^2, P_2^2 - X_2^2)$. The differential-algebraic system defined by J has two solutions satisfying initial conditions $\mathbf{z}(0) = (x_1, x_2)$ if $x_1 x_2 = 1$, namely, $\mathbf{z} = (x_1 e^{\pm t}, x_2 e^{\mp t})$. The system has index 1, its completion $\widetilde{J} = (P_1^2 - X_1^2, P_2 + X_2^2 P_1, X_1 X_2 - 1)$ and $\widetilde{W}$ is the hyperbola $X_1 X_2 = 1$. The associated quasi-linear ideal $\widetilde{J}^\ell = (P_2 + X_2^2 P_1, X_1 X_2 - 1)$. We have $\operatorname{rank} \widetilde{J}^\ell(\mathbf{x}) = 1$ for any $\mathbf{x} \in \widetilde{W}$. This quasi-linear system $\widetilde{J}^\ell$ is underdetermined, with solution $\mathbf{z} = (\varphi_1, \varphi_2)$ where φ_2 is an arbitrary function of t satisfying $\varphi_2(t) \neq 0$ for all t, and where $\varphi_1 = 1/\varphi_2$. The quasi-linearization of J is generated by

$$q\ell(J) = (Q_1 - Y_3, Q_2 - Y_4, Y_1 Y_2 - 1, Y_3^2 - Y_1^2, Y_4^2 - Y_2^2)$$

which also has index 1. The completion is

$$\widetilde{q\ell(J)} = (Q_4 - Y_2, Q_3 - Y_1, Q_2 + Y_3 Y_2^2, Q_1 - Y_3, Y_1 Y_2 - 1, Y_4 + Y_3 Y_2^2, Y_3^2 - Y_1^2).$$

We now have $\operatorname{rank} \widetilde{q\ell(J)}(\mathbf{y}) = 4$ for any $\mathbf{y} \in \widetilde{q\ell(W)}$ which is the algebraic set defined by the constraints

$$Y_1 Y_2 - 1 = 0, \quad Y_4 + Y_3 Y_2^2 = 0, \quad Y_3^2 - Y_1^2 = 0. \tag{7.5}$$

The system can be explicitly represented and has a unique solution for any $\mathbf{y} \in \widetilde{q\ell(W)}$. In effect, we have to give initial conditions $(\varphi_1(0), \varphi_2(0), \dot{\varphi}_1(0), \dot{\varphi}_2(0))$ to decide the branch in the original system. But even in this simple example, we learn that it is necessary to decide the branch using $\dot{\varphi}_1(0)$ only, and $\dot{\varphi}_2(0)$ is determined by (7.5). The point of the example, however, is that by quasi-linearizing the ideal, we can apply symbolic methods of this paper. For a more complicated system, the final algebraic constraints on the first order derivatives may not be easy to compute by hand.

Lemma 7.1.5 *Let J and $q\ell(J)$ be as above. Then $q\ell(J^*) \subseteq q\ell(J)^*$.*

Proof. The prolongation J^* is generated by J, $R = \sqrt{J \cap \mathbb{C}[\mathbf{X}]}$, and ∇R. Thus $q\ell(J^*)$ is generated by (7.1), together with $\lambda(g)$ for $g \in J \cup R \cup \nabla R$. Similarly, $q\ell(J)^*$ is generated by $q\ell(J)$, $R' = \sqrt{q\ell(J) \cap \mathbb{C}[\mathbf{Y}]}$, and $\nabla R'$. Now $q\ell(J)$ is generated by (7.1) and $\lambda(g)$ for $g \in J$. If $g \in J \cap \mathbb{C}[\mathbf{X}]$, then $\lambda(g) \in \lambda(J) \subseteq q\ell(J) \cap \mathbb{C}[\mathbf{Y}]$. Thus $\lambda(J \cap \mathbb{C}[\mathbf{X}]) \subseteq q\ell(J) \cap \mathbb{C}[\mathbf{Y}]$. Taking radical on both sides, we have $\lambda(R) \subseteq R'$. When $g \in \mathbb{C}[\mathbf{X}]$, it is easy to see that $\lambda(\nabla g) \equiv \nabla(\lambda(g))$ modulo (7.1). Thus $g \in R$ implies $\lambda(\nabla g) \in q\ell(J)^*$, proving the lemma. $\qquad\square$

Remark 7.1.6 Since the completion of an ideal can be computed using an ascending chain, the inequality in Lemma 7.1.5 shows that it is more economical to perform quasi-linearization and then prolong than the other way around. However, the inequality in Lemma 7.1.5 can be strict, even when J is quasi-linear.

Example 7.1.7 Let $n = 2$. Let $J = (X_1 + X_2, \ P_1 - X_2)$. The system is quasi-linear. We have $\tilde{J} = J^* = (X_1 + X_2, \ P_1 + X_1, \ P_2 - X_1)$. The quasi-linearizations are

$$
\begin{aligned}
q\ell(J) &= (Q_1 + Y_1, \ Q_2 - Y_4, \ Y_3 + Y_1, \ Y_2 + Y_1), \\
q\ell(J^*) &= (Q_1 + Y_1, \ Q_2 - Y_1, \ Y_4 - Y_1, \ Y_3 + Y_1, \ Y_2 + Y_1).
\end{aligned}
$$

The prolongation of $q\ell(J)$ is

$$
q\ell(J)^* = (Q_3 - Y_1, \ Q_2 - Y_1, \ Q_1 + Y_1, \ Y_4 - Y_1, \ Y_3 + Y_1, \ Y_2 + Y_1)
$$

and clearly $q\ell(J^*) \neq (q\ell(J))^*$. If we prolong once more, we obtain

$$
q\ell(J^*)^* = q\ell(J)^{**} = (Q_4 + Y_1, \ Q_3 - Y_1, \ Q_2 - Y_1, \ Q_1 + Y_1, \ Y_4 - Y_1, \ Y_3 + Y_1, \ Y_2 + Y_1)
$$

which is complete. For this example, $\operatorname{ind} J = 1$, $\operatorname{ind} q\ell(J) = 2$, and the use of quasi-linearization is unnecessary and does not yield any extra information.

The above example may suggest, as does our intuition, that the index of $q\ell(J)$ is one more than the index of J. However, this is not true, and in general, the index of J may increase by more than one after quasi-linearization, even when J is quasi-linear.

Example 7.1.8 Consider the quasi-linear system given by $J = (X_1^2 - X_2^3, \ P_1 - 3X_2)$. The index of J is 1, with $\tilde{J} = (X_1^2 - X_2^3, \ P_1 - 3X_2, \ X_2^2 P_2 - 2X_2 X_1)$. The quasi-linearization $q\ell(J)$ is given by $q\ell(J) = (Q_2 - Y_4, \ Q_1 - Y_3, \ Y_3 - 3Y_2, \ Y_1^2 - Y_2^3)$. The ideal $q\ell(J)$ has index 3, with successive prolongations given by

$$
\begin{aligned}
q\ell(J)^* = q\ell(\tilde{J}) = (&Q_3 - 3Y_4, \ Q_2 - Y_4, \ Q_1 - 3Y_2, \\
&Y_4 Y_2^2 - 2Y_2 Y_1, \ Y_4 Y_1^2 - 2Y_2^2 Y_1, \ Y_3 - 3Y_2, \ Y_2^3 - Y_1^2) \\
q\ell(J)^{**} = (&Y_2 Q_4 + Y_4^2 - 6Y_2, \ Y_1 Q_4 - 2Y_1, \ Q_3 - 3Y_4, \ Q_2 - Y_4, \ Q_1 - 3Y_2, \\
&Y_4^3 - 8Y_1, \ Y_4 Y_2 - 2Y_1, \ Y_4 Y_1 - 2Y_2^2, \ Y_3 - 3Y_2, \ Y_2^3 - Y_1^2) \\
q\ell(J)^{***} = \widetilde{q\ell(J)} = (&Y_4 Q_4 - 2Y_4, \ Y_2 Q_4 - 2Y_2, \ Y_1 Q_4 - 2Y_1, \ Q_3 - 3Y_4, \ Q_2 - Y_4, \\
&Q_1 - 3Y_2, \ Y_4^2 - 4Y_2, \ Y_4 Y_2 - 2Y_1, \ Y_4 Y_1 - 2Y_2^2, \ Y_3 - 3Y_2, \ Y_2^3 - Y_1^2).
\end{aligned}
$$

Note that we do gain information on the initial domains by quasi-linearization. The initial domain of $\widetilde{J}$ where uniqueness of solution holds is defined by

$$\{\,(x_1, x_2) \mid x_1^2 = x_2^3, x_2 \neq 0\,\}.$$

Given $x_2 \neq 0$, we have two choices for x_1. The initial domain of $\widetilde{q\ell\,(J)}$ is defined by

$$\{\,(y_1, y_2, y_3, y_4) \mid 4y_2 = y_4^2, 2y_1 = y_2 y_4, y_4 y_1 = 2y_2^2, y_3 = 3y_2, y_2^3 = y_1^2, y_4 \neq 0\,\}.$$

Thus given $y_4 \neq 0$, y_1, y_2, y_3 are uniquely determined. In addition, the explicit system defined by $\widetilde{q\ell\,(J)}$ is actually linear and easy to solve. The initial value problem at the singular point $\mathbf{x}_0 = (0, 0)$ (or $\mathbf{y}_0 = (0, 0, 0, 0)$) allows the equilibrium solution and the solution $\mathbf{z} = (t^3, t^2)$ and is excluded since $\operatorname{rank} \widetilde{J}(\mathbf{x}_0) = 1$ (or $\operatorname{rank} \widetilde{q\ell\,(J)}(\mathbf{y}_0) = 3$).

8   Differential Algebraic Methods

In this chapter, we shall address a concern raised by the referees of this paper, that has to do with differential algebraic methods. For lack of space, we shall not make this chapter self-contained. We will instead give an intuitive introduction to enable the readers unfamiliar with this subject to follow the main discussion. References to tutorial articles, classic text, and current research are given for the interested readers.

8.1   Differential Polynomials

We begin with a discussion from the perspective of differential algebra. In the introduction to Chapter 5, we mentioned that the explicit forms are perhaps differential algebraic consequences. The field of differential algebra was founded by Ritt and Kolchin and it studies differential polynomial systems of arbitrary order, with coefficients in any differential field $\mathcal{F}$ of arbitrary characteristics, and the systems may be either ordinary or partial[1]. A system Φ consists of a finite set of equations of the form $f_j(z_1, \ldots, z_n) = 0$, $(1 \leqslant j \leqslant m)$, where each f_j is a polynomial in $z_1, \ldots, z_n$ and their derivatives (implicitly notated). As placeholders for unknown functions and their derivatives, the infinitely denumerable family

$$Z = (z_1, \ldots, z_n, \dot{z}_1, \ldots, \dot{z}_n, \ldots, z_1^{(k)}, \ldots, z_n^{(k)}, \ldots)$$

are algebraic indeterminates with the obvious differentiation

$$\frac{dz_i^{(k)}}{dt} = z_i^{(k+1)}.$$

---

[1] For this paper, we stick to only the ordinary case with one derivation d/dt and $\mathcal{F}$ is of characteristic zero and contains t.

We say $z_1, \ldots, z_n$ are *differential indeterminates* and the f_j are *differential polynomials* in $z_1, \ldots, z_n$. Thus, f_j is a polynomial in Z with finite support.

The system Φ is different from a system of differential algebraic equations as defined and studied by Rabier and Rheinboldt [48], where each $f_j(t, \mathbf{x}, \mathbf{p})$ is a sufficiently smooth mapping from $\mathbb{R} \times \mathbb{R}^n \times \mathbb{R}^n$ to $\mathbb{R}$, and the Jacobian with respect to $\mathbf{p}$ has constant rank $\rho < n$, which implies the system includes (possibly as a consequence) algebraic constraints. In differential algebra, there are no restrictions, other than that f_j be polynomials in Z. In particular, some polynomials may involve derivatives and some may not. Each time a differential polynomial of order k is differentiated, we get a differential polynomial of order $k + 1$, which means the number of algebraic indeterminates involved keeps on increasing when we differentiate. This brings us to consider the *differential polynomial ring* $\mathcal{R} = \mathcal{F}\{z_1, \ldots, z_n\}$, which is just a polynomial ring $\mathcal{F}[Z]$ in the family Z equipped with its differentiation as above. Since we can perform arithmetic as well as differential operations on Φ, the polynomials corresponding to all possible equations derivable by these operations form a *differential ideal* $\mathfrak{a}$ (that is, an ideal closed under differentiation with respect to t). This differential ideal $\mathfrak{a}$ is the intersection of all differential ideals containing $f_1, \ldots, f_m$, and is denoted by $\mathfrak{a} = [f_1, \ldots, f_m]$. The polynomials representing the algebraic constraint equations belong to the ideal $\mathfrak{a}_0 = \mathfrak{a} \cap \mathcal{F}[z_1, \ldots, z_n]$, consisting of zero-th order differential polynomials.

Recall that an ideal I is *radical* if $f^k \in I$ for some $k \in \mathbb{N}$ implies $f \in I$. The intersection of all radical differential ideals in $\mathcal{R}$ containing $f_1, \ldots, f_m$ is the radical differential ideal generated by the f_k's, which is denoted by $\mathfrak{r} = \sqrt{[f_1, \ldots, f_m]}$ (the classical and more common notation is $\{f_1, \ldots, f_m\}$). From a geometric view point, if we are interested only in the set of zeros[2] of the given system Φ, we should consider the radical differential ideal $\mathfrak{r}$. Now the algebraic constraints are given by polynomials in the ideal

$$\mathfrak{r}_0 = \mathfrak{r} \cap \mathcal{F}[z_1, \ldots, z_n].$$

Typically, the generators of $\mathfrak{r}_0$ have lower degree than those of $\mathfrak{a}_0$ (unless $\mathfrak{a}_0$ is radical). Thus one of the problems concerning implicit differential algebraic systems is to compute a set of generators for $\mathfrak{a}_0$ or $\mathfrak{r}_0$. The other problems, of course, are to decide whether there is a unique solution when the initial conditions satisfy the algebraic constraints associated with generators of $\mathfrak{a}_0$ or $\mathfrak{r}_0$, and in the affirmative case, to compute an explicit vector field suitable for numerical methods.

8.2   Characteristic Set Methods

There are various strategies for solving the first problem. In the literature we have reviewed, prolongation and elimination take place in the differential ideal $\mathfrak{a}$. We believe ours is the first to involve the radical differential ideal $\mathfrak{r}$. Of course, recently, there has been a lot of research articles in representing $\mathfrak{r}$ as the intersection of a finite set of what are called characteristic or regular differential ideals (see Boulier [4], Boulier *et al.* ([5, 6, 7], Hubert [29], and Bouziane, Rody and Maârouf [8]). These are based

---

[2] A *zero of a differential polynomial* $f(z_1, \ldots, z_n)$ is an n-tuple $(\eta_1, \ldots, \eta_n)$ of elements η_i in some extension differential field of $\mathcal{F}$ that is a solution to the corresponding equation $f(z_1, \ldots, z_n) = 0$.

on the Ritt reduction algorithm using characteristic sets, the definition of which is fairly involved, and its computation even more so and continues to be a research topic. Besides the classic references of Ritt [54], Kolchin [35], and Kaplansky [36], there now exist tutorial articles (see Sit [59], Hubert [30, 31]).

Let us consider the case of a finite system Φ of ordinary differential polynomials. The problem is to compute all zeroth order polynomials derivable from Φ. Given any positive integer h, there are systems for which such a polynomial can only be obtained by repeated differentiation to order h before all the derivatives can be eliminated. For example, the system $z^{(2h)}$, $z^{(h)} + z$ requires h differentiations to show that the associated ideal $\mathfrak{a}_0$ contains z. While this example is easy to compute, the general problem is a difficult one. One approach, suggested by a referee, would be to express $\mathfrak{r}$ as the intersection of its (differential) prime components $\mathbf{p}_k$ and find a characteristic set A_k for each $\mathbf{p}_k$. This is no easy task even when Φ consists of a single differential polynomial, and in general, computing precisely the prime components of a radical differential ideal is an unsolved problem.[3] The characteristic system Φ_k of differential equations $A = 0$, $A \in A_k$ in such a decomposition, in general, is not equivalent to the original system, because each component $\mathbf{p}_k$ is given by the saturation ideal $[A_k] : H_k^\infty$, where H_k is the product of initials and separants of $A \in A_k$. The product H_k represents a differential inequation that, if satisfied by a zero of the corresponding characteristic system Φ_k, will ensure that it is also a zero of the component system $\mathbf{p}_k$. The set of zeros of $\mathbf{p}_k$ may, however, contain some that are also zeros of H_k. Notice that when Φ is first order, the characteristic systems Φ_k may involve higher order (see Example 8.2.1 below). Moreover, since a characteristic set A_k of a differential ideal $\mathbf{p}_k$ need not generate $\mathbf{p}_k$, it is unlikely that the set of differential polynomials in A_k of order zero generates the ideal $\mathbf{p}_{k,0}$ of zeroth order differential polynomials in $\mathbf{p}_k$. Even if it does, it is unclear how to use the characteristic sets A_k (which still represent *implicit* and non-linear differential equations in general) to help solve the system (or even the component system $\mathbf{p}_k$) in a numerically simple way. We do not have a mechanism to pass from one component to another in a numerical scheme. It would seem that there is nothing to be gained by using these characteristic sets rather than the original system. If the aim is just to find the radical ideal $\mathfrak{r}_0$, then we also need to find the intersection of ideals $\mathbf{p}_{k,0}$, which would require Gröbner basis methods.

We may, instead of using a prime decomposition, use a decomposition of $\mathfrak{r}$ into regular systems or characteristic systems (as in references cited earlier), which are only known to generate radical differential ideals. But most of the problems discussed above still remain. Some problems *can* be solved, but the algorithms are not efficient. Assuming that the characteristic sets A_k of the components (be they prime, regular, or characteristic) have been computed, we know by Rosenfeld's Lemma, that the corresponding $\mathbf{p}_{k,0}$ is the intersection of $(A) : H_k^\infty \cap \mathcal{F}[z_1, \ldots, z_n]$, where the *ideal* $(A_k) : H_k^\infty$ is taken in a suitable polynomial subring of $\mathcal{F}\{z_1, \ldots, z_n\}$ generated by the derivatives of z_j that appear in A_k. This ideal and the intersection can be computed by Gröbner basis methods. The other problems, such as existence and uniqueness of solutions have not

---

[3] It is possible to compute a list of prime differential ideals, in the form of their characteristic sets, whose intersection is the given radical differential ideal, and the list contains all the prime components. However, we do not know yet how to compute generators for these primes, or to determine which ones are the minimal primes.

been dealt with in this context. One should note that *all* zeros of Φ are included in the union of the zeros of the components $\mathbf{p}_k$ and even though initial data of a solution must satisfy the algebraic equations of some component, there is no criteria to determine which data satisfying the constraints will lead to a solution (or a unique one) with the data as initial conditions. It would require discovering algorithmically decidable conditions that are sufficient to guarantee existence and uniqueness as well as the computation of the vector field.

Finally, the characteristic decomposition of radical differential ideals is implemented for partial differential polynomials. It would seem to be an overkill to use these packages for systems of ordinary differential polynomials (even though they work just as well).

Our method of prolongation and completion avoids differentiating polynomials of order greater than 0. Since we start with a first order system, we never increase the order of our system in this process. The term[4] *prolongation* as used in this paper is different from the usual usage in that we only prolong zero[th] order polynomials and the completion process as described here attempts to obtain as much information on zero[th] order polynomials in the system as possible in a finite number of steps (including possible lowering of the degrees of generators by means of radical ideal computations). If the system is not underdetermined in the sense of Chapter 7, then we are done (Theorem 6.2.5). Most well-posed scientific problems should fall into this case. If the system is underdetermined, we apply quasi-linearization, which is simply a disguise to differentiate the first order polynomials (getting closer to the usual meaning of *prolongation*). In carrying out the completion process on the quasi-linearization, it is actually more efficient to do so in the ring $\mathcal{F}[z_1, \ldots, z_n, \dot{z}_1, \ldots, \dot{z}_n, \ddot{z}_1, \ldots, \ddot{z}_n]$ than in $\mathcal{F}[\mathbf{Y}, \mathbf{Q}]$; for theoretical development, however, it is more convenient to use our setup.

We see now that we can repeatedly carry out quasi-linearization and completion. Our method tries to find the algebraic constraints and explicit form in a greedy manner without increasing the order (at the expense of more indeterminates), and if necessary, when the system is underdetermined or unconstrained in the sense of Chapter 7, increase the order one at a time, to find additional algebraic constraints, and to resolve possible ambiguities of the initial conditions by finding algebraic relations among higher order initial conditions. Of course, our method does not compute $\mathfrak{r}_0 = \mathfrak{r} \cap \mathcal{F}[z_1, \ldots, z_n]$ (rather, only a piece of it at a time, see Example 7.1.3), and for those systems where $\mathfrak{r}_0 = (0)$ (*a fortiori*, unconstrained), all computational methods will fail to find any zero[th] order polynomials. Applying our method to the quasi-linearization, we proceed to find differential polynomials in $\mathfrak{r}_1 = \mathfrak{r} \cap \mathcal{F}[z_1, \ldots, z_n, \dot{z}_1, \ldots, \dot{z}_n]$ and stop once existence and uniqueness is guaranteed. As proved in Proposition 7.2, the quasi-linearization automatically yields a $\mathbf{Q}$-strong essential basis. For some systems, after completion, we may obtain new zero[th] order and/or first order polynomials (that is, polynomials in $\mathbf{Y}$, which is an alias for $(\mathbf{X}, \mathbf{P})$, see Examples 7.1.3 and 7.1.8) that we would not have obtained without quasi-linearization. Any algebraic constraints involving $\mathbf{Y}$ are *not* interpreted as differential equations or inequations to be satisfied by a *solution* $(\varphi_1, \ldots, \varphi_n)$ and its derivative $(\dot{\varphi}_1, \ldots, \dot{\varphi}_n)$, but rather as algebraic constraints that must be satisfied by the *initial data* $\varphi_1(0), \ldots, \varphi_n(0), \dot{\varphi}_1(0), \ldots, \dot{\varphi}_n(0)$. We then

---

[4] Perhaps *semi-prolongation* is a more accurate term.

compute a vector field which can be integrated to yield solutions that we prove will satisfy the differential constraints. These detailed results are missing in the characteristic set approach.

We do not claim that our method is efficient, but the alternative using characteristic sets does not seem to provide any obvious improvements for the problems we are interested in, namely, initial value problems.

Example 8.2.1 The above points may be illustrated by the well-known system $A = 0$, where $A = (\dot{z})^2 - 4z$. The radical differential ideal $\sqrt{[A]}$ decomposes into two prime components $\sqrt{[A, \ddot{z} - 2]}$ and $\sqrt{[A, \dot{z}]}$, which is an easy consequence of $\dot{A} = 2\dot{z}(\ddot{z} - 2)$. The first component is $\sqrt{[A, \ddot{z} - 2]} = \sqrt{[A]} : \dot{z}$, and its characteristic system Φ_1 is $A = 0$. The other component is $\sqrt{[A, \dot{z}]} = [z]$ and its characteristic system Φ_2 is $z = 0$. There is no zero[th] order polynomial in Φ_1. The zero[th] order polynomial $z \in \Phi_2$ has only the trivial solution, which is a singular solution (in the sense of Ritt). The component systems are $A = 0, \ddot{z} = 2, \dot{z} \neq 0$ and $z = 0$. These systems provide differential constraints on their solutions. In our notation, the system is given by $J = (P_1^2 - 4X_1)$. It is clear (and easy to compute) that $J \cap \mathbb{C}[X_1] = (0)$ and hence J is complete, unconstrained, and has algebraic index 0. The quasi-linearization is

$$q\ell(J) = (Y_2^2 - 4Y_1, Q_1 - Y_2)$$

and its completion is

$$\widetilde{q\ell(J)} = (Y_2^2 - 4Y_1, Q_1 - Y_2, Y_2 Q_2 - 4Y_2),$$

which shows (through change of notations) that for initial conditions $\varphi(0) = a, \dot{\varphi}(0) = b$ satisfying $b^2 - 4a = 0, b \neq 0$, there is a unique solution $\varphi(t)$ that can be numerically integrated from $\ddot{\varphi} = 2$. The case $b = 0$ is ruled out because it does not satisfy the rank condition $\mathrm{rank}\,\widetilde{q\ell(J)}(\mathbf{y}_0) = 2$, where $\mathbf{y}_0 = (a, b)$.

9   Implementation and Examples

In this chapter, we give a few well-known examples as well as one pedagogical example to illustrate the applicability of our algorithms. Our implementation is done in *Axiom*, a scientific computation system first developed in 1971 as a research project by a team at I. B. M. led by Richard Jenks [34]. It was commercially available from 1992 to October, 2001 and distributed by the Numerical Algorithm Group. After it was taken out of the market, it has been revitalized as an open source project headed by Tim Daly [18]. *Axiom* is now freely available (with full source code) on many operating system platforms and hardware. The packages used in this Chapter are available from the second author. They are not yet part of *Axiom*.

For simplicity, we do all the computations over the field $\mathbb{Q}$ of rational numbers. The examples in this Chapter were run on a Windows XP system using the open source version of *Axiom* (Windows version dated November, 2004) with an Intel Pentium 4 (duo

core) processor at 2.8GHz (so only one core is used by *Axiom*), on an ABIT IS7 moth-
erboard with 2GB RAM installed (which is not a requirement). All programs were
compiled with the *Axiom* compiler (that is, not the $A^{\#}$ or Aldor compiler). Timings are
given by the evaluation (EV) portion as reported by *Axiom*. The (EV) time, given in
seconds, excludes input, output, and gabbage collection times, which can be relatively
substantial, but which only reflect performance of the system and not the algorithm.
Typically, most steps require insignificant time, and thus timings are only given for the
more time-consuming routines that force a computation of the completion of a given
ideal for the first time (results are then cached). For example, if the index is computed
first, a subsequent request for the completion will require no evaluation time. Similarly, if a completion is computed first, a subsequent request for the index is answered
instantaneously. In our design, every prolongation is also cached.

9.1   Domain Constructors in *Axiom*

The domain constructor `DistributedMultivariatePolynomial`, for readers
unfamiliar with *Axiom*, is abbreviated by DMP. It takes two parameters, a commutative ring R and a list of symbols V, and returns a domain `DMP(V,R)` that represents
the polynomial ring with coefficient ring R and indeterminates V. The term ordering
used in DMP is the pure lexicographic order. Similarly, the domain `HDMP(V, R)` is
a polynomial ring where the term ordering is degree-lexicographic. We added two
packages in our implementation. The package `EORDER` provides a routine to implement the elimination ordering required in computing essential **P**-degrees. The
domain `GeneralFirstOrderOrdinaryDifferentialEquation`, abbrevi-
ated by GFOODE, constructs the representation $S = S(J)$ for an ideal J through the
`makeSystem` routine. This routine and some others are described below:

makeSytem(F): creates the representation S for the ideal J generated by a finite set
of input polynomials $F = \{ f_1, \ldots, f_m \} \subset \mathbb{Q}[\mathbf{X}, \mathbf{P}]$ as in (2.1); the data structure
supports several view points (as polynomials in an essential **P**-basis, or as matrix
equations if J is quasi-linear, for example), as well as caching of prolongations
of J for computational efficiency.

matrixView(S): displays system S in the form (5.3) if the ideal represented by S is
quasi-linear.

algebraicSystem(S): displays a generating set of polynomials in $J \cap \mathbb{Q}[\mathbf{X}]$, when the
ideal J has representation S.

linearize(S): computes a system S^{ℓ} representing the associated linear ideal J^{ℓ} of the
ideal J, where J is represented by the system S.

prolong(S): computes the system S^* representing the prolongation J^* of the ideal J
where J is represented by S; a Gröbner basis of the radical $R(J)$ is cached in the
data structure of S.

complete?(S): returns `true` if the ideal J represented by S is complete, and `false` otherwise. (The entire sequence of prolongations is cached in the representation S).

completion(S): computes the representation $\widetilde{S}$ for the completion $\widetilde{J}$ of J, when J is represented by S. (The entire sequence of prolongations is cached in the representation S).

index(S): computes the index of J, when J is represented by S. (The entire sequence of prolongations is cached in the representation S).

parSolve(S): computes the algebraic conditions for the matrix $A(\mathbf{X})$ associated with a quasi-linear ideal J represented by S to have rank n and computes the rational functions $r_i(\mathbf{X})$ which represent coordinates of $\mathbf{P}$ such that when $\mathbf{x} \in W$ and satisfies the algebraic conditions, $\mathbf{p} = (r_1(\mathbf{x}), \ldots, r_n(\mathbf{x}))$ satisfies the linear system (5.3) with coefficient matrix $A(\mathbf{x})$. This routine uses the `PLEQN`, or `ParametricLinearEquations`, package.

Other routines include **essentialDegree**, **essentialBasis**, **differentialSystem**, **quasiLinear?**, and **completionSequence**.

9.2 Sample Session

The following set up in *Axiom* is typical ($n = 4$, as in Clairaut's equation below). The purpose is to load the packages, define abbreviations for them, dynamically construct the polynomial domains, and initiate related packages. In what follows, transcripts from *Axiom* have been slightly edited.[1] The ==> notation indicates the left hand side is an abbreviation for the right hand side.

```
)clear all
DMP ==> DistributedMultivariatePolynomial
HDMP ==> HomogeneousDistributedMultivariatePolynomial
NNI ==> NonNegativeInteger
PI ==> PositiveInteger
ODP ==> OrderedDirectProduct
DIRPROD ==> DirectProduct
HDP ==> HomogeneousDirectProduct
OVAR ==> OrderedVariableList
PLEQN ==> ParametricLinearEquations
IDECOMP ==> IdealDecompositionPackage
IDEAL ==> PolynomialIdeals
GBP ==> GroebnerPackage
OUT ==> OutputForm
M ==> Matrix
V ==> Vector
```

---

[1] Some obvious outputs were suppressed (commands ending with a semicolon), and all mathematical expressions were in traditional notations. Algebraic indeterminates are set in lower case rather than upper case.

```
L ==> List
S ==> Symbol
)library EORDER GFOODE -- note order of variables
xvar:List Symbol:=[ x3,x4,x1,x2 ];
pvar:List Symbol:=[ p3,p4,p1,p2 ];
XVAR := OVAR xvar;
PVAR := OVAR pvar;
DPX := DIRPROD(#xvar, NNI);
DPP := HDP(#pvar, NNI);
R := Integer;
F := Fraction R;
GR := DMP(xvar, R);
FGR := Fraction GR;
GF := DMP(xvar, F);
PGR := HDMP(pvar, GR);
gfoode:=GFOODE(xvar,pvar);
pleqn := PLEQN(R, XVAR, DPX, GR);
pideal := IDEAL(F,DPX,XVAR,GF);
idecomp := IDECOMP(xvar, #xvar);
```

Example 9.2.1 Clairaut's Equation. We illustrate with an example of a Clairaut's equation (here y is a function of x and y' is the derivative of y with respect to x):

$$y = xy' - \frac{1}{4}y'^2.$$

This is a non-linear first order equation, algebraically irreducible in the differential polynomial ring $\mathbb{Q}(x)\{y\}$. The radical differential ideal is the intersection of two prime differential ideals $\sqrt{[g(y), x - \frac{1}{2}y']}$ and $\sqrt{[g(y), y'']}$, where

$$g(y) = y - xy' + \frac{1}{4}y'^2$$

(see Buium and Cassidy [10, p. 575]; see also Ince [32] for a more general form of Clairaut's equation). A direct application of the our algorithm (after making the system autonomous) does not provide any new information about the equation. However, by using the quasi-linearization as described earlier, the equation is practically solved, and we obtained the two components. In the *Axiom* setting, we use

$$x_1 = x, \; x_2 = y, \; x_3 = x', \; x_4 = y', \text{ and } p_i = x_i' \text{ for } i = 1, \ldots, 4$$

and the quasi-linearization of the system lives in $\mathbb{Q}[x_1, x_2, x_3, x_4, p_1, p_2, p_3, p_4]$.

```
-- Clairaut Equation
f1:PGR:=  x3 − 1;              -- make system autonomous
f2:PGR:=  4x2 − 4x1x4 + x4²;   -- Clairaut's Equation
f3:PGR:=  p1 − x3;             -- quasi-linearization
f4:PGR:=  p2 − x4;             -- quasi-linearization
F :List PGR:=[ f1,f2,f3,f4 ];
```

S :=makeSystem(F)

$$[p_1 - 1, p_2 - x_4, x_3 - 1, x_4^2 - 4x_4x_1 + 4x_2]$$

$\widetilde{S}$:= completion S (EV) = 0.03 sec

$$[p_3, (x_4 - 2x_1)p_4, (x_1^2 - x_2)p_4, p_1 - 1, p_2 - x_4, x_3 - 1, x_4^2 - 4x_4x_1 + 4x_2]$$

matrixView $\widetilde{S}$

$$\begin{pmatrix} 1 & 0 & 0 & 0 \\ 0 & x_4 - 2x_1 & 0 & 0 \\ 0 & x_1^2 - x_2 & 0 & 0 \\ 0 & 0 & 1 & 0 \\ 0 & 0 & 0 & 1 \\ 0 & 0 & 0 & 0 \\ 0 & 0 & 0 & 0 \end{pmatrix} \begin{pmatrix} p_3 \\ p_4 \\ p_1 \\ p_2 \end{pmatrix} = \begin{pmatrix} 0 \\ 0 \\ 0 \\ 1 \\ x_4 \\ -x_3 + 1 \\ -x_4^2 + 4x_4x_1 - 4x_2 \end{pmatrix}$$

parSolve $\widetilde{S}$ (EV) = 0.03 sec

$$
\begin{aligned}
&[[\quad \texttt{eqzro} = \qquad\qquad [-x_3 + 1,\ x_4 - 2x_1,\ x_1^2 - x_2], \\
&\qquad \texttt{neqzro} = [], \qquad\qquad \texttt{wcond} = [], \\
&\qquad \texttt{bsoln} = \qquad\qquad [\texttt{partsol} = [0, 0, 1, 2x_1]. \\
&\qquad\qquad\qquad\qquad\qquad \texttt{basis} = [[0, 1, 0, 0]]]], \\
&[\quad \texttt{eqzro} = \qquad\qquad [x_3 - 1,\ x_4^2 - 4x_4x_1 + 4x_2], \\
&\qquad \texttt{neqzro} = [x_1^2 - x_2], \quad \texttt{wcond} = [], \\
&\qquad \texttt{bsoln} = \qquad\qquad [\texttt{partsol} = [0, 0, 1, x_4], \\
&\qquad\qquad\qquad\qquad\qquad \texttt{basis} = [[]]]], \\
&[\quad \texttt{eqzro} = \qquad\qquad [x_3 - 1,\ x_4^2 - 4x_4x_1 + 4x_2], \\
&\qquad \texttt{neqzro} = [x_4 - 2x_1], \quad \texttt{wcond} = [] \\
&\qquad \texttt{bsoln} = \qquad\qquad [\texttt{partsol} = [0, 0, 1, x_4], \\
&\qquad\qquad\qquad\qquad\qquad \texttt{basis} = [[]]]] \\
&]
\end{aligned}
$$

We recall that the ParametricLinearEquations solver on which parSolve is based takes as input a linear system $A(\mathbf{X})\mathbf{P}^T = c(\mathbf{X}, \mathbf{W})$ in the unknown $\mathbf{P}$, with coefficients polynomial in the parameters $\mathbf{X}$ and possibly additional parameters $\mathbf{W}$ that only occur on the right hand sides. It returns a list of regimes representing the set of all parametric values for which the system is solvable. Each regime is a quasi-algebraic set U defined by the equations $f = 0$ for $f \in$ eqzro, the inequations $f \neq 0$ for $f \in$ neqzro, and the equations $g = 0$ for $g \in$ wcond. For any point $(\mathbf{x}, \mathbf{w}) \in U$, a particular solution (partsol) for the linear system $A(\mathbf{x})\mathbf{P}^T = c(\mathbf{x}, \mathbf{w})$ is given (*the coordinates of* partsol *are given in the order specified by* pvar). In case the rank

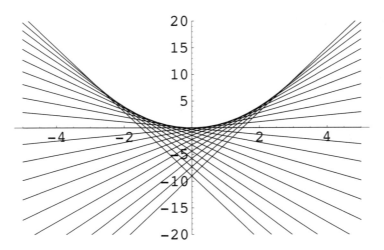

Figure 9.1 Solutions of Clairaut's Equation

of the system is not full, a non-empty basis (`basis`) for the solution space of the homogeneous system associated with the linear system $A(\mathbf{x})\mathbf{P}^T = c(\mathbf{x}, \mathbf{w})$ is also given. For the version of `ParametricLinearEquations` we are using, the regimes need not be irredundant. By using a new domain `QuasiAlgebraicSets` (see Sit [58]) and related packages that implement general algorithms to compare two regimes returned by the parametric linear equation solver, an irredundant decomposition can be computed.

For our application, we are mainly interested in the regimes corresponding to the non-degenerate cases, which have empty `basis`. Since we do not have additional parameters appearing only on the right hand sides, `wcond` is always empty. These regimes U_k returned by `parSolve` $\widetilde{S}$ with empty `basis` then form a cover for $\widetilde{W}^0$ of the complete quasi-linear ideal $\widetilde{J}$ represented by $\widetilde{S}$. The `partsol` in the k-th regime provides an explicit expression for $\mathbf{p}$ when $\mathbf{x}$ belongs to the corresponding U_k. In this example, there are two regimes (the last two are equivalent, as can be checked easily by hand or by the package `QuasiAlgebraicSet`). Using the original notations, the algebraic conditions of the first regime tells us that $y' = 2x$ and $y = x^2$. The explicit form of the derivatives $\mathbf{p}$ tells us that y'' is a constant, which must therefore be 2. This regime consists of a single singular solution. In the second regime, when $y \neq x^2$, we know that $y'' = 0$ or $y' = c$, where c is an arbitrary constant. This regime gives the general component of the differential system. Any initial condition $(x_1, x_2, x_3, x_4) = (0, y(0), 1, c)$ satisfying the algebraic conditions $c^2 + 4y(0) = 0, c \neq 0$ will have a unique solution $y = cx - c^2/4$. See Figure 9.1 above.

Example 9.2.2 We consider next the example of a planar pendulum, which was studied in Rabier and Reinholdt. A mass m is attached to the end of a rigid and massless wire of length ℓ hanging from the origin. Let λ be the tension of the wire at time t, g be the gravity constant, and (x_1, x_2) the coordinates of the mass at time t. The system can be described by the second order system:

$$
\begin{aligned}
x_1^2 + x_2^2 &= \ell^2, \\
\ddot{x}_1 &= -(\frac{\lambda}{m})x_1, \\
\ddot{x}_2 &= -(\frac{\lambda}{m})x_2 - g.
\end{aligned}
$$

The goal of Rabier and Reinholdt is to derive algebraic constraints for the system and also the rate of change in the tension λ. The system may be transformed into a quasi-linear system with auxiliary variables

$$ x_3 = \dot{x}_1, \quad x_4 = \dot{x}_2, \quad \text{and} \quad x_5 = \lambda/m. $$

Substituting p_i for $\dot{x}_i$ for $i = 1, \dots, 5$, we get the following system:

$$
\begin{aligned}
x_1^2 + x_2^2 - \ell^2 &= 0, \\
p_1 - x_3 &= 0, \\
p_2 - x_4 &= 0, \\
p_3 + x_1 x_5 &= 0, \\
p_4 + x_2 x_5 + g &= 0.
\end{aligned}
$$

The parameters ℓ and g are treated as variables with derivatives p_ℓ and p_g set equal to zero. The variable orderings are chosen to eliminate λ and the higher order derivatives of x_1, x_2 first. The system has index 3 and thus generally considered to be difficult.

```
xvar:List Symbol:=[ x5,x4,x3,x2,x1,ℓ,g ];
pvar:List Symbol:=[ p5,p4,p3,p2,p1,pℓ,pg ];
f1 :PGR:=x1² + x2² - ℓ²;
f2 :PGR:=p1 - x3;
f3 :PGR:=p2 - x4;
f4 :PGR:=p3 + x1x5;
f5 :PGR:=p4 + x2x5 + g;
f6 :PGR:=pℓ;
f7 :PGR:=pg;
F :List PGR:=[ f1, f2, f3, f4, f5, f6, f7 ];
S :=makeSystem( F );                            (EV) = 0.05 sec
matrixView S
```

$$
\begin{bmatrix}
0 & 1 & 0 & 0 & 0 & 0 & 0 \\
0 & 0 & 1 & 0 & 0 & 0 & 0 \\
0 & 0 & 0 & 1 & 0 & 0 & 0 \\
0 & 0 & 0 & 0 & 1 & 0 & 0 \\
0 & 0 & 0 & 0 & 0 & 1 & 0 \\
0 & 0 & 0 & 0 & 0 & 0 & 1 \\
0 & 0 & 0 & 0 & 0 & 0 & 0
\end{bmatrix}
\begin{bmatrix}
p_5 \\ p_4 \\ p_3 \\ p_2 \\ p_1 \\ \ell_1 \\ g_1
\end{bmatrix}
=
\begin{bmatrix}
-x_5 x_2 - g \\
-x_5 x_1 \\
x_4 \\
x_3 \\
0 \\
0 \\
-x_2^2 - x_1^2 + \ell^2
\end{bmatrix}
$$

```
index S                                        (EV) = 0.68 sec
      3
```

```
S̃ := completion S;
algebraicSystem S̃
```

$$[x_5\ell^2 - x_4^2 - x_3^2 + x_2 g, x_4 x_2 + x_3\, x_1, x_4 x_1^2 - x_4\ell^2 - x_3 x_2 x_1, x_2^2 + x_1^2 - \ell^2]$$

```
parSolve S̃                                     (EV) = 9.13 sec
```

$$
\begin{aligned}
[[\; \texttt{eqzro} = \quad & [\, x_4^2 + x_3^2 - x_2\, g,\ x_4\, x_3\, x_1 - x_3^2\, x_2, \\
& -\, x_4\, x_2 - x_3\, x_1, -\, x_4\, x_1^2 + x_3\, x_2\, x_1, -3\, x_4\, g, x_3^2\, g - x_2\, g^2, \\
& x_3\, x_2\, g,\ x_2^2 + x_1^2,\ x_1\, g,\ \ell\,], \\
\texttt{neqzro} = [],\quad & \texttt{wcond} = [], \\
\texttt{bsoln} = [\quad & \texttt{partsol} = [0, -x_2\, x_5 - g, -x_1\, x_5, x_4, x_3, 0, 0], \\
& \texttt{basis} = [[1, 0, 0, 0, 0, 0, 0]]]]], \\
[\; \texttt{eqzro} = \quad & [\, x_5\,\ell^2 - x_4^2 - x_3^2 + x_2\, g,\ x_4\, x_2 + x_3\, x_1, \\
& x_4\, x_1^2 - x_4\,\ell^2 - x_3\, x_2\, x_1,\ x_2^2 + x_1^2 - \ell^2\,], \\
\texttt{neqzro} = [\ell],\quad & \texttt{wcond} = [], \\
\texttt{bsoln} = [\quad & \texttt{partsol} = [-\tfrac{3g\, x_4}{\ell^2}, -x_2\, x_5 - g, -x_1\, x_5, x_4, x_3, 0, 0], \\
& \texttt{basis} = [[]]], \\
\cdots\ \;]
\end{aligned}
$$

The first regime is a degenerate case ($\ell = 0$) which may be ignored for this physical problem. In this example, three other unlisted regimes are covered by the second one, that is, the particular solutions are all equivalent rational maps, and the domains of definition for these maps are subsets of the domain of definition of the second one (these verifications can be done using routines from **QuasiAlgebraicSet** and related packages, see Sit [58]). These maps looked different because of non-unique representations of rational functions defined on quasi-algebraic sets. The algebraic constraints **algebraicSystem**($\tilde S$) for the completion are identical to the results of Rabier and Rheinboldt, except for an additional constraint equation of degree 3, which is part of a Gröbner basis. The explicit system is given by the entries in **partsol** above.

Example 9.2.3 This example is a pedagogical one constructed to demonstrate the power of the algorithms. We begin with a space curve C represented by the vector function (e^t, e^{2t}, e^{3t}) for $t > 0$. We then construct, rather arbitrarily, differential equations satisfied by the coordinate functions and their derivatives. We input the following quasi-linear system:

$$
\begin{aligned}
-\; X_2 P_2 &+ X_1 P_3 &=& \quad X_1^4 \\
-X_2 P_1 \qquad\quad &+ \quad 2P_3 &=& \quad 5X_3 \\
X_3 P_1 + X_1^2 P_2 \quad &\qquad &=& \quad 3X_2^2 \\
-\; X_1 P_2 &+ \quad P_3 &=& \quad X_3
\end{aligned}
\tag{9.1}
$$

We use a variable ordering (defined by `xvar`, `pvar`) with $X_3 > X_2 > X_1$ and $P_3 > P_2 > P_1$. An essential **P**-degree basis of the ideal ((EV) = 0.07 sec) shows there is one algebraic constraint of total degree 7 in **X**:

$$3X_3^2 X_2 - 5X_3^2 X_1^2 - X_3 X_2 X_1^3 + 2X_3 X_1^5 + 3X_2^4 - 3X_2^3 X_1^2 + X_2 X_1^6 = 0. \quad (9.2)$$

The ideal has index 3 ((EV) = 5.68 sec) and an essential **P**-degree basis of the completion ideal gives the following equivalent system:

$$
\begin{aligned}
P_3 - 3X_1^3 &= 0 \\
P_2 - 2X_1 P_1 &= 0 \\
X_1^2 P_1 - X_1^3 &= 0 \quad (9.3) \\
X_3 - X_1^3 &= 0 \\
X_2 - X_1^2 &= 0
\end{aligned}
$$

and using `parSolve`, we obtain two explicit systems ((EV) = 0.03 sec). The first is the degenerate case when $X_1 = X_2 = X_3 = 0$ in which case, $P_3 = P_2 = 0$ while P_1 is arbitrary (that is, indeterminate). The second gives the vector field

$$P_3 = 3X_1^3, \quad P_2 = 2X_1^2, \quad P_1 = X_1, \quad (9.4)$$

valid under the algebraic constraints $X_2 = X_1^2, X_3 = X_1^3$ and $X_1 \neq 0$. It should be noted that this is a non-trivial example as some statistics for intermediate computations show. The expressions involved in computing the first and second prolongations are very big (see Table 9.1).

Example 9.2.3 $X_3 > X_2 > X_1$	first prolongation	second prolongation
max deg in algebraic constraints	24	14
max coefficient in constraints	21 digits	30 digits
max **P**-degree in system	2	2

Table 9.1

Note that if we had viewed the original system as a purely algebraic system in 6 variables, we would have gotten both P_2 and P_3 as arbitrary in the degenerate case. Indeed, we also used `psolve` from the package `PLEQN` to solve for P_1, P_2, P_3 directly as a parametric linear system with parameters X_1, X_2, X_3 ((EV) = 0.12 sec), and we got the degenerate case with P_2, P_3 arbitrary, plus four other regimes in which Eqn. (9.2) appeared as a constraint with other inequation constraints. The vector fields for these regimes are more complicated, as rational functions involving terms up to degree 6. Thus our approach catches this "error" in the degenerate case, and simplifies the vector field presentation.

We also experimented with changing the ordering of the variables from $X_3 > X_2 > X_1$ to $X_1 > X_2 > X_3$. The essential **P**-degree basis ((EV) = 0.03 sec) again contains only one algebraic constraint: Eqn. (9.2). The index is 3 ((EV) = 71.08 sec) and an essential **P**-degree basis of the completion ideal gives the following system:

$$
\begin{aligned}
-\quad X_1^2 \ +\quad X_2 &= 0, & -\quad X_1X_2 \ +\quad X_3 &= 0, \\
-\quad X_1X_3 \ +\quad X_2^2 &= 0, & -\quad X_2^3 \ +\quad X_3^2 &= 0, \\
2X_1P_1 \ -\quad P_2 &= 0, & X_2P_1 \ -\quad X_3 &= 0, \\
X_3P_1 \ -\quad X_2^2 &= 0, & X_2P_2 \ -\quad 2X_2^2 &= 0, \\
X_1P_2 \ -\quad 2X_3 &= 0, & P_3 \ -\quad 3X_3 &= 0. \\
X_3P_2 \ -\quad 2X_2X_3 &= 0, &
\end{aligned}
\tag{9.5}
$$

While Eqn. (9.5) is more complicated than Eqn. (9.3), applying `parSolve` ((EV) = 0.02 sec) gives the degenerate case before and the vector field

$$
P_1 = \frac{X_3}{X_2}, \ P_2 = 2X_2, \ P_3 = 3X_3
\tag{9.6}
$$

under the algebraic constraints

$$
X_2 = X_1^2, \ X_3 = X_1X_2, \ X_2^2 = X_1X_3, \ X_2^3 = X_3^2, \ X_2 \neq 0.
\tag{9.7}
$$

Note how the explicit system Eqn. (9.6) forces the initial condition to satisfy the constraint $X_2 \neq 0$ and consequently $X_1 \neq 0, X_3 \neq 0$ by Eqns. (9.7). Had the explicit system been $P_1 = X_1$ (as in Eqn. (9.4)) rather than $P_1 = X_3/X_2$, such constraints will not be reflected. Moreover, two of the equations in (9.6) are decoupled and the system can be solved completely by integration. For this variable ordering, the computation (as reflected also by the computation times) is more involved (see Table 9.2).

Example 9.2.3 $X_1 > X_2 > X_3$	first prolongation	second prolongation
max deg in algebraic constraints	36	10
max coefficient in constraints	95 digits	30 digits
max **P**-degree in system	4	4

Table 9.2

To gauge how our algorithms work with non-quasi-linear systems, we replaced the first equation of Eqns. (9.1) by

$$
-2X_2P_1^2 + X_1P_3 = X_1^4.
$$

We used a variable order $X_3 > X_2 > X_1$. The system then has essential **P**-degree 2 ((EV) = 0.02 sec), An essential **P**-degree basis includes one algebraic constraint of total degree 8. Computing the completion took 10.42 seconds. The completion is given by the equations

$$
P_3 = 3X_1^3, \ P_2 = 2X_1P_1, \ X_1^2P_1 = X_1^3, \ X_3 = X_1^3, \ X_2 = X_1^2
$$

Example 9.2.3, $X_3 > X_2 > X_1$ Non-quasi-linear	first prolongation	second prolongation
max deg in algebraic constraints	24	13
max coefficient in constraints	48 digits	45 digits
max **P**-degree in system	2	2

Table 9.3

The system has index 3 and is eventually quasi-linear. From this, the parametric linear equations solver gives the explicit system ((EV) = 0.03 sec) as

$$P_1 = X_1, \; P_2 = 2X_1^2, \; P_3 = 3X_1^3,$$

and the algebraic constraints

$$X_2 = X_1^2, \; X_3 = X_1^3, X_1 \neq 0.$$

The statistics for intermediate computations are given in the Table 9.3.

10 Conclusion

We have studied initial value problems for a generalized class of quasi-linear systems of differential equations from a point of view of computational commutative algebra and algebraic geometry. We introduce ideal theoretical notions of essential **P**-degree basis, linear rank, first jet domain and initial domain, and 0-points using the fiber of the projection map. We explore extensively several new operators on ideals such as prolongation, completion, associated linearization, and quasi-linearization. We give an algorithm to compute the subset M^0 of the initial domain such that every solution curve starting at a point in M^0 lies in the initial domain. For each **x** in M^0, we guarantee that the solution curve through **x** is unique. We also give algorithms to compute an explicit system of differential equations in a neighborhood of **x** that is equivalent to the original system of differential equations. These equivalent explicit systems can then be used to perform numerical computations. All our algorithms are effective using any set of generators, with no *a priori* assumptions. Maximum flexibility is thus allowed during implementation.

In practice we will start with a system of differential equations (2.5), which we transform (see Chapter 2) into a first order polynomial system with the corresponding algebraic polynomials $f_1, \ldots, f_m$ generating an ideal J. By repeated prolongations, we compute the completion $\tilde{J}$ of J. From this, we can decide if J is eventually quasi-linear (that is, if $\tilde{J}$ is essentially quasi-linear). If this is not the case, we can either use the associated quasi-linear ideal $\tilde{J}^\ell$, or we can perform quasi-linearization and replace J by $q\ell(J)$ and complete again. In either case, we can then apply Theorem 6.2.3 to $\tilde{J}$.

Note that by Propositions 4.1.3 and 2.2.2, the solution set for the initial value problem using the generating system of equations for J will be the same as the solution set for the initial value problem using the generating system of equations for $\tilde{J}$.

Our ideal-theoretic approach provides the ground work for any future improvement in computation. While we have given algorithms to prove that all our theoretical constructions are effective, it is possible that future research may develop better algorithms. To prove a new algorithm, all it takes is to show that the constructed ideals have the theoretical defining properties. There is no need to redevelop a new theory each time a construction depends on some new choice of generators.

Our method simplifies the theory for an important class of differential algebraic equations (typically of high differentiation index) and provides conceptually simple algorithms for deciding algebraic conditions under which the system may have an existence and uniqueness theorem. When the conditions are satisfied, we can compute equivalent explicit systems. While we did not study how well numerical software may be adapted to our explicit systems, such software are far more efficient generally than software for implicit systems. It would be of great interest to explore the numerical implications of our method.

In conclusion, we end with a recent (August 2006) comment by Gear [23] on the origin of differential algebraic equations that may shed some light on the relevance of this article. Citations in the quote refer to those in Gear's article.

> Interestingly, electrical networks were originally modelled with ODEs and a lot of sophisticated techniques were developed to reduce a network to ODEs. (See, for example, [7, 17]). However, the ODEs were generally stiff and it was found that implicit methods had to be used for the stiff equations. It was also found that other types of networks could not be reduced to ODEs so we arrived at the idea of solving DAEs directly [11] and the tableau approach to network problems [14]. Initially these were index one problems, so did not present the DAE difficulties of higher index problems, but newer modelling approaches have lead [sic] to index two problems (e.g., [19]) and the problems have become so large and non-linear that some aspects of them, such as finding consistent initial conditions, are extremely challenging.
>
> Mechanical systems with constraints usually lead to index three problems that cannot be solved directly. The constraints in a DAE restrict the solution to a manifold and usually we cannot easily find the ODE on that manifold.

Acknowledgement. We thank the editors for their encouragement to submit this work and their patience and assistance throughout. We appreciate very much the critical comments from the anonymous referees and their careful reading of the paper. Our algebraic proof of the Theorem 6.2.3 avoids separately dealing with singular and non-singular points, as would be the case if we had presented the complex analytic manifold approach. We thank Edward Grossman of City College for suggesting ideas used in the proofs of Lemma 5.2.6, and Theorems 6.1.1, 6.2.3. We also thank Isaac Chavel (City College) and Richard Churchill (Hunter College) for advice in our original geometric proof of Theorem 6.2.3.

Bibliography

[1] V. Arnol'd, *Ordinary Differential Equations*, 3rd ed., Springer Verlag, Berlin; New York, 1992.

[2] U. M. Ascher, L. R. Petzold, *Computer methods for ordinary differential equations and differential-algebraic equations*, SIAM, Philadelphia, 1998. Second printing with corrections, June 2003. Errata, July 2006: http://www.cs.ubc.ca/spider/ascher/apbook/errata.pdf

[3] T. Becker, V. Weispfenning, *Gröbner Bases: A Computational Approach to Commutative Algebra*, Graduate Texts in Mathematics 141, Springer Verlag, New York, 1993.

[4] F. Boulier, *Étude et Implantation de Quelques Algorithmes en Algèbre Différentielle*, Thése, L'Université des Sciences et Technologies de Lille, 1994.

[5] F. Boulier, D. Lazard, F. Ollivier, M. Petitot, *Representation for the radical of a finitely generated differential ideal*, Proc. ISSAC 1995 (Montreal)(A. Levelt, ed.), ACM Press, New York, 1995, pp. 158–166.

[6] ———, *Computing Representations for Radicals of Finitely Generated Differential Ideals*, Technical Report IT-306, LIFL, 1997.

[7] F. Boulier, F. Lemaire, M. Moreno Maza, *PARDI!*, Proc. ISSAC 2001 (London, Ontario)(B. Mourrain, ed.), ACM Press, New York, 2001, pp. 38–47.

[8] D. Bouziane, K. A. Rody, H. Maârouf, *Unmixed-dimensional decomposition of a finitely generated perfect differential ideal*, J. Symb. Comput. 31 (2001), pp. 631–649.

[9] P. N. Brown, A. C. Hindmarsh, L. R. Petzold, *Consistent initial condition calculation for differential-algebraic systems*, SIAM J. Sci. Comput. 19 (5) (1998)(electronic), pp. 1495–1512.

[10] A. Buium, P. J. Cassidy, *Differential algebraic geometry and differential algebraic groups: from algebraic differential equations to Diophantine geometry*, Selected Works of Ellis Kolchin, with Commentary (H. Bass, A. Buium, P. J. Cassidy, eds.), A.M.S., 1999, pp. 567–636.

[11] S. L. Campbell, *Singular Systems of Differential Equations*, Research Notes in Mathematics 40, Pitman, New York, 1980.

[12] ———, *The numerical solution of higher index linear time varying singular systems of differential equations*, Siam. J. Sci. Stat. Comput. 6 (1985), pp. 334–348.

[13] ———, *A general form for solvable linear time varying singular systems of differential equations*, Siam. J. Math. Anal. 18 (1987), pp. 1101–1115.

[14] S. L. Campbell, C. W. Gear, The index of general nonlinear DAE's, Numerische Mathematik 72 (1995), pp. 173–196.

[15] S. L. Campbell, E. Griepentrog, *Solvability of general differential equations*, Siam. J. Sci. Comp. 16 (2) (1995), pp. 257–270.

[16] D. Cox, J. Little, and D. O'Shea, *Ideals, Varieties, and Algorithms: An Introduction to Computational Algebraic Geometry and Commutative Algebra*, Springer Verlag, Berlin, Heidelberg, New York, 1992.

[17] L. D'Alfonso, G. Jeronimo, P. Solerno, *A linear algebra approach to the differentiation index of generic DAE systems*, `arXiv:cs/0608064v1`, Aug 15, 2006.

[18] T. Daly, *Axiom Volume 1: Tutorial*, Lulu.com, 2005. See also http://wiki.axiom-developer.org/.

[19] J. A. Dieudonné, *Foundations of Modern Analysis*, Academic Press, New York and London, 1960.

[20] M. Fliess, *Automatique et corps différentiels*, Forum Mathematicum 1 (3) (1989), pp. 227–238.

[21] M. Fliess, J. Lévine, P. Martin, P. Rouchon, *Index and decomposition of nonlinear implicit differential equations*, System Struct. and Control, Proc. IFAC Conf. 1995 (Nantes, France), Pergamon Press, 1996, pp. 37–42.

[22] C. W. Gear, *Differential-algebraic equation index transformations*, Siam. J. Sci. Stat. Comput. 9 (1) (1988), pp. 39–47.

[23] _____ , *Towards explicit methods for differential algebraic equations*, BIT Numerical Mathematics 46 (3) (2006), pp. 505–514.

[24] C. W. Gear, L. R. Petzold, *Differential-algebraic systems and matrix pencils*, Matrix Pencils (B. Kagstrom, A. Ruheemph, eds.), Lecture Notes in Math, Springer-Verlag, Berlin, Heidelberg, New York, 1983, pp. 75–89.

[25] _____ , *ODE methods for the solution of differential/algebraic systems*, SIAM J. Numer. Anal. 21 (1984), pp. 367–384.

[26] P. Gianni, B. Trager, G. Zacharias, *Gröbner bases and primary decomposition of polynomial ideals*, Computational Aspects of Commutative Algebra (L. Robbiano, ed.), Academic Press, San Diego, 1989, pp. 15–33.

[27] G. Greuel, *Description of SINGULAR: a Computer Algebra system for singularity theory, algebraic geometry and commutative algebra*, Euromath Bull. 2 (1) (1996), pp. 161–172.

[28] M. Hanke, R. Lamour, *Consistent initialization for nonlinear index-2 differential-algebraic equations: large sparse systems in MATLAB*, Numerical Algorithms 32 (2003), pp. 67–85.

[29] E. Hubert, *Factorization-free decomposition algorithms in differential algebra*, J. Symbolic Comp. 29 (2000), pp. 641–662.

[30] _____ , *Notes on triangular sets and triangulation-decomposition algorithms I: polynomial systems*, Symbolic and Numerical Scientific Computing 2001 (F. Winkler, U. Langer, eds.), LNCS 2630, Springer-Verlag, Berlin, Heidelberg, New York, 2003, pp. 1–39.

[31] _____ , *Notes on triangular sets and triangulation-decomposition algorithms II: differential systems*, Symbolic and Numerical Scientific Computing 2001 (F. Winkler, U. Langer, eds.), LNCS 2630, Springer-Verlag, Berlin, Heidelberg, New York, 2003, pp. 40–87.

[32] E. L. Ince, *Ordinary Differential Equations*, Dover, New York, 1956.

[33] A. Iserles, A. Zanna, *Preserving algebraic invariants with Runge-Kutta methods*, J. Comp. Appld. Maths. 125 (2000), pp. 69–81.

[34] R. D. Jenks, R. S. Sutor, *Axiom, The Scientific Computation System*, Springer-Verlag, Berlin, Heidelberg, New York, 1992.

[35] E. R. Kolchin, *Differential Algebra and Algebraic Groups*, Academic Press, New York, 1973.

[36] I. Kaplansky, *An Introduction to Differential Algebra*, Hermann, Paris, 1957.

[37] P. Kunkel, V. Mehrmann, *Canonical forms for linear differential-algebraic equations with variable coefficients*, J. Comput. Appl. Math. 56 (1994), pp. 225–251.

[38] _____ , *A new class of discretization methods for the solution of linear differential-algebraic equations with variable coefficients*, SIAM J. Numer. Anal. 33 (5) (1996), pp. 1941–1961.

[39] _____ , *Local and global invariants of linear differential-algebraic equations and their relation*, Electronic Trans. Num. Anal. 4 (1996), pp. 138–157.

[40] _____ , *Differential-Algebraic Equations, Analysis and Numerical Solution*. EMS Textbooks in Math., European Math. Soc., Zürich, 2006.

[41] S. Lang, *Algebra*, Addison Wesley, Reading, Mass., 1965.

[42] _____, *Differential Manifolds*, Addison Wesley, Reading, Mass., 1972.

[43] G. Matera, A. Sedoglavic, *Fast computation of discrete invariants associated to a differential rational mapping*, J. Symbolic Comput. 36(2003, pp. 473–499.

[44] H. Matsumura, *Commutative Algebra*, Benjamin, Inc., New York, 1970.

[45] E. H. Nielsen, *The implementation of SIRK methods for differential algebraic equations*, Department of Mathematics Research Report 378, University of Auckland, Auckland, New Zealand, 1997.

[46] L. R. Petzold, *Numerical solution of differential-algebraic equations*, Theory and Numerics of Ordinary and Partial Differential Equations 1994 (Leicester), Adv. Numer. Anal., IV, Oxford Univ. Press, New York, 1995, pp. 123–142.

[47] P. J. Rabier, W. C. Rheinboldt, *A general existence and uniqueness theory for implicit differential-algebraic equations*, Differential Integral Equations 4 (1991), pp. 563–582.

[48] _____, *A geometric treatment of implicit differential-algebraic equations*, J. Differential Equations, 109 (1994), pp. 110–146.

[49] _____, *On impasse points of quasilinear differential algebraic equations*, J. Math. Anal. Appl. 181 (1994), pp. 429–454.

[50] _____, *Discontinuous solutions of semilinear differential-algebraic equations–I*, Nonlinear Anal.,Theory, Methods & Appl. 27 (1996), pp. 1241–1256.

[51] _____, *Discontinuous solutions of semilinear differential-algebraic equations–II*, Nonlinear Anal.,Theory, Methods & Appl. 27 (1996), pp. 1257–1280.

[52] S. Reich, *Differential-Algebraic Equations and Vector Field on Manifolds*, Sekt. Informationstechnik, Nr. 09-02-88, Technische Universität, Dresden, 1988.

[53] _____, *Beitrag zur Theorie der Algebrodifferentialgleichungen*. Ph. D. Dissertaion in Engineering, Technische Universität, Dresden, 1989.

[54] J. F. Ritt, *Differential Algebra*, Dover, New York, 1950.

[55] D. E. Schwarz, *Consistent Initialization for Index 2 Differential Algebraic Equations and Its Application to Circuit Simulation*, Ph.D. Thesis, Humboldt-University, Institute of Mathematics, (2000).

[56] D. E. Schwarz, R. Lamour, *The computation of consistent initial values for nonlinear index 2 DAEs*, Numerical Algorithms 28(1)(2001), pp. 49–75.

[57] W. Y. Sit, *An algorithm for solving parametric linear systems*, J. Symbolic Computation 13 (1992), pp. 353–294.

[58] _____, *Computations on quasi-algebraic sets*, Electronic Proceedings of the Fourth International IMACS Conference on Applications of Computer Algebra 1998 (Prague), http://www-troja.fjfi.cvut.cz/aca98/proceedings.html.

[59] _____, *The Ritt-Kolchin theory of differential polynomials*, Differential Algebra and Related Topics 2000 (Newark, N.J.) (L. Guo, P. J. Cassidy, W. F. Keigher, and W. Y. Sit eds.), World Scientific, New Jersey, London, Singapore, Hong Kong, 2002, pp. 1–70.

[60] M. Spivak, *Differential Geometry Volume I*, Publish of Perish, Inc., Houston, 1970.

[61] G. Thomas, *Symbolic computation of the index of quasi-linear differential-algebraic equations*, Proc. ISSAC'96 (Y. N. Lakshman, ed.), ACM Press, New York, 1996, pp. 196–203.

[62] _____, *The problem of defining the singular points of quasi-linear differential-algebraic systems*, Theoret. Comput. Sci. 187 (1–2) (1997), pp. 49–79.

[63] J. Tuomela, *On singular points of quasilinear differential and differential-algebraic equations*, BIT Numerical Mathematics 37 (1997), pp. 968–977. Also in Research Report A369, Helsinki University of Technology.

[64] _____ , *On the resolution of singularities of ordinary differential systems*, Proc. Differential Algebraic Equations 1997 (Grenoble), Numer. Algorithms 19 (1998), pp. 247–259.

[65] F. W. Warner, *Foundations of Differential Manifolds and Lie Groups*, Scott, Foresman & Company, Glenview, Illinois, 1971. Graduate Texts in Mathematics 94, Springer-Verlag, New York, 1983.

Author information

F. Leon Pritchard, Department of Mathematics and Computer Science, York College, The City University of New York, Jamaica, NY 11451, USA.
Email: pritchard@york.cuny.edu

William Y. Sit, Department of Mathematics, City College, The City University of New York, New York, NY 10031, USA.
Email: wyscc@sci.ccny.cuny.edu

Index

Symbols

$(<_1, <_2)$-reduction, 61
$(\sigma_1, \ldots, \sigma_p)$-dimension polynomial, 72
Δ-ideal, 86
δ_1-fixed, 95
δ-lexicographic, 89
Δ-reduced, 93
δ-regular, 190, 195, 207
Δ-ring, 80
Δ-stable, 88
π_*, 257
π-submanifold of a zero set, 286
σ^*-dimension polynomial, 54
σ, symbol map, 255
σ-algebraic difference field extension, 46
σ-algebraic element, 46
σ-algebraic s-tuple, 48
σ-algebraically dependent family, 46
σ-generators, 45
σ-quasi-linear, 95
σ-transcendental element, 46
ΘU, 112

A

admissible ordering, 88
algebra of difference (σ-) polynomials, 46, 47
algebra of inversive difference (σ-) polynomials, 47
algebraic analysis approach to linear control systems, 163
algebraic index
    of an ideal or system, 299
    algorithm to compute $\sim$, 299
algebraic set
    defining ideal of an $\sim$, 293
    (Krull) dimension of an $\sim$, 293
    effectively computable $\sim$, 293
    in algebraic geometry, 293
    tangent variety of an $\sim$, 302

B

basic set, 44
basic transformations of differential equations, 291
BLAD, 119, 125
block of parameters, 124
boundary problem operator, 273
    normalised, 274
        regular, 275
    regular, 275
Buchberger, Bruno, 110
Buchberger's algorithm, 32, 39, 77

C

canonical differential term ordering, 91
canonical state representation, 31
Carrà Ferro, Giuseppa, 111
Cartan, E., 39
Cartan
    character, 181, 193
    test, 183, 186, 189, 193, 197, 206
Casimir operator, 263
Castelnuovo–Mumford regularity, 192, 196
categorical quotient, 246
Cauchy problem, 34, 37, 38
character, 13
characteristic
    of a vector field, 231
    set, 94, 110, 115, 118
circadian clock, 126

algorithm
    to compute algebraic index, 299
    to compute completion ideal, 299
    to compute **P**-strong essential **P**-degree basis, 295
annihilator subbundle, 235
autonomous element, 163
autonomous observable, 8
autoreduced set, 84

W

Wang, Dongming, 110, 119
weight, 86
Wen-Tsün, Wu, 110
Weyl algebra, 81, 253, 256
White, N., 34, 39

X

$X//G$, categorical quotient, 246
X^{ss}, semi-stable points, 247

Z

Zariski topology, 293
zero of a differential polynomial, 322